*Edited by*
*Avraam I. Isayev*

**Encyclopedia of Polymer Blends**

## Related Titles

Isayev, Avraam I. (ed.)

### Encyclopedia of Polymer Blends

**Volume 1: Fundamentals**

2010
PRINT ISBN: 978-3-527-31929-9

Isayev, Avraam I. (ed.)

### Encyclopedia of Polymer Blends

**Volume 2: Processing**

2011
PRINT ISBN: 978-3-527-31930-5

Elias, H.

### Macromolecules

**Volume 1: Chemical Structures and Syntheses**

2005
Print ISBN: 978-3-527-31172-9

Elias, H.

### Macromolecules

**Volume 2: Industrial Polymers and Syntheses**

2007
Print ISBN: 978-3-527-31173-6

Elias, H.

### Macromolecules

**Volume 3: Physical Structures and Properties**

2007
Print ISBN: 978-3-527-31174-3

Elias, H.

### Macromolecules

**Volume 4: Applications of Polymers**

2008
Print ISBN: 978-3-527-31175-0

Edited by Avraam I. Isayev

# Encyclopedia of Polymer Blends

Volume 3: Structure

WILEY-VCH

Verlag GmbH & Co. KGaA

**The Editor**

*Prof. Avraam I. Isayev*
The University of Akron
Department of Polymer Engineering
250 South Forge Street
Akron, OH 444325-0301
USA

■ All books published by **Wiley-VCH** are carefully produced. Nevertheless, authors, editors, and publisher do not warrant the information contained in these books, including this book, to be free of errors. Readers are advised to keep in mind that statements, data, illustrations, procedural details or other items may inadvertently be inaccurate.

**Library of Congress Card No.:** applied for

**British Library Cataloguing-in-Publication Data**
A catalogue record for this book is available from the British Library.

**Bibliographic information published by the Deutsche Nationalbibliothek**
The Deutsche Nationalbibliothek lists this publication in the Deutsche Nationalbibliografie; detailed bibliographic data are available on the Internet at <http://dnb.d-nb.de>.

© 2016 Wiley-VCH Verlag GmbH & Co. KGaA, Boschstr. 12, 69469 Weinheim, Germany

**Print ISBN:** 978-3-527-31931-2
**ePDF ISBN:** 978-3-527-65399-7
**ePub ISBN:** 978-3-527-65398-0
**Mobi ISBN:** 978-3-527-65397-3
**oBook ISBN:** 978-3-527-65396-6

**Cover Design** Adam-Design, Weinheim, Germany
**Typesetting** Thomson Digital, Noida, India
**Printing and Binding** Markono Print Media Pte Ltd, Singapore

Printed on acid-free paper

# Contents

# Preface

Encyclopedia of polymer blends will include scientific publications in various areas of blends. Polymer blends are mixtures of two or more polymers and/or copolymers. Polymer blending is used to develop new materials with synergistic properties that are not achievable with individual components without having to synthesize and scale up new macromolecules. Along with a classical description of polymer blends, articles in the encyclopedia will describe recently proposed theories and concepts that may not be accepted yet but reflect future development. Each article provides current points of view on the subject matter. These up-to-date reviews are very helpful for understanding the present status of science and technology related to polymer blends.

The encyclopedia will be the source of existing knowledge related to polymer blends and will consist of five volumes. Volume 1 describes the fundamentals including the basic principles of polymer blending, thermodynamics, miscible, immiscible, and compatible blends, kinetics, and composition and temperature dependence of phase separation. Volume 2 provides the principles, equipment, and machinery for polymer blend processing. Volume 3 deals with the structure of blended materials that governs their properties. Volume 4 describes various properties of polymer blends. Volume 5 discusses the blended materials and their industrial, automotive, aerospace, and other high technology applications. Individual articles in the encyclopedia describe the topics with historical perspective, the state-of-the-art science and technology and its future.

This encyclopedia is intended for use by academicians, scientists, engineers, researchers, and graduate students working on polymers and their blends.

Volume 3 is devoted to the structure of blended materials that governs their properties and consists of seven chapters. These chapters cover glass transition phenomena, crystallization and melting behavior, structure–property relationship, morphology and structure of polymer blends, and blends containing various nanofillers. Existing theoretical approaches to describe morphology and structure of blends are extensively discussed. The importance of flow, rheology of components, and rheological aspects of blends is emphasized. These aspects are detailed below and build on each other.

Chapter 1 addresses a number of topics, including general phenomenology, theories, and metrology of the glass transition of polymer blends. The theoretical foundations and practical examples from the analysis of experimental data for miscible systems including binary polymer blends, oligomer/polymer mixtures, and copolymers are critically reviewed. The applicability ranges of important theoretical, semi-empirical, or purely phenomenological mixing rules used for describing the compositional dependence of the glass transition are explored. Examples demonstrating the physical meaning of model parameters are given. A number of case studies involving hydrogen-bonding binary polymer blends and ternary polymer systems are presented. The chapter ends by summarizing general rules that relate the results of glass-transition studies with structural characterizations and miscibility evaluations of polymer blends.

Chapter 2 deals with crystallization and melting behavior of crystalline/amorphous and crystalline/crystalline polymer blends that are strongly influenced by the miscibility and morphology of the polymers. The interspherulitic and intraspherulitic segregations are considered in the case of crystalline/amorphous polymer blends. The crystallization and morphology of crystalline/crystalline polymer blends related to differences in the melting points of each of component, thermodynamic, and kinetic factors during crystallization are discussed. The influence of the composition, rheological characteristics, the interfacial tension, and processing conditions on the superstructures of immiscible polymer blends is presented. Immiscible blends with a crystallizable matrix and an amorphous dispersed phase, and blends with an amorphous matrix and a crystallizable dispersed phase are discussed. The effect of the addition of a copolymer as a compatibilizer that decreases or increases the tendency for crystallization of polymer blends is considered. Reactive compatibilization of polymer blends and its influence on crystallization and morphology are discussed. In addition, the effect of the fillers on the crystallization of immiscible polymer blends is presented.

Chapter 3 is devoted to the morphology and structure of crystalline/crystalline polymer blends with strong emphasis to the recent progress in this field. It focuses mainly on the influence of crystallization on the microphase separation and the effect of phase separation on the crystallization of the blending components. Special attention is given to the various possible crystalline morphologies and phase structures formed in different crystalline/crystalline polymer blends under controlled crystallization conditions. The mechanism of the formation of specific morphologies, such as interpenetrating spherulites, is discussed with elaborately selected model systems. Also, examples of specific polymer blends are presented along with their morphology and structure.

Chapter 4 describes the physics and chemistry of rubber–plastic blends and their structure–property relationship. A greater attention is paid to understanding the interface and the role of the physical process in enabling and extending the interfacial effects of rubber–plastic blends. Various models for the rubber toughening of plastics are described. Numerous techniques that are for the characterization of rubber–plastic blends are provided. Many industrially important examples of many rubber–plastic blends are given along with their structure and morphology.

Chapter 5 deals with the current state of knowledge on the structure and morphology of rubber–rubber blends. Characterization techniques suitable for the study of these blends are introduced. The effect of material parameters and processing conditions on the structure and morphology of rubber–rubber blends is discussed along with the issues related to the filler distribution and curative migration in blends. Various blends containing different pairs of rubbers are presented. When dealing with specific rubber–rubber blends, the characteristics of each rubber component in the blends, such as the crystallization behavior, curing state, and preference of filler distribution is considered, since all these factors influence the blend morphology and structure.

Chapter 6 deals with the miscibility, phase morphology, and properties of ternary polymer blends. A number of interesting cases of miscibility and immiscibility in ternary blends are examined. It is stressed that a simple summation of the contributions of binary interactions to the free energy of mixing in ternary polymer blends is a simplification. The review is devoted to the prediction and formation of phase morphologies in immiscible binary and ternary polymer blends. The theory of spreading coefficients is analyzed in detail and the formation of all the possible morphological types is discussed. Particularly, phase morphologies with separated, fully encapsulated, and partially capsulated dispersed phases are described. Effects of blend composition, kinetic factors as well as the interaction of droplets upon mixing cycle are discussed in detail. An attempt is made to provide the understanding of the principles of the blend formation influencing the mechanical properties of ternary blends. Several cases of property–composition relationships for ternary composition are revised. A hypothesis is offered claiming that the experimental values of the properties of the ternary blends are much closer to those calculated by the additivity properties of the corresponding binary blends.

Chapter 7 deals with the morphology and structure of polymer blends containing various nanofillers. This subject area is increasingly growing due to the interest in polymer nanocomposite indicating that small addition of nanoparticles can dramatically change various properties of a polymer matrix including electrical and thermal conductivity, dielectric and magnetic permeability, gas barrier properties, and mechanical performances. A combination of polymer blending and nanoscale filler reinforcement has received a special attention due to the fact that the addition of nanofillers into multiphase polymer blends is proved to be an efficient strategy to develop a new family of polymer nanocomposites with a great tailoring potential for producing products with a combination of prescribed properties. Among various nanofillers considered in this chapter are silica nanoparticles (hydrophilic and hydrophobic), layered silicate, surface modified nanosilicates (nanoclays), single and multiwalled carbon nanotubes, and graphene along with their surface modification.

There are many people who contributed to the completion of this volume. I wish to express my profound appreciation to the contributors of the various chapters for being patient with my requests for revisions and corrections. I would also like to thank Dr. David Simmons for providing excellent review. I am thankful to Wiley-VCH Publishers for undertaking this project and for their patience, understanding, and cooperation with the authors at all stages of preparation. Finally, the support and patience of my family and the families of all the chapter authors contributed to the completion of this volume.

Akron, OH, USA
October 2015

*Avraam I. Isayev*

# List of Contributors

**Sudhin Datta**
ExxonMobil Chemical Co.
GCR-Product Fundamentals
5200 Bayway Drive
Baytown, TX 77520
USA

**Avraam I. Isayev**
The University of Akron
Department of Polymer Engineering
250 South Forge Street
Akron, OH 444325-0301
USA

**Saleh A. Jabarin**
University of Toledo
Polymer Institute
Department of Chemical and
Environmental Engineering
2801 W. Bancroft Street
Toledo, OH 43606-3390
USA

**Ioannis M. Kalogeras**
University of Athens
Faculty of Physics
Department of Solid State Physics
15784 Zografos, Athens
Greece

**V.N. Kuleznev**
Lomonosov State University of Fine
Chemical Technology
Prospekt Vernadskogo 86
119571 Moscow
Russia

**Tian Liang**
The University of Akron
Department of Polymer Engineering
250 South Forge Street
Akron, OH 444325-0301
USA

**Elizabeth A. Lofgren**
University of Toledo
Polymer Institute
Department of Chemical and
Environmental Engineering
2801 W. Bancroft Street
Toledo, OH 43606-3390
USA

**Kazem Majdzadeh-Ardakani**
University of Toledo
Polymer Institute
Department of Chemical and
Environmental Engineering
2801 W. Bancroft Street
Toledo, OH 43606-3390
USA

**Yu. P. Miroshnikov**
Lomonosov State University of Fine
Chemical Technology
Prospekt Vernadskogo 86
119571 Moscow
Russia

**Hossein Nazockdast**
Amirkabir University of Technology
Department of Polymer Engineering
424 Hafez Ave.
15875-4413 Tehran
Iran

**Zhaobin Qiu**
Beijing University of Chemical
Technology
State Key Laboratory of Chemical
Resource Engineering
15 North Third Ring Road East
Chaoyang District
100029 Beijing
China

**Shouke Yan**
Beijing University of Chemical
Technology
State Key Laboratory of Chemical
Resource Engineering
15 North Third Ring Road East
Chaoyang District
100029 Beijing
China

# 1
## Glass-Transition Phenomena in Polymer Blends

*Ioannis M. Kalogeras*

*University of Athens, Faculty of Physics, Department of Solid State Physics, Zografos 15784, Greece*

## 1.1
## Introduction

The ever-increasing demand for polymeric materials with designed multi-functional properties has led to a multiplicity of manufacturing approaches and characterization studies, seeking proportional as well as synergistic properties of novel composites. Fully integrated in this pursuit, blending and copolymerization have provided a pair of versatile and cost-effective procedures by which materials with complex amorphous or partially crystalline structures are fabricated from combinations of existing chemicals [1–3]. Through variations in material's composition and processing, a subtle adaptation of numerous chemical (corrosion resistance, resistance to chemicals, etc.), thermophysical (e.g., thermal stability, melting point, degree of crystallinity, and crystallization rate), electrical or dielectric (e.g., conductivity and permittivity), and manufacturing or mechanical properties (dimensional stability, abrasion resistance, impact strength, fracture toughness, gas permeability, recyclability, etc.) can be accomplished effortlessly. It is therefore not surprising the fact that related composites have been widely studied with respect to their microstructure (e.g., length scale of phase homogeneity in miscible systems, or type of the segregation of phases in multiphasic materials) and the evolution of their behavior and complex relaxation dynamics as the material traverses the glass transformation range [4–7].

The reversible transformation of amorphous materials (including amorphous regions within semicrystalline polymers) from a molten or rubber-like state into a stiff and relatively brittle glassy state is denoted as "glass transition" (or "liquid–glass transition"). Originally, this term was introduced to describe the striking changes in thermodynamic derivative properties (e.g., heat capacity, compressibility, and thermal expansivity) that normally accompany the solidification of a viscous liquid, such as a polymer melt, during cooling or even compression. In time course, however, the term "glass transition" acquired a broader meaning and is now frequently

*Encyclopedia of Polymer Blends: Volume 3: Structure*, First Edition. Edited by Avraam I. Isayev.
© 2016 Wiley-VCH Verlag GmbH & Co. KGaA. Published 2016 by Wiley-VCH Verlag GmbH & Co. KGaA.

employed for describing "any phenomenon that is caused by a timescale (on which some interesting degree of freedom equilibrates) becoming longer than the timescale on which the system is being observed" [6]. The conventional route to the glassy state of matter is the (rapid) cooling of a melt, provided that crystalization is bypassed. Interestingly, melt mixing provides one of the most common techniques for the large-scale preparation of compression or injection molded polymer blends. The freezing-in of a structural state during cooling, commonly referred to as glassification or vitrification, corresponds to a loss of the state of internal equilibrium possessed by the initial liquid. The vitrification process occurs over a narrow temperature interval, the so-called glass transformation range, over which the characteristic molecular relaxation time of the system changes by some 2–2.5 orders of magnitude, reaching the order of 100 s (the laboratory timescale). In macromolecular substances, this relaxation time is connected with the time response of cooperative (long-range) segmental motions. For convenience, the glass transformation region is traditionally represented by a single value, denoted as the "glass-transition temperature" ($T_g$) [7]. Because of the range of temperature involved and also because of its history, path, and cooling (or heating) rate dependences, assigning a characteristic $T_g$ to a system becomes frequently a problematic task. Nonetheless, when appropriately measured, the glass-transition temperature is very reproducible and has become recognized as one of the most important material properties, directly relating to several other thermophysical and rheological properties, processing parameters, and fields of potential application [8].

Nowadays, polymer engineering largely relies on chemical and compositional manipulation of $T_g$, in an attempt to target particular technological or industrial requirements. A notable paradigm provides meticulous studies on the physical stabilization of active pharmaceutical ingredients (typically, poorly soluble drugs, but potentially also of proteins or other compounds) in the form of binary or ternary solid dispersions/solutions with biologically inert glassy polymers, with the aim of increasing their solubility, dissolution rate, bioavailability, and therapeutic effectiveness [9]. Significant improvements in the performance of related systems are frequently accredited to a combination of factors, including the effects of hydrogen-bonding networks or ion–dipole intercomponent interactions (i.e., stabilizing enthalpic contributions), and strong crystallization-inhibitory steric effects owing to the high viscosity of the polymeric excipient [10]. The implementation of solid-state glassy formulations as a means to preserve the native state of proteins (biopreservation) entails a higher level of complexity in behavior, since chemical and physical stabilization heavily relies on manipulating the local anharmonic motions of the individual protein molecules (fast dynamics) in addition to their slow (glass-transition) dynamics [11]. The established practice in the preservation of proteins primarily focuses on hydrated solid-state mixtures of proteins with glass-forming disaccharides (e.g., trehalose or sucrose) or polyols (e.g., glycerol), serving as lyoprotectants. Still, processing problems such as surface denaturation, mixture separation, and pH changes that lead to physical and chemical degradations, in addition to degradations occurring on storage, make clear that manufacturing of alternative solid-state formulations remains a challenging issue. In this pursuit, however, one has to take into account the complex internal protein dynamics and the fact that in order to successfully maintain protein's structural integrity the selected glass-forming polymer will have to sustain a strong and complicated hydrogen-bonding network (around and to the protein) that will effectively couple matrix dynamics to the internal dynamics of the protein molecules [11]; the adaptability in local structures and chemical environments, offered by polymers' blending, might provide a viable solution to the problem.

While general consensus exists as to the usability of glass transitions in exploring molecular mobilities, molecular environments, and structural heterogeneities in segmental length scales, conflicting arguments still appear regarding the interpretation of fundamental phenomenological aspects of the transition itself. The nature of the glass transition remains one of the most controversial problems in the disciplines of polymer physics and materials science, and that in spite of the in-depth experimental and theoretical research conducted hitherto [12–14]. The difficulty in treating glass transitions even in relatively simple linear-chain amorphous polymers is caused by the almost undetectable changes in static structure, regardless of the qualitative changes in characteristics and the extremely large change in the timescale. Given the significance of this subject, this chapter begins with an overview of important aspects of the phenomenology of glass formation. To address the perplexing behavior encountered when a system passes through its glass transformation range, a number of theoretical models approach this phenomenon using arguments pointing to a thermodynamic or a purely dynamic transition. Although we have not arrived at a comprehensive theory of supercooled liquids and glasses, it is frequently recognized that the observed glass transitions are not *bona fide* phase transitions, but rather a dynamical crossover through which a viscous liquid falls out of equilibrium and displays solid-like behavior on the experimental timescale. Basic notions and derivations of common theoretical models of the glass transition are presented in Section 1.2, with particular reference to early free volume and configurational entropy approaches, in view of their impact on the development of "predictive" relations for the compositional dependence of the $T_g$ in binary polymer systems. Note that regardless of the multitude of treatments we clearly lack a widely accepted model that would allow *ab initio* calculation of the glass-transition temperature. Most theoretical approaches simply allow a prediction of changes in $T_g$ with – among other variables – applied pressure, degree of polymerization (molar mass) or curing (cross-linking), and composition. Important chemical factors that influence the affinity of the components and the magnitude of $T_g$ in polymer blends, in addition to manufacturing processes or treatments that are typically used for manipulating the glass-transition temperature of polymer composites, are briefly reviewed in Section 1.3.

The largely experimentally driven scientific interest on glass transitions in composite materials is equipped with a collection of experimental methods and measuring techniques, with the ability to probe molecular motions at distinctly different length scales [15–17]. Most of them introduce different operational definitions of $T_g$, and some of them are endorsed as scientific standards. In Section 1.4, fundamental aspects of important experimental means are presented, with emphasis placed on the relative sensitivity and measuring accuracy of each technique, and the proper identification and evaluation of glass transitions in multicomponent systems. Typically, the broadness of the glass-transition region is indicative of structural (nano)heterogeneities, whereas its location can be adjusted by appropriate variations in composition and the ensuing changes in the degree of interchain interactions or material's free volume. Given the diversity of chemical structures of the polymers used and the wide range of dynamic asymmetries and molecular affinities explored, it is not surprising the wealth of information on polymers' miscibility and the number of phenomena revealed in related studies. Two general cases are distinguished experimentally: *single-phase* and *phase-separated* systems. In the first case, for instance in a pair of miscible polymers or in random copolymers (i.e., with random alternating blocks along the macromolecular chain), a single – although rather broad – glass-transition region is recorded by most experimental techniques. The glass-transition behavior associated with phase separation, a situation characterizing the vast majority of engineering polymer blends, their related graft, and block copolymers, as well as interpenetrating polymer networks, demonstrates an elevated level of complexity; multiple transitions, ascribed to pure component phases and regions of partial mixing, are common experimental findings. Issues related to miscibility evaluations of polymer blends, such as the determination of the length scale of structural heterogeneity using different experimental approaches, are discussed in Section 1.5. The theoretical foundations and practical examples from the analysis of experimental data for miscible systems are critically reviewed in Section 1.6. The applicability ranges of important theoretical, semiempirical, or purely phenomenological mixing rules used for describing the compositional dependence of $T_g$ are explored, with examples demonstrating the physical meaning of their parameters. A number of case studies, involving intermolecularly hydrogen-bonded binary blends and ternary polymer systems, are presented in the Section 1.7. This chapter ends with a summary of general rules relating the results of glass-transition studies with structural characterizations and miscibility evaluations of polymer blends, as well as typical requirements for reliable determinations of $T_g$ in polymeric systems.

## 1.2
## Phenomenology and Theories of the Glass Transition

### 1.2.1
### Thermodynamic Phase Transitions

What seems to be a long-standing and exceptionally puzzling question is whether the physics of glass formation can be understood considering a purely dynamical origin with no thermodynamic signature, or necessitates thermodynamic or structural foundation. Customarily, the apparent glass-transition phenomenon in appropriately prepared amorphous materials (e.g., several oxides, halides, salts, organic compounds, metal alloys, and numerous polymeric systems) is considered to be a kinetic crossover, and the most remarkable phenomena – structural arrest and dynamic heterogeneities – are strongly linked to molecular dynamics. Purely kinetic explanations, however, overlook the thermodynamic aspects of the phenomenology of glass formation and its deceptive resemblance to a second-order phase transition. Formally, as "phase transition" we consider the transformation of a thermodynamic system from one phase or state of matter to another, produced by a change in an intensive variable. The traditional classification scheme of phase transitions, proposed by Paul Ehrenfest, is based on the behavior of free energy ($F$) as a function of other state variables (e.g., pressure, $P$; volume, $V$; or temperature, $T$). Under this scheme, phase transitions are labeled by the lowest derivative of the free-energy function that is discontinuous at the transition. Thus, *first-order* phase transitions exhibit a discontinuity in the first derivative of the free energy with respect to some thermodynamic variable. In the course of heating, during a first-order transition the material absorbs a certain amount of heat (called the *latent heat of transition*) and undergoes a change in its constant-pressure heat capacity $C_P$. Typical examples are various crystal–liquid–gas phase transitions (e.g., melting or freezing, boiling, and condensation), which involve a discontinuous change in density ($\rho$), the first derivative of the free energy with respect to the chemical potential. On the other hand, *second-order* phase transitions are continuous in the first derivative of the free energy, but exhibit discontinuity in a second derivative of it. The order–disorder transitions in alloys and the ferromagnetic phase transition are typical examples. Passing through such transitions the material will undergo a change in its heat capacity, but no latent heat will be present.

The order of a phase transition can be defined more systematically by considering the thermodynamic Gibbs free-energy function, $G$. In a first-order phase transition, the $G\,(T,\,P)$ function is continuous, but its first derivatives with respect to the relevant state parameters,

$$S = -\left(\frac{\partial G}{\partial T}\right)_P, \quad V = \left(\frac{\partial G}{\partial P}\right)_T \quad \text{and} \quad H = -\left(\frac{\partial (G/T)}{\partial (1/T)}\right)_P, \quad (1.1)$$

are discontinuous across the phase boundary (Figure 1.1; the symbol $S$ denotes entropy, and $H$ stands for enthalpy). In a similar way, in a second-order phase transition the above functions are continuous, but their derivatives with respect to the relevant state parameters, isobaric heat capacity, compressibility, $\kappa_T$, and isobaric expansivity (also called the coefficient of thermal expansion), $\alpha$,

$$C_P = -T\left(\frac{\partial^2 G}{\partial T^2}\right)_P = T\left(\frac{\partial S}{\partial T}\right)_P = \left(\frac{\partial H}{\partial T}\right)_P \quad (1.2a)$$

$$\kappa_T = -\frac{1}{V}\left(\frac{\partial^2 G}{\partial P^2}\right)_T = -\frac{1}{V}\left(\frac{\partial V}{\partial P}\right)_T \quad (1.2b)$$

$$a = \frac{1}{V}\left(\frac{\partial}{\partial T}\left(\frac{\partial G}{\partial P}\right)_T\right)_P = \frac{1}{V}\left(\frac{\partial V}{\partial T}\right)_P, \quad (1.2c)$$

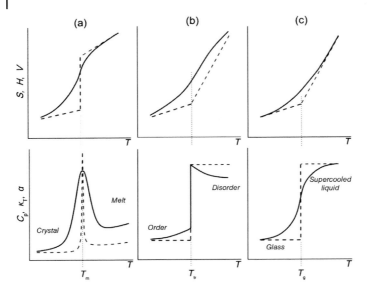

**Figure 1.1** Schematic representation of thermodynamic responses. (a) First-order phase transition: consider, for example, melting of a crystal with defects (—) or of a perfect infinite crystal (--). (b) Second-order transition: transition dominated by intermolecular cooperative phenomena (—), or having only intermolecular cooperative phenomena (--). (c) Glass transition: experimental response (—), and ideal response in an infinately slow experiment (--).

are discontinuous across the phase boundary. Figure 1.1 compares the thermodynamic signatures of the classic first- and second-order transitions with the response of the experimental glass transition. Parameters such as $\alpha$ and $C_p$, and many other properties of inorganic glasses and glassy polymers exhibit a gradual change in slope in the glass-transition temperature, and any such (rounded) step or kink can be used to define $T_g$ (see Section 1.4). This behavior is to be contrasted with the peak or the discontinuity expected, respectively, for genuine first- or second-order phase transformations. The latter is also characterized by a single ordering parameter determining the position of equilibrium in the relaxing system, with the jumps of the above parameters connected via the Ehrenfest ratio

$$\Pi_E \equiv \frac{1}{VT}\frac{\Delta C_p \Delta\kappa_T}{(\Delta\alpha)^2} = 1. \tag{1.3}$$

At the glass-transition temperature, however, the same ratio (known as the Prigogine–Defay ratio) is greater than unity [18,19]. Arguments like the above cast considerable doubt on the validity of models (e.g., configurational entropy models or the random first-order transition [RFOT] theory), postulating the existence of some type of an underlying thermodynamic transition (see Section 1.2.4).

## 1.2.2
## Structural, Kinetic, and Thermodynamic Aspects

One of the most intriguing questions in theoretical physics today is whether the glass is a new state of matter or just a liquid that flows too slowly to observe. The defining property of a structural glass transition for a polymer melt, observed on cooling from a sufficiently high temperature, is the increase of shear viscosity ($\eta$) by more than 14 orders of magnitude, without the development of any long-range order in structure. The typical X-ray or neutron diffraction studies of glassy solids, for example, reveal broad spectra of scattering lengths with no clear indication of primary unit cell structures. The "amorphous halo" of the static structure factor assessed by scattering experiments, or calculated via Monte Carlo and molecular dynamics computer simulations of amorphous cells, also shows insignificant changes when the material crosses the glass transformation range. Voronoi–Delaunay structural analyses of model amorphous systems have provided some means for distinguishing subtle differences between the rigid glass and liquid states of matter [12]. Relevant studies, however, are inconclusive as to the existence of some type of universally accepted geometric descriptor of the feeble structural changes occurring during the transition.

Contrary to the above findings, the marked change in behavior observed for thermodynamic derivative properties or physical quantities during the glass transition has provided the venue for "quantitative" descriptions of the process. Consider, for example, the shape of a typical thermal expansion curve (Figure 1.2a) [20]. In the polymer melt, the thermal expansion coefficient is almost constant, as it is again so in the glassy state but with a smaller value, similar to that of a crystalline solid. At the glass transition, there is therefore a pronounced change in the dependence of density or specific volume on

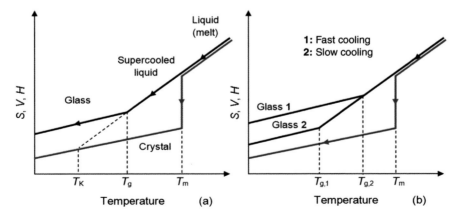

**Figure 1.2** (a) Schematic illustration of the typical temperature dependence of configurational entropy, volume, or enthalpy of glasses and crystals. The glass-transition temperature ($T_g$), the Kauzmann temperature ($T_K$), and the crystal melting point ($T_m$) are indicated on the plot. (b) Schematic illustration of the cooling rate effect on $T_g$.

temperature; this constitutes the foremost identifier of the glass-transition temperature in all glass formers. Interestingly, the density of the glassy state and the location of the glass transition depend on the rate of temperature change, $q = |dT/dt|$. With reference to polymers, the sequence of chain conformation states traversed during a slow cooling process exhibit reduced apparent volume (i.e., higher density), and this behavior extends to lower temperatures, relative to the sequence of conformational states traversed at faster cooling. In parallel, since slower cooling rates allow for longer time for polymer chains to sample different configurations (i.e., increased time for intermolecular rearrangement or structural relaxation), the $T_g$ decreases. The inflection point observed in the apparent volume, enthalpy, or entropy versus temperature plots of glass-forming materials marks the glass-transition temperature and demonstrates a similar cooling-rate dependence (Figure 1.2b). It is well known that variations in the heating rate produce similar effects, which are further complicated by additional aspects of the kinetics of glassy behavior (*structural recovery* effects). All these features reveal that the value of $T_g$, unlike the melting temperature $T_m$, is a rate-dependent quantity, and that the transition defines a kinetically locked thermodynamically unstable state [21], or, otherwise, a metastable state of matter [22].

Among other observations, the shape of the experimentally determined $S(T)$ dependences has provided the stimulus for early studies toward the development of an equilibrium thermodynamic framework for the description of the glass transition. With reference to the generalized behavior already depicted in Figure 1.2a, it becomes clear that in the course of supercooling the difference in entropy between the liquid and crystal phase decreases, with a precipitous decrease in heat capacity at $T_g$. The latter reflects the annihilation of the configurational degrees of freedom that the material possesses in the supercooled liquid state, besides the vibrational contributions found in both the crystalline and glassy ($T < T_g$) states of most materials. If the experimental curve for the entropy or heat capacity of the supercooled liquid is extrapolated to temperatures below the glass transition, it appears that there exists a temperature (the "Kauzmann temperature," $T_K$) at which the *configurational (excess) entropy*, $S_c$, that is the difference between the glass and crystal entropies, will become zero (Figure 1.3) [23,24]. Following the same extrapolation, a further reduction in temperature toward absolute zero would find the noncrystalline state to possess entropy lower than that of the stable crystal phase at the same temperature, which constitutes a violation to Nernst's theorem (the third law of thermodynamics). This paradoxical situation was first pointed out by Walter Kauzmann in 1948 [23]. If the extrapolation is valid, one is forced to admit that even for an infinitely slow cooling process, in which the liquid can reach equilibrium at any temperature, the liquid phase cannot persist below $T_K$. A means to sidestep the so-called Kauzmann paradox, or entropy catastrophe problem, is to consider that a thermodynamic transition to a new state of matter occurs at $T_K$, the *ideal glass transition*, with $T_g \rightarrow T_K$ as the rate of cooling approaches zero. The temperature $T_K$ would thus mark a *divergence* of viscosity and the structural relaxation time of the liquid, and a breaking of the ergodicity, which might be connected with the postulated thermodynamic

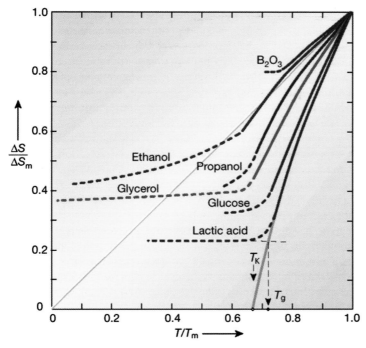

**Figure 1.3** Temperature dependence of the entropy difference between several supercooled liquids and their stable crystals at ambient pressure ($\Delta S/\Delta S_m$, $\Delta S_m$ being the entropy of fusion). The thick lines correspond to experimental data in the range between $T_m$ (normal melting point) and $T_g$. Extrapolation of the curve for lactic acid to lower temperatures is used to show the glass transition and the Kauzmann temperatures (at the point of intersection with the horizontal axis). From ref. [24], with permission, © 2001 Nature Publishing Group.

transition. Experimental manifestation of the phenomenon is presumably masked by the fact that, before getting to this temperature, the liquid falls out of equilibrium. Even so, it is a priori difficult to unequivocally interpret the glass-transition phenomenon as a kinetic manifestation of a second-order transition due to the absence of clear evidence showing growing thermodynamic or structural correlations as the system approaches the transition. Compelling evidence on the existence of a static correlation function that displays a diverging correlation length related to the emergence of "amorphous order," which would classify the glass transition as a standard second-order transition, is still lacking [25]. Recent experimental results on equilibrated structures (see Section 1.2.3.1) cast doubt on the validity of the expectation of a dynamic divergence response, diverging timescales, and a concomitant singularity in the thermodynamics at some temperature well below laboratory $T_g$s.

Considering the glass as a *nonergodic, nonequilibrium,* but slowly evolving *metastable* state of matter, it is expected that its structure will undergo physical processes that will progressively decrease its specific volume, enthalpy, or entropy, until an equilibrated structural state is attained. The principle of the minimization of the Gibbs free energy provides the thermodynamic driving force necessary for the eventual change. The underlying process of slow spatial reorganization of the polymer chains, without irreversible chemical changes, is referred to as *physical aging* (or *structural relaxation*), when it takes place at the use temperature of the polymer, and as *annealing*, when performed at a higher temperature (but below $T_g$). Structure equilibration is achieved quite rapidly at $T \geq T_g$, while, at considerably lower temperatures, glass configurations remain sensibly stable over extremely long periods of time. Physical aging and annealing affect all the temperature-dependent properties that change more or less abruptly at $T_g$ [26,27]. The kinetic attribute of the glass transition is evident in the aging behavior of volumetric or enthalpic data in the glassy state. Figure 1.4 demonstrates the results of a benchmark experiment performed by Kovacs [20], involving the temperature variation of the isobaric (one atmosphere) specific volume data of poly(vinyl acetate) (PVAc). In that study, the sample was initially equilibrated volumetrically at a high temperature. Subsequently, the temperature was stepped to a lower value and the volume was measured at a specified time ($\Delta t = 0.02$ or 100 h) after quenching. Glass densification accomplished in the course of physical aging was found to produce a reduction in the specific volume of about 0.5% for aging time of 100 h at $T \approx T_g - 40\,°C$ ($T_g \approx 27\,°C$), while the longer equilibration times before the temperature change resulted in a lower $T_g$.

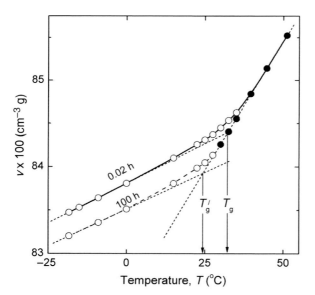

**Figure 1.4** Temperature variation of the isobaric volume of PVAc. The filled symbols represent equilibrium volumes, while open symbols correspond to the volumes observed $\Delta t = 0.02$ and 100 h after quenching of the melt. The intersection of the dashed extrapolated lines marks the glass transition (fictive) temperature. Replotted data from ref. [20]; with permission © 1958 John Wiley & Sons.

Since the measured volumes depend on the temperature and rate of cooling one can also talk about the "time" and "rate" of volume recovery. The rate of volume (or enthalpy) recovery depends on the magnitude and sign of the deviation from the reference equilibrium state, and on how long the sample was allowed to remain at the preceding temperature (memory effect). Plain kinetic effects can be described using a simple kinetic theory, like the single-parameter volume (or enthalpy) relaxation model proposed by Tool [28], and Davies and Jones [29],

$$-\frac{d\delta_v}{dt} = q\delta\alpha + \frac{\delta_v}{\tau_v}, \tag{1.4a}$$

where $\delta_v = (V - V_{eq})/V_{eq}$, $V_{eq}$ is the equilibrium state volume, and $\tau_v$ is the isobaric volume relaxation time. This model was later improved using the Doolittle equation, resulting in

$$-\frac{d\delta_v}{dt} = q\Delta\alpha + \frac{\delta_v}{\delta_{v_g}} \exp\left(\frac{B}{f_g} - \frac{B}{f}\right), \tag{1.4b}$$

where $\tau_{vg}$ is a reference relaxation time and $f_g$ the free-volume fraction at $T_g$. To account for the memory effects, the superposition of a number of elementary relaxation processes has been considered in the multiparameter Kovacs–Aklonis–Hutchinson–Ramos approach [30],

$$-\frac{d\delta_{v,i}}{dt} = q\Delta\alpha_i + \frac{\delta_{v,i}}{\tau_{v,i}} \tag{1.4c}$$

with $\delta_v = \sum_{i=1}^{N} \delta_{v,i}$ and $\Delta\alpha = \sum_{i=1}^{N} \Delta\alpha_i$. Enthalpy relaxation effects on differential scanning calorimetry (DSC) heating scans (Section 1.4.1) are most prominent when the material is isothermally held in the glassy state (20°–50° below $T_g$), for a sufficient duration of time. In addition to rate effects, the glass transition is pressure and path dependent. McKenna and Simon [14] reviewed studies related to the path dependence of $T_g$ and the kinetics of glass formation. Their survey clearly demonstrates that the glass-transition temperature and its pressure dependence are functions of whether the PVT surface of the glass is obtained isobarically, by pressurizing the liquid and cooling from above $T_g$, or isochorically, using variable pressure to retain liquid volume constant until a low temperature is reached at a constant rate.

## 1.2.3
### Relaxation Dynamics and Fragility

The relaxation dynamics in many different kinds of materials encompass contributions from various types of motional processes spanning a range of length scales, which become prominent at different temperature ranges and/or timescales. By virtue of their high densities, supercooled liquids exert strong frustration constraints on the dynamics of individual atomic/molecular entities or "particles" (e.g., atoms, oligomeric molecules, pendant groups, short-chain segments, or even bigger parts of the chain). As temperature decreases toward $T_g$, a tagged particle is most likely trapped by neighbors (i.e., caged) given that the amplitude of thermodynamic fluctuations decreases following a decrease in temperature. Near the glass-transition temperature, several groups of particles may remain caged for relatively long times. For them, liberation from the cage requires collaborative rearrangement of several other particles in its environment, which themselves are also imprisoned. The volume over which cage restructuring – by cooperative motions – must occur presumably increases as the molecular packing increases (with decreasing temperature). Considering the complexity of the systems involved and the diversity of configurational changes that may cause relaxation of a polymeric material, fundamental research in each system often focuses on the description of the time evolution of the relaxation dynamics (i.e., plots of the time dependence of the relaxation or response functions), at a constant temperature, and the creation of relaxation maps (i.e., plots of the temperature dependence of the relaxation times of distinct groups of particles). Some of these issues and the pertinent concept of liquid fragility will be briefly discussed.

### 1.2.3.1 Relaxations in Glass-Forming Materials

For over half a century, the nature of the relaxational response of supercooled liquids and glasses has been extensively explored, in an effort to expand our understanding of the structure–property relationships in the rapidly evolving collection of glassy materials, and at the same time establish connections among experimental responses and theoretical predictions. Out-of-equilibrium studies of glassy dynamics reveal a collection of modes, extending over a broad temperature-, frequency-, or time range. At short times of observation at constant temperature, the approach to equilibrium after a given perturbation is dominated by very fast to moderately fast motions of small parts of the macromolecular chain. The picosecond dynamics of disordered materials include a *fast secondary relaxation* process, which appears as an anharmonic relaxation-like signal (a broad quasielastic scattering) in the GHz–THz region of excitation spectra [31]. This contribution is commonly ascribed to caged molecular dynamics (i.e., cage rattling) with relaxational activity displaying gas-like power-law temperature dependence [32]. Close to it, Raman and neutron scattering inelastic studies reveal a rather controversial lower frequency vibrational mode, or group of modes, known collectively as the "boson peak." Potential correlations between these early-time modes and the long-time dynamics of glass-forming materials emerge from studies relating characteristics (e.g., its intensity and frequency) of the nearly temperature-independent boson peak with the concept of liquid fragility, Kohlrausch's exponent ($\beta_{KWW}$) [33], and the cooperativity length scale ($\xi_\alpha$) [34], all strongly linked to the glass-transition dynamics of disordered media. Several other important secondary processes occur on timescales much slower than cage rattling, but much faster than the structural ($\alpha$) relaxation. These are related to complicated, though local, non or not fully cooperative [35] dynamics. A number of physical origins have been proposed for the principal *slow secondary relaxation* process (in the kHz region of isothermal relaxation spectra [15–17]), the so-called Johari–Goldstein (JG) β-process, a process widely recognized as an intrinsic feature of the glassy state and frequently deemed to originate from the same complicated frustrated interactions leading to the glass transition. Of the alternative attributions proposed hitherto, it is worth mentioning its correlation with molecular motions occurring in "islands of mobility" (i.e., regions of relatively loose structure [36]), the highly restricted stepwise reorientation of practically all molecules in a system [37], and its discussion in terms of intermolecular [38] degrees of freedom or even intramolecular [39] ones. Other secondary relaxations (γ or δ, in the accepted notation for amorphous materials [15,16]) entail more trivial and system-specific motions of structural entities, usually connected with intramolecular degrees of freedom, such as simple bond rotations of lateral groups (including rotations within side groups). With the exception already noted for the modes contributing to the fast (picosecond) dynamics of disordered materials, all other secondary mechanisms are commonly regarded to involve relaxation jumps over asymmetric double-well potentials (e.g., Gilroy–Phillips model [40]). The temperature dependence of the respective relaxation times can thus be well described by a simple exponential Arrhenius-type equation, that is,

$$\tau(T) = \tau_0 \exp\left(\frac{E_{act}}{k_\beta T}\right), \tag{1.5}$$

where $\tau_0$ is the pre-exponential factor (or Arrhenius prefactor) and $k_\beta$ is the Boltzmann constant. The apparent activation energy, $E_{act}$, is typically determined by internal rotation barriers (intramolecular part) and the environment of the relaxing unit (intermolecular part, linked to the stereochemical configuration of the chains). Broad distributions of relaxation times and a strongly temperature-dependent width (presumably due to a Gaussian distribution of barrier heights) are common features of signals related to the JG mode [41]. In several cases, by extrapolating to high temperatures the Arrhenius line, the slow β-mode seems to result from bifurcation of the structural relaxation mode (α-relaxation, Figure 1.5), which encompasses cooperative segmental motions on much longer length scales.

The abrupt retardation of molecular mobility in the course of vitrification is an important facet of the relaxation dynamics of disordered systems. Various experimental results and simulations indicate that the structural relaxation of a supercooled liquid is a dynamically and spatially heterogeneous process with a strongly non-Arrhenius relaxation behavior. Dynamic heterogeneity describes the spatial heterogeneity of the local relaxation kinetics, manifested by the coexistence of "slow" and "fast" mobility regions of limited length within a material [42–44]. Different assumptions that introduce heterogeneity in supercooled liquids exist, including the old concept of liquid-like cells that create liquid-like clusters (Cohen–Crest model), the conception of a solid glass with a small fraction of fluidized domains of extremely high mobility (Stillinger–Hodgdon

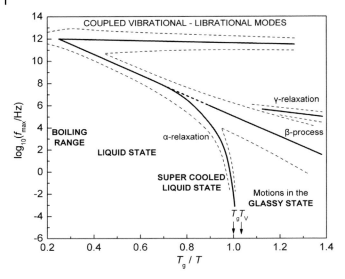

**Figure 1.5** Schematic illustration of the temperature dependence of characteristic relaxation frequencies of representative oscillatory and relaxational modes of motion found in glass-forming polymer liquids. The thick lines indicate the trend of the most probable frequencies, while thin dashed lines indicate typical half-widths of the respective bands. Note the increasing separation (decoupling) of the vibrational and structural relaxation times as the temperature decreases approaching $T_g$. A further decoupling of characteristic motions occurs near and below $T_g$, giving rise to the JG β-process.

model), and the hypothesis of the existence of small distinguishable subvolumes in the system that relaxes statistically independent of their environment. The terminology germane to these regions includes the influential "cooperatively rearranging region(s)" (CRR(s)) introduced in the Adam–Gibbs model, the concept of "entropic droplets," and the "domains" of locally preferred structures, advocated by thermodynamic treatments of glass formation (see Section 1.2.4.3). So far, most techniques provide indirect estimates of the relevant cooperativity length scale, by invoking thermodynamic fluctuation formulae, combined with an appropriate set of *ad hoc* assumptions, to obtain $\xi_\alpha$ from the available experimental data, or by simply introducing external perturbations (e.g., confinement in finely regulated nanometer-sized geometries) [43,45–47]. In the subvolume of each CRR, for example, the density, the temperature, the entropy, and the energy ($E$) are somewhat different, and their mean square fluctuations $\langle \Delta \rho^2 \rangle$, $\langle \Delta T^2 \rangle$, $\langle \Delta S^2 \rangle$, and $\langle \Delta E^2 \rangle$, respectively, are given by standard relations of statistical thermodynamics. Among others [47], Donth proposed to correlate these relations with the width of relaxation time distribution of the so-called α-relaxation process [41]. Each subvolume can be then considered as a thermodynamic system in metastable equilibrium with fluctuating variables having a Gaussian distribution, and a distinct glass-transition temperature ($T_\alpha$) and relaxation time ($\tau_\alpha$). Accordingly, the relaxation time distribution can be related to the glass transition one, with $\langle T_\alpha \rangle$ assumed to be the conventional glass-transition temperature of the system. In Donth's approach, the characteristic volume of cooperativity at $T_g$ ($V_\alpha$) and the number of segments in the cooperative volume ($N_\alpha$) can be estimated by

$$V_\alpha \propto \xi_\alpha^3 = \frac{\Delta(1/C_V)}{\rho(\delta T)^2} k_B T_g^2 \tag{1.6a}$$

and

$$N_\alpha = \frac{\rho N_A \xi_\alpha^3}{M_0}, \tag{1.6b}$$

respectively, with $N_A$ the Avogadro number, $(\delta T)^2$ the mean square temperature fluctuation related to the dynamic glass transition of a CRR, $C_v$ the isochoric heat capacity with $\Delta(1/C_v) \approx \Delta(1/C_p) = (1/C_p)_{glass} - (1/C_p)_{liquid}$, and $M_0$ the molar mass of the repeat unit (monomer). Theoretical claims [46], thermodynamic treatments starting from volume, temperature, and entropy fluctuations [41,47], and extensive use of sensitive experimental probes of dynamic and spatial heterogeneity [43,47] – including results based on the Boson peak frequency [34] – have provided values for $\xi_\alpha$ in the range ~1–4 nm [35]. In parallel, the cooperativity length scale and $N_\alpha$ (of the order of 100 near $T_g$) appear to increase as temperature decreases near $T_g$ [31].

The strongly non-Arrhenius temperature behaviors for the structural relaxation dynamics and the viscosity of glass-forming liquids at temperatures exceeding $T_g$ are well documented (Figure 1.5). Both are frequently interpreted in terms of the empirical Williams–Landel–Ferry (WLF) [48] or Vogel–Fulcher–Tammann–Hesse (VFTH) [49] equations, which were later rationalized in terms of free-volume (or configurational-entropy) concepts. The WLF equation is expressed as

$$\log_{10} \alpha_T = \frac{-C_1(T - T_r)}{C_2 + (T - T_r)}, \tag{1.7a}$$

where $\alpha_T$ is called the temperature shift factor (generally known as the reduced variables shift factor), $T_r$ is a reference temperature, and $C_1$ and $C_2$ are constants. The shift factor is related to the viscosity, $\alpha_T = \eta(T)/\eta(T_r)$, relaxation times, $\alpha_T = \tau(T)/\tau(T_r)$, and several other mechanical (e.g., tensile strength and compliance) or dielectric (e.g., permittivity and electric modulus) relaxing quantities. When $T_g$ is taken as the reference temperature, the following form is obtained:

$$\log_{10} \alpha_T = \frac{-17.44(T - T_g)}{51.6 + (T - T_g)}. \tag{1.7b}$$

By averaging data for various types of synthetic high polymers, the values of $C_1 = 17.44$ and $C_2 = 51.6$ K were derived and applauded as "universal" constants for linear amorphous polymers of any chemical structure. Their usability in complex polymeric systems must be treated with cautiousness, since a different set of values is to be expected when distinct dynamic processes and/or substances are explored. Despite these shortcomings, Eq. (1.7b) introduces some important *kinetic* aspects of the glass-transition phenomenon. For instance, if the time frame of an experiment is decreased by a factor of 10 near $T_g$, this equation reveals an increase of the glass-transition temperature of about 3°. Of particular interest in dynamic experiments is the temperature dependence of the structural α-relaxation times, $\tau_\alpha(T)$, for which the WLF temperature dependence is expressed as

$$\tau_\alpha(T) = \tau(T_r)\exp\left(\frac{-C_1(T - T_r)}{C_2 + (T - T_r)}\right) \tag{1.8}$$

and the mathematically equivalent VFTH equation has the form

$$\tau_\alpha(T) = \tau(\infty)\exp\left(\frac{C}{T - T_V}\right), \tag{1.9}$$

where $C = C_1(T_r - T_V)$ is a material parameter and $T_V = T_r - C_2$ denotes the so-called Vogel temperature (at which the relaxation time is *extrapolated* to diverge). In the absence of deep arguments regarding the underlying physics of glasses, the VFTH equation is mostly regarded an entirely heuristic modification of the Arrhenius rate law to include a finite divergence temperature. Even though the physical meaning of the Vogel temperature has not been clearly defined [41], the universality of the VFTH equation in a wide temperature range ($T_g$ to $T_g + 100$ K) makes clear that $T_V$ is a significant parameter for the dynamics of the glass transition. A survey of the literature provides evidence of a weak connection between $T_V$ and $T_K$ ($T_V \approx T_K$ [50]), with $T_V$ generally found to be approximately 30°–50° (depending on system's fragility) below conventional laboratory $T_g$s. In practice, the glass-transition temperature can be obtained by extrapolating the Arrhenius plot constructed for the structural relaxation times or characteristic frequencies (plots of $\log \tau$ or $\log f_{max}$ versus $10^3/T$, with $f_{max} = 1/2\pi\tau$), using Eq. (1.8) or (1.9) along with the usual convention $\tau(T_g) = 100$ s [51].

Theoretical treatments, computer simulations, and a number of experimental results strongly argue in favor [52] or against [53–55] the existence of a dynamic divergence phenomenon – a behavior also referred to as "super-Arrhenius" – at some temperature above absolute zero. The "geological" ages required for a material to attain equilibrium far below $T_g$ preclude, in general, extensive testing of the above conjecture. Recent data on the temperature dependence of the shift factor obtained by dielectric spectroscopy for PVAc [54], using samples aged to equilibrium as much as 16° below the calorimetric glass-transition temperature, demonstrate, for example, an Arrhenius sub-$T_g$ behavior in contradiction to the predictions made by classic theories. Further work on a Cenozoic (20 million years old) Dominican amber [55] was able to probe equilibrium dynamics nearly 44° below $T_g$, and subsequently present more stronger experimental evidence of nondiverging dynamics at far lower temperatures than previous studies. Notice that several other functions may well provide adequate description of the super-Arrhenius behavior of glass-forming liquids, showing either a divergence at zero temperature (e.g., the Bässler-type law, $\tau_\alpha(T) \sim \exp(A/T^2)$ [56]) or no divergence at all (e.g., the DiMarzio–Yang formula, $\tau_\alpha(T) \sim \exp(-AF_c/k_\beta T)$, where $F_c$ is the configurational part of the Helmholtz free energy [57]).

### 1.2.3.2 The Concept of Fragility

In an attempt to establish some link between the observed thermodynamic behaviors of glass-forming systems and the temperature dependences of several dynamical quantities, Angell introduced the concept of *liquid fragility* [58]. In Angell's classification scheme, the word "fragility" is used to characterize the rapidity with which a liquid's properties (such as $\eta(T)$ or $\tau(T)$) change as the glassy state is approached, in contrast to its colloquial meaning that most closely relates to the brittleness of a solid material. Over the years, fragility has become a useful means of characterizing supercooled liquids, despite the criticism on some inferences of the concept [59]. The term "strong" liquid suggests a system with relatively stable structure and properties (such as the activation energy barriers associated with viscosity or the structural relaxation time) that do not change dramatically in going from the liquid into the glass, while a "fragile" liquid behaves in a reverse manner. Formally, fragility reflects to what degree the temperature dependence of a dynamic property of the glass former deviates from the Arrhenius behavior. One way of evaluating fragility is to construct fragility plots (Angell plots), where the logarithm of a dynamical quantity is plotted versus $T_g/T$ (Figure 1.6) [60]. Several parameters have been introduced, at different theoretical contexts and with varying level of success, for characterizing quantitatively the fragility of glass-forming liquids (e.g., see treatments of Doremus [59], Bruning and Sutton [61], and Avramov [62]). The most common definition of fragility is the fragility parameter (or steepness index), $m$, which characterizes the slope of a dynamic quantity ($X$) with temperature as the material approaches $T_g$ from above [63],

$$m = \left(\frac{\partial \log_{10}X}{\partial(T_g/T)}\right)_{T=T_g}.\tag{1.10}$$

Bearing in mind the non-Arrhenius temperature dependence of $\tau_\alpha$, for example, Eq. (1.10) takes the form

$$m_{\text{VFTH}} = \left(\frac{\partial \log_{10}\tau_\alpha}{\partial(T_g/T)}\right)_{T=T_g} = \frac{C/T_g}{(\ln 10)\left(1 - \dfrac{T_V}{T_g}\right)^2},\tag{1.11a}$$

which provides an estimate of the fragility index in terms of the Vogel temperature. Considering the expression of the relaxation time given by the Tool–Narayanaswamy–Moynihan formula [28,64], another dynamic estimate of the fragility index can be obtained from DSC experiments with different heating rates, through a relation that links $m$ with the apparent activation energy of structural relaxation $\Delta h^*$,

$$m_{\text{DSC}} = \frac{\Delta h^*}{RT_g \ln 10}.\tag{1.11b}$$

The fragility can be intuitively related to the cooperativity of atomic motions in the glassy state (anticipating an increase in cooperativity with increasing $m$)

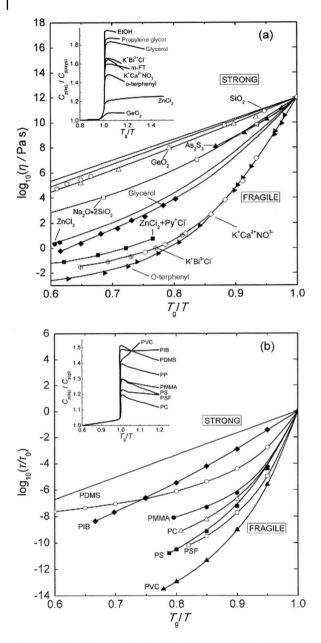

**Figure 1.6** Fragility plots of glass-forming liquids (Figure 1.6a) and polymers (Figure 1.6b). Following the established practice, in the first plot $T_g$ is roughly defined for each system by the relation $\eta(T_g) = 10^{13}$ poise ($10^{12}$ Pa·s), while in the second plot, $T_g$ is defined as the temperature at which the segmental relaxation time $\tau(T_g)$ equals 100 s. Insets show the heat capacity changes at the glass transition for selected systems (replotted from data appearing in ref. [60], with permission © 2001 AIP Publishing LLC).

[41,45], as well as to the breakdown of the Stokes–Einstein relation between viscosity and diffusion coefficient [65]. In that respect, several studies explore the validity of empirical relationships among components of the fragility parameter [34,66] and the length scale of cooperativity or spatial variations in dynamics. For example, by applying the chain rule of differentiation

$$
m = \left. \left( \frac{\partial \log_{10} \tau_\alpha}{\partial \left( \frac{T_g}{T} \right)} \right) \right|_{P} \Bigg|_{T=T_g}
$$

$$
= \left. \left( \frac{\partial \log_{10} \tau_\alpha}{\partial \left( \frac{T_g}{T} \right)} \right) \right|_{V} \Bigg|_{T=T_g} + \left. \left( \frac{\partial \log_{10} \tau_\alpha}{\partial P} \right) \right|_{T} \Bigg|_{T=T_g} \times \left. \left( \frac{\partial P}{\partial \left( \frac{T_g}{T} \right)} \right) \right|_{V} \Bigg|_{T=T_g}
$$

one separates the conventional (isobaric) fragility parameter into two terms

$$
m = m_V + \frac{\Delta V^{\#}}{k_B \ln 10} \times \frac{\alpha_T}{\kappa_T} \tag{1.12}
$$

with $\Delta V^{\#}$ denoting the activation volume at $T_g$. Following the above treatment, Sokolov and coworkers [34] suggested that the isochoric (constant density) fragility, $m_V$, characterizes the pure thermal contribution to fragility that bears no connection to the length scale of dynamic heterogeneities. The second term of Eq. (1.12) encompasses the volume contribution to fragility, that is, the effect due to the temperature-induced change of density under isobaric conditions. Given the correlation evidenced between the cooperativity volume $\xi_\alpha^3$ and $\Delta V^{\#}$ [34], this term is likely to depend directly on cooperativity. A relationship between parameters characterizing the stretching of the relaxation function and isochoric fragility is also probable [66]. Other studies provide pieces of evidence for possible correlations between the conventional (atmospheric pressure) estimates of the fragility parameter and molecular or structural properties of the material, such as its chemical composition, the type of bonding, intermolecular interactions, or the degree of microphase separation [67].

The values of $m$ range from $\sim 250$ [68] for very fragile glass-forming liquids (e.g., ionic systems, organic materials, or polymers with nondirectional intermolecular bonding) to the theoretically low limit of $\sim 16$ [69] for very strong glass formers (the network oxides $SiO_2$ and $GeO_2$, $BeF_2$, etc. [70]). Highly fragile materials demonstrate narrow transitions, while those with lower fragility indices have relatively broader transitions. The roles of chemical structure, composition, and main-chain stiffness in the glass-forming tendency of polymers [71] and polymer blends [51], as well as possible correlations among the "dynamic fragility" index and thermodynamic measures of liquid fragility ($m_T = \Delta C_p$, $C_{p(liq)}/C_{p(gl)}$, $C_{p(liq)}/C_{p(crys)}$, or $1 + \Delta C_p/S_c$, all determined at $T = T_g$, typically used to assert "thermodynamic fragility" [60,72]), are issues in debate [73]. As suggested by the Adam–Gibbs theory (see Section 1.2.4), kinetically fragile liquids are expected to have large configurational heat capacities [6], resulting from their configurational entropy changing rapidly with temperature. Strong glass-forming liquids are resistant to temperature-driven changes in the medium-range order. Therefore, the amount of configurational entropy in the liquid is small, as is the change in heat capacity at $T_g$. Even though the heat capacity changes shown in the inset of Figure 1.6a support the positive correlation between $m$ and $\Delta C_p$, more recent data appear contradicting. Huang and McKenna [60] classified the experimental $m$ versus $\Delta C_p$ dependences into three groups: polymeric glasses with a negative correlation (Figure 1.6b) [72], small-molecule organics and hydrogen-bonding small molecules with no correlation, and inorganic glass formers with a positive one [74]. There are also several reports demonstrating that thermodynamic and kinetic fragilities are not strongly correlated [75], especially when polymeric systems are considered. In view of that, a system concluded to be kinetically fragile will not necessarily be also thermodynamically fragile. Finally, a direct correlation between fragility indices and the average size of the CRRs is frequently considered [41,45,76].

## 1.2.4
### Theoretical Approaches to the Glass Transition

#### 1.2.4.1 General Overview
The intriguing phenomenology of the glass transition has been the driving force of intense efforts aiming to establish firm theoretical perspectives with wide quantitative support for the microscopic and relaxational behavior of liquids in the glass transformation range. The marked decrease in molecular mobility as a system passes through its glass-transition temperature has led several researchers to construct early theories of glass transition based on concomitant changes of conjugate thermodynamic variables, such as the free volume and the configurational entropy. The defect diffusion [77], free volume [78], and configurational entropy [44,79,80] approaches, all dating back to the 1960s, remain in the forefront of current interest about the glass transition. While these early theories fall short in properly defining – among other properties and phenomena – the molecular motions involved in the glass-transition mechanism [13], they are still frequently invoked in interpretations of experimental results. A number of more recent theories and elaborate concepts, including the potential energy landscape (PEL) picture [24,81,82], the coupling model (CM) [13], the mode-coupling theory (MCT) [83–85], and the RFOT [86] theory, the configuron percolation model (CPM) [87–89], as well as the concepts of kinetic constraints [90,91] and geometric frustration [92], have provided an amplification of our perceptions on the glass-transition phenomenon and more plausible explanation of certain experimental observations. Still, irrespective of the intense theoretical efforts to handle the glass-transition phenomenon employing arguments resembling thermodynamic or purely dynamic transitions, we have not yet arrived at a comprehensible theory of supercooled liquids and glasses. Their behavior near the glass transition has been described, but not all that behavior is thoroughly explained by a single theoretical concept [85]. Elements of certain theoretical frameworks

and some insight into their strengths, flaws, capabilities, and limitations will be provided in the following paragraphs; the reader is referred to a − regretfully condensed − list of review papers [13,25,85,93–95] for a comprehensive account of the available theoretical approaches.

### 1.2.4.2 Energy Landscapes and Many-Molecule Relaxation Dynamics

In a seminal 1969 paper, Goldstein [81] put forward the notion that atomic motions in a supercooled liquid comprise high-frequency vibrations in regions of deep potential energy minima in addition to less frequent transitions to other such minima. In an amplification of this concept, Stillinger and coworkers [24,82] formulated the PEL picture of glassy systems, a multidimensional surface describing the dependence of the potential energy on the coordinates of the atoms or molecules. Their conception provided a "topographic" view of phenomena associated with glass formation, along with a valuable theoretical background in the pursuit of distinguishing among vibrational and configurational contributions to the properties of a viscous fluid.

In the phenomenological PEL approach of Stillinger and Weber [24], an $N$-particle system is represented by a potential energy function $U(\vec{r}_1, \vec{r}_2, \ldots, \vec{r}_N)$ in the $3(N − 1)$-dimensional configuration space. The energy of the disordered structure is partitioned into a discrete set of "basins" connected by saddle points − a picture that represents the complicated dependence of potential energy (or enthalpy) on configuration [96]. Each basin contains a metastable local (single) energy minimum and each corresponds to a mechanically stable arrangement of the system's particles. In terms of PEL, relaxations ascribed to short-time molecular motions are considered to occur via intrabasin vibrations (harmonic oscillations) about a particular structure, while long-time molecular motions are considered to take place via occasional activated jumps over saddle points into neighboring basins. In an amplification of this concept, the picture of "metabasins" has been introduced [97], with each metabasin consisting of several local minima ("inherent structures") separated by similar low-energy barriers. Jumps within the superstructure of a metabasin are connected with secondary relaxation events (Figure 1.7a), while much slower collective molecular motions (i.e., "ergodicity restoring" processes related to the glass transition) are considered to proceed via infrequent jumps between neighboring metabasins, separated by large barriers relative to $k_B T$. While the PEL is suitable for modeling glass-transition behavior under isochoric conditions, almost all experimental studies of glass formation are performed under constant pressure conditions. To that end, an enthalpy landscape approach [98] has conveyed an extension of PEL to an isothermal–isobaric ensemble, allowing for changes in both particle positions and the overall volume of the system. In all energy landscape models (potential-energy, free-energy, or enthalpy variants [97,98]), the dynamics are to a certain degree cooperative, since transitions between two minima engage the coordinates of all particles of some localized region. Related frameworks have contributed a certain degree of understanding of the nature of the glass transition and the glassy state of matter. It has been suggested, for example, that it is not possible for the

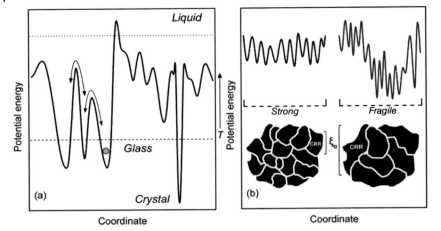

**Figure 1.7** (a) Schematic illustration of a 1D PEL. At a given temperature, the system can visit the configurations between the dashed line and the solid curve. In the glassy state, the system is trapped in a metastable state performing harmonic oscillations. Jumps between neighboring basins (arrows) are related to the β relaxation, while α relaxation involves jumps into neighboring metabasins. (b) Hypothetical forms of the PEL and their relation to the strong/fragile character of a material described in terms of the average size of the CRRs.

configurational entropy of a supercooled liquid to vanish at a finite temperature, since some transitions are always allowed between basins, but only at absolute zero temperature, when it is confined to just a single microstate. Stillinger further extended the concept of a single Kauzmann temperature to multiple "Kauzmann points" in the temperature–pressure plane of a system [82,99]. Although the above statements suggest that the ideas of the "Kauzmann paradox" and the purported "ideal glass transition" are illusive, mere results of improper extrapolations [99] and lacking experimental corroboration [100], several counterarguments cannot be ignored [46]. Worthy of note is also the proposed connection between liquid fragility (as well as between the sharpness and strength of the glass transition [98]) and the topography of the underlying potential energy or enthalpy landscapes, which relates the behavior of strong or fragile liquids to landscapes of rather uniform roughness or highly nonuniform topography, respectively (Figure 1.7b) [82].

Energy landscapes contribute to our understanding of the processes observed during the evolution of molecular dynamics from short to very long times [97]. These encompass contributions from a range of modes, starting from fast vibrations and localized motions and progressively entering the time window of cooperative molecular motions. General characteristics of this behavior are schematically illustrated in Figure 1.8, for the case of the relaxation function, $\Phi(t)$, a pattern closely resembling the time dependence of the density fluctuation autocorrelation function at constant temperature. A phenomenological description of the distinctly nonexponential time dependence of the long-times primary

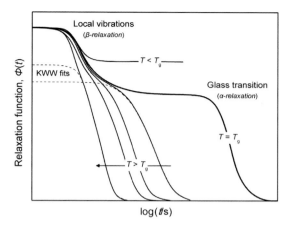

**Figure 1.8** Schematic illustration of the time variation of the relaxation function. Note the different response in the ergodic (supercooled liquid, $T > T_g$) and the nonergodic (glassy, $T < T_g$) states of the system.

relaxation (glass transition) is provided by a function having the *Köhlrausch–Williams–Watts* (KWW) stretched exponential form [101]

$$\Phi_\alpha(t) = \exp\left[-\left(\frac{t}{\tau_\alpha}\right)^{\beta_{\mathrm{KWW}}}\right], \tag{1.13}$$

with $0 < \beta_{\mathrm{KWW}} \leq 1$. Parameter $\beta_{\mathrm{KWW}}$ delivers a convenient measure of the width of the dispersion of the $\alpha$-relaxation and the extent of the many-molecule relaxation dynamics, both critically controlled by intermolecular interactions and spatial constraints. Nonexponentiallity is typically accounted for by two fundamentally different – yet, experimentally indistinguishable – scenarios: the material is considered either to comprise a set of heterogeneous environments with exponential relaxation activity but distinctly different characteristics among different environments (e.g., a scenario equivalent to a distribution of relaxation times), or to be entirely homogeneous with each molecule relaxing nearly identically with an intrinsically nonexponential manner.

Many-molecule relaxation is a central element of the CM, a general theory of relaxation presented by Ngai [13]. Even though this approach is clearly not a complete theory of the glass transition, it has successfully tackled problems originating from prior oversimplified or even illusive considerations of nontrivial interactions between relaxing units in glass-forming materials and the inadequate description of their many-molecule relaxation dynamics. The idea behind CM is the picture of a cooperative system of identical relaxing species (such as ions in a viscous conductor or entangled polymer chains). At short times, the particles can be considered to be noninteracting and thus the relaxation rate is constant, while after some critical time, $t_c$, the molecules interact more strongly and thus the relaxation becomes slowed down. In terms of the

CM, the characteristics of the structural relaxation are correlated with the aspects of the processes that have transpired before it, which include the caging of the molecules (picosecond dynamics) and the universal JG β-process [36]. A rational outcome of the notion that the α-relaxation process originates from the relaxation of individual molecules is to consider that, at sufficiently short times, the many-molecule relaxation is reduced to isolated local motions independent of each other. These correspond to the primitive relaxation of the CM, which can be seen as part of the faster processes in the relaxation spectrum. The model sets forth a relation between $\tau_\alpha$ and the primitive relaxation time ($\tau_p$), or equivalently the JG relaxation time ($\tau_{JG}$), via the coupling parameter $n$ (with $n = 1 - \beta_{KWW}$) characterizing the primary relaxation, of the form [13]

$$\tau_\alpha(T, P) = \left[ t_c^{-n} \tau_p(T, P) \right]^{1/(1-n)} \approx \left[ t_c^{-n} \tau_{JG}(T, P) \right]^{1/(1-n)}. \tag{1.14}$$

The CM predicts the short-time behavior to be essentially Debye. Although the temperature and pressure dependences of $\tau_\alpha$ and $\tau_p$ (or $\tau_{JG}$) are not given or derived, the applicability of Eq. (1.14) has been successfully tested for a wide range of glass formers. The stronger dependence of $\tau_\alpha(T, P)$ compared to that of $\tau_p(T, P)$ or $\tau_{JG}(T, P)$ is expressed by the superlinear factor $1/(1-n)$ and relates to the longer length scale of the motions involved. The CM also provides an explanation of changes in the relaxational behavior of glass formers – including the component dynamics of mixtures or the dynamics of nanoconfined polymers – in terms of the temperature dependence of $n$ or the width of the α-relaxation. Of the cases compiled by Ngai [13], here is only mentioned the projected crossover of the temperature dependence of $\tau_\alpha$ from one VFTH equation to another, at some temperature $T_B > T_g$, coincident with the apparent onset of bifurcation of $\tau_{JG}$ from $\tau_\alpha$, and the onset of decoupling of the translational and orientational motion, which are all related to the small values of $n$ at $T > T_B$ and its more rapid increase with decreasing temperature below $T_B$. Ngai and Rendell [13] mention that an explanation of the heterogeneous picture of relaxation in terms of spectra or distributions of relaxation times is incompatible with the model. They argue that interactions perturb the relaxation in a way as to make it inherently nonexponential and not that it arises from superposition of single exponential (Debye) processes. A main limitation of the CM is connected with the absence of a detailed explanation of the potential relaxation mechanisms in molecular level and how these exactly contribute to the overall macroscopic behavior.

### 1.2.4.3 Approaches with an Underlying Avoided Dynamical Transition

The most famous purely dynamical approach to the glass transition is the MCT, a mean-field treatment of the phenomenon based on a microscopic theory of the dynamics of fluids [83–85]. The theory exploits the idea of a nonlinear feedback mechanism in which strongly coupled microscopic density fluctuations lead to structural arrest and diverging relaxation time at a critical point, with no singularity observed for the thermodynamics. The physical picture adopted by the originally developed scenario of the *idealized* MCT (*i*MCT) attributes the viscous slow-down with decreasing temperature to a so-called cage effect, that is, the assumption that each particle in a dense fluid is kinetically constrained (confined) in a cage formed by neighboring particles. At low temperatures, the probability of occurrence of a strong spontaneous density fluctuation, large enough to liberate a particle from its cage, appears insignificant. In consequence, as the temperature gets lower and the system gets denser, structural arrest occurs because particles can no longer leave their cage even at infinite time. Within the MCT, fast secondary relaxations are related to relatively rapid local motions of molecules trapped inside cages, while the slow process of the breakup of a cage itself contributes to the structural relaxation. Large-scale spatial motion typical of a fluid can only proceed cooperatively, that is, one of the caging particles has to make way, which can only happen if one of its neighbors moves, and so on. One of the main predictions of the *i*MCT is that dynamical freezing and a transition from an ergodic to a nonergodic state occurs at a critical temperature $T_{MCT}$ (~$1.2T_g$). Above $T_{MCT}$, where ergodicity is obeyed, all regions of phase space are accessible, while below $T_{MCT}$, where structural arrest occurs, parts of phase space remain inaccessible. At $T = T_{MCT}$, the *i*MCT visualizes the "self" part of the intermediate scattering function, $F_s(k, t)$, to decay (in the limit $t \to \infty$) to a finite, nonzero, number called the *nonergodicity parameter*. For temperatures exceeding $T_{MCT}$, the *i*MCT predicts that the scattering function decays to zero in two steps (β- and α-regimes), with the decay of the correlation function at long times approximated by the stretched exponential KWW function (see Figure 1.8). Approaching $T_{MCT}$ from above, the structural relaxation time (and viscosity) scales in a power-law fashion

$$\tau_\alpha(T) \propto (T - T_{MCT})^{-\gamma}, \tag{1.15}$$

where $T_{MCT}$ is a critical temperature for the onset of the glass transition, and the exponent $\gamma > 1.5$.

The *idealized* mode-coupling approach successfully describes key aspects of the relaxation dynamics of moderately supercooled liquids, with its main achievement involving the prediction of the two-step relaxation process that emerges as temperature decreases, in agreement with experimental studies and simulation results. Nevertheless, the dynamic arrest and the predicted singularity at the purported critical temperature of the model bear no connection to the laboratory glass transition or a transition to an "ideal" glass state of matter. Experiments clearly provide no evidence of critical singularities above $T_g$ in real systems (e.g., molecular liquids and colloids), while at the shortcomings of this theory one has to count the complete neglect of heterogeneities [47]. Later revisions offered an *extended* version of the theory (*e*MCT), in which inclusion of flux correlators, besides the density correlators, introduced "phonon-assisted hopping processes" that can restore ergodicity below $T_{MCT}$. These changes generated a "rounding" of the singularity, due to the existence of secondary couplings that allow activated processes to occur lower than $T_{MCT}$. The debatable robustness of the *e*MCT to describe dynamical correlations and some aspects of dynamic heterogeneities in the regime $T_g \leq T < T_{MCT}$ suggests that at least in its present form it does not provide a complete theory of the glass transition and, therefore, a particular means of predicting the transition from liquid to glass.

Even so, the mathematical formalism and analysis offered by *e*MCT is acknowledged as a useful starting point in the description of novel systems with unknown behavior (for a review, see Berthier and Biroli [25]).

A different approach offers a group of simple lattice models of glasses, collectively described as kinetically constrained models (KCMs), which are characterized by a trivial equilibrium behavior, but interesting slow dynamics due to restrictions on the allowed transitions between configurations. These models rely on a Hamiltonian for noninteracting entities (spins or particles on a lattice) combined with specific constraints on the permitted motions of any such entity. Their perspective on the glass-transition problem assumes that most of the interesting properties of glass-forming systems are dynamical in origin, and all explanations develop without recourse to any complex thermodynamic behavior. This viewpoint contradicts essential thermodynamic arguments adopted by a number of theoretical treatments (see the following section). Central physical assumptions in most KCMs appear to be the supposition of sparse mobility for the particles (i.e., the atomic motions are deemed to principally involve small amplitude vibrations and not diffusion steps) and the notion of insignificant contribution of static correlations in system dynamics. With appropriate choices of the constraints and explicit mechanisms (e.g., taking into consideration "facilitation" processes), several KCMs provide a natural explanation of the super-Arrhenius slowdown of dynamics and dynamical heterogeneities (e.g., nonexponentiallity) as a consequence of local, disorder-free interactions, notably without the emergence of finite temperature singularities [90]. The super-Arrhenius temperature dependence of the relaxation time is often described by a Bässler-type expression (see Section 1.2.3.1), for temperatures much below an "onset" that marks the beginning of facilitated dynamics with sparse mobile regions. Despite their reliance on local constraints, the implementation of a form of dynamical frustration enables the KCMs to describe cooperative dynamics, aging phenomena, and ergodicity breaking transitions [91]. At low temperatures (or high densities), a struggle between the scarcity of mobility defects (excitations, vacancies) and their need to facilitate motion at neighboring regions is taken into account, leading to a hierarchical collective relaxation. Whether the KCMs provide the correct theoretical framework to explain the glass transition is highly debatable. Among their serious shortcomings stands out the fact that neither glass-transition's phenomenology related to thermodynamics is acknowledged nor are the nontrivial static correlations (argued to accompany the increase of relaxation time in fragile glass formers) properly addressed. Furthermore, these models provide no information either on the slow $\beta$-relaxation or on fast relaxation processes and pertinent anomalous vibrations, and, more importantly, on their acknowledged ties to the structural relaxation mechanism.

### 1.2.4.4 Models Showing a Thermodynamic (or Static) Critical Point

Several statistical–mechanical or mean-field treatments of glass formation build on the premises of the existence of an avoided, or unreachable, thermodynamic (e.g., configurational entropy and random first-order theories), or static (free-volume theories) critical point. Probably the oldest related phenomenological treatment is provided by free-volume theories, which consider that molecular mobility is controlled by the free volume, while the glass is regarded a frozen metastable state of matter, described by an additional kinetically controlled internal order parameter [78,102,103] and a $P-V-T$ equation of state. Many different models of the glass transition that rely on the concept of free volume [78,104] exist, including the simple kinetic (hole) theories [66]. While not identical, these models consider molecular motions in the bulk state of polymeric materials to depend on the presence of structural voids, also known as "vacancies" or "holes" of molecular size ($\sim$0.02–0.07 nm$^3$), or imperfections in the packing order of molecules. These holes are collectively described as "free volume," a term also used to describe the excess volume that can be redistributed freely without energy change [78]. (It is worth noticing, however, that the free volume available for molecular movement does not coincide with the total empty space in the material, which corresponds to the difference between the geometric volume of all segments and the total volume.) The concept that local rearrangement motions in dense systems require some empty space, which can be taken by atoms involved in this motion, is intuitively appealing: in the liquid state, where the free volume is large, molecular movements occur easily and the rearrangement of chain conformations is practically unconstrained, while, following a decrease in temperature, the free volume shrinks until it is too small to allow large-scale molecular motions. As thermal expansion and viscoelastic relaxation of a solid or rubber-like material can be rationalized in terms of changes in the temperature-dependent free volume, the viscoelastic behavior of polymers and related composites has been extensively studied – with variable success – in relation to free-volume variations [105–107]. Evidence on the significance of free-volume theories and support of the hypothesis that $T_g$ is inversely proportional to free volume is often encountered in the studies of binary polymer systems (e.g., see Figure 1.9 for miscible polyethylene oxide [PEO] + phenolic blends [108]).

   Most theories based on the free-volume concept state that the glass transition is characterized by an iso-free-volume fraction state, that is, they consider that "the glass transition temperature is the temperature below which the polymers have a certain universal free volume" [109]. The total volume of the material, $V$, obeys the relation

$$V = V_0 + V_f, \tag{1.16}$$

where the limiting or occupied volume, $V_0$, is associated with the hardcore or incompressible molecular volume (molecular volume at zero thermodynamic temperature or extremely high pressure) and significant volume fluctuations (from thermal motion; i.e., bond vibrations and librations). The free volume at temperatures below $T_g$ (denoted by $V_f^*$) is considered nearly constant, and increases only as the glass-transition temperature is exceeded. In the latter temperature range, free volume can be expressed as

$$V_f = V_f^* + (T - T_g)\frac{\delta V_f}{\delta T}, \tag{1.17}$$

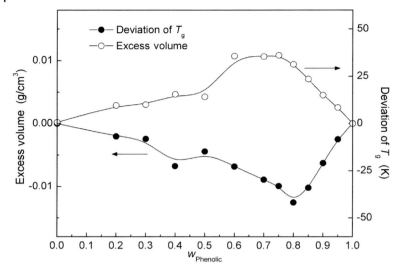

**Figure 1.9** Composition-dependent deviations from the linear mixing rules for the glass-transition temperature (positive deviation) and mixture's volume (negative deviation) in miscible PEO + phenolic blends. Adapted and replotted from ref. [108], with permission © 2000 Elsevier.

and by dividing by $V$, one gets the relation

$$f = f_g + \alpha_f(T - T_g), \tag{1.18}$$

where $f = V_f/V$ is the fractional free volume at some temperature, $f_g$ is the fractional free volume at $T_g$, and $\alpha_f$ is the thermal expansion coefficient of the free volume. Studying a wide range of polymers, Simha and Boyer [110] supported the iso-free-volume hypothesis for the glass transition and derived the value of $f_g = 0.113$. The starting point of the treatment provided by Williams, Landel, and Ferry [48] is that at $T > T_g$ the internal mobility of the system, represented by viscosity, is related to the fractional free volume by an empirical relationship, which is based on the Doolittle equation [102]

$$\eta = A \exp\left(B\frac{V_0}{V_f}\right), \tag{1.19}$$

with $A$ and $B$ material specific constants. With the approximation of $f = V_f/V \approx V_f/V_0$, valid given that $V_0 \gg V_f$, Eq. (1.19) can be written in the form $\ln \eta = \ln A + B/f$. If $\eta(T)$ denotes viscosity at a temperature $T$ and $\eta(T_g)$ corresponds to the viscosity at the glass-transition temperature, the temperature shift factor $(\alpha_T)$ is given by

$$\ln a_T = \ln\left(\frac{\eta(T)}{\eta(T_g)}\right) = B\left(\frac{1}{f} - \frac{1}{f_g}\right). \tag{1.20}$$

By combining Eqs. (1.18) and (1.20), and after appropriate rearrangement in a $\log_{10}$ functional form, one obtains

$$\log_{10} a_{\mathrm{T}} = \frac{-\dfrac{B}{2.303 f_{\mathrm{g}}}(T - T_{\mathrm{g}})}{\dfrac{f_{\mathrm{g}}}{\alpha_{\mathrm{f}}} + (T - T_{\mathrm{g}})}. \tag{1.21}$$

The constant $B$ was found by Doolittle to be of the order of unity. A direct comparison between Eqs. (1.21) and (1.7b) suggests the existence of *universal* values for both the fractional free volume at $T_{\mathrm{g}}$ ($f_{\mathrm{g}} \approx 0.025$) and the thermal expansion coefficient of the free volume ($\alpha_{\mathrm{f}} = f_{\mathrm{g}}/51.6 \approx 4.8 \times 10^{-4}\,\mathrm{K}^{-1}$). An immediate inference of the above treatment is that the glass-transition temperature is reached when the fractional free volume attains the limiting low value of 0.025 (more precisely $0.025 \pm 0.009$ for most polymers). Several equation-of-state models, such as the statistical–mechanical Simha–Somcynsky hole theory [103], permit the determination of $V_{\mathrm{f}}$. The latter, however, is better determined experimentally, given the serious discrepancies often encountered in theoretical derivations [111]. Positron annihilation lifetime spectroscopy (PALS) [112], for example, provides values for the free-volume fraction in the range 0.02–0.11, in reasonable agreement with theoretical estimates [104,110].

Free-volume theories lack the ability to provide connections among relaxation characteristics of glass-forming liquids and important macroscopic thermodynamic properties, notably the configurational entropy and its temperature derivative, the isobaric heat capacity. A thermodynamic perspective to the glass-transition phenomenon may be provided by considering the glass as a thermodynamic phase, an assumed "fourth state of aggregation of matter" [8]. The assumption of an underlying true second-order phase transition (in the Ehrenfest notation), due to a change in the configurational entropy, has led to various phenomenological entropic treatments, most notably the Gibbs–DiMarzio [79] and Adam–Gibbs [44] theories. Strong motivation for their development provided the crisis that emerges when the entropy of a supercooled liquid is extrapolated toward low temperatures. Recall that the so-called Kauzmann's paradox is avoided by the glass transition, since the entropy has a smooth change (as experimentally evidenced) due to an underlying phase transition. The theoretical derivation of the statistical–mechanical theory of Gibbs and DiMarzio was based on the Flory [113] and Huggins [114] lattice model. The model develops by first calculating the number of ways that $n_y$ linear polymer chains of $y$ monomer segments each ($y$-mers) can be placed in a diamond lattice with $n_0$ unoccupied holes. Each chain has a lowest energy shape and the more the shape deviates from it, the greater the internal energy of the molecule. The ensuing configurational partition function, describing the location of holes and polymer molecules, has the form

$$Q = \sum_{f, n_y, n_0} \Omega\big(f_1 n_y \ldots, f_i n_y \ldots, n_0\big) \exp\left[ -\frac{E\big(f_1 n_y \ldots, f_i n_y \ldots, n_0\big)}{k_{\beta} T} \right], \tag{1.22}$$

where $f_i n_y$ is the number of molecules packed in conformation $i$, $E$ in the internal energy of the system, and $\Omega$ is the microcanonical partition function (i.e., the total number of ways that the $n_y$ molecules can be packed into $yn_y + n_0$ sites on the quasi-lattice). Knowledge of the partition function allows the configurational entropy to be determined from the relation

$$S_c(T) = k_\beta T \left( \frac{\partial \ln Q}{\partial T} \right)_{V,n} + k_\beta \ln Q, \tag{1.23}$$

through which several derivations and property calculations become possible. While several of its predictions are in agreement with experimental data (e.g., the molecular mass dependence of $T_g$), the validity of the theory remains rather questionable in the absence of rigorous mathematical justification of critical assumptions (e.g., the actual form of the microcanonical partition function). In an effort to reconcile thermodynamic arguments with purely dynamical aspects of the phenomenon, it has been postulated that, although the observed glass transitions show clear manifestations of a kinetic phenomenon, the underlying true transition features equilibrium properties that are difficult to observe experimentally. This aspect was considered in the Gibbs–DiMarzio approach by defining a new transition temperature, $T_2$, at which the configurational entropy of the system reaches a critically low value $S_{c,0}$ [57] ($S_{c,0}(T_2) = 0$ in ref. [79]). $T_2$ is clearly not an experimentally measurable quantity; calculations place it approximately 50° below the glass-transition temperature observed at ordinary measuring times. In effect, $T_2$ is considered to be the lower limit value that $T_g$ would reach in a hypothetical experiment of infinite timescale.

A somewhat different phenomenological approach to an "entropy-vanishing" glass-transition model has been presented by Adam and Gibbs [44]. The idea behind their molecular-kinetic theory is that the liquid consists of regions that rearrange as units, independently of their environment, when experiencing a sufficient fluctuation in enthalpy. Each related subsystem of the material is referred to as a cooperatively rearranging region (CRR) and has a size determined by the number, $z$, of molecules included (monomeric segments in polymers' case). The temperature-dependent relaxation times for a viscous material are determined from the probabilities for cooperative rearrangements to take place. A structural rearrangement is activated with a barrier height ($\Delta\mu$) proportional to the lower limit number $z^*(T)$ of segments within the hypothetical critically sized CRR with a nonvanishing transition probability. The structural relaxation time is inversely proportional to the average transition probability, providing the relation

$$\tau_\alpha(T) = \tau_\infty \exp\left( \frac{z^* \Delta\mu}{k_\beta T} \right), \tag{1.24}$$

where $\tau_\infty$ is a constant. At high temperatures, the molecular displacements are taken to be entirely noncollective, and the barrier height reduces to a constant Arrhenius activation energy. The size of the CRR depends on temperature and is determined by the configurational entropy of the liquid, while an increase in the dynamic cohesive length, $\xi_\alpha$, is anticipated with decreasing temperature. A link to the Gibbs–DiMarzio theory is provided by the hypothesis that $z^*(T)$ is simply inversely proportional to the (molar) configurational entropy, $z^*(T) = N_A s_c^* / S_c(T)$, where $s_c^*$ is the configurational entropy of the critically sized CRR. By combining the latter equations, configurational entropy and structural relaxation time (or the shear viscosity) in the liquid state are now connected through the relation

$$\tau_\alpha(T) = \tau_\infty \exp\left(\frac{A}{TS_c}\right). \tag{1.25}$$

As temperature decreases and larger units progressively form, $S_c$ decreases until it approaches zero at some temperature $T_0$ (generically close to $T_2$ and $T_K$ [54]). At this point, relaxation times and viscosity diverge, and a second-order thermodynamic transition is suggested to occur. The validity of Eq. (1.25) has been tested in numerous experimental studies (which, typically, use the excess entropy of the liquid over the crystal in the place of $S_c$) and computer simulations (where $S_c$ is directly assessed) [115]. Adam and Gibbs [44] used their entropic theory to derive a WLF-type formula for the temperature shift factor. Furthermore, considering that the configurational entropy can be calculated from the relation

$$S_c(T) = S_c(T_0) + \int_{T_0}^{T} \frac{\Delta C_p}{T} \, dT \tag{1.26a}$$

and roughly assuming a hyperbolic form for the configurational part of the specific heat ($\Delta C_p \sim 1/T$), it follows that

$$S_c(T) \propto \frac{T - T_0}{T}, \tag{1.26b}$$

from which Eq. (1.25) further derives the VFTH equation (with $T_0$ replacing $T_V$). One of the limitations of the Adam–Gibbs theory is that it provides no information regarding the size or number of CRRs in the system, because $s_c^*$ is not specified, except as a formal lower limit that provides, in practice, no useful insight at the microscopic level. As an extreme limit, the entropy of the smallest region capable of undergoing a rearrangement is obtained by the Boltzmann relation as $s_c^* = k_\beta \cdot \ln 2$, given that a minimum of two configurations must be available for a rearrangement to take place. Adopting this value, $\Delta\mu$ and $z^*$ may be evaluated from experimental data. However, the low values deduced for $z^*(T_g)$ in a number of glass formers suggest that the Adam–Gibbs theory, if valid, might be more appropriate for local processes transpiring before the glass transition [13].

An elaborate mean-field approach to glass formation is given by the RFOT theory [86]. Within this framework, vitrification is described analogous to crystallization, with the difference that the system is frozen into a set of aperiodic structures instead of a periodic crystalline structure. This theoretical ensemble ties together intelligibly aspects of several prior concepts, notably, the *unreachable* thermodynamic transition postulated by the configurational entropy theories, the emergence of a complex free-energy landscape with numerous energy minima, and the *avoided* dynamical singularity of the MCT. In this way, a low-temperature thermodynamic transition, at $T_K$, to an ideal glass phase (accompanied by a discontinuous jump in the order parameter, but no latent heat) is postulated to exist along with a high-temperature ergodicity breaking dynamical transition, at $T_{MCT}$. These two coexisting critical points are separated by a regime in which an exponentially large (in system size) number of metastable free-energy states (configurations) dominate the thermodynamics while trapping the dynamics. Support for the above phenomenology has been found in several standard liquid models when treated within mean field-like approximations (see reviews by Berthier and Biroli [25], and Cavagna [116]). The thermally activated hopping mechanisms that take control in the regime $T_K < T < T_{MCT}$ are described in a nonperturbative way. Phenomenological arguments only exist [117], backed by microscopic computations [118], which describe liquid dynamics below the crossover at $T_{MCT}$ as a "mosaic state," while a dynamical scaling theory based on "entropic droplets" (domains of synergetic molecular motions, with linear size $\xi^*$) is considered as temperature approaches $T_K$. The entropic droplets are formed and stabilized by the competition between a favorable driving force (i.e., an increase of the configurational entropy) and an unfavorable surface mismatch penalty. The mosaic length scale $\xi^*$, characteristic of the mosaic cells and entropic droplets, represents the length scale above which any consideration of metastable states becomes inappropriate. Assuming that thermal activation over energy barriers grows with size as $(\xi^*)^\psi$, with $\psi \geq 0$, the super-Arrhenius $\tau_\alpha(T)$ dependence is given by [118]

$$\ln\left(\frac{\tau_\alpha}{\tau_0}\right) = C\,\frac{Y}{k_\beta T}\left(\frac{Y}{T \cdot S_c(T)}\right)^{\frac{\psi}{d-\theta}}, \tag{1.27}$$

where $C$ is a constant, $Y$ is a "bare" energy scale (in appropriate units), $d$ is the dimension of space, and $\psi$ and $\theta$ are two critical exponents. The latter are predicted to obey the relation $\psi \approx \theta = 3/2$, in $d = 3$ [117], providing an Adam–Gibbs-type formula for the relaxation time with the configurational entropy per particle, $s_c(T)$, vanishing at the ideal glass transition $T_K$. Several other nontrivial derivations of the RFOT, frequently based solely on phenomenological arguments [119], address issues such as the decoupling between translation and rotation, nonexponential relaxation (i.e., $\beta_{KWW} < 1$) and its connection with fragility, the specific heat jump at the glass transition, and the accelerated segmental dynamics close to free surfaces [120]. The reader is referred to an excellent review by Berthier and Biroli [25] for a detailed discussion of several limiting approximations and missing links between the pieces of this "patchwork" theory, as well as potential weaknesses of the RFOT construction in providing the description of finite-dimensional systems.

Reference to some form of space fragmentation into dynamically and/or spatially heterogeneous zones (described, for example, as "regions," "spheres," "domains," or "droplets") appears also in the frustration-limited domains (FLDs) theory. Through a series of tentative assumptions [121], this approach directly addresses glass formation in terms of the real space at the molecular level [92]. Its critical assumptions include the existence of a locally preferred structure in a liquid (i.e., a domain-level molecular arrangement that minimizes some local free energy), the unfeasibility for this local structural order to expand over the entire system (i.e., the domain size is limited by frustration, or equivalently, the ordered domains are deemed separated by domains of topological defects), and the possibility to build an conjectural system in which the effect of frustration can be suppressed or disabled. In the context of liquids and glasses, frustration is attributed to a competition between a short-range propensity for expansion of a locally preferred order and global constraints that prohibit the periodic tiling of space with the particular local structure. Considering a system that organizes itself, at low temperatures and finite frustration level, into a mosaic of domains corresponding to some local order, its dynamics will involve restructuring of the domains in a thermally activated manner, using arguments similar to those adopted within the mosaic picture of RFOT. Under rather generic conditions, frustration produces a "narrowly avoided critical point," that is, the ordering transition that may possibly exist at some temperature $T^*$ (above $T_g$) in the absence of frustration may disappear when an infinitesimal degree of frustration is introduced. The crossover temperature, $T^*$, marks the onset of an anomalous supermolecular (collective) behavior and it can be used to establish a scaling description of several collective properties of glass-forming liquids. In relation to the existence of a pertinent heterogeneity length scale, for example, let us consider the free energy of a domain (of volume $L^3$), given by

$$F(L, T) = c_1(T)L^2 - c_2(T)L^3 + c_3(T)L^5. \tag{1.28}$$

The bulk free-energy gain inside the domain (second term) is modified by the strain free energy due to frustration (last term), and this strain induces breaks in the order that give rise to surfaces with free energy $c_1 \cdot L^2$ (first term). By minimizing the free energy per unit volume, $F(L,T)/L^3$, one finds that the characteristic radius of the domains scales as $(c_1/c_3)^{1/3}$, and that their size increases when temperature decreases without showing any divergence [121]. The presence of structured domains, whose size is finely tuned by the amount of frustration, readily connects to fundamental phenomena occurring in glass formers, such as cooperativity, dynamic heterogeneity, and spatial fluctuations, which directly rationalize – at least qualitatively – nonexponential relaxation, decoupling phenomena, and liquid fragility. Relaxation times (and shear viscosity), for example, are predicted to exhibit a distinctive super-Arrhenius temperature dependence at $T < T^*$, of the form

$$\tau_\alpha(T) = \tau_\infty \exp\left(\frac{E_\infty}{T} - \frac{\Delta E(T)}{T}\right). \tag{1.29}$$

The term $\Delta E(T) = BT^*(1 - T/T^*)^{8/3}$ describes the energetic requirement for restructuring the FLDs, and is thus assumed to be zero above $T^*$, yielding a simple exponential behavior. The four adjustable parameters ($\tau_\infty$, $E_\infty$, $B$, $T^*$) of Eq. (1.29) can be obtained by fitting of experimental data. The fragility of a glass former is quantified by the value of parameter $B$, which is inversely proportional to the degree of frustration [93]. A major weakness of this theory is the missing identification of the locally preferred structure for molecules of nonspherical shapes, the structural elements of most real fragile materials. It is very likely that the postulated formation – on supercooling – of tetrahedral or icosahedral local order (or even of hexagonal bond orientational local order) in three-dimensional one-component liquids with atoms interacting via spherically symmetric pair potentials [122], has little to do with the behavior of ordinary molecular liquids and polymers.

### 1.2.4.5 Percolative Phenomena in Glass Formation

Both liquids and glasses have a topologically disordered distribution of elementary particles, without perceptible differences in their translation-rotation symmetry. Their structure cannot be described in terms of repeating unit cells, as the unit cell of any amorphous material would comprise all *particles* (atoms or molecules) due to nonperiodicity. A number of approaches to the glass-transition phenomenon build on alternative structure descriptions [12], such as the consideration of the bond system rather than the set of particles. For each state of matter, one may define a set of bonds, that is, introduce a bond lattice model, which is the congruent structure of its chemical bonds. The bond lattices of glasses and liquids have different symmetries, in contrast to the symmetry similarity of particles in the liquid and glassy states of matter. Structural signatures in the form of percolation thresholds of the Delaunay networks [123] and an increase in icosahedral ordering near $T_g$ [124] have been observed by computer simulations of amorphous solids and simple liquids. Interesting observations have emerged from studies of the percolation thresholds of networks of the Delaunay simplices of different "coloring," with each color denoting the Delaunay sites of identical form (i.e., with identical metric properties). For example, in a molecular dynamics study of the configurations of liquid, supercooled, and quenched (glassy) rubidium, Medvedev and coworkers [123] showed that the Delaunay simplices develop macroscopic aggregates in the form of percolative clusters. In the liquid state, low-density atomic configurations form a cluster that goes across the whole material. This macroscopic structural organization permits extensive motions, like those observed in shear flow. In contrast, in the glassy state nearly tetrahedral high-density configurations contribute to the formation of clusters that percolate across the whole glass.

The CPM explains the glass transition in terms of a percolation-type phase transition, a percolation effect in the system of broken bonds connected with the formation of percolation clusters made of *configurons* [87]. A "configuron" is defined as a fundamental configurational excitation in an amorphous material that involves breaking of a chemical bond and the concomitant strain-releasing local adjustment of centers of atomic vibration. At absolute zero temperature broken bonds are absent, while the concentration of thermally activated broken bonds (configurons) is expected to increase with temperature, accompanied by a loss in material's rigidity. Since configurons weaken the bond system, the higher their content the lower the viscosity of the system becomes. Based on the percolation theory, when the concentration of configurons exceeds the threshold level, these will form a macroscopic so-called percolation cluster of broken bonds, which penetrates the whole volume of the disordered network. The percolation cluster is a dynamic structure postulated to develop at the glass-transition temperature. The critical temperature at which the percolation level is achieved can be found assuming that the configurons achieve their universal critical density, $p(T_g) = p_c$, given by the percolation theory. Although no symmetry changes are evidenced in the atomic distribution, there is a symmetry breaking expressed by a stepwise variation of the Hausdorff–Besikovitch dimensionality of bonds at the glass transition: from 3 (canonical Euclidean space), in the glassy state, the dimension of bonds reduces to $2.55 \pm 0.05$ in the liquid state (fractal network geometry).

Within the CPM, the resemblance of the glass-transition phenomenon to a second-order phase transformation, as a consequence of a change in symmetry, is treated in terms of the Landau–Ginsburg theory. Important role in the latter theory plays the order parameter, which equals zero in the disordered phase and has a finite value in the ordered phase. The density of the percolation cluster of configurons changes from a nonzero value in the liquid state to zero in the glassy state, and thus offers a suitable order parameter. As temperature approaches $T_g$, the CPM reveals diverging behavior for several parameters, such as the correlation length $[\xi(T) = \xi_0/|p(T) - p_c|^n$, with a critical exponent $n = 0.88]$, the thermal expansion coefficient, and the heat capacity (both proportional to $\sim 1/|T - T_g|^{0.59}$). A direct anticorrelation between the fragility ratio, introduced by Doremus [59], and the configuron percolation threshold has been postulated to exist [87], and explained considering that in fragile materials the configurons are larger (delocalized). Important thermodynamic parameters of the configurons (e.g., the configuron formation enthalpy and entropy, $H_d$ and $S_d$, respectively) can be extracted from the temperature dependence of the shear viscosity of amorphous materials [88]. These can be further used to predict the glass-transition temperature from the relation

$$T_g = \frac{H_d}{S_d + R \ln\left[(1 - p_c)/p_c\right]} \tag{1.30}$$

and also to numerically access $T_g$ at arbitrary cooling rates [89]. Unfortunately, while a number of successful predictions have been reported for oxide glasses, the effectiveness of Eq. (1.30) is restricted due to the large uncertainties in the determination of the required thermodynamic parameters (five coefficients concurrently determined in fits of $\eta(T)$) and specific model approximations. Interestingly, the configuron model of viscosity results in a two-exponential $\eta(T)$ equation, a functional form similar to that originally proposed by Douglas [125] for the universal description of viscosity data at all temperatures, which can be readily approximated (within narrow temperature intervals) by known theoretical or heuristic functions.

An alternative approach offers the two-order-parameter model of Tanaka [126]. This model considers the glass transition as being controlled by the competition (due to the incompatibility in their symmetry) between long-range density ordering toward crystallization and short-range bond ordering toward the formation of long-lived rigid structural elements (designated as "locally favored structures"). The supercooled liquid is described as a frustrated metastable liquid state consisting of metastable solid-like islands, in a sea of short-lived random normal-liquid structures, which exchange each other dynamically at the rate of the structural relaxation time. Depending on the type of measurement, $T_g$ can be defined either as the temperature at which the average lifetime of the metastable islands exceeds the characteristic observation time or as the temperature where percolation of long-lived metastable islands occurs. The average fraction of locally favored structures is regarded as a suitable measure of fragility (with a higher fraction indicating a stronger liquid), while extensive reasoning addresses the crossover from the noncooperative to the cooperative regime and the origin of the non-Arrhenius behavior of the structural relaxation.

At present, approaches like those mentioned above simply provide an interesting, yet incomplete, picture of glass formation, given the fact that they can address only a narrow range of its plethoric phenomenology.

## 1.3
## Manipulating the Glass Transition

The glass-transition temperature reflects the ease by which polymer chains commence waggling and break out of the rigid glassy state into the soft rubbery state in the course of heating. Its location is thus primarily regulated by *intrinsic characteristics* of the macromolecular system (Section 1.3.1), typically related to the chemical structure (main-chain structure, tacticity, type of pendant groups, etc.), chain conformations, and the degree of polymerization (molecular mass). Knowledge of the potential influence of each of these factors is a prerequisite – but clearly not the only – for appropriate selection of the components in binary and ternary mixtures and the preparation of polymer composites with finely adjusted properties and structural characteristics. Besides blending, which is comprehensively treated later in this chapter, reference should also be made to a number of technologically important *externally controllable processes* (application of pressure, orientation processes, presence of additives, electron-beam irradiation, etc.), *chemical reactions* (copolymerization, curing), or *physical phenomena* (aging, crystallization), with a substantial bearing on the glass structure and the temporal response (during storage or service life) of engineering polymeric materials (Section 1.3.2).

Guidance for targeted polymeric molecular design may come from explicit theoretical and simulation methodologies that aim to predict material's properties from its molecular details. Freed and coworkers [127], for example, developed a generalized entropy theory that combines the lattice cluster theory – for a semiflexible-chain polymer fluid – with the Adam–Gibbs model for the structural relaxation time. Their model provided a rational, predictive framework for calculating the essential properties (including $T_g$ and fragility) of glass-forming fluids as a function of their molecular architecture, bond stiffness, cohesive interaction energy, pressure, molar mass, concentration and structure of additives, and so forth. The estimates appear to corroborate several of the experimentally established general trends discussed below. Simulation approaches, on the other hand, usually employ either quantitative structure–property relationship (QSPR) models (where structural and quantum-chemical descriptors of perceived significance along with group additive methods are typically used in relevant estimations [128,129]) or atomistic modeling techniques (where full atomic detail of the polymers is considered, as in molecular dynamics simulations [130]). For selected polymers, interesting information on various structure–property dependences, such as the variation of $T_g$ with chain stiffness and substituent volume [131] or the type of the polymerization initiator [132], has been derived from QSPR studies.

## 1.3.1
### Effects of Chemical Structure

The location of the glass transformation range is representative of the flexibility of polymer chains, which in turn is determined by the degree of freedom with which their segments rotate along the backbone. As a result, low $T_g$s are typical of linear polymers with single covalent bonds and a high degree of rotational freedom about $\sigma$ bonds in the main chain, while stiffening groups along the backbone (e.g., aromatics and cyclic structures) reduce chain flexibility and increase intermolecular cohesive forces. For instance, the incorporation of a $p$-phenylene ring (Ph) into polyethylene's (PE) monomeric unit gives poly($p$-xylylene) $[(-CH_2-Ph-CH_2-)_n]$, with a glass-transition temperature ($\approx +80\,°C$) significantly elevated compared to that observed in various commercial PEs ($-130\,°C$ to $-80\,°C$). Tacticity has considerable bearing on the rotational energy requirements of the backbone. In the case of mono- and disubstituted vinyl polymers, $(-CH_2-CXY-)_n$, for example, Karasz and MacKnight [133] indicate that steric configuration affects $T_g$ only when the substitutes are different, and neither of them is hydrogen. Based on the Gibbs–DiMarzio theory of the glass transition, the increase observed going from isotactic and highly syndiotactic polymers can be related to the larger difference among the energy levels between rotational isomers in syndiotactic chains. In monosubstituted vinyl polymers, where the other substitute is hydrogen, the energy differences between the rotational states of the two pairs of isomers are comparable and the effect of tacticity is weak (e.g., polystyrene[PS] and poly(vinyl acrylate)s; Table 1.1). If the different tactic configurations of a single disubstituted vinyl polymer are considered, the glass-transition temperature appears to increase with the increasing content of syndiotactic triads. This trend is clearly demonstrated by poly($N$-vinyl carbazole) (PVK) (i.e., $T_g = 126\,°C$, $227\,°C$, and $276\,°C$, for isotactc (i-), atactic (a-), and syndiotactic (s-) PVK, respectively) [134]), poly(methyl methacrylate) (PMMA) [16,135], and several other poly(vinyl methacrylate)s (Table 1.1).

The chemical nature (e.g., polarity and ionicity), bulkiness, and flexibility of the groups attached to a polymer chain are often used to adjust its glass-transition

**Table 1.1** Effect of tacticity on the glass-transition temperatures for various poly(vinyl acrylate)s and poly(vinyl methacrylate)s [1,136].

| | $T_g$ (°C) | | | | |
| | Poly(vinyl acrylate)s | | Poly(vinyl methacrylate)s | | |
| Side chain | Isotactic | Atactic[a] | Isotactic | Atactic[a] | Syndiotactic |
| --- | --- | --- | --- | --- | --- |
| Methyl | 10 | 8 | 43 | 105 | 160 |
| Ethyl | −25 | −24 | 8 | 65 | 120 |
| *n*-Propyl | | −44 | | 35 | |
| Isopropyl | −11 | −6 | 27 | 81 | 139 |
| *n*-Butyl | | −49 | −24 | 20 | 88 |
| Isobutyl | | −24 | 8 | 53 | 120 |
| *sec*-Butyl | −23 | −22 | | 60 | |
| Cyclohexyl | 12 | 19 | 51 | 104 | 163 |

a) Atactic specimens with high syndiotactic content.

temperature. Pendant groups endorsing stronger intermolecular forces are responsible for polymers of higher $T_g$. The polar carbon chloride bond in poly(vinyl chloride) (PVC), for example, is the source of stronger intermolecular (dipole–dipole) interactions, compared to the relatively weak van der Waals forces present in polypropylene (PP); the large difference of $T_g^{(PVC)} - T_g^{(PP)} \approx +100\,°C$, among the atactic forms of these polymers, is an immediate consequence of the substitution. The upshift is even stronger when the Cl— groups are completely substituted by OH— (+105 °C) or CN— (+117 °C) groups, owing to the establishment of an extensive network of hydrogen-bonding interactions. Large and inflexible groups normally bring about an increase in chain's rigidity. Bulky groups, such as the benzene ring, may even hook up on neighboring chains and restrict rotational freedom (physical cross-linking), with a concomitant increase in polymer's $T_g$. Ample experimental evidence exists in the case of monosubstituted vinyl polymers [5,7,134], with the glass-transition temperatures following an increasing trend after substituting the hydrogen atom by progressively bulkier, stiffer, and/or more polar pendant groups: $T_g \leq -80\,°C$ (for X = H), $T_g = -10\,°C$ (X = —CH$_3$), 31 °C (X = —O—CO—CH$_3$), 85 °C (X = —OH), $T_g = 100$–130 °C in *p*-methyl, 2-methyl, or *p*-chloro styrene containing polymers, $T_g = 130$–150 °C when biphenyl or napthalene pendant groups are introduced, and $T_g > 170\,°C$ in poly(*n*-vinyl pyrrolidone) and PVK [7,134]. Moreover, changes in the number of successive methanediyl (—CH$_2$—) or methyl (—CH$_3$) groups in the aliphatic sequence of the flexible pendant groups in polyvinyl acrylates [137], methacrylates [133], and several other linear-chain polymers [138] have been extensively used to regulate chain's packing and system's $T_g$. Typically, by increasing the length of the aliphatic chain, free volume (at a given temperature) increases and the frictional interaction between chains is lowered [16]. Examples illustrative of the above behavior are presented in Table 1.2, for several

**Table 1.2** Effect of the pendant group length of the glass-transition temperature of typical linear-chain thermoplastic polymers.

| Alkyl | Poly(n-alkyl methacrylate)s TD (V − T) [133] | Poly(vinyl n-alkyl ether)s | | |
| --- | --- | --- | --- | --- |
| | | DMA (1 Hz) [106] | TD (L − T) [140] | TD (V − T) [139] |
| Methyl | 105 | −10 | −22 | |
| Ethyl | 65 | −17 | −33 | −42 |
| n-Propyl | 35 | −27 | | |
| n-Butyl | 20 | −32 | −56 | −54 |
| n-Pentyl | | | | −66 |
| n-Hexyl | −5 | | −74 | |
| n-Octyl | −20 | | −80 | −80 |
| n-Dodecyl | −65 | | | |

poly(n-alkyl methacrylate)s [133], of the general formula $(-CH_2-CCH_3(OX)-)_n$, and poly(vinyl n-alkyl ether)s [106,139,140], $(-CH_2-CH(OX)-)_n$, with X representing the n-alkyl group. The predictions of the generalized entropy theory [127] for polymer chains with stiff backbones and flexible side groups (e.g., poly(n-alkyl methacrylate)s and poly(n-α-olefin)s [141]) are in line with the experimental trends.

It is also worth stressing the modulation of the flexibility of a pendant group by the compactness of its isomers. The different isomeric forms of the butyl radical in poly(vinyl butyl ether)s is an interesting example: $T_g$ increases from −32 °C in the case of poly(vinyl n-butyl ether) to −1 °C in poly(vinyl isobutyl ether), and, finally, to 83 °C in poly(vinyl tert butyl ether), which accommodates the most compact isomer [142]. Mixed dependences may appear in more complex chain structures, attributable to coalescent counteracting effects. In different classes of polyphosphazenes [143,144], for example, in the region of short alkyl groups the response is regulated by an increase in free volume, with the short side groups acting as chain ends [144]. Above the side-chain length associated with each $T_g$ minimum, a further addition of methylene groups enhances intermolecular interactions, presumably due to a physical cross-linking action of the longer alkyl side groups [144,145]. For chains with flexible backbones and stiff side groups, theoretical predictions reveal a strong increase in the glass-transition temperature with increasing length of the side groups [127].

Main chain's length and polarity have also significant bearing on the glass-transition temperature. The latter is very dependent on the degree of polymerization up to a value known as the critical $T_g$ or the critical molar mass [1]. In most cases, the strong dependence persists only up to $M_n \sim 10^4$, with no appreciable effect being seen for longer chains. The theoretical treatments of Fox and Flory [109] and Somcynsky and Patterson [146] suggest a linear dependence of

$T_g$ on the inverse molar mass of a homopolymer, expressed as

$$T_g = T_g^\infty - \frac{K}{M_n}.$$ 

(1.31)

In this relation, $K$ is a constant depending on the polymer, $M_n$ is the number-averaged molecular mass of the homopolymer, and $T_g^\infty$ is the glass-transition temperature of a linear chain of "infinite" length. For PVC, for example, Pezzin and coworkers [147] reported values of $T_g^\infty = 78\,°C$ and $K = 8.05 \times 10^4$ (for $3 \times 10^3 \le M_n \le 10^5$). Equation (1.31) can be deduced from the free-volume theory, taking into account that terminal (end) groups bring about more free volume than the internal ones [148]. More satisfactory description of the experimental pattern has been reported using functional forms deriving from the statistical–mechanical Gibbs–DiMarzio [80] theory. For commercial polymers, the influence of $M_n$ variation on $T_g$ is insignificant and is almost always overtaken by other factors. Note also that a decrease in chain's flexibility frequently stems from an increase in the density and/or the strength of interchain interactions. The first effect is illustrated in the gradual change of the glass-transition temperature of PE from $-7\,°C$ to $137\,°C$, with chlorination levels increasing from 28.2% to 77.4% [149].

### 1.3.2
**Externally Controlled Processes or Treatments**

#### 1.3.2.1 **Pressure Effects**
The measurement of changes in the glass-transition temperature with pressure variations has become an important topic of polymer science, and new experimental studies and theoretical interpretations of pressure effects are in development [150]. In practical terms, interest is primarily directed toward the studies of product failure at high-pressure applications, while polymer engineering seeks information on the pressure dependence of the glass transition of polymers – in part – as a result of its involvement in the commercial large-scale production or treatment of polymer mixtures (e.g., in hot-melt extrusion and compression molding). In general, $T_g$ in polymers increases with increasing pressure (Figure 1.10 [150,151]), as expected from the generalized Ehrenfest relations applied to the glass–liquid transformation [152] and in accordance with predictions of free-volume theories of the glass transition [1]. The change is roughly described by the relation deriving if one considers the continuity of volume at the transition,

$$\left(\frac{\partial T_g}{\partial P}\right)_T = \frac{\Delta \kappa_T}{\Delta \alpha},$$ 

(1.32)

while a much better description of a wider collection of experimental data [153] is provided by the second Ehrenfest relation, which considers continuity of entropy at the transition,

$$\left(\frac{\partial T_g}{\partial P}\right)_T = VT\frac{\Delta \alpha}{\Delta C_p}.$$ 

(1.33)

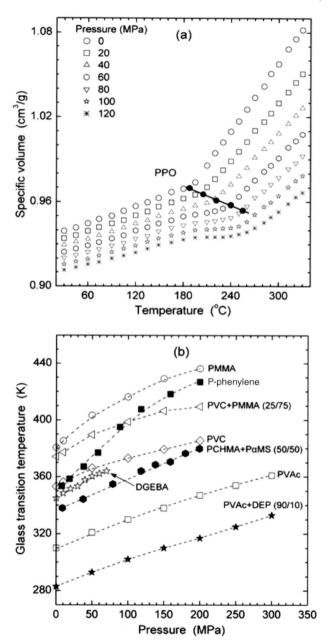

**Figure 1.10** (a) Dependence of the specific volume versus temperature and pressure, reported for PPO. The solid circles represent $T_g(P)$ as determined from the intersection of the liquid and glassy $V(T)$ data [150]. (b) Pressure dependence of $T_g$ reported for various glass formers [151], with permission © 2007 American Chemical Society.

The latter relation has been recently shown by Schmelzer [19] to be derived using an entropy-based approach for viscous flow and relaxation (Adam–Gibbs model). For all glass formers, the change in $T_g$ with pressure is linear at low pressures, but the pressure coefficient of the glass-transition temperature decreases with increasing pressure. The empirical relation proposed by Andersson and Andersson [154]

$$T_g = k_1 \left(1 + \frac{k_2}{k_3}P\right)^{1/k_2}$$
(1.34)

with $k_1$, $k_2$, and $k_3$ material constants, is frequently used to describe experimental data. A rough estimate considers an increase of the glass-transition temperature at a rate of around 20°–25° per kbar of pressure [1]. In view of that, the stiffening effect of pressure (vitrification) becomes important only in the case of applications at very high pressures, as well as in engineering operations where the polymer is processed near $T_g$.

### 1.3.2.2 Crystallization Effects

Polymers with a high tendency for crystallization are anticipated if their molecular geometry permits the formation of specific molecular orientations. As the glass-transition phenomenon activates only in the amorphous regions of a partially crystalline material, the intensity of the signal is controlled by both the degree of crystallinity and the formation of a "rigid amorphous fraction" (RAF) [155,156]. The latter term is often used to distinguish strongly hindered (often presumed immobile) chains at the amorphous–crystalline interface from mobile chains in the remaining amorphous phase (the "mobile amorphous fraction" (MAF); see also Section 1.7.2). As a consequence of the need to accommodate flexible molecules of typically 1–100 μm length into micro- or nano-sized intercrystallite regions, a crossing of the interface by the long molecules is rather common. This produces a strong coupling between crystalline and amorphous phases and a dependence of molecular mobility on segment's proximity to crystallite surfaces. Because of that, the main glass-transition signal in semicrystalline polymers is broader than that of the amorphous ones and extends toward the high-temperature side [27]. Surprisingly, the transition temperature may either increase [5] or decrease with increasing degree of crystallinity. In poly(ethylene terephthalate) (PET), for example, the amorphous specimen $T_g$ of 65 °C changes to 92 °C in a highly crystalline material (relaxational DSC data [157]), while, in the case of poly(4-methyl-1-pentene), an increase in crystallinity from 0% to 76% is accompanied by a drop in $T_g$ from 29 °C to 18 °C (specific volume versus temperature thermomechanical data [158]). A broadly accepted explanation for these opposite shifts is still missing. A plausible explanation, however, considers the difference in the relative densities of the amorphous and crystalline phases as the determining factor [7]. In many polymers, the coupling observed at the interface also causes a separate glass transition for the RAF. This portion of the polymer may lose its rigid character below the melting temperature of the surrounding crystals, within the melting region, or even above the melting temperature [156].

When a molten polymer is subjected to stretch during processing, orientation of the chains occurs accompanied by a significant rise of its glass-transition temperature. Highly oriented materials, such as films or fibers, can yield an apparent $T_g$, that is, by as much as 30° higher than that of the unoriented amorphous material [159,160]. Even stronger dependences are observed when orientation coexists with, or promotes, crystallization [159]. A common interpretation of these changes is based on the decrease of the free volume with increasing orientation, as chains in the amorphous phase are constrained in relatively close-packed elongated forms.

### 1.3.2.3 Plasticizer Effects

In order to improve the flexibility, processability, and utility of a given polymer, it is often necessary to decrease its glass-transition temperature. This can be accomplished by the addition of low-molecular-mass chemical compounds ("diluents"), mostly nonvolatile and chemically inert liquids, referred to as *plasticizers* [161]. A plasticizer may be considered as a substance, breaking intermolecular bonds in a polymer network, or even as a lubricant, reducing intermolecular friction and by this increasing deformability of the structure [162]. The ensuing reduction of the cohesive forces between neighboring polymer chains facilitates molecular rotations with a concomitant decrease in $T_g$. From the perspective of free-volume theories, plasticizer molecules may be considered to increase free volume by pushing apart neighboring chains, permitting the translational and rotational mobility of their segments to be retained to temperatures lower than those in the pure polymer. In contrast to molecular plasticizers that decrease the stiffness of the glassy polymer (lowering modulus and tensile strength, followed by an increase in elongation), some chemical compounds act as *antiplasticizers*, increasing the material's stiffness while they produce a (softer) depression of $T_g$. The addition of an antiplasticizer has been determined by Riggleman and coworkers [163] to cause significant changes in the long-wavelength properties, which are associated with an enhanced packing efficiency in the glass state (e.g., a decrease in $\kappa_T$) and increased material stiffness (i.e., an increase in shear modulus, bulk modulus, and Poisson ratio). The efficiency of any potential plasticizer/antiplasticizer is expected to depend on its polarity, solubility parameter, stiffness, density, and loading (weight, volume, or molar fraction). These factors manipulate the relative importance of the enthalpic and entropic contributions to the glass-transition temperature. An analysis of the issue, based again on the generalized entropy theory, exemplifies the significance of diluent's molecular properties [141]: plasticization is favored by small additives whose cohesive energies are less than the cohesive energy of the host polymer (i.e., only weak attractive interactions between the diluent and the polymer matrix are present), while antiplasticization is promoted in the opposite case. Absorbed water often functions as a plasticizer on many hydrophilic materials (e.g., polyamides, starches, and sugars), while the simultaneous occurrence of plasticization and antiplasticization effects (yet in different concentration ranges) is not uncommon in hydrated amorphous food matrices [164]. Weakly polar esters are good plasticizers because they tend to be miscible with many polar and nonpolar polymers. The most commonly used plasticizers are obtained from phthalic acid and include diethyl, dibutyl, and n-dioctyl phthalates. Smaller and/or more flexible diluents generally depress $T_g$ more than the larger ones. In terms of application, however, low-molecular-mass plasticizers present some disadvantages, caused by their volatility and tendency for diffusion within the final product (and subsequent leaching), potentially posing environmental dangers. This fact has provoked the development of polymeric plasticizers, formed by polymers of low $T_g$ and miscible with the base polymer [2], which provide materials with longer service times. A typical example is PVC plasticized with acrylonitrile butadiene rubber or copolymers of ethylene vinyl acetate [165].

### 1.3.2.4 Filler Effects

Several properties, such as the thermal stability, mechanical strength, or electrical conductivity of a polymeric material, can also be regulated by incorporating microscopic or nanoscopic inorganic fillers. Particulate systems typically involve polymers with finely dispersed clays, alumina, silica, silver or gold nanoparticles, carbon blacks, carbon nanofibers, and single-wall or multiwall carbon nanotubes [166]. The level of adjustment in the properties of the host matrix is determined by a number of factors, including the nature, size, amount (load), and surface chemistry of the filler, as well as the level of interaction between the components. In a number of studies [166,167], a loosened molecular packing of the polymer chains in the presence of the nanoparticles results in enhanced molecular mobility and a lower $T_g$ for the composite. In contrast, wetted nanoparticle interfaces experiencing strong attractive interactions (as a rule, hydrogen bonding) with the polymer bring about moderate-to-strong upshifts in the glass-transition temperature [168]. The presence of a rigid amorphous polymer fraction around the nanoparticles has been well documented [39,169,170]. Calorimetric and dielectric results from different filler–polymer combinations suggest that the constraint in segmental mobilities does not extend throughout the material but affects only an interfacial layer with a thickness of a few nanometers. In some cases, this interfacial layer has been identified as totally immobilized [169], while in others, an additional glass transition emerges as a separate signal at higher temperature, or as shoulder at the high-temperature flank of the relaxation peak (e.g., in poly(dimethyl siloxane) + $SiO_2$ nanocomposites [171]). Interparticle spacing efficiently modulates the apparent glass-transition temperature of the nanocomposite [172–174], with different shifts recorded even in the same particulate system depending on the state of dispersion. This is clearly illustrated in the current thermograms of natural rubber–silica nanocomposites (Figure 1.11), where finely dispersed silica produces a small systematic increase of $T_g$ compared to the neat rubber, while no change or a marginal decrease is observed for aggregated silica nanoparticles [175].

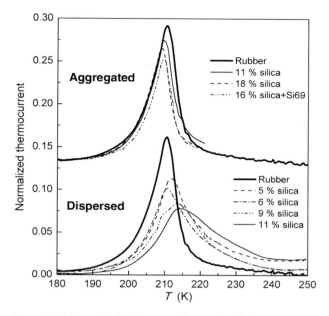

**Figure 1.11** Normalized TSC spectra of natural rubber-silica nanocomposites, in the temperate range of the glass transition of the polymer matrix. For clarity, the group of spectra referring to systems with aggregated silica nanoparticles is vertically upshifted. After ref. [175], with permission © 2011 Elsevier.

### 1.3.2.5 Cross-linker Effects

Cross-linking of neighboring chains can be performed in a number of ways. The widely used "normal" cross-links involve strong covalent bonding of the polymer chains together into one molecule, while the weaker "reversible" cross-links found, for example, in several thermoplastic elastomers, rely on noncovalent or secondary interactions (typically hydrogen bonding and ionic bonding, respectively) between neighboring polymer chains. The stabilizing action of noncrystalline (e.g., in styrene-butadiene block copolymers) or crystalline (e.g., in thermoplastic copolyesters) domains within a composite material also provides a type of "physical" cross-linking. When normal cross-links are present, with an increasing cross-link density of the material the glass-transition temperature increases (and $\Delta C_p$ decreases) since the segmental mobility becomes progressively hindered to a higher degree. The effect of cross-linking on $T_g$ bears some analogies to that imposed by an increase in molecular mass. Typically, the introduction of cross-links into a polymer is accomplished by the addition of a cross-linking agent, which can be regarded as a comonomer. The copolymer effect and the effect of cross-linking itself were combined by Fox and Loshaek [176] into the form of the equation

$$T_g = T_g^\infty - \frac{K}{M_n} + \frac{K^*}{M_c}, \tag{1.35}$$

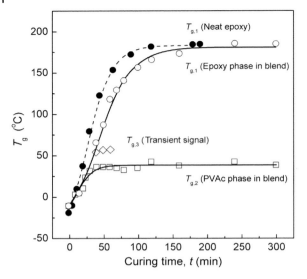

**Figure 1.12** Glass-transition temperatures versus curing time for immiscible PVAc + epoxy resin blends, cured at 180 °C: (●) $w_{PVAc} = 0$ (neat epoxy resin), (○, ◇, □) blend with $w_{PVAc} = 0.05$. Phase separation after the first ~30 min of curing is evident from the appearance of multiple $T_g$s and the opacity in the blends. After ref. [180], with permission © 2007 John Wiley & Sons.

in which $K^*$ is a constant and $M_c$ is the number-average molar mass of the chains between cross-links, a parameter inversely proportional to the number of cross-links per gram of material [148]. Cross-linking is a rather common route for the preparation of molecular structures intended for applications where strength and rigidity are important. In thermosetting polymers, for example, a prepolymer in a soft solid or viscous state changes irreversibly into an infusible, insoluble polymer network by a chemical reaction known as "curing." The degree of cure ($\alpha_c$), also known as fractional conversion, is a key parameter in determining the end-use properties (stiffness, creep, etc.) of the material. The glass-transition temperature is highly sensitive to the degree of cure: its value increases significantly as the curing reaction proceeds to completion (Figure 1.12), due to the progressive establishment of a three-dimensional network of bonds. Several theoretical approaches to modeling the $T_g$ versus conversion relationship during isothermal cure have been proposed for thermosetting materials [177–179]. For example, based on thermodynamic considerations, Venditti and Gillham [179] suggested the relation

$$\ln(T_g) = \frac{(1 - \alpha_c)\ln(T_{g,0}) + \lambda\alpha_c\ln(T_{g,\infty})}{(1 - \alpha_c) + \lambda\alpha_c},$$

(1.36)

where $T_{g,0}$ and $T_{g,\infty}$ are the glass-transition temperatures of the uncured mixture ($\alpha_c = 0$) and the fully cured network, respectively, and $\lambda = \Delta C_{p,\infty}/\Delta C_{p,0}$ ($\Delta C_{p,0}$ and $\Delta C_{p,\infty}$ are the heat capacities of the uncured monomer mixture and the fully cured network, respectively).

### 1.3.2.6 Geometric Confinement Effects

Parallel to sophisticated model simulations, the glass-transition dynamics and the characteristic transition temperature of nanoscopically confined glass-forming materials have been intensively studied with the purpose of elucidating the characteristic length scale of cooperativity experimentally [181]. A strong body of scientific literature [182] has been accumulated for a number of systems, including – but not limited to – amorphous layers confined between neighboring crystal lamellae [181], polymers in clay galleries, intercalated or exfoliated clay–polymer hybrid systems [183], colloidal systems, micro/nanoemulsions, ultrathin films, multilayers, organic glass formers within controlled pore inorganic glasses, as well as binary systems (e.g., filler + polymer, diluent + polymer, and binary polymer blends [184–186]) and copolymers [187]. Of particular merit appear to be studies on the behavior of organic liquids and polymers confined in nanopores and of polymers in the form of thin films, for which striking changes are often recorded for confining dimensions below 10 nm in porous environments, and for even higher thicknesses in free-standing polymer film geometries [181,188,189]. Excellent reviews on the topic [181,182,189,190] clearly demonstrate that we are far from having achieved a final, complete, and self-consistent, picture for the behavior of glass-forming materials in confined states. Confinement-induced perturbations of the molecular relaxation dynamics and the apparent glass-transition signal, usually quantified by the shift of the transition temperature ($\Delta T_{\mathrm{g}}^{*} \equiv T_{\mathrm{g}}^{\mathrm{conf.}} - T_{\mathrm{g}}^{\mathrm{bulk}}$) and changes in the relative strength and breadth of the respective transitions, clearly demonstrate remarkable, irregular, and often contradictory dependences on the surface chemistry of the confining system (i.e., the interfacial energy controlled by the nature of the repulsive/soft or attractive/rough surfaces), compositional, and structural characteristics of the confined phase (e.g., polarity, molecular mass, chain conformation/stereoregularity, steric hindrance, and fragility), and the topology (fractality, porosity) or degree of confinement [182,189,190]. Attempts to unify interpretations and resolve pending questions stumble on the diversity of features which real systems demonstrate (depending, among other parameters, on the type and strength of interactions between the components), and to a lower degree on conflicting results that arise from intrinsic differences between the experimental techniques employed (e.g., wide variations in sensitivity and length scale of the probe) or even subtle variations in thermal histories and sample pretreatments [182]. A matter of intense research debate, consonant with the intriguing nature of the glass-transition phenomenon and the diversity of factors controlling it, constitutes reports of different trends in the variation of the transition temperature ($\Delta T_{\mathrm{g}}^{*}$) with changes in the confining length obtained using different experimental techniques for nearly identical systems [189,191]. Inconclusive arguments also appear with respect to the potential influence of the molar mass of the confined polymer on the type and strength of the $\Delta T_{\mathrm{g}}^{*}$ variation [189].

Since Jackson and McKenna [192] first observed the depression of the glass-transition temperature of organic liquids (of *ortho*-terphenyl and benzyl alcohol) into the pores of a controlled pore glass, a number of studies on hydrogen bonded glass formers have demonstrated the glass-transition temperature to decrease [189,190], remain unchanged, or even increase with increasing confinement. This complex behavior is in part related to the interplay among *surface effects* (e.g., strong hydrogen-bonding interactions between the confined molecules and the natural (uncoated) pores) and the intrinsic or *confinement effect* (e.g., breakdown of the cooperative motion, which translates into a change in the cooperativity length scale). Hydrogen-bonded liquids confined in silanized nanopores generally display accelerated relaxation dynamics, a broadening of the relaxation function, and lower glass-transition temperatures compared to those recorded in the bulk or when confined in unsilanized pores. Matrices with hydrophilic (hard) surfaces have been found to produce diverse responses, beyond the abovementioned rather typical trend. Examples provide the bulk-like behavior observed for salol in unsilanized MCM-41 nanoporous glass [193], and the elevation of the glass-transition temperature – in selected confining dimensions only – of glycerol, ethylene glycol, and oligomeric propylene glycol in unsilanized nanoporous glasses [193,194].

The majority of published results on thin polymer films suggest that the dynamics of thin polymer layers with free surfaces (free-standing films) or in contact with repulsive/soft surfaces (softly supported films) are governed by entropic effects, including chain-end segregation, density anomalies, and disentanglement. The different energetic states of the molecules located on the free surface and in the internal (bulk) regions of any material, and in particular the existence of an enhanced mobility layer at the free surface of polymers, remain a subject of intense debate [195]. The diversity of results obtained in several investigations of ultrathin or multilayer films, binary polymer blends [186], and other systems is compounded by the fact that different length scales (or extends of cooperativity) are impacted by confinement differently [196]. Several results support the idea of Keddie *et al.* [197] that a sufficiently thin liquid-like layer adjacent to the free surface exists, with reduced $T_g$ due to the reduced requirements for cooperative segmental mobility. de Gennes [198] has further suggested a "sliding-motions" mechanism for propagating the mobility of the near-surface segments to depths comparable to the overall size of the polymer molecules. The perturbations caused by the surface layer are thus allowed to propagate some tens of nanometers into the film interior, usually resulting in a strong reduction in $T_g$ with decreasing film thickness ($D$). This variation can be described by the relation

$$T_g(D) = \frac{D(2k + D)}{(\xi + D)^2} T_g(\infty),$$ (1.37)

where $T_g(\infty)$ denotes the thickness-independent value determined in sufficiently thick samples, and $k$ and $\xi$ are model-specific constants [199]. For $2k = \xi$, Eq. (1.37) reduces to a Michaelis–Menten-type function $[T_g(D) = T_g(\infty)/(1 + \xi/D)]$, which provides a satisfactory description of several experimental dependences. In contrast, strong enthalpic forces in the vicinity of the polymer–substrate interface (e.g., hydrogen bonds between films of PMMA, poly(2-vinyl pyridine) [P2VP], or

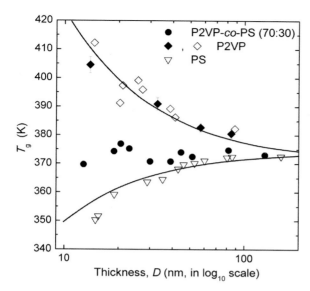

**Figure 1.13** Measured and fitted glass-transition temperatures against film thickness for thin films of P2VP, PS, and poly(2-vinyl pyridine)-*co*-polystyrene (70:30), on Si substrates. Strong interaction among P2VP and the substrate is evident. Lines are data fits to Eq. (1.37). After ref. [199], with permission © 2004 Elsevier.

PVAc and native $SiO_2$ substrates) lead to a partial immobilization of polymer chains segments. In such cases, the local increase of the glass-transition temperature [189] is likely to produce positive values for $\Delta T_g^*$. Reports indicate that the thickness dependence of the effective $T_g$ of polymer blend [184,185] and copolymer [199] films (Figure 1.13) is intermediate to that recorded for neat component thin films. Data collected for miscible PS + poly(2,6-dimethyl-1,4-phenylene oxide) (PPO) [184] and PS + tetramethyl polycarbonate (TMPC) [185] mixtures on $SiO_x$/Si substrates manifest a strong dependence of the sign and magnitude of $\Delta T_g^*$ from blend composition. Accordingly, the increasingly positive $\Delta T_g^*$ with decreasing film thickness observed for neat TMPC gradually transforms to a negative $\Delta T_g^*(D)$ pattern, which becomes stronger as the PS content increases in the blend [185].

When a particular polymer structure is considered, and with the assumption of only marginal confinement-induced changes in its tacticity and degree of polymerization, there are strong indications for the existence of a principal underlying mechanism that operates similarly in all different types or topologies of confinement. A paradigm offers the similarities (notably observations of $\Delta T_g^* > 0$) observed in the behavior of atactic or predominantly syndiotactic PMMAs under different conditions of geometric restriction: for example, polymer in the form of supported ultrathin films (on an unsilanized/attractive $SiO_2$ surface) [200], polymerized *in situ* 5-nm-diameter gel-silica glass nanopores [188] or in hydrophobic and hydrophilic 13-nm-diameter controlled pore glasses [201], containing finely dispersed $SiO_2$ [202] or metal (Ce, Co) nanoparticles [203], restricted between layers of organophilic montmorillonite [204], or loaded with organic bentonite [205].

## 1.4
### Experimental Means of Determination

For a number of methods and experimental techniques, the temperature variation of different thermodynamic, physical, mechanical, or electrical properties of a material provides the means for identifying the glass transition and other transitions. From rheometry emerges the time-honored definition of $T_g$ as the temperature at which the viscosity of the internally equilibrated supercooled liquid reaches the value of $10^{12}$ Pa s. Electron spin resonance (ESR) and nuclear magnetic resonance (NMR) spectroscopies permit the observation of glass transformations and $T_g$ determinations from changes in molecular mobility (Figure 1.14 [206]). Fourier transform infrared (FTIR) and Raman spectroscopies monitor changes in molecular bonding, occurring within the temperature range of the phenomenon. Sensitive determinations of thickness-dependent glass transitions in thin homopolymer or polymer blend films are

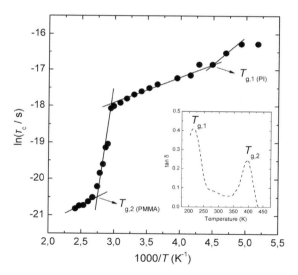

**Figure 1.14** Glass transitions and ESR: Arrhenius plots of nitroxide radical (probe) rotational correlation times and loss tangent spectrum (inset) of an immiscible semi-inter-penetrating polymer network based on PI and PMMA ($w_1/w_2 = 1$; 4% dicumyl peroxide, PI cross-linker). Three crossover points, the extreme ones associated with the glass transition temperatures of PI and PMMA and the intermediate related to a secondary relaxation, are indicated. The dynamic mechanical spectrum (tan $\delta$ versus $T$ plot) is included for comparison. Compiled from plots appearing in ref. [206], with permission © 2010 Elsevier.

frequently accomplished using ellipsometry [207], X-ray or neutron reflectivity, fluorescence spectroscopy [208], Brillouin light scattering [209], and PALS [210]. The latter techniques are all sensitive to properties related to the change in polymer density at $T_g$. Significant results can also be extracted by combining microscopy with thermal analysis techniques (e.g., local thermal analysis [211]).

Modern thermal analysis comprises a number of sophisticated techniques, particularly efficient in performing – among other characterizations – a quantitative description of several types of thermal events and of their dependence on molecular architecture, chemical composition, or processing details. Studies devoted to glass-transition phenomena in polymer-based systems constitute an important area of related experimentation [5,17,106,107,212,213]. Nevertheless, the variability of measuring approaches and probes, the multitude of operational definitions for the glass-transition temperature, and the interrelated temperature change rate and oscillating frequency dependences of the respective signals (Table 1.3) continue to create sources of misconception in relevant

**Table 1.3** Characteristics of routine thermoanalytical techniques used in studies of glass-transition phenomena in polymer-based materials.

| Technique[a] | Property measured | Operational definitions of $T_g$ | Resolution, sensitivity |
|---|---|---|---|
| DTA | Temperature difference | Onset, midpoint, or endpoint of step change in $\Delta T$ versus $T$ plots[b] | Low to moderate |
| DSC/ MTDSC | Heat flow (differential heat flow, heat capacity) | Onset or midpoint of step change in $C_p$ versus $T$ plots; peak maximum in the derivative of the (reversing) heat flow versus $T$ plots[b] | Moderate/ high |
| TMA/ MTTMA | Dimension changes (thermal expansion coefficients) or softening | Point of intersection of the glassy and rubbery expansion versus $T$ curves[b] | Moderate/ high |
| DMA | Viscoelastic properties (mechanical strength, energy loss) | Temperature of the α-relaxation peak in loss modulus ($E''$) or damping factor ($\tan\delta$) versus $T$ plots.[c] Onset temperature of the storage modulus ($E$) drop at the transition[c] | High |
| DEA | Dielectric properties and electrical relaxation (permittivity, dielectric loss, polarization change rate) | Temperature of the α-relaxation peak in loss factor ($\varepsilon''$) or loss tangent ($\tan\delta$) versus $T$ plots.[c] Onset temperature of the real part of relative permittivity ($\varepsilon'$) rise at the transition.[c] Temperature of the α-peak in depolarization current versus $T$ plots[b] | Moderate to high |

a) DTA = Differential thermal analysis, TMA = Thermomechanical analysis, DMA = Dynamic mechanical analysis, DEA = Dielectric analysis.
b) Temperature change rate dependent.
c) Oscillation frequency dependent.

determinations or data comparisons. In the following sections, theoretical concepts and experimental aspects of thermal analysis techniques, emphasizing in problems encountered in glass-transition determinations in polymer blends, will be briefly presented.

## 1.4.1
### Calorimetric Techniques

The science of calorimetry is associated with determinations of the changes in the energy of a system by measuring the heat exchanged with its surroundings in the course of physical phenomena or chemical reactions. Although somewhat dated, differential thermal analysis (DTA) is still used for heat exchange measurements, phase diagram determinations, and thermal decomposition recordings in various atmospheres of materials used in mineralogical research, environmental sciences, as well as in pharmaceutical and food industries. In the course of a typical DTA experiment, the difference in temperature between a substance (the "sample") and a reference material (a thermally inactive material, such as $Al_2O_3$), both placed in the same furnace and in a specified atmosphere, is monitored against time or temperature, while the temperature of the reference ($T_r$) and the sample ($T_s$) is subjected to identical linear heating cycles. The differential temperature ($\Delta T = T_s - T_r$) is plotted against program temperature (or time), providing the DTA curve. Changes in the sample, either exothermic or endothermic, can thus be detected relative to the inert reference material providing data on the transformations (e.g., glass transitions, crystallization, melting, sublimation, oxidation, and thermal degradation) that have occurred [17,214]. The DTA curve is often treated only as a *fingerprint* for identification purposes (i.e., determination of characteristic temperatures alone); quantitative results are produced only when calibration with a standard material is performed, allowing for the quantifiable conversion of $\Delta T$ to heat flow and, eventually, to heat of transition ($\Delta H$) or the constant pressure heat capacity. Except for some high-temperature applications, in recent years DTA has been largely displaced in the field of polymer science by the more sophisticated analytical technique of DSC. In DSC, the heat flow rate difference into a substance and a reference material is measured as a function of temperature while both are subjected to a controlled temperature program. The "apparent" heat capacity of the substance, $C_{app}$, is related to the differential heat flow and the heating rate through the relation $dQ/dt = C_{app} \cdot (dT/dt)$, assuming that the weights of the sample and reference pans are identical. The term "apparent" is implemented here because $C_{app}$ comprises the true heat capacity of the sample along with kinetic (time-dependent) contributions from various physical or chemical processes.

The typical DSC plots of heat flow as a function of temperature usually reveal a series of thermal effects, and the actual temperature (or temperature range) at which each thermal event appears is primarily determined by the polymer's structure. With reference to Figure 1.15, and starting from the lowest temperature onward, the first discontinuity usually observed signifies the glass transition; this appears as a rounded step or a shift of the base line, corresponding to the

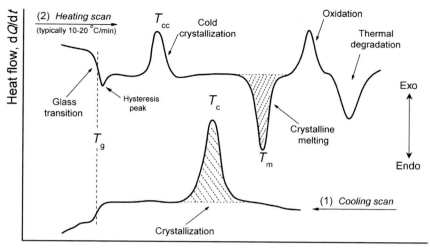

**Figure 1.15** Schematic cooling (1) and heating (2) DSC curves showing a range of different transitions (glass transition, crystallization, melting) and reactions (oxidation, thermal degradation) of a typical polymer. A hysteresis peak appears in the high-temperature side of the glass transition. Possible signals from mesophase transitions (e.g., transitions between the crystalline, smectic, nematic, or isotropic phases) in polymorphic materials (liquid-crystalline polymers) are not considered here.

heat capacity difference ($\Delta C_p$) of the sample before and after the transition. Such a step does not occur for local or normal mode relaxation processes because of the missing contribution from entropy fluctuation. Enthalpy relaxation is typically evidenced by an endothermic hysteresis peak in the high-temperature side of the glass transition, with a magnitude that strongly depends on the thermal history of the material; under some circumstances, it may even make the glass transition appear to be a melting transition (a problem particularly intense in DSC analyses of polymer blends). A controversial and rather weak liquid–liquid transition (at $T_{LL}$, with $T_{LL}/T_g \approx 1.1$–$1.2$ [215]), related to increased chain mobility and segment–segment associations, may also be present. As temperature increases, there may be a cold crystallization peak (at $T_{cc}$, exothermic), followed by a crystalline melting peak ($T_m$, endothermic). The intensity of the cold crystallization peak depends on the sample's history and ability to crystallize in the timescale of the experiment. In the course of heating, oxidation reactions (and even oxidation degradation) may appear if the experiments are not performed in inert atmosphere, in addition to thermal events related to curing chemical reactions (e.g., in elastomer formulations or thermosets). At much higher temperatures, the polymer undergoes thermal degradation, resulting in main-chain scission, cross-linking, cyclization, or loss of volatile fragments. In an inert atmosphere, the degradation pattern may be endothermic, exothermic, or both, whereas in oxygen or air, it is always exothermic.

The DSC peak area provides quantitative calorimetric information for each thermal event. The curve can be used to calculate heats (enthalpies) of transitions or reactions by integrating the peak corresponding to a given transition (e.g., the crystalline melting peak in curve 2 of Figure 1.15). There is no such heat of transition at the glass transition, since only first-order transitions have a heat of transition. The heat capacity change at the glass transition is a characteristic constant number for a given amorphous polymer. An empirical rule proposed for amorphous polymers – often called Wunderlich's rule – suggests that $\Delta C_p$ at the glass transition is around 11 J/°C·mol per mobile unit of the polymer main chain in the case of relatively small units, while for larger mobile units (such as the phenylene rings) $\Delta C_p$ may be double or triple this value [156]. It should be noted, however, that $\Delta C_p$ is strongly influenced by crystallinity and the so-called RAF in semicrystalline polymers. Detection of the temperature interval where vitrification modifies thermodynamic parameters and measurement of $T_g$ and $\Delta C_p$ provides the most important results of a DSC scan on an amorphous or partially crystalline polymeric material. It has been suggested [212] that $T_g$ should be measured during sample cooling, rather than in the subsequent heating run, from the intersection of the extrapolated equilibrium glass and liquid lines obtained in the enthalpy versus temperature graph. Since the sample exists in thermal equilibrium at the start of the measurement, enthalpy relaxation that often complicates $T_g$ measurements is avoided. Instrumental drawbacks, such as poor control of the cooling rates in some differential scanning calorimeters and difficulties in performing calibration on cooling, preclude the general use of cooling curves for determining $T_g$. Therefore, when the glass-transition temperature is to be obtained by progressive heating of a cooled sample, and in order to minimize enthalpy relaxation effects, a cooling rate somewhat faster than that of the subsequent heating rate is recommended. A high heating rate is beneficial in detecting $T_g$, because the heat flow signal associated with the transition enhances, with very little corresponding increase in noise, thereby increasing resolution. Such a change produces a shift of $T_g$, due to a combination of thermal gradient and kinetic effects, in addition to a broadening of the transition range. In view of the above facts, and for comparison purposes, the heating and cooling rates, the breadth of the transition signal and the methodology used to extract the glass-transition temperature from the curve need to be reported along with its estimate.

Difficulties in DSC determinations of the glass-transition temperature of semicrystalline polymers or polymer blends are common since the transition can be very broad and smeared out. In such cases, derivative curves (e.g., d$C_p$/d$T$ versus $T$ plots) become useful as the heat capacity change is replaced by a more visible peak. Several methods are available for marking the "exact" location of the glass-transition temperature in a typical heating DSC curve. Figure 1.16a demonstrates five characteristic temperatures that are frequently cited throughout the literature as $T_g$, often without mention of the specific location actually picked. These are (1) $T_b$, the "onset" temperature (usually difficult to determine), which defines the point at which the first deviation from the base line on the low-temperature side is

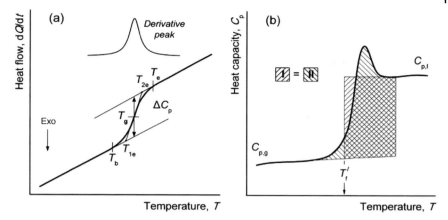

**Figure 1.16** Determination of the glass-transition temperature from a heating DSC scan: (a) various $T_g$ estimates from an idealized dQ/dt versus T curve. The height of the marker (double arrow) is proportional to $\Delta C_p$ at the glass transition. (b) Graphical representation of the procedure used to determine the limiting fictive temperature from $C_p$ heating data. The shaded areas I and II correspond, respectively, to the left and right integrals appearing in Eq. (1.38).

observed; (2) $T_{1e}$, the frequently reported extrapolated onset temperature (reproducible), which is the temperature at the intersection of the extrapolated base line and the tangent taken at the point of maximum slope; (3) $T_g$, the temperature at the half-height of the heat capacity increase (highest reproducibility; preferred point), also called the "temperature of half-unfreezing," which is very close to the temperature at the inflection point (temperature peak in the first derivative of the $C_p$ versus T graph); (4) $T_{2e}$, the extrapolated "end" temperature of the glass transition; and finally, (5) $T_e$, the end temperature of the glass transition, where the heat capacity dependence becomes linear again. $T_{1e}$ is very close to the fictive temperature, $T_f$, which characterizes the glass. Tool [28] defined the $T_f$ of a material in a nonequilibrium (glassy) state as the actual temperature of the same material in the equilibrium (liquid) state whose structure is similar to that of the nonequilibrium state. In that respect, the fictive temperature corresponds to the temperature at which a property of interest (enthalpy, specific volume, refractive index, logarithm of shear viscosity, etc.) when extrapolated along the glass line intersects the equilibrium liquid line [216]. The important limiting value of the fictive temperature, $T'_f$, is obtained if the extrapolation is performed from a point deep in the glassy state after cooling at a given rate. Both $T'_f$ (measured on heating) and $T_g$ (measured on cooling) depend only on the cooling rate, with experimental evidence supporting their close proximity [217]. A convenient way to measure $T'_f$ is through the area matching method from Moynihan *et al.* [216] (Figure 1.16b),

$$\int_{T'_f}^{T \gg T_g} \left(C_{p,l} - C_{p,g}\right) \mathrm{d}T = \int_{T \ll T_g}^{T \gg T_g} \left(C_p - C_{p,g}\right) \mathrm{d}T. \tag{1.38}$$

As indicated by Chartoff and Sircar [17], for amorphous thermoplastic polymers it is technically more significant to use the temperature at the extrapolated onset, $T_{1e}$ ($\approx T'_f$) as the glass-transition temperature, since $T_{1e}$ defines the initial temperature for the loss of structural properties (e.g., modulus) as the polymer softens through the glass transformation range. Therefore, $T_{1e}$ defines both a low-temperature limit for the processing of amorphous thermoplastic polymers and an upper-use temperature. In the case of elastomers, the temperature for useful elastomeric properties lies above the glass-transition region, that is, at $T > T_e$. If the elastomer is cooled beyond $T_e$, it enters the glass-transition region and starts losing its elasticity, becoming progressively stiffer as the temperature decreases.

Modulated-temperature DSC (MTDSC) [218] is an important advancement of DSC. In MTDSC, the same heat-flux DSC cell arrangement is employed, but with a nonlinear heating profile applied across the sample and reference material. The heating profile results from the sinusoidal modulation (oscillation) overlaid on the traditional linear ramp, that is, temperature changes in the form

$$T(t) = T_0 + qt + A_T \sin(\omega t), \tag{1.39}$$

where $T_0$ is the starting temperature, $q$ is the linear heating rate, and $A_T$ is the amplitude of temperature modulation. The instantaneous heat flow in a modulated DSC experiment can be written as

$$\frac{dQ}{dt} = C_p(q + A_T \omega \cos \omega t) + f'(t, T) + A_k \sin \omega t, \tag{1.40}$$

where the term $q + A_T \omega \cos \omega t$ is the measured heating rate, $f(t, T)$ is the kinetic response without temperature modulation, and $A_k$ is the amplitude of kinetic response to temperature modulation. Using Fourier's transformation analysis, the complex heat flow signal can be separated into reversing heat-capacity-related effects, which are in-phase with the temperature changes (such as glass transitions and crystalline melting), and nonreversing effects that are out of phase with the changes in temperature (e.g., cold crystallization, enthalpic relaxation, oxidation, evaporation, and thermal decomposition). Heat capacity can be calculated from the ratio of modulated heat flow amplitude to the product of amplitude of modulation temperature and frequency, while the reversing heat flow is determined by multiplying the heat capacity with average heating rate. Preference to MTDSC is supported by its ability to measure heat capacity in a simple run with increased sensitivity (five times greater than that of conventional DSC), its higher resolution (due to the very low underlying heating rate), and unique ability to separate overlapping thermal effects (due to the superimposed modulated heating profile). Examples provide the multiple glass transitions of nanoheterogeneous amorphous phases created in PEO + PMMA blends [219]. The efficacy of MTDSC to extract glass transitions masked by the rapid cold crystallization occurring in the same temperature range is also extremely useful in miscibility evaluations of polymer blends. Figure 1.17 shows, as an example, the total heat flow profile of a PET + acrylonitrile/butadiene/styrene (ABS) blend, and its separation into reversing and

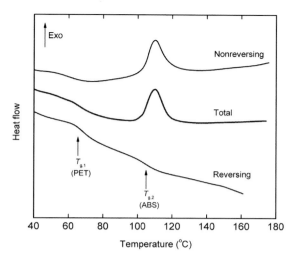

**Figure 1.17** Total, reversing and nonreversing heart flow curves of an immiscible PET + ABS blend composition. Adapted and replotted from ref. [220].

nonreversing components [220]. Complete phase separation is evidenced by the existence of two well-separated glass-transition steps that appear in the curve of the reversing component of heat flow and are located within the region of the glass-transition temperatures of the constituting polymers ($T_g \approx 65\,^\circ$C for PET and $\approx 105\,^\circ$C in the case of ABS). The crystallization of PET is indicated by an exothermic peak in the nonreversing heat flow curve; the fact that the glass-transition signal of ABS is submerged in the low-temperature flang of the crystallization peak precludes its detection in a conventional DSC experiment.

### 1.4.2
### Thermomechanical Analysis (TMA)

Thermomechanical analysis (TMA) is an established thermoanalytical technique based on the measurement of changes in sample length ($L$) or volume ($V$) as a function of temperature or time, under load at an atmospheric pressure. The technique is also referred as thermodilatometry (TD) if dimensions are measured with negligible force acting on the testing material while it is subjected to a controlled temperature program. Thermomechanical studies are usually performed under static load with a variety of probes for measuring dimensional changes in expansion/compression, penetration, tension, flexure, or shear test modes [106]. By applying special modes and different attachments, stress relaxation, creep, parallel-plate rheometry, and volume dilatometry measurements are also feasible. Polymeric materials are usually studied in the form of rigid or nearly rigid solids, or in the liquid (melt) state using special accessories. These can also be used to measure the volume changes in irregularly shaped samples or powders submerged in an inert liquid (e.g., in mercury dilatometers). The basic physical quantities utilized in TMA is *stress*, $\sigma$, which is defined as the force applied per unit area of the material, $\sigma = F/A$, with $A$ denoting sample's cross-sectional area, and the deformation per unit dimension caused by the applied stress and measured by the *strain*, $\varepsilon$. For a simple tensile experiment, strain is defined as $\varepsilon = \Delta L/L$, where $\Delta L$ is the change in length and $L$ is the original length. When subjected to a mechanical force, materials may behave in a variety of ways. A brittle material will deform reversibly to a small amount and then fracture, while a ductile material also deforms reversibly up to a certain amount and then yield and flow under the applied force until it begins to harden under load and then fail. Up to the elastic limit, the material will return to its former shape and size when the force is removed. The slope of this linear region corresponds to Young's modulus, $E$, also known as *elastic modulus, tensile modulus*, or simply *modulus*, defined as

$$E = \frac{\sigma}{\varepsilon} = \frac{L}{A}\left(\frac{\Delta F}{\Delta L}\right)_T, \tag{1.41}$$

with $\Delta F/A$ representing the change in stress and $\Delta L/L$ the change in strain. The magnitude of $E$ is a measure of material stiffness.

The identification of glass-transition signals in polymer-based systems (in filled, crystalline, or cross-linked materials particularly) has evolved into a routine application of TMA. The measurement of $T_g$ is achievable with a single experiment and from the same data used to determine the linear isobaric expansivity, also known as the coefficient of linear thermal expansion ($\alpha_L$),

$$\alpha_L = \frac{1}{L}\left(\frac{\Delta L}{\Delta T}\right)_p, \tag{1.42}$$

or the (volumetric) coefficient of thermal expansion $\alpha$ ($\alpha = 3\alpha_L$ for isotropic materials). The schematic curves of expansion and penetration TMA runs are shown in Figure 1.18. During expansion measurements of amorphous and semicrystalline polymers that are not oriented, a sudden increase in expansion rate is observed above $T_g$, as the material shifts from a structural configuration of limited or no chain mobility to an increased chain mobility state. The point of intersection, seen as an inflection or bend, of the glassy and rubbery linear or volume expansion curves typically defines the glass-transition temperature. Note, however, that a somewhat different $T_g$ value is seen for each mode of testing, as they each measure a different effect. The change in the slope of the expansion curves below and above $T_g$ is related to the expansion of free volume, since the actual volume of the molecules (the hardcore or incompressible volume, corresponding to zero thermodynamic temperature or extremely high pressure) does not change appreciably around $T_g$. In the case of semicrystalline systems only, a further increase in temperature may result in the penetration of the probe into the sample, even with a negligible load on it. This abrupt decrease in the probe position (onset temperature), as illustrated in Figure 1.18, can in most cases be assigned to crystal melting, and the temperature of the break on the curve represents the melting point.

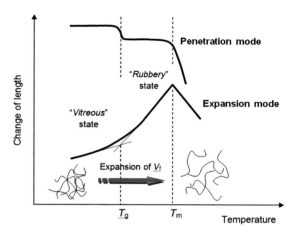

**Figure 1.18** Schematic representation of TMA curves for a semicrystalline specimen in expansion or penetration-mode experiments. In the first testing mode, the coefficient of thermal expansion can be determined in certain temperature ranges in the linear parts of the $L - T$ or $V - T$ graph, using Eq. (1.42) for $\alpha_L$, and Eq. (1.2c) for $\alpha$, respectively.

Ideally, linear thermal expansion and subsequent contraction of a specimen to its original dimensions are totally reversible phenomena. Nonetheless, if the material softens as it is heated while subjected to a mechanical load it will flow and creep, resulting in a nonreversible dimensional deformation. Moreover, if the material was stretched (oriented) when soft and then cooled before the experiment, residual stresses will remain in the sample (memory effects). Within the conditions of the thermomechanical experiment, stress relaxation on heating will cause irreversible morphological changes (i.e., randomization of the orientation achieved during manufacture) along with shrinkage of the testing material. The length changes typically measured by conventional TMA are thus a convolution of the above effects, unless the specimen is completely isotropic and measurements are made under zero loads. A separation of overlapping thermodynamic and kinetic, reversing and nonreversing, dimensional changes is achieved in modulated temperature TMA (MTTMA) experiments [106]. In this case, the dependent physical quantity measured is the length, with the sample exposed to a sinusoidal temperature modulation superimposed on a linear underlying heating, cooling, or isothermal profile, similar to MTDSC. The modulation conditions are different from MTDSC, since the sample and test fixture and enclosure are larger in the mechanical test, thus requiring longer equilibration (and scan) times.

Several studies demonstrate the usability of TMA as a complementary technique to DSC, DTA, or thermogravimetric analysis studies, in performing structural and thermal characterization of homogeneous polymer systems (e.g., mixtures of poly(vinyl phenyl ketone hydrogenated (PVPhKH) with poly(2-ethyl-2-oxazoline) or poly(styrene-co-4-vinylpyridine) (PS4VP) [221]), mixtures showing elevated structural heterogeneity at selected blend compositions (e.g.,

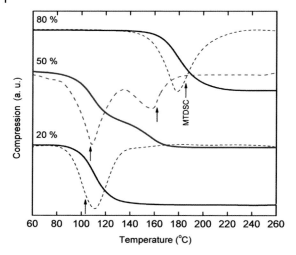

**Figure 1.19** The (—) TMA curve and (– –) derivative curve of the sample thickness changes of PVPhKH + PPO blends with 20, 50, and 80 wt. % PPO. Vertical arrows show the location of $T_g$ as determined by MTDSC. Note the double glass-transition signal only in the intermediate blend composition. Replotted data from ref. [222], with permission © 2004 John Wiley & Sons.

PVPhKH + PPO blends, Figure 1.19 [222]), or phase-separated systems (polyurethane + PMMA, PVC + PMMA [223]).

### 1.4.3
### Dynamic Mechanical Analysis (DMA)

There are three fundamental test methods for characterizing the viscoelastic behavior of polymers: stress relaxation, creep, and dynamic mechanical analysis (DMA) [15,107]. DMA methods, in particular, are most popular among thermal analysts for measuring mechanical properties and glass-transition-related relaxation modes, due to their high sensitivity, versatility, and increased resolution of overlapping mechanisms. Successful applications of DMA and related techniques for the study of polymeric solids and liquids are well documented [15,107], with particularly important results extracted in evaluations of orientation processes, of the effects of additives, and structural modifications or phase separation processes in composites. The relatively short measuring times (important when curing, aging, and crystallization processes are of interest), compared to those required in stress relaxation or creep experiments, and the wide range of information extracted for complex materials are noteworthy advantages of DMA.

In dynamic mechanical spectroscopy, one measures the mechanical properties of materials typically by imposing in a sinusoidal fashion a small strain (or, equivalently, stress) on a sample and measuring the resulting stress (strain) response as a function of temperature. For a perfectly elastic material (Hookean solid), stress and strain are in phase. In an ideal liquid behavior (Newtonian fluid), however, instead of deforming reversibly under load the material will flow; the strain is now proportional to the rate of change of the stress. The phase angle difference of $\delta = 90°$ between sinusoidal stress and strain in liquids is the key for the use of DMA as a tool in the characterization of viscoelastic polymers. Since such materials have properties intermediate between those of an ideal solid and an ideal liquid, they will exhibit a phase lag ($\delta$) somewhere between 0° (ideal solid) and 90° (ideal liquid). Along these lines, an application of a sinusoidal force (strain) to the testing material,

$$\varepsilon(t) = \varepsilon_{max} \cdot \sin(\omega t), \tag{1.43a}$$

where $\omega$ denotes the angular frequency, produces a stress of the form

$$\sigma(t) = \sigma_{max} \cdot \sin(\omega t + \delta). \tag{1.43b}$$

The relative amplitudes of stress and strain and the phase lag, which reflects the relative degree of viscous character to elastic character, can thus be obtained. DMA data are commonly reported using the complex modulus function

$$E^* = \frac{\sigma(t)}{\varepsilon(t)} = \frac{\sigma_{max} e^{i(\omega t + \delta)}}{\varepsilon_{max} e^{i\omega t}} = \left(\frac{\sigma_{max}}{\varepsilon_{max}}\right) e^{i\delta} = \frac{\sigma_{max}}{\varepsilon_{max}} (\cos(\delta) + i \sin(\delta)) = E' + iE''. \tag{1.44}$$

Its real component, $E'$, is known as the *storage modulus* and is a measure of the elastic character or solid-like nature of the material. The imaginary part, $E''$, is known as the *loss modulus* and is a measure of the viscous character or liquid-like nature of the material. In a physical sense, the first is related to the stiffness of the material, and the latter reflects the damping capacity of the material. For an ideal elastic solid, $E'$ is simply Young's modulus of the material and $E''$ is zero, while for an ideal viscous liquid $E'$ is zero and the loss modulus is related to the viscosity of the material. The ratio of the loss modulus to the storage modulus is known as the *damping factor* or *loss tangent*, or, more commonly, as

$$\tan \delta = \frac{E''}{E'}. \tag{1.45}$$

Tan $\delta$ ranges from zero for an ideal elastic solid to infinity for an ideal liquid, and represents the ratio of energy dissipated to energy stored per cycle of deformation. Analogous definitions apply for the parameters used in tests under a shear mode of deformation (i.e., the complex shear modulus $G^*$ and $\tan \delta = G''/G'$) and rheological measurements in polymer liquids [107].

The dynamic mechanical data obtained for polymer-based systems convey information on a broad range of relaxation processes, that goes from very local motions (secondary relaxations, generally due to branching chains or side chains) to segmental mobility exhibiting cooperativity ($\alpha$-relaxation or dynamic glass transition), or even large-scale relaxation processes (Rouse dynamics or reptation) [224]. Thermal analysts usually perform a series of temperature scans at a constant oscillatory frequency, covering a narrow range of frequencies

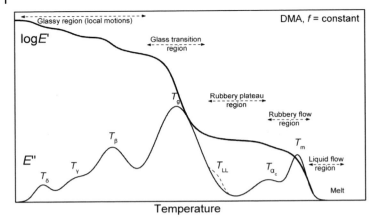

**Figure 1.20** An idealized isochronal DMA scan of a partially crystalline polymer, showing the effect of various molecular relaxations on the storage modulus, $E'$, and loss modulus, $E''$, curves. Peaks ascribed to the various transitions are indicated in the loss modulus spectrum.

(*isochronal* scans) (Figure 1.20). The mechanically active relaxation modes populate the spectrum in a number dependent on the complexity of the system and the magnitude of the temperature window. In amorphous polymers, for example, loss peaks at $T/T_g < 1$ are related to secondary relaxation processes. For semi-crystalline polymers, one should also note the presence of the melting peak, and a secondary relaxation ($\alpha_c$-peak) related to the slippage of the crystallites past each other. The liquid–liquid transition (at $T_{LL}$) may also be present within the rubbery plateau region.

Identifying the glass transition and how various system modifications affect $T_g$ is a major application of DMA. The glass transition is generally easily detected from isochronal dynamic mechanical data because of the sharp decrease in storage modulus $E'$ (or, equivalently, in shear storage modulus $G'$), and the corresponding loss dispersion in $E''$ (or in shear loss modulus $G''$) or $\tan \delta$, which peak at $T_g$. Either estimate is valid, but the values attained from each plot are different: the $T_g$ value obtained from the $\tan \delta$ peak is several degrees higher than that from the peak in $E''$, as shown in Figure 1.21 for PVC [225]. The loss modulus peak more closely denotes the initial drop of $E'$ from the glassy state into the transition. In this respect, the $T_g$ value based on the $E''$ peak is generally close to the intersection of the two tangents to the log storage modulus curve originating from both the glassy region and the transition region, the so-called onset temperature; it is therefore regarded as the most appropriate value. For most linear amorphous polymers, the glass-transition region is relatively narrow (width $< 20\,°C$) and thus the difference between the $E''$ and $\tan \delta$ peak temperature is of minor importance. There are cases, however (e.g., partially crystalline polymers, cross-linked thermosets, miscible polyblends, and heterophase polymers), where the transition region is very broad ($50$–$60\,°C$) and none of the two

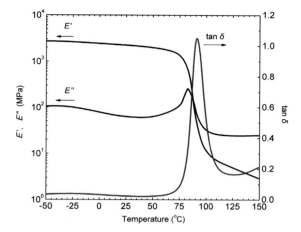

**Figure 1.21** Temperature dependence of the storage modulus $E'$, loss modulus $E''$, and mechanical loss factor $\tan \delta$, of PVC. Runs were conducted at a heating rate of $2\,°C$ $min^{-1}$ at an oscillating strain frequency of 1 Hz. Adapted data from ref. [225], with permission © 2011 John Wiley & Sons.

definitions may be entirely suitable for specifying $T_g$. Note that the frequency of oscillation needs to be taken into account in $T_g$ determinations using dynamic mechanical (or dielectric) relaxation studies, just as the heating rate intervenes in calorimetric results. The frequency of 1 Hz is usually chosen as a compromise value that provides $T_g$ values comparable to other thermal methods, while allowing collection of mechanical data at a sufficient rate to permit reasonable experimental times. DMA experiments using multiple frequencies deliver additional information (relaxation maps, activation energies, etc.) useful for defining kinetic parameters for the glass transition. The latter permit the prediction of material properties over broader frequency ranges (time-temperature superposition principle), such as those the material could encounter in actual end use.

## 1.4.4
### Dielectric Analysis (DEA)

The application of dielectric analysis (DEA) in the characterization of many different kinds of polar macromolecular materials and polymer-based composites is a steadily growing field of research [213,226]. Related studies have provided detailed insight into the molecular relaxation dynamics at various time and length scales. The properties and phenomena usually investigated by dielectric techniques include transition temperatures, chemical relaxation, cold crystallization and stabilizer effects, influence of space charges, microstructure, plasticization and oxidation phenomena, water–polymer interactions and hydration properties, physical aging, weathering, or effects of gamma and neutron radiation [16]. The remarkable sensitivity of several dielectric techniques to the presence of structural nanoheterogeneities is related to their small probe length scales (~3 nm), as opposed to conventional DSC that can only sense movements at considerably larger length scales (~20–60 nm). The application of DEA even in apolar substances has become feasible by introducing polar groups in their structure, either by chemical modification (labeling) of the polymer chain (chlorination, oxidation, etc.) or by dissolving of suitable polar probe molecules, which act as dielectric probes of polymer dynamics. In the latter case, strong coupling of the probe molecule fluctuations with the molecular motions in its environment is essential [227]. A successful application of this approach has been presented for binary blends comprising PE, PS, or PP, using low concentrations of the highly soluble and polar probe molecule 4,4'-(N,N-dibutylamino)-(E)-nitrostilbene [228].

High-precision measurements of complex physical quantities, connected with the orientational fluctuations of permanent dipoles or the translational mobility of charges, are now possible at an extremely wide frequency range (10 µHz, . . . , 30 GHz) through a variety of sophisticated experimental methods and measuring techniques; all these are collectively described as DEA techniques [213,226]. In dielectric relaxation spectroscopy (DRS), for example, an alternate current bridge or similar device is typically used to measure the equivalent conductance, $G$, and capacitance, $C$, of a material as a function of frequency $f$ (or $\omega = 2\pi f$). All the experimental information regarding electrical relaxation at a given temperature is contained in the functions

$$G(\omega) = \varepsilon''(\omega)\varepsilon_0\omega\frac{A}{d} \qquad (1.46a)$$

$$C(\omega) = \varepsilon'(\omega)\varepsilon_0\frac{A}{d}, \qquad (1.46b)$$

where $A$ and $d$ are sample's cross-sectional area and thickness, respectively, and $\varepsilon_0$ is the permittivity of free space. In that way, one obtains the frequency dependence of the components, $\varepsilon'(\omega)$ and $\varepsilon''(\omega)$, of the complex relative permittivity,

$$\varepsilon^*(\omega) = \varepsilon'(\omega) - i\varepsilon''(\omega), \qquad (1.47)$$

and other important dielectric functions [229]. The real part of relative permittivity expresses the ability of the dielectric medium to store energy and consists of the contributions of free space and the real part of the susceptibility of the material itself, while the imaginary component describes the energy losses entirely due to the material medium. The dielectric data (commonly permittivity, but also dielectric loss tangent $\tan\delta = \varepsilon''/\varepsilon'$ or electric modulus $M^* = 1/\varepsilon^*$) are usually presented in *isothermal plots*, that is, as a function of frequency at a constant temperature (Figure 1.22). In the frequency domain, $\varepsilon'(\omega)$ due to dipoles' relaxation exhibits dispersion, falling from $\varepsilon_r$ (relaxed) to $\varepsilon_u$ (unrelaxed) with increasing frequency. The dielectric loss band $\varepsilon''(\omega)$ of secondary relaxations displays a broad and generally symmetric shape, with a half-width of 3–6 decades (in a logarithmic scale) that usually narrows as temperature rises. The broad shape of the respective bands, compared with the width of the single relaxation time (Debye-type) mechanism (1.14 decades), is usually explained in terms of

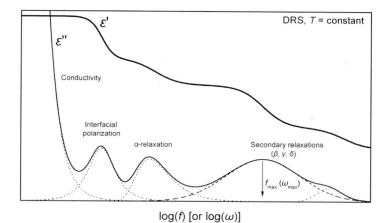

$$\log(f) \ [\text{or} \ \log(\omega)]$$

**Figure 1.22** Generic behavior of frequency dependence of permittivity components ($\varepsilon'$ and $\varepsilon''$) for the typical dielectric relaxation modes in glassy polymers and related composites. Interfacial polarization effects are common in heterogeneous materials (e.g., immiscible or partially miscible polymer blends), while crystalline phases and additives are potential sources of additional relaxations.

a distribution of environments in which the molecular entities relax. The α-relaxation (glass transition) displays a relatively broad but asymmetric shape, with a half-width of 2–5 decades. Permittivity data collected at some temperature are typically analyzed by a superposition of several Havriliak and Negami (HN) functions [230] and a conductivity term,

$$\varepsilon^*_{\text{exp}}(\omega) = \varepsilon_\infty + \sum_i \frac{\Delta\varepsilon_i}{\left[1 + (i\omega\tau_{\text{HN}})^{1-\alpha_{\text{HN},i}}\right]^{\beta_{\text{HN},i}}} - i\frac{\sigma_{\text{dc}}}{\varepsilon_0\omega^p}, \tag{1.48}$$

where $\sigma_{\text{dc}}$ is the dc conductivity of the material ($p = 1$ for purely ohmic conductivity). The shape parameters $0 \leq \alpha_{\text{HN}} < 1$ and $0 < \beta_{\text{HN}} \leq 1$ describe the width and the asymmetry of the loss peak, respectively, and with the addition of the dielectric strength $\Delta\varepsilon = \varepsilon_r - \varepsilon_u$ and $\tau_{\text{HN}}$ (a parameter directly related to the relaxation time $\tau$) provide a complete description of each relaxation process. The dielectric glass-transition temperature ($T_{\text{g,diel}}$) is defined by the convention $\tau_\alpha(T_{\text{g,diel}}) = 100 \, \text{s}$, and corresponds to the temperature at which the maximum of the loss factor band is located at the frequency of $f_{\text{max},\alpha} = 1.6 \times 10^{-3} \, \text{Hz}$. As with DMA, dielectric relaxation data can also be plotted as a function of temperature at a constant frequency (*isochronal plots*). Current-temperature ($I(T)$) spectra, resembling the isochronal $\varepsilon''(T)$ plots, are collected using the thermally stimulated (depolarization) currents (thermally stimulated currents [TSC] or thermally stimulated depolarization currents [TSDC]) technique. The equivalent frequency of this technique is in the range $10^{-4}$–$10^{-2}$ Hz, depending on the rate used for sample heating [213], and is sufficiently low to permit increased resolution of the complex signals recorded in polymer composites. In both cases, an

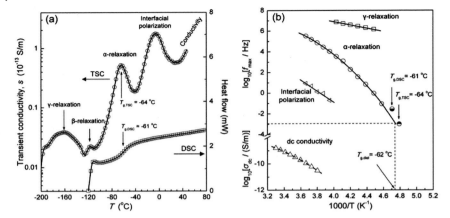

**Figure 1.23** (a) TSC (heating rate 5 °C min⁻¹, equivalent frequency ~10⁻³ Hz) and DSC (20 °C min⁻¹, ~2.6 10⁻² Hz) spectra of a segmented polyurethane. (b) Arrhenius diagram of the various relaxation modes and dc conductivity. The lines correspond to the fittings of the data to the Arrhenius (straight lines) and VFTH (curved lines) equations. Partly based on data appearing in ref. [231], with permission © 2005 Springer.

oscillation frequency (or heating rate) dependent estimate of $T_g$ is readily identified by the position of the peak corresponding to the α-relaxation mechanism.

Dielectric experiments performed at different temperatures allow analytical recording of the temperature dependences of the relaxation times or frequencies (Arrhenius plots). These can be further analyzed to provide, for example, the glass-transition temperature, dynamic estimates of the fragility index, data about the degree of curing and polymerization (by online dielectric monitoring [213]), an insight into the factors controlling polymers' miscibility and morphology, quantitative information about the degree of intermolecular and interface interactions in blends and composites, as well as conductivity percolation thresholds. The information extracted from dielectric studies is complementary to that obtained by DMA, DSC, and sensitive molecular spectroscopies. The data shown in Figure 1.23 for a thermoplastic segmented polyurethane reveal a close proximity for the dielectric and calorimetric estimates for the glass-transition temperature [231]. The limited agreement often found in $T_g$ estimates of different thermoanalytical techniques, even in cases of comparable frequencies, is in part related to the different modes of segmental motion activated by the thermal, electrical, or mechanical stimulus used in each experiment.

## 1.5
## Blend Morphology and Glass Transitions

The most frequent result of mixing of two polymers is an immiscible system that exhibits complete phase separation due to the chemical incompatibility (i.e., the repulsive interaction) between the components. Compatibilization via chemical, structural, or compositional modifications is often necessary to achieve some level of molecular mixing and better response to selected applications. A miscible binary system, on the other hand, is purportedly homogeneous down to the molecular level, with a domain size comparable to the dimensions of the statistical chain segment. The latter describes a rather limiting situation, since our current perception of miscibility also includes systems with compositional nanoheterogeneities. Poor consideration of the extent and relative strength of inter- and intramolecular interactions and of the impact of excess mixing volume effects often result in rather inaccurate theoretical predictions of the phase diagrams and miscibility windows of several polymer blends, calling for independent experimental validation. Glass-transition temperature determinations currently provide the most popular miscibility test. However, the effectiveness of thermal analyses and of several other relaxational (or vibrational) approaches in determining miscibility or phase behavior in multicomponent systems heavily relies on the nature of the technique and the probe length scale (i.e., the minimal domain size capable of producing experimentally resolvable behavior). Related issues will be critically examined in the following paragraphs.

## 1.5.1
### Miscibility and Phase Boundaries in Polymer Blends

Complete miscibility and phase stability in a binary mixture, of composition $\varphi$ at fixed temperature and pressure, requires that the following thermodynamic conditions,

$$\Delta G_{\mathrm{mix}} = \Delta H_{\mathrm{mix}} - T\Delta S_{\mathrm{mix}} < 0 \tag{1.49a}$$

$$\left(\frac{\partial^2 \Delta G_{\mathrm{mix}}}{\partial \varphi_i^2}\right)_{P,T} > 0, \tag{1.49b}$$

are satisfied [4]. If the latter condition is fulfilled merely in a limited range of compositions, then the blend is partially miscible, with stable one-phase mixtures expected only at the ends. At a given temperature, the sign of Gibb's free energy of mixing function, $\Delta G_{\mathrm{mix}}$, almost exclusively depends on the value of the enthalpy (heat) of mixing $\Delta H_{\mathrm{mix}}$, since $T\Delta S_{\mathrm{mix}}$ attains always positive values due to the increase in entropy on mixing. The chemical nature of the polymers and their molecular mass are the main characteristics that affect miscibility. The former determines the strength of the cohesive forces between the components, with the most favorable conditions attained in the presence of strong specific interactions (e.g., hydrogen bonds; Section 1.7.1), while the molecular mass influences $\Delta G_{\mathrm{mix}}$ in two different ways. If the mixing is endothermic ($\Delta H_{\mathrm{mix}} > 0$), then the higher the molecular mass the lower is the entropy change, rendering miscibility less probable. The opposite happens if $\Delta H_{\mathrm{mix}}$ is negative, since the number of intermolecular interactions increases by increasing the molecular mass, even if these interactions imply a reduction of $\Delta S_{\mathrm{mix}}$.

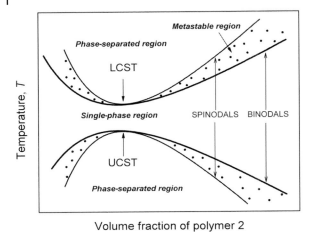

Volume fraction of polymer 2

**Figure 1.24** Phase diagram showing the LCST and UCST, respectively, behavior of polymer blends. Nonsymmetric phase diagrams are common in binary blends of polymers with large differences in their molecular masses.

For a given combination of polymers, blend composition and temperature have substantial bearing on system's morphology. Figure 1.24 shows a schematic phase diagram of a binary system, with the three regions of different degree of miscibility: the single-phase miscible region between the two binodals, the four fragmented metastable regions between binodals and spinodals, and the two phase-separated regions of immiscibility, bordered by the spinodals. The binodal curve is related to the equilibrium phase boundary between miscible (one-phase) and metastable regions. In general, the binodal is defined by the condition at which the component chemical potentials are equal, and is determined by the points of common tangent to the free energy curve (Figure 1.25a). The spinodal curve separates metastable and unstable (unconditionally two-phase) regions, where the curvature of the free energy versus composition graph changes from positive to negative and the second derivative of $\Delta G_{mix}$ is zero. The point of intersection of these curves is denoted as the critical point. Phase separation takes place when a single-phase system suffers a change of composition, temperature, or pressure that forces it to enter either the metastable or the spinodal region (e.g., a shift from the single-phase system at $T_A$ to a phase-separated one at $T_B$, Figure 1.25b). When the binary system enters from the single-phase region into the metastable region, the phase separation occurs by a mechanism resembling crystallization – slow nucleation, followed by the growth of phase-separated domains. Inside the spinodal, the system is unstable to all concentration fluctuations and the blend spontaneously separates into coexisting phases via the mechanism known as spinodal decomposition. Given sufficient time for the process, structure rearrangement will eventually lead to very large regions of the two coexisting phases; however, the spinodal structure can be frozen-in by rapidly cooling the mixture below its glass-transition temperature or by triggering

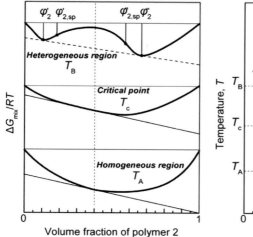

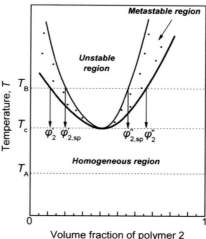

**Figure 1.25** Generalized behavior of $\Delta G_{mix}/RT$ versus volume fraction of polymer 2 ($\varphi_2$) plot (Figure 1.25a), for various positions on the phase diagram (Figure 1.25b) of a binary polymer blend with LCST behavior. The dashed line on the upper site of Figure 1.25a is the common tangent that connects the binodal points.

a chemical reaction between the components. The diagram also shows two critical solution temperatures, the lower, lower critical solution temperature (LCST) (at a higher temperature), and the upper, upper critical solution temperature (UCST) (at a lower temperature). Phase diagrams with two critical points are typical for mixtures of low molar mass components [232], whereas polymer blends usually display only one of them (e.g., LCST [233–238] or UCST [239–241]). An LCST is common in enthalpically driven demonstrations of polymers' miscibility, as is the case of mixtures of poly(vinyl methyl ether) (PVME) with PS [233] or poly(2-chloro styrene) (P2CS) [234]. Here, favorable enthalpic interactions promote mixing of phases at the molecular level, while immiscibility at higher temperatures is entropically driven, with the unfavorable component of entropy of mixing emerging from either equation-of-state effects (e.g., differences in free volume) or the collapse of existing interassociations (e.g., strong intercomponent hydrogen bonding). In several cases, such as in blends of PS with polyisoprene (PI) [239], polybutadiene (PB) [240], or poly(α-methyl styrene) (PαMS) [241], entropics provide a stabilizing contribution strong enough to offset destabilizing enthalpic interactions; these systems exhibit an UCST behavior, with a phase separation observed on cooling driven by unfavorable energetics [104].

Theoretical assessments of miscibility windows are commonly based on the compositional dependence of $\Delta G_{mix}/RT$ (e.g., see Figure 1.25). The lattice theory for the enthalpy of mixing in polymer solutions, developed by Flory [113] and Huggins [114], finds frequent application for modeling the free energy of binary polymer mixtures. The assumption of random mixing of the two polymers and volume additivity ($\Delta V_{mix} = 0$) leads to the well-known expression for the combinatorial entropy of mixing,

$$\Delta S_{mix} = -R \left[ \frac{\varphi_1}{N_1} \ln \varphi_1 + \frac{\varphi_2}{N_2} \ln \varphi_2 \right], \quad (1.50)$$

where $\varphi_i$ and $N_i$ are the volume fraction and the number of segments of the $i$th polymer, respectively, and $R$ is the gas constant. Applying the concept of regular solutions and assuming all pair interactions within the framework of mean-field theory yield for the enthalpy of mixing the approximate relation [3]

$$\Delta H_{mix} = BV\varphi_1\varphi_2 = \chi_{12}RT\varphi_1\varphi_2, \quad (1.51)$$

where $V$ is the total volume of the mixture, $B$ is the interaction energy density, and $\chi_{12}$ is the so-called Flory–Huggins binary interaction parameter. The term "parameter" is widely used to describe $\chi_{12}$ but it is definitively better characterized by the term "function," given its dependence on quantities such as temperature, composition, pressure, molar mass, and related distribution, and even on model parameters such as the coordination number of the lattice and the length of the segment. For incompressible binary systems, the free energy of mixing can thus be expressed as

$$\frac{\Delta G_{mix}}{RT} = \frac{\varphi_1}{N_1} \ln \varphi_1 + \frac{\varphi_2}{N_2} \ln \varphi_2 + \chi_{12}\varphi_1\varphi_2. \quad (1.52)$$

It can be seen that the combinatorial entropy of mixing decreases with increasing molar mass ($N_i$ is proportional to the degree of polymerization) and practically vanishes for high molar mass polymers. The state of mixing is thus highly dependent on the nature and magnitude of the contribution of the enthalpic term. Positive values of $\chi_{12}$ necessarily lead to immiscibility for mixtures of polymers with high molecular weight. Specific interactions, such as ionic or hydrogen bonds, are implicitly eliminated from the Flory–Huggins model, as $\Delta H_{mix}$ is derived only for the Van der Waals interactions. However, experimentally determined $\chi_{12}$ values can include specific interactions, with negative interaction parameters often obtained from melting point depression or inverse gas chromatography studies. Note that depending on the functional form of the temperature dependence of $\chi_{12}$ a wide variety of phase diagrams is possible. For example, the UCST behavior is well accounted for by the Flory–Huggins theory, with $\chi_{12} = C_1 + C_2/T$, but the theory fails to predict LCST.

In the association model approach, Painter and Coleman [242] suggested adding to the simple Flory–Huggins expression an additional term, $\Delta G_H/RT$, to account for the free-energy changes corresponding to specific interactions, most commonly – but not necessarily – hydrogen bonds. The equation, also modified to account for same-chain contacts and screening effects, takes the form

$$\frac{\Delta G_{mix}}{RT} = \frac{\varphi_1}{N_1} \ln \varphi_1 + \frac{\varphi_2}{N_2} \ln \varphi_2 + (1 - \gamma_{ii})\chi_{12}\varphi_1\varphi_2 + \frac{\Delta G_H}{RT}, \quad (1.53)$$

where $\gamma_{ii}$ is an intramolecular screening parameter, defined as the fraction of same-chain contacts. Ordinary screening effects in mixtures containing linear polymer structures lead to $\gamma_{ii}$ values between 0.25 and 0.35, while higher values are obtained when hyperbranched components are used (dendrimer-like polymers, polymers with chain-end or side-chain tethered polyhedral oligomeric silsesquioxanes, etc.) [243]. The association model considers $\chi_{12}$ to represent only unfavorable "physical" forces (dispersion and weak polar forces), which provide a positive contribution to $\Delta G_{mix}$. It is therefore determined using the non-hydrogen-bonded Hildebrand solubility parameters ($\delta_i$, defined as the square root of the cohesion energy function), through the relation

$$\chi_{12} = \frac{V_{ref}}{RT}(\delta_1 - \delta_2)^2, \tag{1.54}$$

with $V_{ref}$ denoting a reference volume. The association model quantitatively determines $\Delta G_H/RT$ and the compositional dependence of blend $T_g$ for selected hydrogen-bonded blend systems (see Section 1.6.2.3). The only necessary parameters to be known are the molar volumes of the individual segments, the relevant enthalpies of hydrogen bond formation, and the equilibrium constants that describe self- and interassociation. These parameters are readily obtained from group contributions and infrared spectroscopy studies.

### 1.5.2
### State of Dispersion and the Glass Transition

The approaches mentioned above bear some major limitations (e.g., by neglecting changes in free volume on mixing) that reduce the effectiveness of related theoretical predictions of polymer–polymer miscibility. It is therefore a common practice to locate the compositional window and the relevant length scale of phase homogeneity by experimental means. Polymer blends are generally assigned as miscible (single phase) or immiscible (multiphase) based on the result of some convenient physical test. Nevertheless, the evaluations are highly dependent on the nature and resolution of the probing technique. The classical turbidity (or haze) measurement, for example, which is used to appraise phase homogeneity based on the turbidity of systems comprising phases with significant differences in their refractive indices, is characterized by a level of resolution on the micrometer's scale (1–100 μm). This is also the threshold of the light-scattering techniques typically used in miscibility evaluations. Related techniques seem to be of limited practical importance for the majority of polymer blends, since their structural heterogeneities have smaller domain sizes (within the range 50 nm to 5 μm). Short-wavelength radiation scattering, such as scattering of X-rays and thermal neutrons with wavelengths in the range of 0.1–0.3 nm, permits much finer structures to be resolved. Phase separation in domain sizes down to 5 and 20 nm can be easily identified with small-angle scattering techniques such as small-angle X-ray scattering (SAXS) and small-angle neutron scattering (SANS), respectively. SAXS has been mainly used in morphological studies of semicrystalline blends, in the determination of spinodal and binodal temperatures in binary polymer systems, as well as to measure $\chi_{12}$. SANS, on the other hand, has been extensively used to study conformation and morphology in single- or multicomponent macromolecular systems, in the molten or liquid states. Among other molecular spectroscopy techniques, ESR spectroscopy and solid-state NMR measurements of the spin-lattice relaxation times are capable of distinguishing motional traces and structural or dynamic heterogeneities at domain sizes down to 2–4 nm. Wide-angle X-ray scattering (WAXS) offers much higher resolution, permitting the identification of structural heterogeneities down to true molecular levels, with scattering phase dimensions in the range 0.1–1 nm. Apparently, the assessment of phase homogeneity by a particular method or technique does not necessarily mean that the assertion is readily extendible to lower length scales.

Without a doubt, glass-transition temperature determinations currently provide the most popular test for miscibility evaluations. Depending on the chemical nature and the morphology of the system, several DSC studies indicate that multiple glass-transition signals are expected to appear at a level of heterogeneity characterized by domain sizes within the range of 15–20 nm; contributions from considerably lower domain sizes are detected using thermal techniques of higher sensitivity and resolving power (Section 1.4). Still, several complications exist that frequently cast doubt on the interpretation of related results. Problems arise from the fact that the glass-transition behavior of any glassy system encompasses the slowly dissipating memories of its preparation procedure, and the effects of any special treatment applied to the material prior to testing. The above factors, in combination with potential differences in the rate at which the system has been brought to the conditions of measurement, affect in a usually poorly predictable way its morphological characteristics and preclude direct comparisons of data obtained for the same material from different laboratories or using different experimental protocols. In view of that, several studies [2,244] clearly demonstrate that a reliable direct relationship between miscibility and the glass-transition phenomenon can be anticipated only in cases where the blend has achieved its equilibrium morphology before the evaluation, and the behavior of the glass-transition signal is unaffected by the experimental method.

On the condition that all abovementioned requirements are fulfilled, the number of transitions and the compositional dependence of the respective glass-transition temperatures offer reliable "indicators" of miscibility, partial miscibility (compatibility), or immiscibility between blend components. Figure 1.26 demonstrates schematically the expected compositional variation of the glass-transition temperatures, and the temperature variation of typical dispersion signals in dynamic mechanical or dielectric relaxation experiments (see Table 1.3), in each of these cases. A classic and simple experimental criterion for determining miscibility is based on the measurement of a single glass-transition signal for all blend compositions, located in between the glass-transition temperatures of the components and consistent with the composition of the blend (Figure 1.26a). The single glass-transition temperature recorded is also frequently referred to as the "blend-average $T_g$"; its position depends on the type and extent of the interactions between the components, as well as on mixing-induced variations in the free

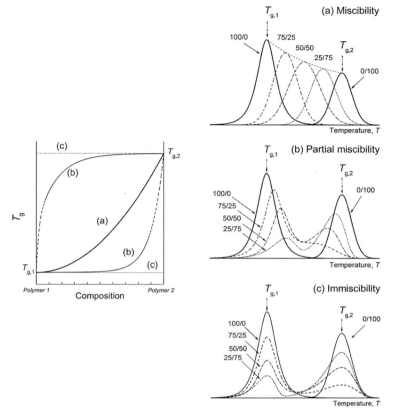

**Figure 1.26** Schematic plot of the expected shifts of glass-transition signals of binary polymer mixtures, in the case where miscible (a), partially miscible (b), or immiscible (c) blends are formed. A behavior analogous to that shown in plot (b) is likely to appear in dynamically heterogeneous miscible blends (see Section 1.6.2.5).

volume (see Section 1.6). In the case of partially miscible (or compatible) polymer blends, the glass transitions of both components are recorded (Figure 1.26b): convergence of the glass-transition temperatures of the two phases is observed, along with a change in the width and the strength of the signals. Completely immiscible polymer blends clearly demonstrate two glass-transition regions for all blend compositions, but in this case peak positions are nearly composition independent (Figure 1.26c).

In the ordinary immiscible polymer blends and related commercial materials, an asymmetric *interface* exists, which may remain indefinitely [1]. Depending on the statistical segment length and the binary interaction parameter of the two polymers, interpenetration at the interface may extent to a depth of a few to several nanometers. The term *interphase* or *interphase region* is used to describe the interpenetration zone, which may have physical properties distinctly

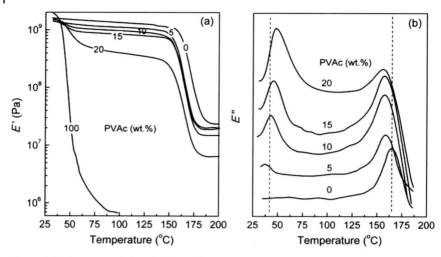

**Figure 1.27** Storage modulus $E'(T)$ (a) and loss modulus $E''(T)$ (b) isochrones (1 Hz) for PVAc + epoxy thermosets with different PVAc contents. After ref. [245], with permission © 2010 Springer.

different from either polymer. The interphase regions in several polymer blends, like those containing a dispersed phase of a low-$T_g$ rubber, provide great toughness and/or impact resistance to the final products. Segmental motions activating within the interphase are likely to produce minor shifts and/or asymmetric broadening in the apparent glass-transition signals of the constituting polymers. Along these lines, the transition temperatures of each component in a phase-separated blend may vary within a window with a width less than one-tenth of the $T_g$ contrast (e.g., see the response of the immiscible PVAc + epoxy resin blends depicted in Figure 1.27) [245]. Such minor changes are frequently reported to arise from morphology changes, physical interactions, mismatch between the thermal expansion coefficients of the components (negative pressure effect), or dilution effects. Note that when chains in the interphase are chemically (as in block or graft copolymers) or physically bonded (e.g., at most, through hydrogen bonding of alike chains) a strengthened interphase is formed, providing a tougher material and much stronger perturbations in the glass transitions of the system.

The immiscible binary blends formed by PP, PE, or PS offer an illustrative example of the relation between the compositional variations of morphology and changes of the glass-transition temperature [228,246]. Figure 1.28 shows cross-sectional views of PP + PS samples prepared by melt blending in a single screw extruder. Fine micron-scale morphologies consisting of PS dispersed throughout a PP matrix, ranging from 0.5 to 20 μm, appear when PS is a minority phase. A co-continuous structure is observed at equal mass fractions (a phase inversion point near $w_{PS} = 0.50$), illustrating a configuration in which both phases surround each other, while a PP dispersed in PS structure appears at

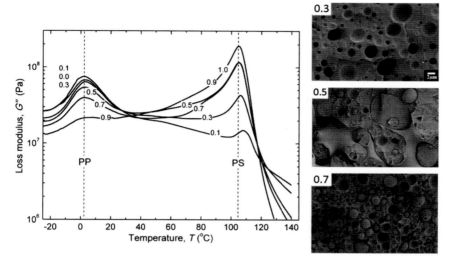

**Figure 1.28** DMA loss modulus curves, showing the glass transitions of PP and PS in immiscible PP + PS blends. Numbers indicate the mass fraction of PS in the blends, and the vertical lines indicate the $T_g$s of neat PP and PS. Insets show scanning electron microscopy (SEM) micrographs of selected blends ($w_{PS} = 0.30$, 0.50, and 0.70). The PS domain morphology in the first image ($w_{PS} = 0.30$) is represented by the dark etched (PS removed) regions. Replotted and adapted from ref. [246], with permission © 2006 Elsevier.

higher PS loadings. Immiscibility is evident in MTDSC and rheological measurements of the glass transitions in all blend compositions. The DMA spectra shown in Figure 1.28 reveal a tendency of the glass-transition temperature of PS to increase as its loading decreases below 50 wt.%, representing a shift opposite to that expected if parts of the constituent polymers were miscible. The gradual upshift is restricted to compositions in which the PP phase surrounds the PS phase either partially or completely. Thirtha *et al.* [246] explained this behavior taking into account the compressive pressure exerted on the amorphous PS domains due to differential shrinkage between the amorphous PS and crystallizing PP. Furthermore, the glass-transition temperature of PP (~3 °C) does not change with composition, suggesting that the phase interactions are weak physical, not chemical, and are dependent on changes in morphology that accompany changes in composition.

The compatibilization of inherently immiscible polymers proffers significant modifications in both the relaxational and glass-transition behavior of the components, as well as in the morphology of the resultant materials. Epoxy resin, for example, appears to be an efficient reactive compatibilizer for immiscible poly (trimethylene terephthalate) (PTT) + polycarbonate (PC) blends [247]. With reference to Figure 1.29, when epoxy content reaches 2.7 phr, or more, the blends cease to exhibit a distinct two-phase morphology, indicating stronger interfacial adhesion of the blend; etched cavities are absent at the fracture surface, and the

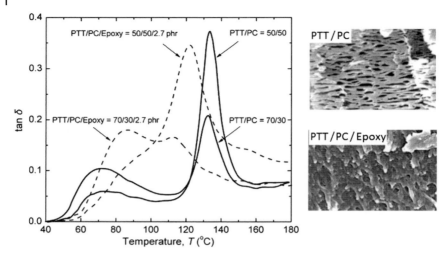

**Figure 1.29** DMA damping factor versus temperature curves, showing the glass transitions of immiscible and compatibilized (with epoxy resin) PTT + PC blends, melt-mixed at 250 °C for 10 min. SEM micrographs compare the morphologies of selected blends (PTT/PC = 50/50 and PTT/PC/epoxy = 50/50/2.7 phr). The PC-rich phase domain in PTT + PC is represented by the dark etched (PC removed) regions with stalactitic morphology. Replotted and adapted from ref. [247], with permission © 2005 Taylor & Francis.

scraggy surface reveals clear miscibility among microdomains, consistent with the changes in the glass transformation range observed in the damping factor versus temperature spectrum.

Miscible polymer blends still have heterogeneity in the segmental length scale because of the chain connectivity (that results in the self-concentration of the segments of respective chains) as well as the dynamic fluctuation over various length scales. As a result, in several cases the blend components feel different dynamic environments that may translate into different temperature dependences of their segmental relaxation rates. This type of dynamic heterogeneity often results in a wide glass transformation range, a broad distribution of the segmental relaxation modes, and the thermorheological complexity of this distribution. Two separate glass transitions may even appear, as has been clearly demonstrated using techniques of increased resolving power in several miscible polymer mixtures with components showing a strong dynamic contrast ($\Delta T_g$ exceeding ~100°) and weak intermolecular interactions (see Section 1.6.2.5). The weakly intermolecularly hydrogen-bonded PMMA + poly(vinyl phenol) (PVPh) blends, for instance, appear homogeneous (fully miscible) when inspected at a domain-size scale exceeding ~20–30 nm, taking into account the single composition-dependent glass transition reported for all blend compositions by conventional DSC (Figure 1.30a [248]). Nevertheless, the binary mixtures appear heterogeneous at lower probing length scales (~2–15 nm), after considering the pair of transitions found in melt-mixed blends by DMA, or

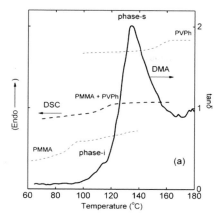

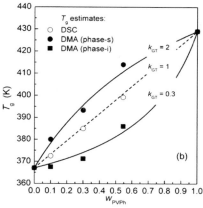

**Figure 1.30** (a) Comparison of the DSC (heating rate 20 °C min⁻¹) and DMA (oscillating frequency of 10 Hz) spectra of a representative PMMA + PVPh blend ($w_{PVPh} = 0.3$). The DSC spectra of the two homopolymers are also shown. (b) Compositional variation of the glass-transition temperatures. DMA estimates are downshifted by 16°. Lines represent data fits based on the GT equation (Section 1.6.2.1). Replotted and adapted from ref. [248], with permission © 1996 American Chemical Society.

observing the proton spin-lattice relaxation time in the rotating frame $[T_{1\rho}(H)]$ in tetrahydrofuran (THF) cast blends by solid-state NMR spectroscopy [249]. With reference to the thermograms shown in Figure 1.30a, Li and Brisson [248] related the major DMA peak to a blend rich in syndiotactic-like PMMA segments (phase-s) and the minor peak (lower strength and $T_g$) to a blend rich in isotactic-like PMMA segments (phase-i), with diffuse boundaries between them and distinctly different compositional dependences of the respective transition temperatures (Figure 1.30b).

Several complications in miscibility evaluations also arise in the case of melt-miscible partially crystalline blends, where multiple glass-transition-like transitions may appear (e.g., contributions from intermediate rigid-amorphous phases with $T_g$s below, at, or above $T_m$ [156]) although the polymers themselves are miscible. Bearing in mind the above findings, the "single-$T_g$" miscibility criterion undoubtedly constitutes an oversimplification of the actual picture, as the validity of related assessments is critically dependent on the preparation process and the experimental technique used for the evaluation. As indicated by Utracki, the presence or absence of a single glass transition in a polymer blend practically provides a means for evaluating "technological miscibility," that is, to distinguish systems that are so well homogenized so that their phase domains will remain unaffected by conventional processing conditions (e.g., mixing methods and subsequent heating–cooling cycles). Homogeneity on the single-segment level is probably unattainable (due to the constraints of chain flexibility), so a characteristic domain size will be present in all blends; this can range from several repeat units (nominally miscible) to several microns (unambiguously immiscible).

Accordingly, blends displaying a single glass transition (one $T_g$) should be deemed miscible only on a scale at or above the total number of segments cooperatively relaxing at the glass transition, and after considering the spatial resolution limit of the employed experimental technique.

## 1.6
## Analyzing Glass Transitions in Single-Phase Systems

Information on the state of mixing and intercomponent interactions in a binary polymer system can be extracted by analyzing the aspects of the experimental glass-transition behavior, with important quantitative results usually obtained for single-phase and compatibilized materials. The main procedures adopted in related studies involve monitoring of changes in the shape (breadth and asymmetry of the respective signal), the strength, and particularly the location (apparent $T_g$) of the glass transformation range, as a function of composition (Section 1.6.1). Without a doubt, when polymers are combined it is essential to understand how the properties of the resulting materials will change with compositional variations. A key objective of the application research is to develop mixing rules for the desired properties, with the glass-transition temperature getting significant attention. Mixing rules are hardly ever linear. They may be synergetic, which means the desired property increases strongly with the mass ($w$), volume ($\varphi$), or molar ($x$) fraction of the minor component, or nonsynergetic when the property deteriorates. For single-phase materials, the typical mixing rule includes the contributions from each component as well as additional interaction terms, while, for partially miscible or immiscible blends, mixing rules can be extremely complex due to the different morphologies that may develop. The following paragraphs (Section 1.6.2) present established theoretical or empirical mixing rules for the glass-transition temperature, in addition to the interpretations of common experimentally observed dependences, with illustrative examples extracted from the available literature on miscible polymer blends, small-molecule + polymer mixtures and interpenetrating polymer networks.

### 1.6.1
### Shape Characteristics and Strength of the Transition

The breadth of the glass transition provides information about the homogeneity of the system, as it reflects the width of the distribution of relaxation times connected to the transition mechanism. The broadness of a transition signal is defined in DSC studies as the difference between the onset and endset temperatures of the $\Delta C_p$ step at $T_g$. In dynamic experiments (DMA and DEA studies), the broadness of the segmental $\alpha$-relaxation mode is commonly represented by the width at half-height of the respective loss modulus, dielectric loss, or loss tangent (tan $\delta$) peaks; the derivative of a DSC curve ($dC_p/dT$ versus $T$ plot) in the glass transformation range may be used in the same way. In binary mixtures of

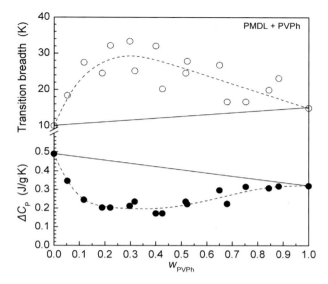

**Figure 1.31** Composition dependence of the heat capacity difference at $T_g$ and of the breadth of the glass transition region of *miscible* poly(*N*-methyldodecano-12-lactam) (PMDL) + PVPh blends. In each plot, linear dependences are depicted by solid lines, while the dashed curves are guides for the eye. After ref. [251], with permission © 2011 John Wiley & Sons.

low-molecular-mass glass formers (e.g., solvent + solvent mixtures), the transition width is approximately composition independent [250], indicating equivalent distributions of local environments. In contrast, in most miscible binary polymer systems, the observation of a single glass transition is accompanied by a *moderate-to-strong* signal broadening, compared with the width of the transitions recorded in the neat components (Figure 1.31) [251,252]. An intermediate behavior is more likely in several small-molecule + polymer mixtures. Most homopolymers typically exhibit glass transition widths of 10–15 °C, while in polymer blends, cross-linked polymers, and other polymer-based composites, the width can be greater by a factor of 3–4. In carefully conducted experiments, this broadening may even translate into a *bimodal pattern* [250]. The latter is usually qualitatively interpreted based on the existence of structural heterogeneities on segmental level, and concentration fluctuations within a local distribution of volumes associated with specific intercomponent interactions [250,253], which may pass undetected in evaluations relying on typical light-scattering techniques.

The intensity of the transition signal is directly related to the mobility of the chains and the fraction of amorphous chains in a partially crystalline blend, excluding the RAF near crystallite faces (see Section 1.7.2). Depending on the type of the probe method, a number of parameters can be used as quantitative measures of the intensity of the process, such as the heat capacity change at $T_g$ ($\Delta C_p$ in DSC) or the strength of the dielectric $\alpha$-transition signals ($\Delta\varepsilon_\alpha$ in DEA).

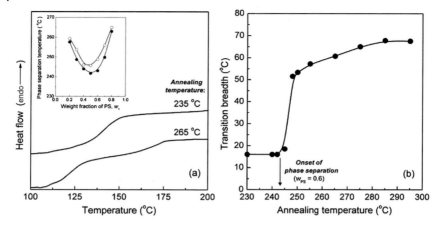

**Figure 1.32** (a) DSC thermograms of a PS + TMPC blend ($w_{PS} = 0.50$) after annealing at 235 °C or 265 °C, below and above, respectively, the respective phase separation temperature (~245 °C). The inset shows the phase separation temperatures obtained by DSC (filled circles) and light transmission (open circles). (a) Change of the transition breadth for a PS + TMPC blend ($w_{PS} = 0.60$) as a function of annealing temperature. In all cases, annealing time was 5 min. After ref. [238], with permission © 1992 Elsevier.

Early studies indicate that the $\Delta C_p(w)$ and $T_g(w)$ dependences show a nonlinear and somewhat reversed behavior in blends that exhibit specific interactions (Figure 1.31), while a linear behavior is expected in systems with noninteracting components [254]. Along these lines a decrease in the strength is usually associated with the formation of more ordered structures, with enhanced restrictions for the segmental motions occurring within the blend environment, as a result of intense specific interactions among functional groups along the chains and a concomitant reduction in free volume.

The phase boundary curve of a binary blend can be determined by studying, as a function of annealing temperature and time, the development of the two phases that form when the miscibility gap is entered [236,255]. The usability of transition breadths in determining phase behavior is illustrated in Figure 1.32, using the calorimetric analysis of the thermally induced phase separation process occurring in PS + tetramethyl bisphenol-A PC blends [238]. Solution-cast samples of the above mixture demonstrate a single-$T_g$ behavior, at all blend compositions. On heating, the blends turn cloudy due to phase separation and the onset temperature of the phase separation can be readily identified by optical methods (e.g., light transmission; open circles in the inset of Figure 1.32a), but also from the glass-transition behavior after heat treatment. The simple experimental procedure employed involves heating the blend at a progressively increasing high annealing temperature for a short period of time, followed by rapid cooling (quenching). In that way, the phase structure at the annealing point is frozen and the nonequilibrium morphology can be explored by subsequent study of thermal events. The glass-transition region recorded by DSC may be either single or bimodal, depending on whether the annealing temperature is below or above, respectively, the phase separation temperature. An LCST-type phase boundary can thus be determined by identifying, for each blend composition, the lowest annealing temperature (cloud point) at which the breadth of the calorimetric glass-transition signal shows a drastic increase (Figure 1.32b). Good agreement between the experimental pattern obtained by DSC and the predicted spinodal curve is often demonstrated [236].

### 1.6.2
### Description and Interpretation of $T_g$ versus Composition Behaviors

What holds for the glass-transition temperatures generally holds for several other properties of binary blends. Mechanical properties, resistance to chemicals, radiation, or heat, they all generally plot the same way as the $T_g$ does with respect to the relative amounts of each polymer in the blend. For that reason, an extensive theoretical and experimental effort has been devoted in attempts to establish relations for the compositional dependence of the glass-transition temperature of amorphous blends. The usability of the ensuing functions is primarily judged by their ability to describe real experimental patterns without resorting to the use of adjustable parameters. In several purely phenomenological approaches, however, a number of heuristic parameters have been introduced to achieve better correlation of models to experimental data, inevitably, at the expense of their predictive power. In view of that, frequently emerges the need to provide a posteriori plausible physical meaning to important empirical parameters, or establish classification schemes for the blend behaviors. Given the multiplicity of approaches and the complexity of the ensuing data interpretations, an overview of the origins and the limits of applicability of selected equations will be discussed in some detail in the following sections.

#### 1.6.2.1  Specific Volumes or Flexible Bonds Additivity Models
Most of the $T_g(w)$ expressions proposed hitherto for binary systems can be represented as variations of the *general* mathematical form:

$$T_g = \frac{w_1 T_{g,1} + k w_2 T_{g,2}}{w_1 + k w_2},$$
(1.55)

with parameter $k$ representing a model-specific constant, which is usually assumed temperature independent. A relation of this form has been proposed by Gordon and Taylor (GT) [256], based on the assumption of *volume additivity* (i.e., ideal volume of mixing) and the existence of a linear change in volume with temperature. Their constant, $k_{GT}$, is given by

$$k_{GT} = \rho_1 \Delta\alpha_2 / \rho_2 \Delta\alpha_1,$$

where $\rho_i$ represents the densities, and $\Delta\alpha_i = (\alpha_{liquid} - \alpha_{glass})_i$ is the change in expansivities at the glass transition of the homopolymers. The GT relation was initially derived for copolymers, and subsequently extensively used in studies of binary mixtures. Equation (1.55) is also analogous to the relation proposed for copolymers by Wood [257], as well as to the Kelley–Bueche [258] relation for the volume fraction dependence of the glass-transition temperatures observed in diluent + polymer systems. In another approach, DiMarzio extended the Gibbs–DiMarzio entropy theory of the glass transition for polymer blends [259,260], assuming that additivity of the flexible bonds is responsible for conformational changes. This treatment provided a description of the change of $T_g$ as a function of the fractions of flexible bonds ($B_i$, with $i = 1, 2$),

$$T_g = B_1 T_{g,1} + B_2 T_{g,2}.$$ (1.56)

By relating the bond fraction to the weight fraction via

$$w_1 = m_1 B_1 \gamma_1 / (m_1 B_1 \gamma_2 + m_2 B_2 \gamma_1),$$

with $w_1 + w_2 = 1$, where $\gamma_i$ is the number of flexible bonds per repeating unit, and $m_i$ is the mass of the monomer unit, the DiMarzio equation becomes identical in form with Eq. (1.55), with its constant written as

$$k_{DM} = m_1 \gamma_2 / m_2 \gamma_1.$$

The GT equation leads to the simple Fox equation [261]

$$\frac{1}{T_g} = \frac{w_1}{T_{g,1}} + \frac{w_2}{T_{g,2}}$$ (1.57)

by invoking the Simha–Boyer rule, $\Delta \alpha_i T_{g,i} = 0.113$ [110], and assuming similar specific volume for the two components ($\rho_1 \approx \rho_2$). In the limiting case, where $k_{GT} = 1$, the GT equation also reduces to the expression for the linear combination (mass additivity)

$$T_g = w_1 T_{g,1} + w_2 T_{g,2}.$$ (1.58)

This provides the simplest estimate of the glass-transition temperature of binary mixtures. The routine application of the above equations in miscible binary systems produces smooth, monotonic, theoretical $T_g(w)$ dependences that frequently either substantially *underestimate* (Figure 1.33) or *overestimate* (Figure 1.34) the experimental patterns [10,262,263]. In view of that, their application for predictive purposes should be considered with cautiousness.

### 1.6.2.2 Additivity of Free Volumes

Several years ago careful analyses of literature data indicated that a singular point, or cusp, may be present in the glass-transition temperatures versus composition curves of some binary mixtures [266–269]. Such a behavior was demonstrated and quantitatively described by Kovacs [266] in the framework of the free-volume theory, assuming a negligible excess volume between the two polymers on mixing. Based on the free-volume approach, if the difference $\Delta T_g = T_{g,2} - T_{g,1}$ between the two homopolymers involved becomes large, the free volume of the high-$T_g$ polymer becomes zero at a critical temperature $T_{crit}$. As a result, equations deriving from the free-volume theory become invalid below

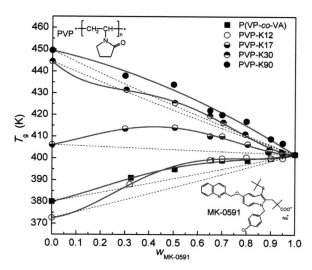

**Figure 1.33** Compositional variation of the $T_g$ of solid dispersions of the model hydrophobic drug MK-0591 with different pharmaceutical grades of PVP [with $T_g$ (°C) = 175 − 9685/$K^2$], or P(VP-*co*-VA) [264]. Solid lines are fits to the BCKV model equation, while dashed lines are theoretical predictions based on the Fox equation. After ref. [10], with permission © 2011 Elsevier.

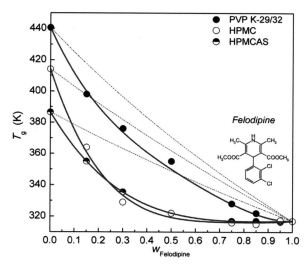

**Figure 1.34** $T_g$ versus composition patterns of miscible solid dispersions of amorphous felodipine in PVP K-29/32, hydroxypropyl methylcellulose (HPMC), or hydroxypropyl methylcellulose acetate succinate (HPMCAS) [265]. Solid lines are fits to the BCKV model equation, while dashed lines are theoretical predictions based on the GT equation. After ref. [10], with permission © 2011 Elsevier.

that point, since in a different case one would stumble on the physically unrealistic situation of a negative free volume. This critical temperature and the corresponding critical volume fraction $\varphi_{crit}$ (relative to polymer 1) were calculated by Kovacs as

$$T_{crit} = T_{g,2} - \frac{f_{g,2}}{\Delta\alpha_2} \tag{1.59a}$$

and

$$\varphi_{crit} = \frac{f_{g,2}}{[\Delta\alpha_1(T_{g,2} - T_{g,1}) + f_{g,2}(1 - \Delta\alpha_1/\Delta\alpha_2)]}, \tag{1.59b}$$

respectively, where $f_{g,2}$ is the free-volume fraction of polymer 2 at its respective transition temperature. Kovacs showed that below $T_{crit}$ the glass-transition temperature of the mixture is given by

$$T_g = T_{g,1} + \frac{f_{g,2}}{\Delta\alpha_1}\left(\frac{\varphi_2}{\varphi_1}\right). \tag{1.60}$$

According to this equation, $T_g$ depends uniquely on the parameters of polymer 1 ($T_{g,1}$, $\Delta\alpha_1$, and $\varphi_1$), if $f_{g,2}$ is given by the universal value of 0.025. Data in the range $T > T_{crit}$ can be described using the classical equations [e.g., Eqs. (1.55), (1.57), or (1.58)]; however, a judicious selection of a particular functional form has to be made given the usually limited width of the pertinent concentration range.

In cases where the excess mixing volume is not negligible, Braun and Kovacs [267] proposed a modified form of Eq. (1.60)

$$T_g = T_{g,1} + \frac{\varphi_2 f_{g,2} + g\varphi_1\varphi_2}{\varphi_1\Delta\alpha_1}, \tag{1.61a}$$

which is again valid below $T_{crit}$. In this equation, $g$ is an interaction term which is defined by

$$g = \frac{V_{ex}}{V}\frac{1}{\varphi_1\varphi_2}, \tag{1.61b}$$

where $V_{ex}$ is the excess volume due to specific interactions. $V_{ex}$ and $g$ are positive if the interactions between the components are stronger than or at least equal to the average interactions between molecules of the same species, and negative otherwise. Typical values of $g$ found in the literature range from $-0.020$ to $+0.020$. The parameter $g$ can be incorporated into the GT equation to provide the relation [267]

$$T_g = \frac{\varphi_1 T_{g,1} + k\varphi_2 T_{g,2} + g\varphi_1\varphi_2}{\varphi_1 + k\varphi_2}, \tag{1.62}$$

where volume fractions are used instead of weight fractions, and $k = \Delta\alpha_2/\Delta\alpha_1$. Equation (1.62) can also be used to fit the data above $T_{crit}$ provided a sufficient number of points is available.

The above treatments give a singular point, or cusp, in the $T_g$-composition curve of miscible polymer blends when $T_{crit} > T_{g,1}$. Typical examples of this

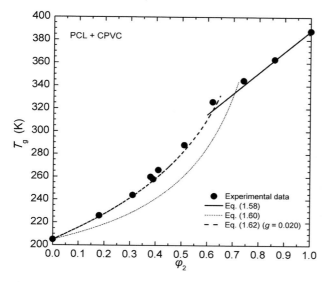

**Figure 1.35** Glass-transition temperature of PCL + chlorinated PVC (CPVC) blends as a function of volume fraction, $\varphi_2$, of CPVC in the amorphous phase (data from ref. [268], with permission © 1988 John Wiley & Sons). Using the iso-free-volume theory predictions of $f_{g,2} = 0.025$ and $\Delta\alpha_2 = 0.00048\,K^{-1}$, $T_{crit}$ is estimated to be $T_{crit} = T_{g,2} - 52\,K = 336\,K$ and $\varphi_{crit} = 0.28$ (relative to polymer 1).

behavior can be found in blends of poly($\varepsilon$-caprolactone) (PCL) with several chlorinated polymers [267,268] (e.g., chlorinated PVC; Figure 1.35), PVAc with aliphatic polyesters [270], PEO + PVPh [271], PVME + PVPh [271,272], poly(vinyl ethyl ether) (PVEE) + PVPh [273], and several other binary mixtures [269,270,274]. As a rule of thumb, Aubin and Prud'homme [268] indicate that a cusp is likely to appear if the difference $\Delta T_g$ is greater than 52 K, in practice higher than at least 70°–80°, and $\Delta\alpha_2$ is sufficiently large.

### 1.6.2.3 Predictions Based on Thermodynamic Considerations

Couchman and Karasz [275] and Couchman [276,277] followed a classical thermodynamic treatment of the glass transition that provided several relations for the compositional dependence of the apparent glass-transition temperature in various single-phased solid solutions. The approach is based on the supposition of continuity of the thermodynamic entropy [275] or enthalpy [276] functions at $T_g$, and equality of the respective excess functions of mixing of the liquid (melt) and glass states. In the entropy model used by Couchman and Karasz, the components of a miscible binary polymer blend were considered to have pure-component molar entropies denoted as $S_1$ and $S_2$, and respective mole fractions $x_1$ and $x_2$. The molar entropy of the mixed system is expressed as

$$S = x_1 S_1 + x_2 S_2 + \Delta S_{mix}, \tag{1.63}$$

where $\Delta S_{\mathrm{mix}}$ incorporates all excess entropy changes (conformational, thermal, etc.) associated with mixing. By considering as $S_1^o$ and $S_2^o$ the pure-component molar entropies at the respective glass-transition temperatures, the mixed system total molar entropy takes the form

$$S = x_1 \left( S_1^o + \int_{T_{\mathrm{g},1}}^{T} C_{\mathrm{p},1} \mathrm{d} \ln T \right) + x_2 \left( S_2^o + \int_{T_{\mathrm{g},2}}^{T} C_{\mathrm{p},2} \mathrm{d} \ln T \right) + \Delta S_{\mathrm{mix}},$$

(1.64)

where $C_{\mathrm{p},1}$ and $C_{\mathrm{p},2}$ are the molar heat capacities of the components. Relations of the above form apply for both the total molar entropies of the glassy ($S^G$) and the liquid ($S^L$) states. At a fixed pressure, the glass transition of the mixture is defined by the continuity condition $S^G(T_g) = S^L(T_g)$, which leads to the relation:

$$x_1^G \left( S_1^{o,G} + \int_{T_{\mathrm{g},1}}^{T_g} C_{\mathrm{p},1}^G \mathrm{d} \ln T \right) + x_2^G \left( S_2^{o,G} + \int_{T_{\mathrm{g},2}}^{T_g} C_{\mathrm{p},2}^G \mathrm{d} \ln T \right) + \Delta S_{\mathrm{mix}}^G$$
$$= x_1^L \left( S_1^{o,L} + \int_{T_{\mathrm{g},1}}^{T_g} C_{\mathrm{p},1}^L \mathrm{d} \ln T \right) + x_2^L \left( S_2^{o,L} + \int_{T_{\mathrm{g},2}}^{T_g} C_{\mathrm{p},2}^L \mathrm{d} \ln T \right) + \Delta S_{\mathrm{mix}}^L .$$

(1.65)

Considering that the excess entropy of mixing is continuous during the glass transition (i.e., $\Delta S_{\mathrm{mix}}^G = \Delta S_{\mathrm{mix}}^L$), and using the (zeroth order) approximation that the transition isobaric capacity increments,

$$\Delta C_{\mathrm{p},i} = C_{\mathrm{p},i}^L - C_{\mathrm{p},i}^G \ (i = 1, 2),$$

are temperature independent, Eq. (1.65) provides the expression

$$\ln T_g = \frac{x_1 \Delta C_{\mathrm{p},1} \ln T_{\mathrm{g},1} + x_2 \Delta C_{\mathrm{p},2} \ln T_{\mathrm{g},2}}{x_1 \Delta C_{\mathrm{p},1} + x_2 \Delta C_{\mathrm{p},2}}.$$

(1.66)

The above equation can be equivalently expressed in terms of mass (weight) fractions, instead of the molar fractions,

$$\ln T_g = \frac{w_1 \Delta C_{\mathrm{p},1} \ln T_{\mathrm{g},1} + w_2 \Delta C_{\mathrm{p},2} \ln T_{\mathrm{g},2}}{w_1 \Delta C_{\mathrm{p},1} + w_2 \Delta C_{\mathrm{p},2}},$$

(1.67)

with $\Delta C_{\mathrm{p},i}$ denoting the heat capacity change per unit mass. This equation reduces to several other common equations, following a certain number of simplifying assumptions. By setting $k = \Delta C_{\mathrm{p},2}/\Delta C_{\mathrm{p},1}$, one obtains the logarithmic GT-like expression

$$\ln T_g = \frac{w_1 \ln T_{\mathrm{g},1} + k w_2 \ln T_{\mathrm{g},2}}{w_1 + k w_2},$$

(1.68a)

a relation initially proposed by Utracki [244]. Furthermore, if it is assumed that $\Delta C_{p,1} = \Delta C_{p,2}$ (a rather crude approximation), then the relation of Pochan *et al.* [278] is derived as

$$\ln T_g = w_1 \ln T_{g,1} + w_2 \ln T_{g,2}. \tag{1.68b}$$

Making use of the expansions of the form $\ln(1+y) = y$, for small $y$, Eq. (1.67) can be written as

$$T_g = \frac{w_1 \Delta C_{p,1} T_{g,1} + w_2 \Delta C_{p,2} T_{g,2}}{w_1 \Delta C_{p,1} + w_2 \Delta C_{p,2}}, \tag{1.69}$$

which corresponds to the general form of $T_g(w)$ equation (Eq. (1.55)), using

$$k_{CK} = \Delta C_{p,2} / \Delta C_{p,1}$$

as a parameter. Assuming that $\Delta C_{p,i} T_{g,i} = $ constant (Boyer criterion: $\Delta C_p T_g = 115\,\mathrm{J\,g^{-1}}$), Eq. (1.69) finally transforms to the Fox equation.

As in the case of the GT function, the applicability of Eq. (1.69) remains low even if its so-called *constant* is used as free fitting parameter, with satisfactory results obtained only for random mixtures (e.g., the commercialized miscible blend of PS and PPO, with weakly interacting components [279]). It has been suggested that the entropy of mixing as well as molecular interactions may be contributing factors for the frequently observed inconsistencies among the theoretical predictions and actual experimental behaviors. Elaborate modifications of the Couchman–Karasz function have been proposed to increase its accuracy [280] and extend its applicability to nonrandom systems [277]. To account, for example, for the effect of the entropy of mixing on the glass-transition temperature, Pinal [280] presented an entropic analysis that extended Eq. (1.66) in the form of

$$\ln T_g = \frac{x_1 \Delta C_{p,1} \ln T_{g,1} + x_2 \Delta C_{p,2} \ln T_{g,2}}{x_1 \Delta C_{p,1} + x_2 \Delta C_{p,2}} - \frac{\Delta S_{c,mix}}{\Delta C_{p,m}}. \tag{1.70}$$

Here, $\Delta C_{p,m}$ is the heat capacity difference between the liquid and the crystalline forms of the material, and $\Delta S_{c,mix}$ is the configurational entropy of mixing that is accessible to the liquid within the timescale of the experiment.

The fitting parameter free method developed by Painter *et al.* [281] to predict the composition dependence of the glass-transition temperature of strongly interacting systems derived an equation,

$$\left[ w_1 \Delta C_{p,1}(T_g - T_{g,1}) + w_2 \Delta C_{p,2}(T_g - T_{g,2}) \right] + w_2 \left[ \left( H_B^{H,L} \right)_{T_g} - \left( H_B^{H,L} \right)_{T_{g,2}} \right] + \Delta H_{mix}^{H,L} = 0, \tag{1.71}$$

separated into three components: a nonspecific interaction term, a term that accounts for that part of the temperature dependence of the specific heat that is due to self-association, and a heat of mixing in the liquid state term. In this relation, $\left( H_B^{H,L} \right)_{T_g}$ stands for the pure state enthalpy of the self-associating polymer at $T_g$, while $\Delta H_{mix}^{H,L}$ is the heat of mixing determined at blend's $T_g$. The thermodynamic Painter–Coleman association model has been found to predict well

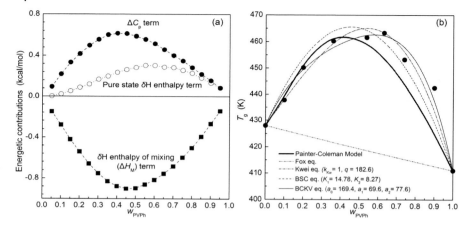

**Figure 1.36** (a) Energetic contributions to the $T_g$ equation, and (b) the predicted blend $T_g$ for mixtures of PVPy with the self-associating PVPh. The prediction of the Fox equation (no specific interactions considered) and the results of various polynomial fitting functions (see Section 1.6.2.4) are included for comparison. After refs [263,281], with permission © 1991 American Chemical Society.

polymer–polymer miscibility and the glass-transition behavior of most hydrogen-bonded polymer blends, and particularly of mixtures characterized by a strongly negative heat of mixing. An illustrative example of this type provides the blends of self-associating PVPh and non-self-associating poly(vinyl pyridine) (PVPy) [281] (Figure 1.36), where strong intercomponent interactions are promoted by the presence of the pyridine group in PVPy, capable of accepting the hydrogen-bonding proton of PVPh. Other examples include mixtures of PVPh with poly(ethyl methacrylate-*co*-methyl methacrylate) [282], PVAc, PVME, P2VP or poly(4-vinyl pyridine) (P4VP) [283], poly(styrene-*co*-vinylphenol) + poly (tert-butyl acrylate) blends [281], PS4VP (15 mol% of 4-vinylpyridine) + poly (styrene-*co*-acrylic acid) (PSAA, 14 mol% acrylic acid) blends [284], poly(styrene-*co*-N,N-dimethylacrylamide) (PSAD, 17 or 25 mol% N,N-dimethylacrylamide) + PSAA (14, 18, 27, or 32 mol% acrylic acid) blends [284], poly(p-(hexafluoro-2-hydroxyl-2-propyl)styrene) with PVAc or poly(ethylene-*co*-vinyl acetate) [285], and PS4VP + poly(styrene-*co*-methacrylic acid) (PSMA, 15 mol% of methacrylic acid) blends [286]. Different types of $T_g(w)$ patterns reported for hydrogen-bonding miscible blends [287] are typically discussed in terms of entropy or enthalpy changes induced by the relative amount of intramolecular (self-association) and intermolecular hydrogen bonding. The Painter–Coleman association model can also produce – with moderate success, however – negative or even sigmoidal deviations of blend $T_g$s from the simple rule of mixtures [287], which may appear depending on the relative strength and the type of the compositional dependence of the energetic terms of Eq. (1.71) [281]. As a result of the relation among fragility and intermolecular interactions, the association model has been successfully applied to also predict blend compositions with maximal fragility [285].

Several other entropy-related models that feature adjustable parameters have also been considered [288,289] to account for interchain interactions that bring on strong entropy of mixing effects on the glass transition (i.e., negative excess mixing volumes) and produce structured mixtures. Kim and coworkers [288], for example, using configurational entropy and the Flory–Huggins theory derived the equation

$$T_g = \exp\left[\frac{z'R}{M_1\Delta C_{p,1}}\left(1 - \gamma\ln\left(\frac{z'-1}{e}\right)\right)\left(\frac{\varphi_1}{r_1}\ln\varphi_1 + \frac{\varphi_2}{r_2}\ln\varphi_2\right)\right.$$
$$\left. + \left(\varphi_1\ln T_{g,1} + \varphi_2\ln T_{g,2}\right)\right], \tag{1.72}$$

with $z'$ being the lattice coordination number. In Eq. (1.72,) $r_1 = v_1/v_0$ and $r_2 = v_2/v_0$, where $v_i$ are the molar volumes of components 1 and 2, respectively, and $v_0$ is the unit lattice volume (normally set for convenience as $v_0 = 1$). In addition, $\gamma$ is a proportional constant, an adjustable parameter, representing the specific interaction between two polymers. Still, for routine applications the usability of the above relation is hampered by the poor knowledge of the required parameters. In another thermodynamic approach, by Lu and Weiss [289], the enthalpy of mixing is represented by a van Laar relationship (Eq. (1.51)), resulting in the relation

$$T_g = \frac{w_1 T_{g,1} + kw_2 T_{g,2}}{w_1 + kw_2} + \frac{Aw_1 w_2}{(w_1 + kw_2)(w_1 + bw_2)(w_1 + cw_2)^2}, \tag{1.73}$$

where $A = \chi_{12}R(T_{g,2} - T_{g,1})c/M_1\Delta C_{p,1}$, $k \approx \Delta C_{p,2}/\Delta C_{p,1}$, $b = M_2/M_1$ (ratio of molecular mass of the repeat units of each polymer), and $c = \rho_1/\rho_2$. Reliable experimental estimates of the Flory–Huggins binary interaction parameter are usually difficult to obtain. Therefore, rather than predicting the $T_g(w)$ dependence for a given miscible polymer blend, this relation probably offers only a mean to calculate $\chi_{12}$ for a limited number of systems from their experimental $T_g(w)$ patterns. An example provides the results for poly(ether ketone ketone) (PEKK) + thermoplastic polyimide (TPI) blends (Figure 1.37). The values of $\chi_{12}$, calculated from Eq. (1.73) for the three PEKK + TPI blends are −0.1, −1.0, and −2.4, respectively. The negative $\chi_{12}$ indicates an attractive interaction between PEKK and TPI, and its decreasing trend (i.e., more negative value) with increasing TPI concentration suggests miscibility improvement as the thermoplastic component increases [290].

### 1.6.2.4 Empirical Concentration Power $T_g$ (w) Equations and Systems' Complexity

Given the complexity of the structures attained and the frequent lack of information from independent studies (e.g., data on the enthalpies of hydrogen-bond formation and the equilibrium constants used in the Painter–Coleman association model), it is often difficult to achieve predictions of blends' behavior based on fixed properties of the starting materials. The inclusion of several fitting parameters is thus considered inevitable in an attempt to alleviate the shortcomings of the established theoretical $T_g(w)$ equations. In the early 1950s, Jenckel

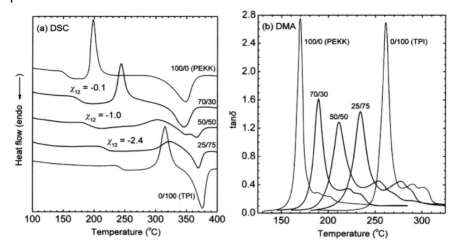

**Figure 1.37** DSC (heating scans at 20 °C min$^{-1}$) and DMA (oscillating frequency 1 Hz, heating rate 2 °C min$^{-1}$) spectra for PEKK, TPI, and selected miscible blends. Both methods provide similar $T_g(w)$ dependences, which can be used in $\chi_{12}$ determinations based on the thermodynamic approach of Lu and Weiss. Replotted and adapted from ref. [290], with permission © 2004 John Wiley & Sons.

and Heusch [291] proposed for plasticized polymer blends an empirical concentration second-order power equation of the form

$$T_g = w_1 T_{g,1} + w_2 T_{g,2} + k_{JH}(T_{g,2} - T_{g,1})w_1 w_2, \tag{1.74}$$

with a parameter, $k_{JH}$, used to characterize the solvent "quality" of the plasticizer molecules [291]. This relation can effectively describe strong, but only monotonic (all positive or all negative) and smooth, deviations from the linear mixing rule. Concentration second-order power equations for the compositional dependence of the blend $T_g$ have also been proposed by Kanig [292], who related the changes in the interaction energies to the respective Gibbs energies for generating one mole of holes in the equilibrium polymer melt, and by DiMarzio [260], who assumed – beside flexible bond additivity – the effect of volume changes due to the different specific volumes of the blend components at $T_g$. In another widely applied approach, Kwei [293] extended the GT equation to a concentration second-order power equation by introducing a quadratic term, $qw_1 w_2$, to read

$$T_g = \frac{w_1 T_{g,1} + k_{Kw} w_2 T_{g,2}}{w_1 + k_{Kw} w_2} + q w_1 w_2. \tag{1.75}$$

The physical meaning of the empirical Kwei parameters, $k_{Kw}$ and $q$, has been the subject of subsequent interpretations based on the strength of specific interactions in the blend and the balance between the breaking of self-association interactions and the formation of interassociation interactions. By this equation, sigmoidal curves can be explained, but only with positive deviations from additivity in the low-$T_g$ range and negative deviations in the high-$T_g$ range; to reproduce data showing the reversed behavior or more complicated dependences, additional correction terms are required. One of the most effective equations in describing complex $T_g(w)$ patterns is the phenomenological virial-like concentration third power function proposed by Brekner, Schneider, and Cantow (BSC equation) [294,295]

$$T_g = T_{g,1} + (T_{g,2} - T_{g,1})[(1 + K_1)w_{2c} - (K_1 + K_2)w_{2c}^2 + K_2 w_{2c}^3]. \quad (1.76)$$

The variable $w_{2,c}$ is the expansivity-corrected mass fraction of the GT expression $w_{2,c} = kw_2/(w_1 + kw_2)$, with $k \approx T_{g,1}/T_{g,2}$. This functional form results directly from the assumption that both the free-volume distribution and the conformational mobility in polymer mixtures are dependent on the specific intercomponent interactions. It has been pointed out that parameter $K_1$ mainly accounts for the differences between the interaction energies of the binary heterointeractions (between different components) and homointeractions (between molecules of the same component), while parameter $K_2$ depends on contributions resulting from conformational entropy changes. Depending on the values of the fitting parameters of Eq. (1.76), $K_1$ and $K_2$, and of their difference, $K_1 - K_2$, it has become possible to categorize the glass-transition behavior of various binary systems into general classes [295].

Interestingly, even the two-parameter empirical equations fall short in describing very complex $T_g(w)$ behaviors that come into sight, when, for instance, at least one of the components partly crystallizes in the blend environment (Figure 1.38) [296], entropic or enthalpic factors prevail at different compositional ranges [251], or the neat components have almost identical segmental mobilities ($\Delta T_g \approx 0$, Figure 1.39). Tentative explanations of these discrepancies are often based on molecular size effects, the diminishing (or enhanced) free volume of one component in the presence of the other or composition-dependent excess mixing volume effects. As an example, the patterns of poly(epichlorohydrine) (PECH) + PVME [297] and PαMS + poly (cyclohexyl methacrylate) (PCHMA) [298] blends can be explained only by bearing in mind the combined effect of heterocontact formation on interchain orientation and the corresponding conformational entropy changes. A negative excess mixing volume – which signifies less space for molecular and macromolecular chain relaxation – is highly probable in the case of PECH + PVME. Here, according to the IR and $^{13}$C NMR data of Alegría et al. [297], the heterocontacts between PECH (with —Cl as electron-acceptor moieties) and PVME (with —OCH$_3$ as electron-donating moieties) are merely slightly favored. Specific volume determinations corroborate an analogous interpretation for the case of the PαMS + PCHMA mixtures: in the intermediate composition ($w_1 = 0.5$), the specific volume calculated for the blend assuming volume additivity is $V_g = 0.966 \text{ mL g}^{-1}$, while the experimental value is only $0.958 \text{ mL g}^{-1}$ [298].

If the effects of the enthalpic and entropic changes are not symmetrical, then irregular patterns with maxima and/or minima deviating from the midpoint of

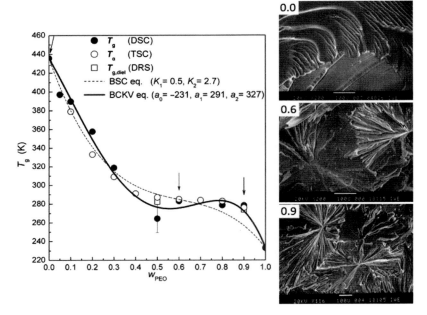

**Figure 1.38** Plots of glass-transition temperatures as a function of blend composition for thermosetting networks prepared with PEO ($M_w = 4.0 \times 10^3$ g mol$^{-1}$). SEM images demonstrate the presence of spherulitic formations of crystalline PEO at $w_{PEO} \geq 0.5$. Fitting lines to the calorimetric data, based on the BSC and BCKV equations, are shown. Data taken from ref. [296], with permissions © 2007 American Chemical Society.

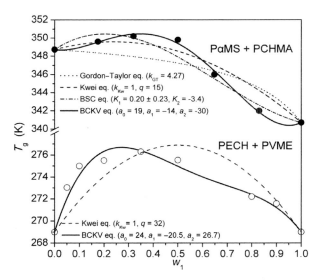

**Figure 1.39** $T_g(w)$ plots of miscible blends of polymers with low $T_g$-contrast: PECH + PVME (dielectric loss peak at 1 kHz, $\Delta T_g \approx 0$) [297], PαMS + PCHMA (DSC data, $\Delta T_g = 8°$) [298]. From ref. [263], with permission © 2010 Elsevier.

the composition $(2w_1 - 1 = 0)$ are usually observed. In view of that, an empirical three-parameter equation has been proposed by Brostow, Chiu, Kalogeras, and Vassilikou-Dova (BCKV equation) [299]

$$T_g = w_1 T_{g,1} + w_2 T_{g,2} + w_1 w_2 \left[ a_0 + a_1 (2w_1 - 1) + a_2 (2w_1 - 1)^2 \right] \qquad (1.77)$$

and successfully tested in binary polymer mixtures, interpenetrating polymer networks, interpolymer complexes, and oligomeric organic + polymer blends [10,262,263]. The quadratic polynomial on the right-hand side of Eq. (1.77), centered around $2w_1 - 1 = 0$, is defined to represent deviations from linearity; that is, with $a_0 = a_1 = a_2 = 0$ the equation leads to the simple rule of mixtures. The number and magnitude of the adjustable parameters required to represent an experimental $T_g(w)$ pattern have been postulated to provide quantitative measures of system's complexity [10]. Based on detailed comparisons between the results obtained for a number of binary systems (Figure 1.40), using established equations and the BCKV function, the empirical parameter $a_0$ and its normalized form, $a_0/\Delta T_g$, have been shown to reflect mainly differences between the strengths of inter- and intracomponent interactions. The magnitude and sign of the higher order parameters are in part related to asymmetric (composition-dependent) energetic contributions of heterocontacts, entropic effects, and structural nanoheterogeneities (e.g., crystalline inclusions) observed in some blend compositions.

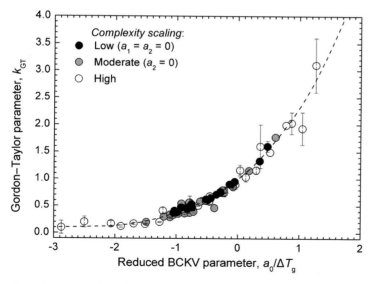

**Figure 1.40** Dependence between the fitting estimates of the parameter of the GT equation ($k_{GT}$) and the reduced principal parameter of the BCKV equation ($a_0/\Delta T_g$). Data for over 80 binary polymer [262,263] and drug + polymer [10] (with permission © 2011 Elsevier) miscible systems are included. The dashed line is a guide for the eye.

### 1.6.2.5 Dynamically Heterogeneous Miscible Blends

In most miscible polymer mixtures, the calorimetric glass transitions are unusually broad (e.g., with asymmetrically broadened derivative peaks) and this feature is frequently accompanied by several well-known anomalies, such as the failure of the time-temperature superposition principle [300]. Display of two concentration-dependent glass transitions – a phenomenon long considered an indication of partial miscibility and large-scale spatial heterogeneities – has been proven to exist in several "dynamically heterogeneous" blends that are miscible on the molecular level. Dynamic heterogeneity is connected with the observation of distinct relaxation (dynamic) behavior for each component in a mixture, despite the existence of phase homogeneity. Traces of this behavior are already known from the late 1960s, as a result of the application of sensitive thermal analysis techniques (e.g., DMA data on PS + PPO blends), with the first report of bimodal calorimetric glass-transition signals in PS + PαMS blends dating back to 1982 [301]. Since then, bimodal transitions have been theoretically justified [302] and experimentally demonstrated (e.g., Figure 1.41) in a large number of molecularly mixed binary systems [252,303–314].

Both in the cases of the partially miscible and the dynamically heterogeneous miscible blends, a number of glass transitions are likely to be seen only in carefully conducted experiments with techniques of higher signal-resolving power, higher sensitivity (i.e., detection of much lower concentrations of relaxing segments of some type), and probes of sufficiently small length scale [300,304]. In studies of the relaxation behavior of PVME + PS blends, for example, Lorthioir

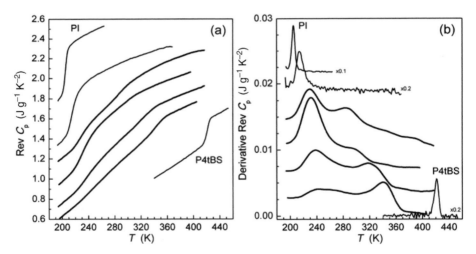

**Figure 1.41** MTDSC thermograms of miscible PI + P4tBS blends: (a) reversible $C_p$ curves, and (b) temperature derivative curves. From pure PI to pure P4tBS, the curves represent blends with PI mass fraction of $w_{PI} = 0.75$, 0.50, 0.40, 0.33, and 0.25, respectively, with two distinct glass transitions recorded at intermediate blend compositions. Vertical shifts and reduced scales (plot (b)) have been used for clarity. From ref. [303], with permission © 2009 American Chemical Society.

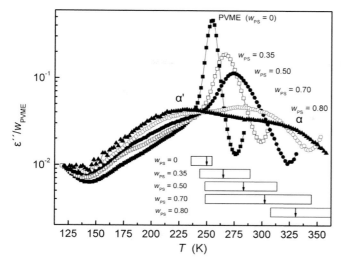

**Figure 1.42** Isochronal $\varepsilon''(T)$ spectra obtained at $f = 1$ Hz on pure PVME and various PVME + PS blends. For each sample, $\varepsilon''$ values have been divided by the PVME weight fraction ($w_{PVME}$). The boxes below the curves indicate the breadth of the single calorimetric transitions recorded for each system, while vertical arrows indicate the midpoint $T_g$ value. Adapted from ref. [304], with permission © 2003 American Physical Society.

*et al.* [304] observed a single glass-transition region in each blend composition by conventional DSC (Figure 1.42), while a bimodal dielectric relaxation signal is readily discernible in several mixtures. With reference to Figure 1.42, in the blend with $w_{PS} = 0.80$ the typical glass-transition mode ($\alpha$-relaxation) appears near blend's calorimetric $T_g$ (at 332 K), while a second glass-transition signal ($\alpha'$-relaxation) forms at a much lower temperature (225 K) and relates to motions of polar PVME chain segments confined in a nonpolar matrix created by frozen PS chains. Note that near the calorimetric glass-transition temperature of the blend the relaxation rate of PVME is over three decades faster compared to that of PS segments.

Several approaches explore the origins of dynamic heterogeneity in polymer blends considering either differences in the intrinsic mobility of the components, thermal concentration fluctuations, the effect of chain connectivity, or combinations of them [304–311]. One approach extends the CM, in which cooperativity between neighboring molecules leads to a broad distribution of local environments (i.e., relaxation times) for each component. In terms of the free-volume theories, one may consider that a distribution of free volumes around component segments is responsible for a wide distribution of $T_g$ values. Another approach highlights the contribution of thermodynamically driven local concentration fluctuations in generating and controlling the unique local environment experienced by each component [308,309]. These fluctuations are manipulated by component molecular masses and the binary interaction parameter $\chi_{12}$.

Related models consider further the local concentration fluctuations to be quasi-stationary near the glass transition, since their average relaxation time is much longer than that of segmental dynamics in the same range of temperatures [308]. This approach has been successfully applied in a limited number of blends, indicating that the relevant length scale is roughly 10 nm near $T_g$.

The methodology developed by Lodge and McLeish (LM model) [311] puts forward an analysis of dynamic heterogeneity and the ensuing pair of effective glass transitions for each blend component ($T_{g,i}^{eff}(\varphi)$, with $i = 1, 2$) considering the "self-concentration" effect. As a consequence of chain connectivity, the average number of nearest neighbors of a given segment that belong to the same component is larger than the number of neighboring segments of the other component. Accordingly, in a region of volume $V$ centered on the basic structural unit of each polymer in the miscible blend, the effective concentration sensed by each polymer segment will be larger than the macroscopic one (i.e., $\varphi_{eff,i} > \varphi_i$). If the typical length scale associated with a relaxational process, as the glass transition, is such that $\varphi_{eff,i}$ is larger than $\varphi_i$, the dynamics will be intermediate between that of the pure polymer and the average dynamics of the blend. In a different perspective, if the length scale of the probe is low enough to effectively sample and resolve segmental motions within such small volumes, then two glass-transition signals will be an immediate result. Based on the LM model, the effective local composition, that is, the local composition of the blend in a volume $V \sim l^3$ around the segment $i$, is described by the relation

$$\varphi_{eff,i} = \varphi_{self,i} + (1 - \varphi_{self,i})\varphi_i, \tag{1.78}$$

where $\varphi_{self,i}$ is the self-concentration of the pertinent polymer segment and $\varphi_i$ the nominal (average) concentration of the same blend component. These local compositions might be quite different from that of the bulk as the flexibility of the polymer chain decreases. The length scale related to the monomeric friction factor is regarded of the same order as the Kuhn length, $l_k$ (the length scale beyond which the conformation of the chain becomes Gaussian), defined as $C_\infty l_b$, where $C_\infty$ is the characteristic ratio, and $l_b$ is the length of the average backbone bond. The relaxation of the Kuhn segment is influenced by the concentration of monomers within a volume $V \approx l_k^3$. The self-concentration for each component is thus calculated as the volume fraction occupied by the Kuhn length's worth of monomers inside $V$, using the relation

$$\varphi_{self} = \frac{C_\infty M_0}{\rho k_b N_A V}, \tag{1.79}$$

where $k_b$ is the amount of backbone bonds per repeat unit. In the framework of the LM model, each blend component experiences a distinct effective glass-transition temperature that depends on $\varphi_{eff}$. Estimates of the self-concentration (between 0.1 and 0.6 based on calculations using Eq. (1.79)) are initially used to determine the effective local concentrations. With the assumption that $T_{g,i}^{eff}(\varphi) = T_g(\varphi)\big|_{\varphi=\varphi_{eff,i}}$, the effective glass-transition temperature of each component can be obtained using the Fox equation, and confronted to the experimental

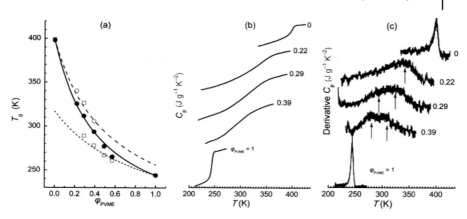

**Figure 1.43** (a) $T_{g,av}$ (circles) and $T_{g,1}$ and $T_{g,2}$ (squares) of PVME + PoCS blends against $\varphi_{PVME}$. Dashed and dotted curves are $T_{g,eff}$ values for the PoCS and PVME components, respectively, calculated by the LM model. The solid curve is determined by the GT equation with $k_{GT} = 0.3$. (b) $C_p$ curves. (c) Derivative $C_p$ curves. $T_{g,1}$ and $T_{g,2}$ are assigned with arrows [306], with permission © 2005 American Chemical Society.

patterns [311–313,315]. Alternatively, the self-concentration can be incorporated as parameter in a suitable $T_{g,av}(\varphi)$ or $T_g^{eff}(\varphi)$ equation, to be indirectly obtained by subsequent fitting of experimental data [262,304–306,312]. In this manner, self-concentration quantifies the extent to which a component has its dynamics perturbed by the blend partner. The upper limit of $\varphi_{self} = 1$ denotes a component showing its pure or neat component dynamics, while $\varphi_{self} = 0$ corresponds to a component whose dynamics are slaved to those of the blend partner. An example of the application of the LM model analysis for miscible PVME + poly($o$-chlorostyrene) (PoCS) blends is shown in Figure 1.43 [306].

Several papers provide comparisons of the LM model predictions to either calorimetric data or dynamic results (Table 1.4). The model is generally successful in predicting component dynamics in athermal polymer blends, though this success usually relies on using $\varphi_{self}$ as a fitting parameter rather than using its theoretic estimate. In the case of PI + poly(vinylethylene) (PVE), for example, the theoretical value of $\varphi_{self,2}$ is very low to account for the observed behavior [316], suggesting that the self-concentration effect predicted by the LM model is not strong enough to account for the tracer dynamics of PVE in a PI matrix. Data evaluations are usually consistent with the model prediction of a smaller self-concentration for the slower component (i.e., $\varphi_{self,1}/\varphi_{self,2} > 1$), which is expected to arise from its stiffer backbone and larger Kuhn length (higher persistence length) [311]. Nevertheless, wide differences often appear between model predictions and experimental derivations of the self-concentration. Typically, theoretical predictions are based on the assumption that $V = l_k^3$, despite the fact that even within the assumptions of the LM model a length scale that is of the order of Kuhn's length, and not necessarily equal to it, is anticipated. Ample experimental evidence suggests that $\varphi_{self}$, for a given polymer, is matrix dependent,

**Table 1.4** Examples of miscible binary polymer systems exhibiting distinct segmental dynamics for the components in the blend environment and results from their analysis in terms of the LM model.

| System | $\Delta T_g$ (°C) | Experimental technique | $\varphi_{self,1}$ | $\varphi_{self,2}$ | Reference |
|---|---|---|---|---|---|
| PVAc + PMMA | 44 | Fluorescence spectroscopy | 0.80[a] | 0.66[a] | [314] |
| PI + PVE | 60–65 | DSC, NMR, DRS | 0.45[b] | 0.25[b] | [311,318] |
| | | | | 0.20[b] | [319] |
| | | | | 0.50[a] | [320] |
| PCHA + PCHMA | 75 | ESR, DSC | 0.21[b] | 0.41[b] | [305] |
| PEO + PVAc | 96 | DSC | 0.26[b] | 0.23[b] | [311] |
| | | | 0.64[a] | 0.16[a] | [313] |
| | | | 0.40[a] | 0.08[a] | [300] |
| PEO + PLA | 110 | DSC | 0.37[a] | 0.19[a] | [313] |
| PI + PS | 130 | $^{13}$C and $^2$H NMR | 0.45[b] | 0.27[b] | [321] |
| | | | 0.33[a] | 0.42[a] | [321] |
| PVME + PS | 130 | MTDSC, DRS, TSC | 0.25[c] | 0.27 | [304,306] |
| | 50–130 | DEA, DSC[d] | | | [239,322] |
| PVME + P2CS | 150 | DSC, DRS, SAXS | 0.25[b] | | [240] |
| | | | 0.62[a] | 0.24[a] | [323] |
| PVME + PoCS | 154 | MTDSC, DRS, TSC | 0.25[b] | 0.22[b] | [304,311] |
| | | | 0.15[c] | 0.20[c] | [262] |
| PαMS (5-mer) + PαMS | 180 | DSC | 0.22 | 0.22 | [324] |
| | | | 0.051[c] | 0.087[c] | [324] |
| PCL + PSMAn(14) | 190 | TSC, DSC, WAXS | 0.33[b] | 0.27 | [325] |
| PEO + PMMA | 200 | DSC | 0.23[b] | 0.25[b] | [302] |
| | | | 0.55[a] | 0.60[a] | [302] |
| PCL + PC | 206 | DSC, DMA, TSC, WAXS | 0.33[b] | 0.05[b] | [312] |
| | | | 0.47[a] | 0.19[a] | [312] |
| | | | 0.20[c] | 0.49[c] | [312] |
| PI + P4tBS | 215 | MTDSC, DSC | 0.45[b] | 0.2[b] | [303] |
| | | | 0.63[a] | 0.03[a] | [303] |

a) Result of fitting using the Fox equation.
b) Model prediction using Eq. (1.79) and $V = l_k^3$.
c) Result of fitting using complex $T_g(\varphi)$ equations.
d) In contrast to DEA, early DSC studies indicate only a single calorimetric signal.

and thus the addition of a geometric factor may be required [300,313,317]. The LM model ignores confinement effects, which may become important below the glass-transition temperature of the blend when the high-$T_g$ component becomes glassy and confines the more mobile low-$T_g$ component. In PI + poly

(4-*tert*-butylstyrene) (P4tBS) [303] and PEO + PMMA [302], for instance, the results reveal enhanced dynamics for the fast component in a miscible blend at temperatures beneath the glass-transition temperature of the slowest component (nonequilibrium or confinement effect). The model disregards concentration fluctuations or strong intermolecular interactions, which can influence both the mean relaxation time and the breadth of the distribution [300,304].

Several instances of borderline miscibility have been reported by examining the glass-transition behavior of binary systems of very weakly interacting polymers, such as PVAc with poly(methyl acrylate) (PMA) [254], PMMA [316], or PEO [313]. Miscibility or immiscibility has been reported for PVAc + PMMA blends, depending on the nature of the solvent [232,316,326], the conditions of mixing, and the type of thermal treatment [327]. It has been suggested that in solution the conformational changes resulting from hydrogen bond interactions between the two chemically similar polymers and the solvent molecules (chloroform) may be liable for miscibility [326]. The extremely weak enthalpic interaction among PEO and PMMA chains and the slightly negative Flory–Huggins interaction parameter are responsible for the marginal miscibility detected for PEO + PMMA blends, based on the single glass-transition temperature following a Fox-type compositional dependence [328]. The strong dynamic heterogeneity of the system (Table 1.4) is reflected in the complex compositional and temperature dependence of component relaxation times, which indicate that each component in the blend retains a separate rheological identity [328]. There are, in fact, frequent reports of two $T_g$s and complicated blend morphologies for compositions with PMMA as a majority component ($w_{PMMA} \geq 0.6$) [329]. Strong intermolecular hydrogen bonding, on the other hand, produces a coupling of the segmental dynamics of the components in blends of polymers with even very large intrinsic mobility differences. As a consequence, suppressed dynamic heterogeneity has been demonstrated in PVME + PVPh (for $w_{PVME} \leq 0.5$) [330], PVEE + PVPh (for $w_{PVEE} \leq 0.4$) [273,331], and poly(ethyl methacrylate) (PEMA) + PVPh [332] mixtures, while the time-temperature superposition principle has been reported to hold for PVEE + PVPh, or the strongly interassociated mixtures of PVPh with P2VP, P4VP, or PVAc [283], for which the Painter–Coleman model is also successfully applied.

## 1.7
## Case Studies

### 1.7.1
### Miscibility Achievement via Chemical Modification

In the absence of specific intermolecular interactions, the Gibbs free energy of mixing is usually positive for polymer blends, due to the small combinatorial entropy of mixing and the positive enthalpy of mixing. Consequently, to exhibit thermodynamic miscibility of the blend, in general, there needs to be some degree of intermolecular interactions (e.g., hydrogen bonding, ion–dipole, $\pi$–$\pi$, and charge-transfer interactions) that will provide a favorable heat of mixing. Hydrogen bonding is a particularly important mechanism to expand the range of miscible polymer pairs since if no favorable interactions are present miscibility is very rare and is only found when solubility parameters match each other ($\Delta\delta_{crit} \leq 0.1$ (MPa)$^{1/2}$). If favorable weak interactions are present (e.g., dipole–dipole forces), miscibility can be found even if the difference in solubility parameters approach $\Delta\delta_{crit} = 1.0$ (MPa)$^{1/2}$, but when hydrogen bonds are established it can go up to ~6.0 (MPa)$^{1/2}$ [3].

Over the past few decades, there has been considerable interest in enhancing the miscibility of polymer blends, either by adding a third component as a compatibilizing agent or by introducing specific functional groups into the polymers to promote exothermic interactions between them. Many novel and useful polymer blends have been produced in this manner, with the styrene-based polymers constituting an important category (Table 1.5). In the absence of specific interactions between segments of PS and PMMA, for example, their blends exhibit a

**Table 1.5** Type of the deviation from the linear mixing rule and description of the $T_g$ ($w$) patterns in selected miscible binary blends containing styrene-type units (DSC data).

| Polymer 1[a] | Polymer 2[a] | $\Delta T_g$ (°C) | $T_g(w)$ patterns | | Ref. |
|---|---|---|---|---|---|
| | | | Deviation | Comments | |
| PαMS | PCHMA | 8 | Positive | Complexity: BSC ($K_1 = 0.20$, $K_2 = -3.4$), BCKV ($\alpha_0 = 19$, $\alpha_1 = -14$, $\alpha_2 = -30$). Solvent: chloroform | [258,263] |
| SAN (17.3) | PMMA | 8 | Positive | BCKV ($\alpha_0 = 12$, $\alpha_1 = -8$, $\alpha_2 = 0$). Solvent: chloroform | [254,263] |
| PSCA(15) | P4VP (16.6) | 13 | Positive | Kwei ($q = 50$, $k_{Kw} = 1$). Hydrogen bonding and partial protonation of the pyridine units (FTIR). Solvent: DMF | [337] |
| PS4VP (5–50) | PSMA(20) | 16...33 | Positive | Kwei ($q = 20...60$, $k_{Kw} = 1$). Hydrogen bonding (FTIR). Solvent: chloroform | [338] |
| PS4VP (15) | PSAA(14) | 15 | Positive | Painter–Coleman model. BCKV ($\alpha_0 = 21$, $\alpha_1 = 5.6$, $\alpha_2 = 2.4$). Solvent: THF | [284] |
| PSAD(17) | PSAA(18) | 17 | Positive | BCKV ($\alpha_0 = 21.6$, $\alpha_1 = 0.1$, $\alpha_2 = 17$). Solvent: THF | [263,339] |
| PSAD(17) | PSAA(27) | 20 | Negative | BCKV ($\alpha_0 = -4.9$, $\alpha_1 = 2.6$, $\alpha_2 = -6.2$). Solvent: THF | [263,340] |
| PSAD(25) | PSAA(14) | 21 | Negative | Painter-Coleman model. BCKV ($\alpha_0 = -38.8$, $\alpha_1 = -6.13$, $\alpha_2 = 67.46$). Solvent: THF | [284] |
| PSAD(17) | PSAA(32) | 26 | Negative | BCKV ($\alpha_0 = -64.9$, $\alpha_1 = 36$, $\alpha_2 = -4.5$). Solvent: THF | [263,340] |

| | | | | | |
|---|---|---|---|---|---|
| PS | PPE | 24 . . . 122 | Negative | BCKV: $\alpha_0/\Delta T_g = -1.04 \ldots -0.20$. Solvent: benzene | [262,341] |
| PVPhKH | PS4VP | 30 | Positive | GT ($k_{GT} = 2.03$). Hydrogen bonding (FTIR). Solvent: chloroform | [221] |
| PSCA(15) | PS4VP(17) | 31 | Positive | Kwei ($q = 22$, $k_{Kw} = 1$). Hydrogen bonding (FTIR). Solvent: DMF | [342] |
| PS4VP (15) | PSMA(15) | 33 | Positive | Painter–Coleman model. BCKV ($\alpha_0 = 77.46$, $\alpha_1 = 0.18$, $\alpha_2 = -63.33$). Hydrogen bonding and partial protonation of the pyridine units (FTIR). Solvent: THF | [286] |
| PIBM4VP (20) | PSMA(12) | 39 | Positive | Kwei ($q = 12.81$, $k_{Kw} = 1$). BCKV ($\alpha_0 = 13.8$, $\alpha_1 = \alpha_2 = 0$). Solvent: heptane | [263,343] |
| PIBM4VP (10) | PSMA(12) | 49 | Negative | BCKV ($\alpha_0 = -64.8$, $\alpha_1 = 74$, $\alpha_2 = -12.6$). Solvent: heptane | [263,343] |
| PSHS(28) | PNB | 43 | Negative | GT ($k_{GT} = 0.33$). Hydrogen bonding (FTIR). Solvent: anisole | [344] |
| PSHS(5) | PNB | 64 | Negative | GT ($k_{GT} = 0.37$). Hydrogen bonding (FTIR). Solvent: anisole | [344] |
| PIBM4VP (20) | PSMA(29) | 76 | Positive | Kwei ($q = 38.6$, $k_{Kw} = 1$). BCKV ($\alpha_0 = 39.6$, $\alpha_1 = -25$, $\alpha_2 = -58$). Solvent: heptane | [263,343] |
| PA-6 | MnSPS | 81 | Positive | Lu–Weiss. $-2 < \chi_{12} < -1.5$. Specific interactions among MnS and amide groups (FTIR). Solvent: *m*-cresol | [289] |
| PS | TMPC | 93 | Negative | Solvent: THF | [236] |
| PS | PPO | 107 . . . 205 | Negative | BCKV: $\alpha_0/\Delta T_g = -0.28 \ldots -0.74$. Solvent: benzene | [263,279] |
| PVME | PS | 50 . . . 135 | Negative | BCKV: $\alpha_0/\Delta T_g = -0.22 \ldots -0.66$. Solvent: toluene | [263,295] |
| PVME | PS | 123 | Negative | BCKV ($\alpha_0 = -142$, $\alpha_1 = 33$, $\alpha_2 = 0$). Dynamically heterogeneous (van der Waals interactions). Solvent: toluene | [262,345] |
| PVME | P2CS | 154 | Negative | BCKV ($\alpha_0 = -150$, $\alpha_1 = 81$, $\alpha_2 = 0$). Dynamically heterogeneous (dipole–dipole interactions). Solvent: toluene | [234,262] |

a) Numbers in parentheses denote the molar percentage of the second monomer in each copolymer, with the exception of SAN for which the number refers to weight percentage.

distinct phase-separated morphology. Miscibility with PMMA is only achieved by chemical modification of the molecules, such as by incorporation of hydrogen-bonding capable functional groups along the chains of PS. The copolymer poly(styrene-*co*-acrylonitrile) (SAN), with 17–24 wt.% acrylonitrile, provides an interesting case [333]. Given the small $T_g$ contrast of the components ($\Delta T_g \leq 10°$), conventional DSC data appear insufficient to differentiate among miscible SAN + PMMA blends and their immiscible physical mixture. In the latter case, the bimodal structure of the glass-transition signal (two $T_g$s) becomes apparent only in differential heat capacity spectra obtained using MTDSC and for systems with $\Delta T_g$ exceeding ~5°. Styrene-hydroxy styrene copolymers (PSHS) also form miscible blends with various polymers, such as a homologous series of poly(alkyl methacrylate)s [334], PCL [335], poly(acrylic acid) [336], and polyethers [271]. Hydrogen bonding between the carbonyl ester groups and the hydroxyl groups in hydroxystyrene units inserted into PS chains provides the driving force for miscibility. Similarly, a number of studies suggest improved miscibility of PS with P4VP, or PS4VP, by incorporating proton-donor monomers (acrylic acid, *p*-vinyl phenol, cinnamic acid, maleic acid, or methacrylic acid, etc.) into PS, to utilize the proton acceptor nature of 4-vinylpyridine.

For the systems included in Table 1.5, a tendency appears to develop for positive departures of the $T_g(w)$ patterns from the linear mixing rule (mass additivity) for low $\Delta T_g$'s (<20°) and negative deviation for blends with intense dynamic asymmetry. The type of the $T_g(w)$ dependences and metric properties of these patterns depend on the interplay of entropic (free volume) and enthalpic (relative strength and extend of intercomponent interactions) factors. The relative significance of each factor bears influences from a number of parameters, such as the molecular mass [279,295,341] and flexibility of the polymer chains, the type and accessibility of the functional groups, and the conditions of blend preparation (melt mixing or solvent cast). As an example, the compositional variation reported for the $T_g$s of the miscible binary blends formed by the weakly interacting PS and PPO ($\chi_{12} = -0.06$) shows a decreasing departure from linearity with increasing molecular mass of PS (Figure 1.44a) [279]. In addition,

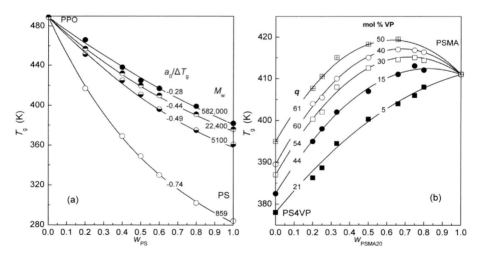

**Figure 1.44** Compositional variation of the glass-transition temperatures recorded for (a) PS + PPO blends, with PSs of different molecular masses: $M_w = 859$, 5100, 22 400, and 582 000 g mol$^{-1}$ [263,279] (with permission © 2010 Elsevier), and (b) PS4VP + PSMA20 blends, with PS4VP of different 4-vinylpyridine loadings: 5, 15, 30, 40, or 50 mol% [338], with permission © 2011 John Wiley & Sons.

depending on the nature of the interacting species, their densities within the polymer chains, and the nature of the solvent [344,346], miscible polymer blends, or interpolymer complexes can be prepared [338,342,343]. Benabdelghani and Etxeberria [338], for example, reported a gradually intensified elevation of the $T_g(w)$ pattern (Figure 1.44b), accompanied by higher thermal stability for the blends and a stronger intercomponent hydrogen-bonding interaction, with an increasing amount of 4-vinylpyridine units in PS4VP + PSMA blends. Conversely, in PSAD + PSAA blends, Djadoun and coworkers [339,340] observed a shift to an increasingly negative deviation as the density of the carboxylic groups within PSAA increases (e.g., by using copolymers with 18, 27, or 32 mol% acrylic acid), due to the occurrence of higher amounts of self-associated carboxyl groups and the corresponding reduction in heterocontacts.

A fair amount of research activity has been devoted particularly to miscible systems comprising PVPh – otherwise referred to as poly(4-vinyl phenol), poly (*p*-hydroxystyrene), poly(4-hydroxystyrene) (P4HS), or poly(*p*-vinyl phenol) – and its copolymers, since the hydroxyl group at para position of the pendent phenyl ring is capable of hydrogen bonding with proton-accepting functional groups (carbonyl, ether) found in several other amorphous (Table 1.6) or partially crystalline (Table 1.7) polymers [347–369]. Thermal analysis studies of the glass transformation range clearly demonstrate miscibility of PVPh with polymers such as poly(vinyl pyrrolidone) (PVP) [347], *polyacrylates* (e.g., PMA and PEA), and *polymethacrylates* (e.g., PMMA, PEMA, and poly(*n*-propyl methacrylate) [P*n*PMA]) [249,352], *polyethers* (e.g., PEO) [348]), *poly(vinyl alkyl ethers)* (e.g., PVME [271,283,330] and PVEE), and *polyketones* (poly(vinyl methyl ketone)). In addition, PVPh can also form miscible blends with PVAc and its random copolymers, a number of *aliphatic polyesters*, such as PCL [271], poly (butylene succinate) (PBSuc) [349], poly(ethylene succinate) (PESuc) [350], poly (*β*-hydroxybutyrate) (PHB) [351], poly(hydroxyvalerate) (PHV) [353], poly(ethylene adipate) (PEA), and poly(butylene adipate) (PBA) [353], poly(L-lactide) (PLLA), poly(D,L-lactide) (PDLLA) [355,356], as well as some *aromatic polyesters* (e.g., PET, poly(butylene terephthalate) (PBT) [354] and PTT [357]). An early review of the glass-transition behavior and miscibility of PVPh with a number of classes of polymers, including polyamides, polyimides, polyurethanes, polyesters, and PCs, has been published by Landry and coworkers [369].

Blends that are prepared by solvent casting often do not represent an equilibrium structure due to varying solvent–polymer interactions among their constituents. This is particularly true for strongly intermolecularly interacting systems, with the limiting case corresponding to solvent-induced phase separation when solvent molecules interact more strongly with one of the blend components. In view of that, several reports indicate that the nature of the solvent has a noticeable effect on the shape of the composition dependence of the glass-transition temperatures. The glass-transition and segmental relaxation dynamics of PVPh + PMMA blends, for instance, have been studied by a number of techniques, which indicate that PMMA and PVPh are miscible at all compositions and over a wide temperature range, due to strong intercomponent hydrogen

**Table 1.6** Type of the deviation from the linear mixing rule and description of the $T_g$ (w) patterns in selected miscible binary blends of PVPh with amorphous polymers (DSC, MTDSC, or DRS data).

| Second component | $\Delta T_g$ (°C) | $T_g(w)$ patterns | | Ref. |
|---|---|---|---|---|
| | | Deviation | Description and comments | |
| PVPy | 17 | Positive | Kwei ($q = 183$, $k_{Kw} = 1$) | [281] |
| PATM | 21 | Positive | Kwei ($q = 86$, $k_{Kw} = 1$), BCKV ($\alpha_0 = 11.5$, $\alpha_1 = 0$, $\alpha_2 = -33$). Solvent: THF | [358] |
| PVP | 25 | Positive | Kwei ($q = 140$, $k_{Kw} = 1$). Solvent: DMF | [347] |
| P4VP | 27 | Positive | Kwei ($q = 96$, $k_{Kw} = 1$), $\chi_{12} = -18.2$, $\Delta\nu = 410\,\text{cm}^{-1}$. Solvent: MEK | [283] |
| PAA | 50 | Negative | Solvent: DMF | [359] |
| Phenoxy | 55 | Negative | Entropy changes due to a decrease in the density of hydrogen bonding. Solvent: THF | [360] |
| PMMA | 66 | s-Shaped | BCKV ($\alpha_0 = 2.3$, $\alpha_1 = 55$, $\alpha_2 = 46$). Solvent: MEK | [249] |
| PTHFMA | 72 | Positive | Kwei ($q = 30.3$, $k_{Kw} = 1.38$), BCKV ($\alpha_0 = 64.1$, $\alpha_1 = -14.9$, $\alpha_2 = -71.2$). Solvent: THF | [361] |
| PEMA | 76 | Positive | BCKV ($\alpha_0 = 35.1$, $\alpha_1 = 7.3$, $\alpha_2 = 6.8$). Solvent: MEK | [282] |
| P2VP | 83 | Positive | Kwei ($q = 58$, $k_{Kw} = 1$), $\chi_{12} = 2.9$, $\Delta\nu = 395\,\text{cm}^{-1}$. Solvent: THF[a)] | [283] |
| Phenolic | 84 | Negative | Deviation result of an entropy change corresponding to a decrease in the density of hydrogen bonding. Solvent: THF | [360] |
| P$n$PMA | 85 | Positive | Kwei ($q = 10$, $k_{Kw} = 1.17$), BCKV ($\alpha_0 = 19.5$, $\alpha_1 = 57.4$, $\alpha_2 = -7.6$). Solvent: 2-butanone | [361] |
| PMTMA | 118 | Positive | Kwei ($q = 86$, $k_{Kw} = 1$), BCKV ($\alpha_0 = 67.9$, $\alpha_1 = 54.4$, $\alpha_2 = 95.7$). Solvent: THF | [358] |
| PMA | 143 | Negative | BSC ($K_1 = -0.36$, $K_2 = -0.90$). Solvent: acetone | [346] |
| PVAc | 152 | Negative | Kwei ($q = -84$, $k_{Kw} = 1$), $\chi_{12} = -2.67$, $\Delta\nu = 72\,\text{cm}^{-1}$. Solvent: MEK | [283] |
| PEEMA | 160 | s-Shaped | Kwei ($q = -144$, $k_{Kw} = 3.2$), BCKV ($\alpha_0 = 34.5$, $\alpha_1 = 101.3$, $\alpha_2 = 0$). Solvent: acetone | [362] |
| PEA | 172 | Negative | BSC ($K_1 = 0.19$, $K_2 = -0.62$). Solvent: acetone | [346] |
| PVEE | 186 | s-Shaped | BCKV ($\alpha_0 = 2.3$, $\alpha_1 = 55.0$, $\alpha_2 = 45.6$). Solvent: MEK | [249,273] |
| PVME | 178 | Negative | Kovacs ($g = -0.011$, $\varphi_{2c} = 0.49$, $T_{crit} = 344\,\text{K}$). Solvent: THF | [271] |
| | 188 | Positive | GT ($k_{GT} = 1.1$), independent of the solvent used (THF, acetone) | [363] |
| | 200 | Negative | Kwei ($q = -152$, $k_{Kw} = 1$), $\chi_{12} = -0.45$, $\Delta\nu = 210\,\text{cm}^{-1}$. Solvent: MEK | [283] |

a) $\Delta\nu$ = difference between the vibrational frequencies of free and hydrogen-bonded (OH $\cdots$ O=C) hydroxyls.

**Table 1.7** Type of the deviation from the linear mixing rule and description of the $T_g$ (w) patterns in selected miscible binary blends of amorphous PVPh with semicrystalline polymers (DSC data). The reported parameters of the Kwei and BCKV equations are assessed from curve fitting of as-received $T_g(w)$ data, without correction for the real composition of the amorphous phase.

| Second component | $\Delta T_g$ (°C) | $T_g(w)$ patterns | | Ref. |
|---|---|---|---|---|
| | | Deviation | Description and comments | |
| PBN | 81 | None | Kwei ($q = 0$, $k_{Kw} = 1$), $\chi_{12} < 0$. Solvent: $n$-hexane | [364] |
| PTT | 105 | None | $B = -7.8\,\text{cal cm}^{-3}$, $\chi_{12} = -0.74$. Melt blending | [357] |
| PLLA | 118 | Negative | Kwei ($q = -78$, $k_{Kw} = 1$), $B = -8.8\,\text{cal cm}^{-3}$, $\chi_{12} = -0.42$. Solvent: dioxane | [355] |
| PDLLA | 120 | Negative | Kwei ($q = -87$, $k_{Kw} = 1$), $B(w) \leq -11.8\,\text{cal cm}^{-3}$. Solvent: THF | [356] |
| PHB | 141 | Negative | Complexity: amorphous PVPh squeezed into the interlamellar region of PHB (BCKV: $\alpha_0 = -70$, $\alpha_1 = -11$, $\alpha_2 = 50$). $\chi_{12} = -1.4$, $B = -12.5\,\text{cal cm}^{-3}$. Solvent: THF + chloroform (1:1) | [351] |
| PHV | 157 | Negative | Kwei ($q = -38$, $k_{Kw} = 0.45$), $\chi_{12} = -1.2$. Solvent: THF | [353] |
| P(3HB-co-3HH) | 168 | Negative | Complexity: BCKV ($\alpha_0 = -143$, $\alpha_1 = 18$, $\alpha_2 = 75$). GT ($k_{GT} = 0.45$). Blends prepared with 20 mol% 3-HH. Strong intermolecular $\delta H$ (FTIR). Solvent: acetone | [365] |
| P(BA-co-BT) | 181 | Negative | Kwei ($q = -82$, $k_{Kw} = 1$), $\Delta\nu = 101\,\text{cm}^{-1}$. Melt blending | [354] |
| PBT | 183 | None | Kwei ($q = 5$, $k_{Kw} = 1$), $\Delta\nu = 103\,\text{cm}^{-1}$. Melt blending | [354] |
| PESuc | 191 | Negative | Complexity: BCKV ($\alpha_0 = -287$, $\alpha_1 = 262$, $\alpha_2 = -201$). Solvent: DMF | [350] |
| PMDL | 206 | s-Shaped | Complexity: BCKV ($\alpha_0 = 28$, $\alpha_1 = -60$, $\alpha_2 = -95$). Miscibility dictated by enthalpic ($\delta H$ at $w_{PVPh} \geq 0.28$) or entropic (random mixing at $w_{PVPh} < 0.28$) factors. Solvent: THF (and precipitation in hexane) | [251] |
| PBSuc | 208 | Negative | Complexity: BCKV ($\alpha_0 = -281$, $\alpha_1 = 175$, $\alpha_2 = -57$), $\chi_{12}(w) = -1.03 \ldots -2.57$. Solvent: DMF | [349] |
| PCL | 210 | Negative | Kwei ($q = -85$, $k_{Kw} = 1$), B $= -9.82\,\text{cal cm}^{-3}$, $\Delta\nu = 85\,\text{cm}^{-1}$, $\Delta\nu* = -65\,\text{cm}^{-1}$. Solvent: THF[b] | [335] |
| PBA | 211 | Negative | Kwei ($q = -225$, $k_{Kw} = 1$), $\Delta\nu = 95\,\text{cm}^{-1}$. Solvent: THF | [354] |
| PBSA | 215 | Negative | Kwei ($q = -160$, $k_{Kw} = 1$), $\chi_{12} = -0.82$. Solvent: THF | [366] |
| PEO | 217 | Negative | Kovacs ($g = -0.020$, $\varphi_{2c} = 0.62$, $T_{crit} = 348\,\text{K}$). $\chi_{12} = -1.5$. Solvent: THF | [271,348,367] |
| PESeb | 229 | Negative | Kwei ($q = -125$, $k_{Kw} = 1$). $\delta H$ interactions (FTIR). $\chi_{12} = -1.3$, B $= -4.7\,\text{cal cm}^{-3}$. Solvent: MEK | [368] |

a) All blends prepared by the solvent-casting method except from PBT or P(BA-co-BT) + PVPh that were prepared by melt blending, and miscible PLLA or PDLLA + PVPh prepared by solution precipitation.

b) $\Delta\nu* = $ difference between the vibrational frequencies of self-associated (OH $\cdots$ OH) and interassociated (OH $\cdots$ O=C) hydrogen-bonded hydroxyls.

bonding. In the miscible blends obtained by solution casting from methyl ethyl ketone (MEK) solution, the breadths of the glass-transition regions in the blends are only slightly broader (15°–20°) compared to those found in neat PMMA or PVPh (~10°) [249]. In contrast, the same polymers are trapped in a phase-separated state when THF is used as solvent [346], since their solubility in this liquid is significantly different and PMMA precipitates first during the solvent evaporation process [370]. The $^1$H spin-lattice relaxation times of PMMA and PVPh show no change in the blend environment [371], a result also advocating immiscibility of the polymers at the length scale of the $^{13}$C cross-polarization/ magic angle sample spinning NMR study. An analogous behavior is found in several other systems, including blends of poly(acrylic acid) (PAA) with PVPh, solvent-cast from $N,N$-dimethylformamide (DMF) (miscible, at a molecular length scale 2–3 nm, as evidenced by NMR data and the single-$T_g$ criterion) or ethanol (immiscible, two composition-independent $T_g$s) [359].

Generally speaking, the irregular $T_g(w)$ patterns observed for several binary mixtures point toward the existence of a complex system, in the sense that at least one important property (e.g., the degree of polymers' mixing, the relative balancing between hetero- and homocontacts or between enthalpic and entropic contributions, and the tendency for crystallization) has a distinctive compositional dependence [263]. Inspecting the data presented in Tables 1.5–1.7, the question that emerges is why both positive and negative deviations in $T_g$ – or, even, both miscibility and immiscibility – occur in systems featuring intermolecular hydrogen bonding. Undoubtedly, factors such as the length of the repeating unit and the degree of polymerization, the mobility of the side chains, and the number (density) and accessibility of the hydrogen-bonding functional groups in each polymer have a drastic effect – in a not always predictable way – on the phase behavior. An example provides the combined DSC and FTIR studies of Lee and Han [364] in blends of PVPh with poly($n$-alkylene 2,6-naphthalate)s containing alkylene units of different lengths. The number of methylene units in the polyester affects chains' mobility and the accessibility of the ester carbonyl functional groups toward the hydroxyl groups of PVPh, which in turn impact glass-transition behavior and miscibility. Accordingly, blends of poly(ethylene 2,6-naphthalate) or poly(trimethylene 2,6-naphthalate) with PVPh demonstrate partial or complete immiscibility, while poly(butylene 2,6-naphthalate) (PBN) appears miscible with PVPh over the whole range of compositions in the amorphous state. Disregarding possible effects from the use of different solvents, the $T_g(w)$ patterns of the various polymethacrylates included in Table 1.6 also indicate a tendency for stronger positive deviation from additivity with increasing number of proton accepting groups in the side chain of the thermoplastic component. In accordance with this, the reduced prime BCKV parameter increases from $\alpha_0/\Delta T_g = 0.035$ in PMMA (one carbonyl group per side chain) to 0.590 in poly(methylthiomethyl methacrylate) (PMTMA) (two proton acceptor groups per side chain) and goes up to 0.890 in poly(tetrahydrofurfuryl methacrylate) (PTHFMA) (three different groups per side chain). For PVPh + PMTMA, in particular, FTIR studies indicate strong hydrogen-bonding interactions between the thioether sulfur atoms of PMTMA and hydroxyl groups of PVPh, while NMR reveals structural homogeneity extending down to ~3 nm.

A number of studies also provide information on differences in the degree and strength of intercomponent hydrogen-bonding interactions present in polymer blends and the corresponding copolymers, and their effect on the glass-transition behavior [287,372,373]. According to the Painter–Coleman association model, the interassociation equilibrium constant of PEMA-co-PVPh ($K_A = 67.4$) is higher than the interassociation equilibrium constant of PEMA + PVPh ($K_A = 37.4$), indicating that the experimental $T_g$'s for copolymers and blends of the same composition should be different. This situation has been experimentally verified by Coleman and coworkers, who demonstrated a higher $T_g$ in copolymer's case [272]. A plausible explanation of the observation considers the different degrees of rotational freedom that arise from intermolecular screening and spacing effects. Furthermore, spectroscopic studies reveal that the number of hydrogen bonds formed between hydroxyl and ester groups in PEMA + PVPh blends is significantly smaller than in random PEMA-co-PVPh copolymers containing the same segments. Similar behavior is found in PVP-co-PVPh copolymers and the corresponding PVP + PVPh blends [287], or PMMA + poly(methyl acrylic acid) (PMAA) and PMMA-co-PMAA (Figure 1.45 [373]), but not in non–hydrogen-bonding systems, such as those comprising isoprene and vinyl ethylene units (PI-block-PVE copolymers and PI + PVE blends [372]).

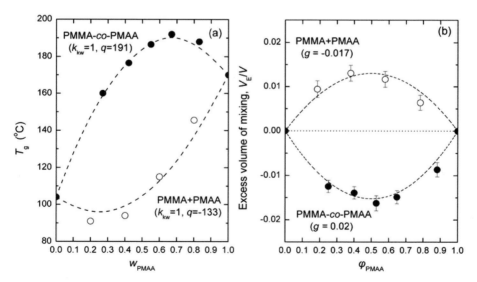

**Figure 1.45** Comparison of the composition dependences of $T_g$ (a) and the excess mixing volume (b) for PMMA + PMAA blends and PMMA-co-PMAA copolymers. The dashed lines show the fitting results of the Kwei equation (plot a), and the excess mixing volume predictions of the Kovacs' free-volume theory (plot b). Replotted data from ref. [373], with permission © 2003 Elsevier.

### 1.7.2
### Microstructure of the Amorphous Phase in Semicrystalline Blends

Blending of crystalline with amorphous polymers is a convenient way for the development of amorphous–crystalline interfaces and various morphological patterns, which may result in an improvement of several physical properties (e.g., toughness, ductility, and impact strength). Although the miscibility window and structural morphologies of related systems have been methodically described in time course, intense research activity is still engaged in studies of aspects of the segregation between the crystalline/amorphous interface and of the possible relationships between intercomponent interactions and domain size of the separated phases. Detection of a single glass transformation range is typical of the nonrigid (bulk-like) miscible amorphous phase, formed by the amorphous fraction of the crystalline polymer and the amorphous component, with miscibility achievement generally interpreted as a result of specific (hydrogen bonding [335]) or nonspecific interchain interactions, or by simply invoking the matched polarity of the blend components [374].

The aliphatic polyester PCL provides an illustrative paradigm of technologically important semicrystalline polymers, offering a potential replacement for conventional polymers due to its biodegradability and low-temperature adhesiveness. PCL is known for its ability to provide miscible blends with polymers, such as chlorinated polyethylenes [375], PVC [376], PC [377], SAN (with 8–28 wt.% acrylonitrile) [378,379], poly(styrene-*co*-maleic anhydride) (PSMAn, with 14 mol% maleic anhydride) [325], poly(styrene-*co*-vinylphenol) (with >13 mol% of VPh in the copolymer) [380], poly(benzyl methacrylate) (PBzMA), and poly(phenyl methacrylate) (PPhMA) [374], PVME, phenolic, phenoxy, and PVPh [335,381–383] (Table 1.8). Several studies reveal interrelations among the extent of partial crystallization and the glass-transition temperature of the amorphous phase. Rim and Runt [380], for example, observed that the greater the SAN concentration in PCL + SAN blends, the higher the $T_g$ and the lower the tendency for PCL crystallization. FTIR spectroscopy provides links between thermophysical properties ($T_g$ and $T_m$) and the structural organization of phases in crystalline-amorphous blends exhibiting specific interactions. The stretching vibrations of the carbonyl group of PCL and proton donating (e.g., hydroxyl) groups in several amorphous counterparts provide excellent probes of intermolecular interactions. For instance, based on estimates of the Kwei parameter, $q$, the average strengths of the intermolecular interactions in blends of PCL with phenolic, PVPh, and phenoxy are weaker than the corresponding self-association for the homopolymers: $q = -10$ (phenolic) $> -85$ (PVPh) $> -100$ (phenoxy) (Kwei fits for $k_{Kw} = 1$) [335]. From a thermodynamic viewpoint, the strength of the specific interaction in a blend can also be described by its interaction energy density parameter, $B$, which is obtained from the depression in the equilibrium melting point based on the Nishi–Wang equation. The negative values of $B$ ($-12.51$, $-9.82$, and $-7.55$ cal cm$^{-3}$ for blends with phenolic, PVPh, and phenoxy, respectively) and the single glass-transition signals obtained are indicative

**Table 1.8** Type of the deviation from the linear mixing rule and description of the $T_g$ (w) patterns in selected miscible binary blends comprising semicrystalline PCL (DSC data).

| Second component | $\Delta T_g$ (°C) | $T_g(w)$ patterns | | Ref. |
|---|---|---|---|---|
| | | Deviation | Comments | |
| PECl (30 . . . 48% Cl) | 50 . . . 67 | Negative | GT ($k_{GT} = 0.26$ and 0.35, for 36 wt.% and 48 wt.% Cl, respectively). Melt mixing | [375] |
| Phenolic | 125 | Negative | Kwei ($q = -10$, $k_{Kw} = 1$), B = −12.51 cal cm$^{-3}$, $\Delta v = 95$ cm$^{-1}$, $\Delta v* = -30$ cm$^{-1}$. Solvent: THF | [335] |
| PBzMA | 126 | Negative | GT ($k_{GT} = 0.25$), nonspecific interactions; miscibility due to matched polarity. Solvent: THF | [374] |
| PVC (56% Cl) | 143 | Negative | GT ($k_{GT} = 0.56$), $\chi_{12} = -0.33$. Solvent: THF | [376] |
| Phenoxy | 159 | Negative | Kwei ($q = -100$, $k_{Kw} = 1$), B = −7.55 cal cm$^{-3}$, $\Delta v = 45$ cm$^{-1}$, $\Delta v* = -105$ cm$^{-1}$. Solvent: THF | [335] |
| SAN (27.5% AN) | 160 | Negative | GT ($k_{GT} = 0.47$). Solvent: THF | [379] |
| SAN (25% AN) | 173 | Negative | GT ($k_{GT} = 0.63$), $\chi_{12} = -0.52$. Solvent: 1,2-dichloroethane | [378] |
| CPVC (63% Cl) | 170 | Negative | GT ($k_{GT} = 0.69$), $\chi_{12} = -0.35$. Solvent: THF | [376] |
| CPVC (67% Cl) | 181 | Negative | GT ($k_{GT} = 0.76$), $\chi_{12} = -0.38$. Solvent: THF | [376] |
| PPhMA | 180 | Negative | Nonspecific interactions; miscibility due to matched polarity. Solvent: THF | [374] |
| PVPh | 206 | Negative | Kovacs ($g = -0.014$; positive excess mixing volume) | [382] |
| | 210 | Negative | Kwei ($q = -85$, $k_{Kw} = 1$), B = −9.82 cal cm$^{-3}$, $\Delta v = 85$ cm$^{-1}$, $\Delta v* = -65$ cm$^{-1}$. Solvent: THF | [335] |
| | 217 | Negative | GT ($k_{GT} = 0.24$). Solvent: THF | [383] |
| PC | 210 | Negative | Kovacs ($g = -0.0227$, $\varphi_{2c} = 0.72$, $T_{crit} = 372$ K). Solvent: CH$_3$Cl (and precipitation in methanol) | [377] |
| P4HS | 197 | Negative | Kovacs ($g = -0.01$, $\varphi_{2c} = 0.58$, $T_{crit} = 375$ K). Solvent: THF[a] | [384] |
| P4HSBr | 213 | Negative | Kovacs ($g = 0.02$, $\varphi_{2c} = 0.51$, $T_{crit} = 387$ K). Solvent: THF[a] | [384] |

a) Analysis performed on $T_g(w)$ data with correction for the real composition of the amorphous phase.

of miscibility, with the relative strength of hydrogen bonding increasing for lower values of B. The frequency difference ($\Delta v$) between the hydrogen-bonded hydroxyl absorption and free hydroxyl absorption bands in FTIR spectra provides an independent verification of the above behavior, by demonstrating that the average strength of the intermolecular hydrogen-bonding interaction decreases in the same order: $\Delta v = 95$ cm$^{-1}$ (phenolic) $> 85$ cm$^{-1}$ (PVPh) $> 45$ cm$^{-1}$ (phenoxy) [381].

Observations of strong elevations in the experimental glass-transition temperatures, with respect to the $T_g$ versus composition patterns of their wholly amorphous mixtures [337,384], are common in blends with some amount of crystallinity. The discrepancy may, to a certain extent, be corrected if the

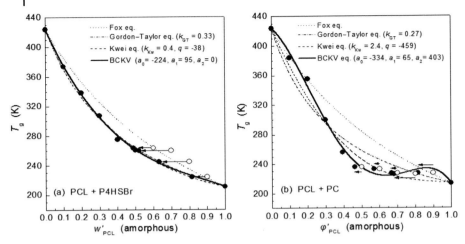

**Figure 1.46** Blend $T_g$ versus amorphous phase composition ($w'$, $\varphi'$) dependences reported for (a) PCL + P4HSBr [384], and (b) PCL + PC [377]. Open symbols refer to the actual experimental data, plotted as a function of the overall weight ($w_1$) or volume ($\varphi_1$) content of PCL in the blend, which were subsequently corrected for blend crystallinity (filled symbols denoted by arrows). From ref. [262], with permission © 2009 John Wiley & Sons.

experimental data are plotted not as a function of the overall weight fraction of the component but in dependence of its real weight fraction in the amorphous phase in the blend (e.g., see Figure 1.46a for PCL + poly(4-hydroxystyrene) brominated (P4HSBr) [384]). In several cases, however, the unusual nonmonotonic $T_g$ versus composition variation persists, even after the necessary corrections for the actual weight fractions of each component in the amorphous phase of the mixture (e.g., see Figure 1.46b for PCL + PC [377]). In terms of microstructure, such rather atypical variations may be – in part – attributed to the formation of complex rigid amorphous phases and different types of segregation of the amorphous polymer. Because of the highly entangled nature of polymer systems, on crystallization and cooling below $T_g$ amorphous layers that are unable to crystallize themselves become entwined with crystalline regions and are constrained in loops and chains connected to the crystal surface (Figure 1.47); this fraction is described as the rigid amorphous phase (or "rigid amorphous fraction"). The physical tethering of amorphous chains progressively diminishes as one moves away from the crystal surface, giving to these chains an increasingly greater degree of mobility. The remaining MAF relates to the unconstrained bulk-like amorphous phase, which exhibits the strong glass transition (at $T_g$) as described earlier. In order of increasing degree of segregation, one may find in partially crystalline materials: interlamellar segregation (the amorphous fraction resides in the interlamellar region within the lamellar stack), interfibrillar segregation (the amorphous chains are placed outside the lamellar stacks of the crystalline component(s), but are still located within the spherulite), and/or interspherulitic segregation (the amorphous phase is expelled from the lamellar stacks and

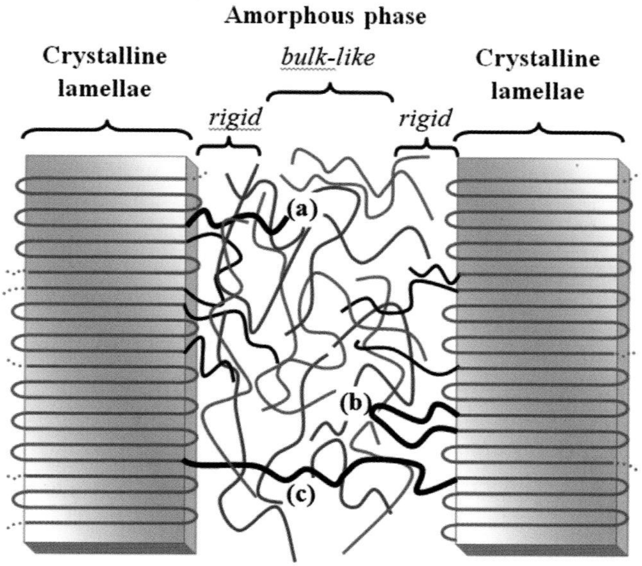

**Amorphous phase**

**Crystalline lamellae**  *bulk-like*  **Crystalline lamellae**

*rigid*  *rigid*

(a)

(b)

(c)

**Figure 1.47** Scheme showing a small section of the well-ordered lattice of crystalline lamellae within a spherulite in a binary polymer blend with a semicrystalline component and the amorphous interlamellar links composed of chains from both polymers. The different zones where a rigid amorphous phase is present or a bulk-like glass transition behavior is likely to appear are shown. Cilia (a), loose loops (b), and tie molecules (c) in the interlamellar region are also indicated.

resides between neighboring spherulites). The structural complexity encountered in many crystalline–amorphous systems is likely to generate highly complicated $T_g(w)$ dependences, which are often only described by heuristic multiparametric mixing rules. In the case of the PCL + PC mixture, in which both components are partially crystalline, one may observe a cusp at a critical composition, above which the Braun–Kovacs equation can be successfully implemented [377]. Similar arguments also apply for the blends of PEO with cellulose acetate butyrate (CAB) [274]. In the latter system, complementary optical microscopy and SAXS experiments verified the complexity of the blend revealing that at low CAB contents the chains of the amorphous component are incorporated into interlamellar regions and commence to segregate to the interfibrillar region with an increase of its weight fraction.

### 1.7.3
### Ternary Polymer Blends: Phase Behavior and Glass Transitions

With the increasing application of multicomponent systems, much interest has been directed toward ternary polymer blends [357,360,385–400]. The thermodynamic phase relationships for ternary mixtures, where one component is

solvent and the other two polymers, were described in the early studies of
Scott [386] and Tompa [387] based on the Flory–Huggins lattice model. Since
then, the majority of miscibility studies on ternary polymer blends remain
focused on the hypothesis that any polymer miscible with any of two other polymers can "compatibilize" their immiscible binary pair. Examples illustrative of
the above behavior provide poly(vinylidene fluoride) (PVDF) [387], PVPh [388],
and SAN [389], which independently act as compatibilizers of the immiscible
PEMA + PMMA mixture. Miscibility evaluations and phase diagram assessments
heavily rely on optical microscopy and calorimetric studies. However, the single-
$T_g$ miscibility criterion for ternary blends is far from being considered
unequivocal, with uncertainties particularly severe for blend compositions rich
in one of the components or for systems comprising components of neighboring
$T_g$s. Questionable assessments are often reported particularly in the case of conventional DSC studies, due to the lack of resolution of this method. For example,
the thermal analysis studies of Ponoit and Prud'homme [391] on melt-mixed
PECH + PVAc + PMMA ternary blends have demonstrated that PECH effectively compatibilizes the immiscible PVAc + PMMA blends only at PECH loadings exceeding 70 wt.%. The higher resolution of DMA has permitted the
identification of structural heterogeneities (immiscibility) – on the basis of system's glass-transition behavior – for several mixtures that would otherwise be
deemed miscible based merely on typical DSC scans (Figure 1.48). Zhang and
coworkers [385] have furthermore demonstrated instances where, after careful
enthalpy relaxation studies, the rather broad glass transformation range recorded

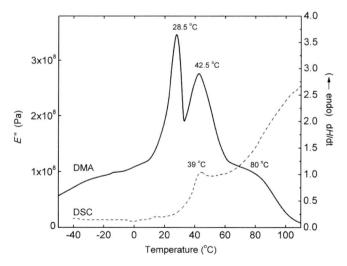

**Figure 1.48** DMA (loss modulus $E''$ at an oscillating strain frequency of 10 Hz) and DSC
curves of a PECH + PVAc + PMMA (8.5/23.5/68)
blend. Numbers indicate the glass-transition
temperatures determined from each thermal
technique. Indications of complex blend morphology are only present in the dynamic
mechanical spectrum. Replotted data from
ref. [391], with permission © 2002 Elsevier.

for particular ternary blend compositions transforms to a multipart signal. Despite the apparent limitations of the glass-transition approach, related studies remain at the frontline of miscibility evaluations of ternary blends. In such cases, however, it seems appropriate to consider the effect of the compatibilizer to involve a reduction of the size of the heterogeneous domains to a value comparable to the probe size of the particular measuring technique.

When all three binary pairs are miscible, completely homogeneous ternary blends or close immiscibility loop phase diagrams are possible. In general, in the absence of strong specific interactions between the components, the ternary phase behavior of polymer blend systems is primarily governed by differences in the physical interaction among the components, which is expressed by the magnitude of the binary interaction parameters, $\chi_{ij}$ (or equivalent solubility parameters, $\delta_i$). If just one of the binary interaction parameters is significantly larger than a critical value [385], a large portion of the ternary phase diagram is predicted to be heterogeneous. In addition, a strong driving force toward phase separation exists if there are significant differences in the solubility or interaction parameter values (i.e., $\chi_{ij} - \chi_{ik} \neq 0$), which produces the so-called "$\Delta\chi$ effect". In terms of the Painter–Coleman association model, the effect of strong specific interactions (hydrogen bonds) on the phase behavior is considered to be controlled by the magnitudes of the equilibrium constants describing self- and interassociation, and the difference between the effective interassociation equilibrium constants of the binary systems ($\Delta K$). The latter is usually called the "$\Delta K$ effect" and reflects the difference in the "chemical" interaction between the self-associating polymer and the other polymers in the mixture; a strong $\Delta K$ effect is responsible for phase separation. Bearing in mind the above restrictions, only very few ternary polymer blends are expected to be homogeneous over their entire range of compositions. These "totally" miscible, ternary blends – including PVDF + PVAc + PMMA [390], PECH + PMMA + PEO [392], PHB + PEO + PECH [393], poly(ether diphenyl ether ketone) + poly(ether ether ketone) + PEI [394], PEI + PET + PBT [395], and PCL + PPhMA + PBzMA [396] – all possess low $\Delta\chi$ effects and no hydrogen-bonding interactions between their segments (i.e., the $\Delta K$ effect can be neglected).

Addition of a polymeric component capable of forming strong hydrogen-bonding interactions (i.e., a proton donor) may potentially act as a mutual common link for the other polymers (containing proton-accepting functional groups), leading to a "bridging effect" and the formation of a more homogeneous ternary mixture. For this to happen, however, it is essential that the interassociation equilibrium constants of the two miscible hydrogen-bonding binaries are comparable. Thermal analyses, for example, have provided compelling evidence of phase homogeneity for all blend compositions in ternary blends of PVPh with PVAc and PMA [385], or with pairs of homologous aryl polyesters (PET, PTT, or PBT) [357]. In all these ternary systems, ultimate miscibility is accredited to negligible differences in the level of the physical interactions among their components (no $\Delta\chi$ effect), and the sensitive balancing of the hydrogen-bonding interactions in the ternary blends between PVPh and either one of the proton accepting polymers (weak $\Delta K$ effect). In the case of the phenoxy + phenolic + PCL [400] and phenoxy + phenolic + PVPh [360] ternary blends, where all binaries interassociate through hydrogen bonds, complete miscibility in the amorphous phase has been ascribed to intermolecular hydrogen bonds that exist within the individual binary blends and create a network-like structure.

Several calorimetric studies of the glass-transition behavior of ternary blends demonstrate a closed loop of a phase-separated region in the phase diagram. Examples provide the ternary blends of PEO + phenolic + PCL [389], PVAc + PVPh + PEO [385], PMMA + phenoxy + PEO [397], PMMA + PVPh + poly(vinyl cinnamate) [398], and PS + poly(cyclohexyl acrylate) (PCHA) + P2CS [399], all consisting of three miscible binaries. In the case of the PEO + phenolic + PCL ternary blend (Figure 1.49), for example, the observed interassociation equilibrium constant between the hydroxyl group of phenolic and the ether group of PEO ($K_A = 264.7$) is found substantially higher than the interassociation equilibrium constant between the hydroxyl of phenolic and carbonyl of PCL ($K_C = 116.8$) and the self-association equilibrium constant of hydroxyl multimer formation ($K_B = 52.3$). This result implies that the tendency toward forming the hydrogen bonding between phenolic and PEO dominates over the interassociation of the phenolic with PCL and the self-association by forming the intramolecular hydrogen bonding of the pure phenolic resin. The fact that the phenolic resin interacts more favorably with PEO than that with PCL ($K_A \gg K_C$) produces a strong $\Delta K$ effect, which provides the driving force for the closed-loop region of immiscibility.

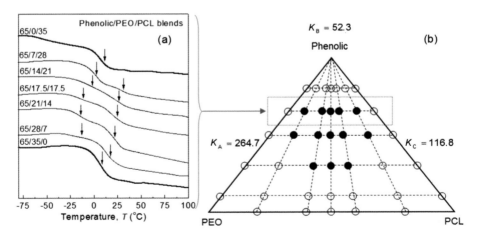

**Figure 1.49** (a) Second-run DSC thermograms and (b) closed-loop immiscibility phase diagram at room temperature of PEO + phenolic + PCL blends. All ternary blends containing 65 wt.% phenolic display two glass transitions, implying that they are immiscible in the amorphous phase. Open and filled circles in the phase diagram denote miscible and immiscible blends, respectively. Replotted from ref. [389], with permission © 2002 American Chemical Society.

## 1.8
## Concluding Remarks

The reversible transformation of amorphous materials from a devitrified or viscoelastic (rubber-like) state into the fairly brittle state of solid glass constitutes one of the most puzzling phenomena in materials science. The distinct kinetic and thermodynamic aspects of the glass transition and the diverse experimental responses recorded hitherto for polymers under different molecular environments and geometric constrictions hamper a unified theoretical interpretation of the event. Regardless of the practically imperceptible structural changes during the transition, numerous experimental techniques employing thermal, mechanical, electrical, or electromagnetic excitation of specific molecular motions have provided compelling evidence of the drastic changes in segmental mobility and molecular bonding that occur in the glass transformation range. Thermal analysis techniques are particularly effective in determining characteristics of the cooperative molecular motions activating in the glass transformation range, with numerous studies indicating that the location and width of the respective transitions are regulated by inherent properties of the material (chemical structure, chain architecture, free volume, etc.), in addition to a number of externally controlled factors (pressure, additives, level of confinement, etc.).

The variability of motional probes and the ensuing differences in the probing length scale of the various measuring techniques, the broadly different oscillation frequencies used in popular isochronal dynamic experiments, along with the different (often inappropriate) procedures adopted for conditioning the testing materials, may become sources of misperception – rather than power tools – in the hands of the novice experimentalist. Notwithstanding the above complications, the glass-transition phenomenon and its defining marker, the glass-transition temperature, have for long been explored and used to evaluate the success of polymers' mixing. The resulting materials exhibit properties intermediate or even superior to those of their pure components, offering a means to improve the poor performances of existing polymers without sacrificing any excellent characteristic. The paradigms illustrated in this chapter make evident that the number of transitions and the compositional dependence of the respective glass-transition temperatures offer indicators of miscibility, partial miscibility (compatibility), or immiscibility between blend components. The conventional – although somewhat misused – experimental criterion for determining miscibility in polymer blends is based on the measurement of a single glass-transition signal for all blend compositions, usually located in between the glass-transition temperatures of the components and consistent with the composition of the blend. For miscible binary polymer systems, in particular, the various theoretical or phenomenological approaches critically examined in this chapter have formulated different analytical $T_g(w)$ functions, endorsing as a minimum qualitative correlations of their parameters with miscibility-controlling factors, such as the type and strength of intercomponent interactions, the relative balancing between inter- and intramolecular interactions, conformational entropy changes, or the magnitude and partition of the excess mixing volume in the system. In the case of the partially miscible or compatibilized polymer blends, the glass transitions of both components are recorded. Convergence of the glass-transition temperatures of the two phases is typically observed, along with a change in the width and the strength of the signals. Interestingly, an analogous behavior often emerges in dynamically heterogeneous blends of weakly interacting components, with the level of homogeneity approaching the segmental length scale. Moreover, dynamic studies of partially mixed systems reveal shifts in the glass-transition temperatures and changes in the width of the transition signals, which provide a measure of the extent of molecular mixing. In contrast, complete segregation of phases (immiscibility) is evidenced by separate glass transitions for all blend compositions, with nearly composition-independent transition temperatures.

The reproducibility and appropriate resolution of the complex glass transition data obtained in several polymer-based composites is prerequisite for achieving unambiguous experimental determinations of structural homogeneity or of the state of dispersion in multiphase systems. The cases discussed in this chapter emphasize the need for testing materials that have attained equilibrium conditions prior to the experiment; that is, the thermal, mechanical or electrical history (memory effects) has to be erased, following standard pretreatments. Moreover, optimal mixing conditions, which include appropriately selected type and amount of solvent(s), temperature of mixing, and method of preparation (i.e., solvent evaporation, coprecipitation or hot-melt extrusion), are essential. Experimental attempts to establish the miscibility window of any pair of polymers via their glass-transition behavior have to consider also the nature, sensitivity, and resolution of the probing technique, the latter being directly related to the limiting domain size with distinguishable thermal response. Along these lines, blends displaying a single glass transition may be deemed miscible only on a scale at or above the total number of segments cooperatively relaxing at the glass transition (i.e., determined level of homogeneity no less than the probe length scale of the particular technique). The issues presented in several sections of this chapter reveal that proper identification of glass transitions entails rigorous separation of the pertinent signals from neighboring transitions (e.g., melting peaks, liquid–liquid transitions), relaxational phenomena (secondary relaxations, interfacial polarizations, enthalpy, or volume relaxation, etc.), or other signals (e.g., noise, background), and subsequent analysis following appropriate models. Unfortunately, despite the advent of modulated temperature techniques in many fields of thermal analysis along with the introduction of novel commercial experimental systems with remarkable accuracy and signal-resolving power, proper consideration of the above issues is often lacking in routine experiments. This situation may cast doubts in several evaluations and precludes straightforward comparisons among seemingly complementary studies of glass formation and polymer dynamics in complex systems.

**Abbreviations**

| | |
|---|---|
| BCKV | Brostow–Chiu–Kalogeras–VassilikouDova (equation) |
| BSC | Brekner–Schneider–Cantow (equation) |
| CM | Coupling model |
| CPM | Configuron percolation model |
| CRR | Cooperatively rearranging region |
| DEA | Dielectric analysis |
| DMA | Dynamic mechanical analysis |
| DRS | Dielectric relaxation spectroscopy |
| DSC | Differential scanning calorimetry |
| DTA | Differential thermal analysis |
| ESR | Electron spin resonance |
| FLDs | Frustration-limited domains (theory) |
| FTIR | Fourier transform infrared |
| GT | Gordon–Taylor (equation) |
| HN | Havriliak–Negami (equation) |
| JG | Johari–Goldstein (model) |
| LCST | Lower critical solution temperature |
| LM | Lodge–McLeish (model) |
| MAF | Mobile amorphous fraction |
| MCT | Mode-coupling theory |
| MTDSC | Modulated temperature DSC |
| MTTMA | Modulated temperature TMA |
| NMR | Nuclear magnetic resonance |
| PALS | Positron annihilation lifetime spectroscopy |
| PEL | Potential energy landscape |
| QSPR | Quantitative structure–property relationships (model) |
| RAF | Rigid amorphous fraction |
| RFOT | Random first-order transition (theory) |
| SANS | Small-angle neutron scattering |
| SAXS | Small-angle X-ray scattering |
| TD | Thermodilatometry |
| TGA | Thermogravimetric analysis |
| TMA | Thermomechanical analysis |
| TSC | Thermally stimulated currents |
| TSDC | Thermally stimulated depolarization currents |
| UCST | Upper critical solution temperature |
| VFTH | Vogel–Fulcher–Tammann–Hesse (equation) |
| WAXS | Wide-angle X-ray scattering |
| WLF | Williams–Landel–Ferry (equation) |
| ABS | Acrylonitrile/butadiene/styrene |
| CAB | Cellulose acetate butyrate |
| DMF | *N,N*-dimethylformamide |
| HPMC | Hydroxypropyl methylcellulose |

| | |
|---|---|
| HPMCAS | Hydroxypropyl methylcellulose acetate succinate |
| MnSPS | Manganese sulfonated PS (10.1 mol% of MnS in PS) |
| PαMS | Poly(α-methyl styrene) |
| P(3HB-*co*-3HH) | Poly(3-hydroxybutyrate-*co*-3-hydroxyhexanoate) |
| P(BA-*co*-BT) | Poly(butylene adipate-*co*-butylene terephthalate)) |
| P(VP-*co*-VA) | Poly(vinyl pyrrolidone-*co*-vinyl acetate) |
| P2CS | Poly(2-chloro styrene) |
| P2VP | Poly(2-vinyl pyridine) |
| P4HS | Poly(4-hydroxystyrene) |
| P4HSBr | Poly(4-hydroxystyrene) brominated |
| P4tBS | Poly(4-*tert*-butylstyrene) |
| P4VP | Poly(4-vinyl pyridine) |
| PA-6 | Polyamide-6 |
| PATM | Poly(*N*-acryloylthiomorpholine) |
| PB | Polybutadiene |
| PBA | Poly(butylene adipate) |
| PBN | Poly(butylene 2,6-naphthalate) |
| PBSA | Poly(butylene succinate-*co*-butylene adipate) |
| PBSuc | Poly(butylene succinate) |
| PBT | Poly(butylene terephthalate) |
| PBzMA | Poly(benzyl methacrylate) |
| PC | Polycarbonate |
| PCHA | Poly(cyclohexyl acrylate) |
| PCHMA | Poly(cyclohexyl methacrylate) |
| PCL | Poly($\varepsilon$-caprolactone) |
| PDLLA | Poly(D,L-lactide) |
| PE | Polyethylene |
| PEA | Poly(ethylene adipate) |
| PECH | Poly(epichlorohydrine) |
| PECl | Chlorinated polyethylene |
| PEEMA | Poly(2-ethoxyethyl methacrylate) |
| PEKK | Poly(ether ketone ketone) |
| PEMA | Poly(ethyl methacrylate) |
| PEO | Polyethylene oxide |
| PESeb | Poly(ethylene sebacate) |
| PESuc | Poly(ethylene succinate) |
| PET | Poly(ethylene terephthalate) |
| PHB | Poly($\beta$-hydroxybutyrate) |
| Phenolic | Phenolic resin |
| Phenoxy | Poly(hydroxylether of bisphenol A) |
| PHV | Poly(hydroxyvalerate) |
| PI | Polyisoprene |
| PLLA | Poly(L-lactide) |
| PMA | Poly(methyl acrylate) |
| PMDL | Poly(*N*-methyldodecano-12-lactam) |

| PMAA | Poly(methyl acrylic acid) |
|------|---------------------------|
| PMMA | Poly(methyl methacrylate) |
| PMTMA | Poly(methylthiomethyl methacrylate) |
| PNB | Polynorbornene |
| PnPMA | Poly(n-propyl methacrylate) |
| PoCS | Poly(*o*-chlorostyrene) |
| PP | Polypropylene |
| PPE | Poly(2,6-dimethylphenylene ether) |
| PPhMA | Poly(phenyl methacrylate) |
| PPO | Poly(2,6-dimethyl-1,4-phenylene oxide) |
| PS | Polystyrene |
| PS4VP | Poly(styrene-*co*-4-vinylpyridine) |
| PSAA | Poly(styrene-*co*-acrylic acid) |
| PSAD | poly(styrene-*co*-N,N-dimethylacrylamide) |
| PSCA | Poly(styrene-*co*-cinnamic acid) |
| PSHS | Poly(styrene-*co*-hydroxystyrene) |
| PSMA | Poly(styrene-*co*-methacrylic acid) |
| PSMAn | Poly(styrene-*co*-maleic anhydride) |
| PTHFMA | Poly(tetrahydrofurfuryl methacrylate) |
| PTT | Poly(trimethylene terephthalate) |
| PVAc | Poly(vinyl acetate) |
| PVC | Poly(vinyl chloride) |
| PVDF | Poly(vinylidene fluoride) |
| PVE | Poly(vinylethylene) |
| PVEE | Poly(vinyl ethyl ether) |
| PVK | Poly(*N*-vinyl carbazole) |
| PVME | Poly(vinyl methyl ether) |
| PVP | Poly(vinyl pyrrolidone) |
| PVPh | Poly(vinyl phenol) |
| PVPhKH | Poly(vinyl phenyl ketone hydrogenated) |
| PVPy | Poly(vinyl pyridine) |
| SAN | Poly(styrene-*co*-acrylonitrile) |
| THF | Tetrahydrofuran |
| TMPC | Tetramethyl polycarbonate |
| TPI | Thermoplastic polyimide |

## Symbols

| $< \ldots >$ | Average value |
|---|---|
| $A, B$ | Material or model-specific constants |
| $a_0, a_1, a_2$ | Parameters of the BCKV equation |
| $B$ | Interaction energy density |
| $B_i$ | Fraction of flexible bonds of the $i$th polymer |
| $A_T$ | Amplitude of temperature modulation |
| $C, C_1, C_2$ | Model-specific constants |

| | |
|---|---|
| $c_i$ | Model-specific parameters ($i = 1, 2, \ldots$) |
| $C_{app}$ | Apparent heat capacity |
| $C_p$ | Heat capacity |
| $D$ | Thickness |
| $d$ | Dimension of space |
| $E*$ | Complex modulus ($E^* = E' + iE''$) |
| $E$ | Energy |
| $F$ | Free energy; acting force |
| $f$ | Free-volume fraction; scanning frequency |
| $f_g$ | Free-volume fraction at $T_g$ |
| $f_{max}$ | Peak frequency of the dynamic glass transition ($\alpha$-relaxation) |
| $E, E_{act}$ | Energy; activation energy |
| $G$ | Gibbs free energy |
| $G*$ | Shear complex modulus ($G^* = G' + iG''$) |
| $g$ | Interaction term (Kovacs' model) |
| $H$ | Enthalpy |
| $I$ | Current |
| $K, K^*, k_1, k_2, k_3$ | Material specific constants |
| $K_A, K_C$ | Interassociation equilibrium constants |
| $K_B$ | Self-association equilibrium constant |
| $K_1, K_2$ | Parameters of the BSC equation |
| $k, k_{CK}, k_{DM}, k_{GT}, k_{JH}, k_{kw}$ | Parameters of various $T_g(w)$ functions |
| $k_b$ | Amount of backbone bonds per repeat unit |
| $k_\beta$ | Boltzmann constant |
| $L, l$ | Length |
| $l_k$ | Kuhn length |
| $l_b$ | Length of the average backbone bond |
| $M*$ | Complex modulus |
| $M_0$ | Monomer (or repeat-unit) molar mass |
| $M_c$ | Molar mass of chains between cross-links |
| $M_n$ | Molar mass (number average) |
| $m$ | Mass; fragility parameter (steepness index) |
| $m_T$ | Thermodynamic fragility (parameter) |
| $N_A$ | Avogadro's number |
| $N_i$ | Number of segments of the $i$th polymer |
| $n$ | Exponent (various models) |
| $P$ | Pressure |
| $P_{av}$ | Average probability |
| $p_c$ | Critical configurons concentration |
| $Q$ | Partition function; heat |
| $q$ | Rate of temperature change; parameter (Kwei equation) |
| $R$ | Gas constant |
| $S$ | Entropy |

| | |
|---|---|
| $S_c$ | Configurational entropy |
| $s_c^*$ | Critical configurational entropy |
| $t$ | Time |
| $T$ | Absolute temperature |
| $T_c$ | Crystallization temperature |
| $T_{cc}$ | Cold crystallization temperature |
| $T_{MCT}$ | Critical temperature (MCT, RFOT) |
| $T_{crit}$ | Critical temperature (Kovacs' model) |
| $T_f$ | Fictive temperature |
| $T_g$ | Glass-transition temperature |
| $T_K$ | Kauzmann temperature |
| $T_{LL}$ | Liquid–liquid transition temperature |
| $T_m$ | Melting temperature |
| $T_r$ | Reference temperature |
| $T_s$ | Sample temperature |
| $T_V$ | Vogel temperature |
| $T_0, T_2$ | $S_c$-vanishing temperatures |
| $T_b, T_{1e}, T_{2e}, T_e$ | Different marks of glass-transition temperature in a DSC curve |
| $T_{1\rho}(H)$ | Proton spin-lattice relaxation time in the rotational frame |
| $\tan \delta$ | Loss tangent ($E''/E'$, $G''/G'$, $\varepsilon''/\varepsilon'$) |
| $V$ | Volume |
| $V_\alpha$ | Cooperativity volume |
| $V_f$ | Free volume |
| $V_f^*$ | Free volume at $T < T_g$ |
| $V_0$ | Hardcore (or incompressible) molecular volume |
| $w$ | Mass (or weight) fraction |
| $x$ | Mole fraction |
| $z$ | Size of CRR (Adam–Gibbs model) |
| $z^*$ | Critical size of CRR |

## Greek Symbols

| | |
|---|---|
| $\alpha$ | Coefficient of volumetric thermal expansion |
| $\alpha$ | Primary relaxation, dynamic glass transition |
| $\alpha_c$ | Degree of cure |
| $\alpha_f$ | Coefficient of free-volume thermal expansion |
| $\alpha_L$ | Coefficient of linear thermal expansion |
| $\alpha_T$ | Temperature shift factor; thermal expansion coefficient |
| $\beta, \gamma, \delta$ | Secondary relaxations |
| $\beta_{KWW}$ | KWW function exponent |
| $\gamma$ | Exponent (MCT) |
| $\gamma_i$ | Number of flexible bonds per repeating unit |
| $\Delta C_p$ | Heat capacity change at the transition |

| | |
|---|---|
| $\Delta H$ | Heat of transition |
| $\Delta T$ | $T_r - T_s$ |
| $\Delta T_g$ | Difference of components' $T_g$ in a binary system ($T_{g,2} - T_{g,1}$) |
| $\Delta T_g^*$ | Shift of polymer's $T_g$ under confinement ($T_g^{conf.} - T_g^{bulk}$) |
| $\Delta V^{\neq}$ | Activation volume (at $T_g$) |
| $\Delta \mu$ | Activation energy (Adam–Gibbs model) |
| $\Delta \nu$ | Difference of OH vibrational frequencies ($\nu_{free} - \nu_{inter\text{-}assoc.}$) |
| $\Delta \nu*$ | Difference of OH vibrational frequencies ($\nu_{self\text{-}assoc.} - \nu_{inter\text{-}assoc.}$) |
| $\delta$ | Phase lag |
| $\delta_i$ | Solubility parameters |
| $\varepsilon$ | Strain, permittivity |
| $\varepsilon_0$ | Permittivity of free space |
| $\varepsilon_\infty$ | Permittivity at very large frequencies |
| $\eta$ | Viscosity |
| $\kappa_T$ | Compressibility |
| $\nu$ | Vibrational frequency |
| $\xi_\alpha$ | Cooperativity length scale of molecular motions |
| $\xi*$ | Length scale of entropic droplets (RFOT) |
| $\Pi_E$ | Ehrenfest ratio |
| $\rho$ | Density |
| $\sigma$ | Stress; conductivity |
| $\sigma_{dc}$ | Direct current conductivity |
| $\tau$ | Relaxation time |
| $\Phi$ | Relaxation function |
| $\varphi$ | Volume fraction |
| $\varphi_{eff}$ | Effective local concentration |
| $\varphi_{self}$ | Self-concentration |
| $\chi_{12}$ | Flory–Huggins binary interaction parameter |
| $\omega$ | Angular frequency |

## References

1 Sperling, L.H. (2006) *Introduction to Physical Polymer Science*, 4th edn, John Wiley & Sons, Inc., Hoboken, NJ.

2 Utracki, L.A. (ed.) (2002) *Polymer Blends Handbook*, vol. **1**, Kluwer, Dordrecht.

3 Robeson, L.M. (2007) *Polymer Blends: A Comprehensive Review*, Hanser, München, Germany.

4 Arrighi, V., Cowie, J.M.G., Fuhrmann, S., and Youssef, A. (2010) Miscibility Criterion in Polymer Blends and its Determination, in *Encyclopedia of Polymer Blends, Volume 1: Fundamentals*

(ed. A.I. Isayev), John Wiley & Sons, Chapter 5, pp. 153–198.

5 Menczel, J.D., Judovits, L., Prime, R.B., Bair, H.E., Reading, M., and Swier, S. (2009) Differential scanning calorimetry (DSC), in *Thermal Analysis of Polymers, Fundamentals and Applications* (eds J.D. Menczel and R.B. Prime), John Wiley & Sons, Inc., Hoboken, NJ, Chapter 2, pp. 7–239.

6 Angell, C.A. (2004) Glass transition, in *Encyclopedia of Materials: Science and Technology*, 2nd edn, (Editors-in-Chief K.H.J. Buschow, R.W. Cahn, M.C.

Flemings, and B. Ilschner) (print) E.J. Kramer, S., Mahajan, and P. Veyssière, (updates), Elsevier, Oxford, pp. 1–11.

7 Plazek, D.J. and Ngai, K.L. (1996) The glass temperature, in *Physical Properties of Polymers Handbook* (ed. J.E. Mark), AIP Press, Woodbury, N.Y., p. 139.

8 Gutzow, I.S. and Schmelzer, J.W.P. (2011) Basic properties and the nature of glasses, in *Glasses and the Glass Transition* (eds J.W.P. Schmelzer and I.S. Gutzow), Wiley VCH Verlag GmBH, Weinheim, Chapter 2.

9 Van den Mooter, G. (2012) *Drug Discov. Today Technol.*, **9**, e79–e85; Meeus, J., Scurr, D.J., Chen, X., Amssoms, K., Davies, M.C., Roberts, C.J., and Van den Mooter, G. (2015) *Pharm. Res.*, **32**, 1407–1416; Demirdirek, B. and Uhrich, K.E. (2015) *J. Drug Target.* **23**, 716–724; Dubois, J.N.L. and Lavignac, N. (2014) *Polym. Chem.* **5**, 1586–1592.

10 Kalogeras, I.M. (2011) *Eur. J. Pharm. Sci.*, **42**, 470–483; Kim, Y., Liemmawal, E.D., Pourgholami, M.H., Morris, D.L., and Stenzel, M.H. (2012) *Macromolecules* **45**, 5451–5462.

11 Hill, J.J., Shalaev, E.Y., and Zografi, G. (2014) *J. Pharm. Sci.*, **103**, 2605–2614.

12 Kalogeras, I.M. and Hagg Lobland, H.E. (2012) *J. Mater. Ed.*, **34**, 69–94.

13 Ngai, K.L. (1979) *Com. Solid State Phys.*, **9**, 127; Ngai, K.L. and Rendell, R.W. (1993) *J. Mol. Liq.*, **56**, 199–214; Ngai, K.L. (2000) *J. Phys.: Condens. Matter*, **12**, 6437–6451; Ngai, K.L. (2007) All standard theories and models of glass transition appear to be inadequate: missing some essential physics, in *Soft Matter under Exogenic Impacts* (eds S.J. Rzoska and V.A. Mazur), Springer, Dordrecht, pp. 91–111.

14 McKenna, G.B. and Simon, S.L. (2002) The glass transition – its measurement and underlying physics, in *Handbook of Thermal Analysis and Calorimetry*, vol. **3**, Applications to Polymers and Plastics (ed. S.Z.D. Cheng), Elsevier, Amsterdam, Chapter 2, pp. 49–110.

15 Menard, K.P. (2008) *Dynamic Mechanical Analysis: A Practical Introduction*, 2nd edn, CRC Press, Boca Raton, FL.

16 Hedvig, P. (1977) *Dielectric Spectroscopy of Polymers*, Adam Hilger, Bristol.

17 Chartoff, R.P. and Sircar, A.K. (2005) Thermal analysis of polymers, in *Encyclopedia of Polymer Science and Technology*, John Wiley & Sons, Inc., Hoboken, NJ.

18 Prigogine, I. and Defay, R. (1954) *Chemical Thermodynamics*, Longmans, Green and Co., New York.

19 Ribeiro, M.C.C., Scopigno, T., and Ruocco, G. (2009) *J. Phys. Chem. B*, **113**, 3099–3104; Schmelzer, J.W.P. (2012) *J. Chem. Phys.*, **136**, 074512.

20 Kovacs, A.J. (1958) *J. Polymer Sci.*, **30**, 131–147.

21 Öttinger, H.C. (2006) *Phys. Rev. E*, **74**, 011113.

22 Cohen, M.H. and Turnbull, D. (1964) *Nature*, **203**, 964; Herlach, D., Holland-Moritz, D. and Gelenko, P. (2007) *Metastable Solids from Undercooled Melts*, Elsevier, Amsterdam.

23 Kauzmann, W. (1948) *Chem. Rev.*, **43**, 219–256.

24 Stillinger, F.H. and Weber, T.A. (1982) *Phys. Rev. A*, **25**, 978–989; Stillinger, F.H. and Weber, T.A. (1983) *Phys. Rev. A*, **28**, 2408–2416; Debenedetti, P.G. and Stillinger, F.H. (2001) *Nature*, **410**, 259–267.

25 Berthier, L. and Biroli, G. (2011) *Rev. Mod. Phys.*, **83**, 577–645.

26 Yun, M., Jung, N., Yim, C., and Jeon, S. (2011) *Polymer*, **52**, 4136–4140.

27 Struik, L.C.E. (1978) *Physical Aging of Amorphous Polymers and Other Materials*, Elsevier, New York.

28 Tool, A.Q. (1946) *J. Am. Ceram. Soc.*, **29**, 240–253.

29 Davies, R.O. and Jones, G.O. (1953) *Adv. Phys.*, **2**, 370–410.

30 Kovacs, A.J., Aklonis, J.J., Hutchinson, J.M., and Ramos, A.R. (1979) *J. Polymer Sci. B*, **17**, 1097–1162.

31 Hong, L., Begen, B., Kisliuk, A., Novikov, V.N., and Sokolov, A.P. (2010) *Phys. Rev. B*, **81**, 104207.

32 Simmons, D.S. and Douglas, J.F. (2011) *Soft Matter*, **7**, 11010–11020.

33 Ngai, K.L., Sokolov, A., and Steffen, W. (1997) *J. Chem. Phys.*, **107**, 5268–5272.

34 Hong, L., Novikov, V.N., and Sokolov, A.P. (2011) *J. Non-Cryst. Solids*, **357**, 351–356.

35 Stevenson, J. and Wolynes, P. (2010) *Nat. Phys.*, **6**, 62.

36 Johari, G.P. and Goldstein, M. (1970) *J. Chem. Phys.*, **53**, 2372; (1970) *J. Phys. Chem.*, **74**, 2034; (1971) *J., Chem., Phys.*, **55**, 4245; (1972) *J. Chem., Phys.*, **56**, 4411.

37 Williams, G. and Watts, D.C. (1971) *Trans. Faraday Soc.*, **67**, 1971–1979.

38 Kudlik, A., Tschirwitz, C., Blochowicz, T., Benkhof, S., and Rossler, E. (1998) *J. Non-Cryst. Solids*, **235–237**, 406–411.

39 Schmidt-Rohr, K., Kulik, A.S., Beckham, H.W., Ohlemacher, A., Pawelzik, U., Boeffel, C., and Spiess, H.W. (1994) *Macromolecules*, **27**, 4733–4745.

40 Gilroy, K.S. and Phillips, W.A. (1981) *Phil. Mag. B*, **43**, 735.

41 Donth, E. (1982) *J. Non-Cryst. Solids*, **53**, 325; Donth, E. (1991) *J. Non-Cryst. Solids*, **131–133**, 204; Donth, E., Kahle, S., Korus, J., and Beiner, M. (1997) *J. Phys. I France*, **7**, 581–598; Donth, E. (2001) *The Glass Transition. Relaxation Dynamics in Liquids and Disordered Materials*, Springer, Berlin.

42 Cohen, M.H. and Crest, G.S. (1959) *Phys. Rev. B*, **20**, 1077; Stillinger, F.H. and Hodgdon, J.A. (1994) *Phys. Rev. E*, **50**, 2064; Stillinger, F.H. and Hodgdon, J.A. (1996) *Phys. Rev. E*, **53**, 2995.

43 Schröter, K. (2009) *J. Therm. Anal. Calorim.*, **98**, 519–599.

44 Adam, G. and Gibbs, J.H. (1965) *J. Chem. Phys.*, **43**, 139–146.

45 Hempel, E., Hempel, G., Hensel, A., Schick, C., and Donth, E. (2000) *J. Phys. Chem. B*, **104**, 2460–2466.

46 Sastry, S. (2006) *J. Indian Inst. Sci.*, **86**, 731–749.

47 Sillescu, H. (1999) *J. Non-Cryst. Solids*, **243**, 81–108.

48 Williams, M.L., Landel, R.F., and Ferry, J.D. (1955) *J. Am. Chem. Soc.*, **77**, 3701–3707.

49 Vogel, H. (1921) *Phys. Z.*, **22**, 645; Fulcher, G.S. (1925) *J. Am. Ceram. Soc.*, **8**, 339, 789; Tammann, G. and Hesse, W. (1926) *Z. Anorg. Allg. Chem.*, **156**, 245.

50 Hodge, I.M. (1997) *J. Res. Natl Inst. Stand. Technol.*, **102**, 195–205.

51 Saiter, J.M., Grenet, J., Dargent, E., Saiter, A., and Delbreilh, L. (2007) *Macromol. Symp.*, **258**, 152–161; Bureau, E., Cabot, C., Marais, S., and Saiter, J.M. (2005) *Eur. Polymer J.*, **41**, 1152–1158.

52 Lubchenko, V. and Wolynes, P.G. (2007) *Annu. Rev. Phys. Chem.*, **58**, 235–266; Richert, R. and Wagner, H. (1998) *Solid State Ionics*, **105**, 167–173; Richert, R. and Wagner, H. (1997) *Polymer*, **38**, 255–261.

53 Ngai, K.L., Capaccioli, S., Paluch, M., and Provosto, D. (2014) *J. Phys. Chem. B*, **118**, 5608–5614; Kossack, W., Adrjanowicz, K., Tarnacka, M., Kipnusu, W.K., Dulski, M., Mapesa, E.U., Kaminski, K., Pawlus, S., Paluch, M., and Kremer, F. (2013) *Phys. Chem. Chem. Phys.*, **15**, 20641–20650; Hecksher, T., Nielsen, A.I., Olsen, N.B., and Dyre, J.C. (2008) *Nat. Phys.*, **4**, 737–741; O'Connell, P.A. and McKenna, G.B. (1999) *J. Chem. Phys.*, **110** 11054–11060.

54 Zhao, J. and McKenna, G.B. (2012) *J. Chem. Phys.*, **136**, 154901.

55 Zhao, J., Simon, S.L., and McKenna, G.B. (2013) *Nat. Commun.*, **4**, 1783.

56 Richert, R. and Angell, C.A. (1998) *J. Chem. Phys.*, **108**, 9016–9026.

57 DiMarzio, E.A. and Yang, A.J.M. (1997) *J. Res. Natl Inst. Stand. Technol.*, **102**, 135–157.

58 Angell, C.A. (1985) Strong and fragile liquids, in *Relaxations in Complex Systems* (eds K.L. Ngai and G.B. Wright), National Technical Information Service, U.S. Department of Commerce, Springfield, VA 22161, pp. 3–11.

59 Doremus, R.H. (2002) *J. Appl. Phys.*, **92**, 7619–7629.

60 Huang, D. and McKenna, G.B. (2001) *J. Chem. Phys.*, **114**, 5621–5630.

61 Bruning, R. and Sutton, M. (1996) *J. Non-Cryst. Solids*, **205–207**, 480–484.

62 Avramov, I. (2005) *J. Non-Cryst. Solids*, **351**, 3163–3173.

63 Moynihan, C.T. and Angell, C.A. (2000) *J. Non-Cryst. Solids*, **274**, 131–138.

64 Narayanaswamy, O.S. (1971) *J. Am. Ceram. Soc.*, **54**, 491; Moynihan, C.T., Macebo, P.B., Montrose, C.J., Gupta, P.K., DeBolt, M.A., Dill, J.F. *et al.* (1976) *N.Y. Ann. Acad. Sci.*, **279**, 15.

65 Angell, C.A., Ngai, K.L., McKenna, G.B., McMillan, P.F., and Martin, S.W. (2000) *J. Appl. Phys.*, **88**, 3113–3157.

66 Niss, K., Dalle-Ferrier, C., Tarjus, G., and Alba-Simionesco, C. (2007) *J. Phys.: Condens. Matter*, **19**, 076102; Zhang, C., Guo, Y., and Priestley, R.D. (2014) *ACS Macro Lett.*, **3**, 501–505.

67 Bohmer, R., Ngai, K.L., Angell, C.A., and Plazek, D.J. (1993) *J. Chem. Phys.*, **99**, 4201; Robertson, C.G. and Wilkes, G.L. (2001) *J. Polymer Sci. B*, **39**, 2118–2129.

68 Böhmer, R. and Angell, C.A. (1994) *Disorder Effects on Relaxational Processes* (eds R. Richert and A. Blumen), Springer, Berlin.

69 Vilgis, T.A. (1993) *Phys. Rev. B*, **47**, 2882–2885.

70 Novikov, V.N., Ding, Y., and Sokolov, A.P. (2005) *Phys. Rev. E*, **71**, 061501; Sokolov, A.P., Novikov, V.N., and Ding Y. (2007) *J. Phys.: Condens. Matter*, **19**, 205116.

71 Kunal, K., Robertson, C.G., Pawlus, S., Hahn, S.F., and Sokolov, A.P. (2008) *Macromolecules*, **41**, 7232–7238.

72 Roland, C.M., Santangelo, P.G., and Ngai, K.L. (1999) *J. Chem. Phys.*, **111**, 5593.

73 Cangialosi, D., Alegria, A., and Colmenero, J. (2006) *J. Chem. Phys.*, **124**, 024906.

74 Ngai, K.L. and Plazek, D.J. (1995) *Rubber Chem. Technol. Rubber Review*, **68**, 376–434; Angell, C.A., Richards, B.E., and Velikov, V. (1999) *J. Phys.: Condens. Matter*, **11**, A75.

75 Johari, G.P. (2000) *J. Chem. Phys.*, **112**, 8958.

76 Saiter, A., Saiter, J.M., and Grenet, J. (2006) *Eur. Polymer J.*, **42**, 213–219.

77 Glarum, S.H. (1960) *J. Chem. Phys.*, **33**, 639.

78 Turnbull, D. and Cohen, M.H. (1959) *J. Chem. Phys.*, **31**, 1164–1169; Turnbull, D. and Cohen, M.H. (1961) *J. Chem. Phys.*, **34**, 120–125.

79 Gibbs, J.H. and DiMarzio, E.A. (1958) *J. Chem. Phys.*, **28**, 373.

80 DiMarzio, E.A. (1981) *Ann. N.Y. Acad. Sci.*, **371**, 1.

81 Goldstein, M. (1969) *J. Chem. Phys.*, **51**, 3728–3739.

82 Stillinger, F.H. and Debenedetti, P.G. (2003) *Biophys. Chem.*, **105**, 211–220.

83 Bengtzelius, U., Gotze, W., and Sjolander, A. (1984) *J. Phys. C*, **17**, 5915–5934; Leutheusser, E. (1984) *Phys. Rev. A*, **29**, 2765–2773.

84 Debenedetti, P.G. (1996) *Metastable Liquids: Concepts and Principles*, Princeton University Press, Princeton.

85 Binder, K., Baschnagel, J., and Paul, W. (2003) *Prog. Polymer Sci.*, **28**, 115–172.

86 Kirkpatrick, T.R. and Wolynes, P.G. (1987) *Phys. Rev. A*, **35**, 3072–3080; (1987) *Phys. Rev. B*, **36**, 8552–8564; Kirkpatrick, T.R. and Thirumalai, D. (1987) *Phys. Rev. Lett.*, **58**, 2091–2094.

87 Ojovan, M.I. (2006) *J. Exp. Theor. Phys. Lett.*, **79**, 632–634; Ojovan, M.I. and Lee, W.E. (2006) *J. Phys.: Condens. Matter*, **18**, 11507–11520.

88 Ojovan, M.I., Travis, K.P., and Hand, R.J. (2007) *J. Phys.: Condens. Matter*, **19**, 415107.

89 Ojovan, M.I. (2008) *Entropy*, **10**, 334–364.

90 Ritord, F. and Sollich, P. (2003) *Adv. Phys.*, **52**, 219–342.

91 Garrahan, J.P., Sollich, P., and Toninelli, C. (2011) Kinetically constrained models, in *Dynamical Heterogeneities in Glasses, Colloids and Granular Media* (eds L. Berthier, G. Biroli, J.-P. Bouchaud, L. Cipelletti, and W. van Saarloos), Oxford University Press, Oxford, Chapter 10, pp. 341–369.

92 Nelson, D.R. (1983) *Phys. Rev. Lett.*, **50**, 982–985; (1983) *Phys. Rev. B*, **28**, 5515–5535.

93 Tarjus, G. (2011) An overview of the theories of the glass transition, in *Dynamical Heterogeneities in Glasses, Colloids and Granular Media* (eds L. Berthier, G. Biroli, J.-P. Bouchaud, L. Cipelletti, and W. van Saarloos), Oxford University Press, Oxford, Chapter 2, pp. 39–67.

94 Ediger, M.D., Angell, C.A., and Nagel, S.R. (1996) *J. Phys. Chem.*, **100**, 13200–13212.

95 Chandler, D. and Garrahan, J.P. (2010) *Annu. Rev. Phys. Chem.*, **61**, 191.

96 Doliwa, B. and Heuer, A. (2003) *Phys. Rev. E*, **67**, 031506.

97 Heuer, A. (2008) *J. Phys.: Condens. Matter*, **20**, 373101.

98 Mauro, J.C., Loucks, R.J., Varshneya, A.K., and Gupta, P.K. (2008) *Sci. Model Simul.*, **15**, 241–281.

99 Stillinger, F.H. (1988) *J. Chem. Phys.*, **88**, 7818–7825.

100 Huang, D., Simon, S.L., and McKenna, G.B. (2003) *J. Chem. Phys.*, **119**, 3590–3593.

101 Kohlrausch, R. (1854) *Pogg. Ann. Phys.*, **91**, 179–213; Williams, G. and Watts, D.C. (1970) *Trans. Faraday Soc.*, **66**, 80–85.

102 Doolittle, A.K. (1951) *J. Appl. Phys.*, **22**, 1471–1479.

103 Simha, R. and Somcynsky, T. (1969) *Macromolecules*, **2**, 342–350.

104 Sanchez, I.C. and Stone, M.T. (2000) Statistical thermodynamics of polymer solutions and blends, in *Polymer Blends Volume 1: Formulation* (eds D.R. Paul and C.B. Bucknall), John Wiley & Sons Inc., New York.

105 Frenkel, J. (1946) *Kinetic Theory of Liquids*, Clarendon, Oxford; Chow, T.S. (1984) *Polymer Eng. Sci.*, **24**, 1079–1086.

106 Bair, H.E., Akinay, A.E., Menczel, J.D., Prime, R.B., and Jaffe, M. (2009) Thermomechanical analysis (TMA) and thermodilatometry (TD), in *Thermal Analysis of Polymers, Fundamentals and Applications* (eds J.D. Menczel and R.B. Prime), John Wiley & Sons, Inc., Hoboken, NJ, Chapter 4, pp. 319–385; Price, D.M. (1998) *Thermochim. Acta*, **315**, 11–18.

107 Chartoff, R.P., Menczel, J.D., and Dillman, S.H. (2009) Dynamic mechanical analysis (DMA), in *Thermal Analysis of Polymers, Fundamentals and Applications* (eds J.D. Menczel and R.B. Prime), John Wiley & Sons, Inc., Hoboken, N J, Chapter 5, pp. 387–495.

108 Chu, P.P. and Wu, H.-D. (2000) *Polymer*, **41**, 101–109.

109 Fox, T.G. and Flory, P.J. (1950) *J. Appl. Phys.*, **21**, 581; (1954) *J., Polymer Sci.*, **14**, 315.

110 Simha, R. and Boyer, R.F. (1962) *J. Chem. Phys.*, **37**, 1003–1007.

111 Dlubek, G., Kilburn, D., and Alam, M.A. (2004) *Electrochim. Acta*, **49**, 5241–5247.

112 Kobayashi, Y., Haraya, K., Kamiya, Y., and Hattori, S. (1992) *Bull. Chem. Soc. Jpn*, **65**, 160–163.

113 Flory, P. (1942) *J. Chem. Phys.*, **10**, 51–61.

114 Huggins, M.L. (1942) *J. Am. Chem. Soc.*, **64**, 1712–1719.

115 Sastry, S. (2001) *Nature*, **409**, 164–167.

116 Cavagna, A. (2009) *Phys. Rep.*, **476**, 51.

117 Kirkpatrick, T.R., Thirumalai, D., and Wolynes, P.G. (1989) *Phys. Rev. A*, **40**, 1045.

118 Bouchaud, J.-P. and Biroli, G. (2004) *J. Chem. Phys.*, **121**, 7347.

119 Lubchenko, V. and Wolynes, P.G. (2007) *Ann. Rev. Phys. Chem.*, **58**, 235.

120 Ashtekar, S., Scott, G., Lyding, J., and Gruebele, M. (2010) *J. Phys. Chem. Lett.*, **1**, 1941.

121 Kivelson, D., Kivelson, S.A., Zhao, X.-L., Nussinov, Z., and Tarjus, G. (1995) *Physica A*, **219**, 27; Tarjus, G., Kivelson, S.A., Nussinov, Z., and Viot, P. (2005) *J. Phys.: Condens. Matter*, **17**, R1143.

122 Coslovich, D. and Pastore, G. (2007) *J. Chem. Phys.*, **127**, 124504; Tanaka, H., Kawasaki, T., Shintani, H., and Watanabe, K. (2010) *Nat. Mater.*, **9**, 324; Pedersen, U.R., Schroder, T.B., Dyre, J.C., and Harrowell, P. (2010) *Phys. Rev. Lett.*, **104**, 105701.

123 Medvedev, N.N., Geiger, A., and Brostow, W. (1990) *J. Chem. Phys.*, **93**, 8337; Naberukhin, Yu.I., Voloshin, V.P., and Medvedev, N.N. (1991) *Mol. Phys.*, **73**, 919.

124 Kondo, T. and Tsumuraya, K. (1990) *J. Chem. Phys.*, **94**, 8220.

125 Douglas, R.W. (1949) *J. Soc. Glass Technol.*, **33**, 138–162.

126 Tanaka, H. (2005) *J. Non-Cryst. Solids*, **351**, 3371–3384, 3385–3395, 3396–3413.

127 Freed, K.F. (2011) *Acc. Chem. Res.*, **44**, 194–203 (and cited references).

128 Katritzky, A.R., Kuanar, M., Slavov, S., Hall, C.D., Karelson, M., Kahn, I., and Dobchev, D.A. (2010) *Chem. Rev.*, **110**, 5714–5789; Le, T., Epa, V.C., Burden, F.R., and Winkler, D.A. (2012) *Chem. Rev.*, **112**, 2889–2919.

129 Liu, W.Q. and Cao, C.Z. (2009) *Colloid Polymer Sci.*, **287**, 811–818.

130 Barrat, J.L., Baschnagel, J., and Lyulin, A. (2010) *Soft Matter*, **6**, 3430–3446.

131 Hamerton, I., Howlin, B.J., and Kamyszek, G. (2012) *PLoS One*, 7, e38424.

132 Mhlanga, P., Hassan, W.A.W., Hamerton, I., and Howlin, B.J. (2013) *PLoS One*, **8**, e53367.

133 Karasz, F.E. and MacKnight, W.T. (1968) *Macromolecules*, **1**, 537.

134 Mark, J.E. (ed.) (1999) *Polymer Data Handbook*, Oxford University Press, New York.

135 O'Reilly, J.M., Bair, H.E., and Karasz, F.E. (1982) *Macromolecules*, **15**, 1083–1088.

136 Rogers, S.S. and Mandelkern, L. (1957) *J. Phys. Chem.*, **61**, 985–990.

137 Shetter, J.A. (1963) *J. Polymer Sci. B Polymer Lett.*, **1**, 209–213.

138 Shen, M.C. and Eisenberg, A. (1970) *Rubber Chem. Technol.*, **43**, 95, 156.

139 Lal, J. and Trick, G.S. (1964) *J. Polymer Sci. A2*, **2**, 4559–4572.

140 Haldon, R.A., Schell, W.J., and Simha, R. (1967) *J. Macromol. Sci. B*, **1**, 759–775.

141 Stukalin, E.B., Douglas, J.F., and Freed, K.F. (2009) *J. Chem. Phys.*, **131**, 114905–114913.

142 Vincent, P.I. (1965) *The Physics of Plastics* (ed. P.D. Ritchie), D. Van Nostrand Co., Inc., Princeton, NJ, Chapter 2.

143 Allcock, H.R., Connolly, M.S., Sisko, J.T., and Al-Shali, S. (1988) *Macromolecules*, **21**, 323–334.

144 Wisian-Neilson, P., Huang, L., Islam, M.Q., and Crane, R.A. (1994) *Polymer*, **35**, 4985–4989.

145 Foucher, D.A., Ziembinski, R., Ben-Zhong, T., Macdonald, P.M., Massey, J., Jaeger, C.R., Vansco, G.J., and Manners, I. (1993) *Macromolecules*, **26**, 2878.

146 Somcynsky, T. and Patterson, D. (1962) *J. Polymer Sci.*, **62**, 151.

147 Pezzin, G., Zilio-Grandi, F., and Sanmartin, P. (1970) *Eur. Polymer J.*, **6**, 1053–1061.

148 Gedde, U.W. (1995) *Polymer Physics*, Chapman & Hall, London.

149 Schmieder, K. and Wolf, D. (1953) *Kolloid Z.*, **134**, 149.

150 Roland, C.M. and Casalini, R. (2005) *Macromolecules*, **38**, 8729–8733; Roland, C.M., Hensel-Bielowka, S., Paluch, M., and Casalini, R. (2005) *Rep. Prog. Phys.*, **68**, 1405–1478; Karlou, K. and Schneider, H.A. (2000) *J. Thermal Anal. Calorim.*, **59**, 59–69.

151 Schwartz, G.A., Paluch, M., Alegría, A., and Colmenero, J. (2009) *J. Chem. Phys.*, **131**, 044906; Roland, C.M. (2008) *Soft Matter*, **4**, 2316–2322; Roland, C.M. and Casalini, R. (2007) *Macromolecules*, **40**, 3631–3639.

152 Speedy, R.J. (1999) *J. Phys. Chem. B*, **103**, 8128–8131.

153 Goldstein, M. (1973) *J. Phys. Chem.*, **77**, 667–673.

154 Andersson, S.P. and Andersson, O. (1998) *Macromolecules*, **31**, 2999–3006.

155 Cheng, S.Z.D. and Jin, S. (2002) Crystallization and melting of metastable crystalline polymers, in *Handbook of Thermal Analysis and Calorimetry, Vol. 3: Applications to Polymers and Plastics* (ed. S.Z.D. Cheng), Elsevier, Amsterdam, Chapter 3, pp. 167–195.

156 Calleja, G., Jourdan, A., Ameduri, B., and Habas, J.-P. (2013) *Eur. Polymer J.* **49**, 2214–2222; Wunderlich, B. (2005) *Thermal Analysis of Polymeric Materials*, Springer Berlin; Wunderlich, B. (2003) *Prog. Polymer Sci.*, **28**, 383–450.

157 Alves, N.M., Mano, J.F., Balaguer, E., Meseguer Duenas, J.M., and Gómez Ribelles, J.L. (2002) *Polymer*, **43**, 4111–4122.

158 Griffith, J.H. and Rånby, B.G. (1960) *J. Polymer Sci.*, **44**, 369–381.

159 Miller, R.W. and Murayama, T. (1984) *J. Appl. Polymer Sci.*, **29**, 933–939.

160 Garrett, P.D. and Grubb, D.T. (1988) *J. Polymer Sci. B*, **26**, 2509.

161 Shen, M.C. and Tobolsky, A.V. (eds) (1965) Glass transition temperature of polymers – effect of plasticizer, chain ends, and comonomer, in *Plasticization and Plasticizer Processes*, vol. **48** (ed. N.A.J. Platzer), ACS, Washington.

162 Shtarkman, B.P. and Razinskaya, I.N. (1983) *Acta Polym.*, **34**, 514–520.

163 Riggleman, R.A., Douglas, J.F., and de Pablo, J.J. (2010) *Soft Matter*, **6**, 292–304.

164 Pittia, P. and Sacchetti, G. (2008) *Food Chem.*, **106**, 1417–1427.

165 Pena, J.R., Hidalgo, M., and Mijangos, C. (2000) *J. Appl. Polymer Sci.*, **75**, 1303–1312.

166 Rothon, R.N. (ed.) (2003) *Particulate-Filled Polymer Composites*, 2nd edn, Rapra Technology Ltd, Shrewsbury, UK.

167 Mohomed, K., Gerasimov, T.G., Abourahma, H., Zaworotko, M.J., and Harmon, J.P. (2005) *Mater. Sci. Eng. A Struct.*, **409**, 227.

168 Ash, B.J., Siegel, R.W., and Schadler, L.S. (2004) *J. Polymer Sci. B*, **42**, 4371–4383.

169 Sargsyan, A., Tonoyan, A., Davtyan, S., and Schick, C. (2007) *Eur. Polymer J.*, **43**, 3113–3127; Davtyan, S.P., Tonoyan, A.O., Varderesyan, A.Z., and Müller, S.C. (2014) *Eur. Polymer J.*, **54**, 182–186.

170 Klonos, P., Panagopoulou, A., Bokobza, L., Kyritsis, A., Peoglos, V., and Pissis, P. (2010) *Polymer*, **51**, 5490–5499.

171 Fragiadakis, D., Pissis, P., and Bokobza, L. (2005) *Polymer*, **46**, 6001–6008.

172 Srivastava, S. and Basu, J.K. (2007) *Phys. Rev. Lett.*, **98**, 165701.

173 Srivastava, S., Kandar, A.K., Basu, J.K., Mukhopadhyay, M.K., Lurio, L.B., Narayanan, S., and Sinha, S.K. (2009) *Phys. Rev. E*, **79**, 021408.

174 Rittigstein, P. and Torkelson, J.M. (2006) *J. Polymer Sci. B*, **44**, 2935–2943.

175 Fragiadakis, D., Bokobza, L., and Pissis, P. (2011) *Polymer*, **52**, 3175–3182.

176 Fox, T.G. and Loshaek, S. (1955) *J. Polymer Sci.*, **15**, 371–390.

177 DiBenedetto, A.T. (1987) *J. Polymer Sci. B*, **25**, 1949–1969.

178 Pascault, J.P. and Williams, R.J.J. (1990) *J. Polymer Sci. B*, **28**, 85.

179 Venditti, R.A. and Gillham, J.K. (1997) *J. Appl. Polymer Sci.*, **64**, 3–14.

180 Prolongo, M.G., Arribas, C., Salom, C., and Masegosa, R.M. (2007) *J. Appl. Polymer Sci.*, **103**, 1507–1516.

181 Schick, C. (2010) *Eur. Phys. J. Special Topics*, **189**, 3–36; Arabeche, K., Delbreilh, L., Adhikari, R., Michler, G.H., Hiltner, A., Bear, E., and Saiter, J.-M. (2012) *Polymer*, **53**, 1355–1361.

182 McKenna, G.B. (2010) *Eur. Phys. J. Special Topics*, **189**, 285–302.

183 Anastasiadis, S.H., Karatasos, K., Vlachos, G., Manias, E., and Giannelis, E.P. (2000) *Phys. Rev. Lett.*, **84**, 915; Vaia, R.A., Sauer, B.B., Tse, O.K., and Giannelis, E.P. (1997) *J. Polymer Sci. B*, **35**, 59.

184 Ao, Z.M. and Jiang, Q. (2006) *Langmuir*, **22**, 1241–1246.

185 Pham, J.Q. and Green, P.F. (2003) *Macromolecules*, **36**, 1665–1669.

186 Boiko, Y.M. (2010) *Colloid Polymer Sci.*, **288**, 1757–1761.

187 Mok, M.M., Kim, J., Marou, S.R., and Torkelson, J.M. (2010) *Eur. Phys. J. E*, **31**, 239–252.

188 Kalogeras, I.M. (2005) *Acta Mater.*, **53**, 1621–1630; Kalogeras, I.M. and Neagu, E.R. (2004) *Eur. Phys. J. E*, **14**, 193–204.

189 Alcoutlabi, M. and McKenna, G.B. (2005) *J. Phys.: Condens. Matter*, **17**, R461–R524; Roth, C.B. and Dutcher, J.R. (2005) *J. Electroanal. Chem.*, **584**, 13–22.

190 Richert, R. (2011) *Ann. Rev. Phys. Chem.*, **62**, 65–84; D. Cangialosi (2015) Effect of confinement geometry on out-of-equilibrium glassy dynamics, in *Non-equilibrium Phenomena in Confined Soft Matter, Soft and Biological Matter* (S. Napolitano, ed.), Springer, Basel, Chapter 12, pp. 265–298.

191 Zheng, W. and Simon, S.L. (2007) *J. Chem. Phys.*, **127**, 194501.

192 Jackson, C.L. and McKenna, G.B. (1991) *J. Non-Cryst. Solids*, **131–133**, 221.

193 Trofymluk, O., Levchenko, A.A., and Navrotsky, A. (2005) *J. Chem. Phys.*, **123**, 194509.

194 Schönhals, A., Goering, H., and Schick, Ch. (2002) *J. Non-Cryst. Solids*, **305**, 140; He, F., Wang, L.-M., and Richert, R. (2005) *Phys. Rev. B*, **71**, 144205; Arndt, M., Stannarius, R., Gorbatchow, W., and Kremer, F. (1996) *Phys. Rev. E*, **54**, 5377.

195 Fakhraai, Z. and Forrest, J.A. (2008) *Science*, **319**, 600–604; Dutcher, J.R. and Ediger, M.D. (2008) *Science*, **319**, 577–578; Xia, W., Mishra, S., Keten, S. (2013) *Polymer* **54**, 5942–5951.

196 Fukao, K. and Miyamoto, Y. (2000) *Phys. Rev. E.*, **61**, 1743–1754; Fakharaai, Z. and Forrest, J.A. (2005) *Phys. Rev. Lett.*, **95**, 025701.

197 Keddie, J.L., Jones, R.A.L., and Cory, R.A. (1994) *Europhys. Lett.*, **27**, 59.

198 de Gennes, P.G. (2000) *Eur. Phys. J. E*, **2**, 201–205.

199 Park, C.H., Kim, J.H., Ree, M., Sohn, B.-H., Jung, J.C., and Zin, W.-C. (2004) *Polymer*, **45**, 4507–4513; Forrest, J.A. and Dalnoki-Veress, K. (2001) *Adv. Colloid Polymer Sci.*, **94**, 167–195.

200 Priestley, R.D., Mundra, M.K., Barnett, N.J., Broadbelt, L.J., and Torkelson, J.M. (2007) *Aust. J. Chem.*, **60**, 765–771; Rittigstein, P., Priestley, R.D., Broadbelt,

L.J., and Torkelson, J.M. (2007) *Nat. Mater.*, **6**, 278–282.

201 Zhao, H.Y., Yu, Z.N., Begum, F., Hedden, R.C., and Simon, S.L. (2014) *Polymer*, **55**, 4959–4965.

202 Priestley, R.D., Rittigstein, P., Broadbelt, L.J., Fukao, K., and Torkelson, J.M. (2007) *J. Phys.: Condens. Matter*, **19**, 205120.

203 Zanotto, A., Spinella, A., Nasillo, G., Caponetti, E., and Luyt, A.S. (2012) *Express Polymer Lett.*, **6**, 410–416.

204 Tran, T.A., Said, S., and Grohens, Y. (2005) *Macromolecules*, **38**, 3867.

205 Ouaad, K., Djadoun, S., Ferfera-Harrar, H., Sbirrazzuoli, N., and Vincent, L. (2011) *J. Appl. Polymer Sci.*, **119**, 3227–3233.

206 John, J., Klepac, D., Didović, M., Sandesh, C.J., Liu, Y., Raju, K.V.S.N., Pius, A., Valić, S., and Thomas, S. (2010) *Polymer*, **51**, 2390–2402.

207 Priestley, R.D., Ellison, C.J., Broadbelt, L.J., and Torkelson, J.M. (2005) *Science*, **309**, 456–459.

208 Brady, R.F. and Charlesworth, J.M. (1994) *Prog. Org. Coat.*, **24**, 1.

209 Forrest, J.A., Dalnoki-Veress, K., and Dutcher, J.R. (1997) *Phys. Rev. E*, **56**, 5705–5716.

210 DeMaggio, G.B., Frieze, W.E., Gidley, D.W., Zhu, M., Hristov, H.A., and Yee, A.F. (1997) *Phys. Rev. Lett.*, **78**, 1524–1527.

211 Gorbunov, V.V., Grandy, D., Reading, M., and Tsukruk, V.V. (2009) Micro- and nanoscale local thermal analysis, in *Thermal Analysis of Polymers, Fundamentals and Applications* (eds J.D. Menczel and R.B. Prime), John Wiley & Sons, Inc., Hoboken, NJ, Chapter 7, pp. 615–649.

212 Turi, E.A. (ed.) (1997) *Thermal Characterization of Polymeric Materials*, 2nd edn, Academic Press, New York.

213 Vassilikou-Dova, A. and Kalogeras, I.M. (2009) Dielectric analysis (DEA), in *Thermal Analysis of Polymers, Fundamentals and Applications* (eds J.D. Menczel and R.B. Prime), John Wiley & Sons, Inc., Hoboken, NJ, Chapter 6, pp. 497–610.

214 Rieger, J. (2001) *Polymer Test.*, **20**, 199–204.

215 Kalogeras, I.M., Pallikari, F., Vassilikou-Dova, A., and Neagu, E.R. (2006) *Appl.*

*Phys. Lett.*, **89**, 172905; (2007) *J. Appl. Phys.*, **101**, 094108.

216 Moynihan, C.T., Easteal, A.J., DeBolt, M.A., and Tucker, J. (1976) *J. Am. Ceram. Soc.*, **59**, 12–16.

217 Badrinarayanan, P., Zheng, W., Li, Q., and Simon, S.L. (2007) *J. Non-Cryst. Solids*, **353**, 2603–2612.

218 Reading, M. (1993) *Trends Polymer Sci.*, **8**, 248–253; Lacey, A.A., Price, D.M., and Reading, M. (2006) Theory and practice of modulated temperature differential scanning calorimetry, in *Modulated-Temperature Differential Scanning Calorimetry, Theoretical and Practical Applications in Polymer Characterisation. Series: Hot Topics in Thermal Analysis and Calorimetry*, vol. 6 (eds M. Reading and D.J. Hourston), Springer, New York, pp. 1–79.

219 Silva, G.G., Machado, J.C., Song, M., and Hourston, D.J. (2000) *J. Appl. Polymer Sci.*, **77**, 2034–2043.

220 www.tainst.com: Applications Library, Thermal Solutions – Determination of polymer blend composition (TS-22A).

221 Maldonado-Santoyo, M., Ortiz-Estrada, C., Luna-Barcenas, G., Sanchez, I.C., Cesteros, L.C., Katime, I., and Nuno-Donlucas, S.M. (2004) *J. Polymer Sci. B*, **42**, 636–645; Maldonado-Santoyo, M., Cesteros, L.C., Katime, I., and Nuno-Donlucas, S.M. (2004) *Polymer*, **45**, 5591–5596.

222 Maldonado-Santoyo, M., Nuno-Donlucas, S.M., Cesteros, L.C., and Katime, I. (2004) *J. Appl. Polymer Sci.*, **92**, 1887–1892.

223 Chakrabarti, R., Das, M., and Chakraborty, D. (2004) *J. Appl. Polymer Sci.*, **93**, 2721–2730; Patricio, P.S.O., Silva, G.G., and Machado, J.C. (2007) *J. Appl. Polymer Sci.*, **105**, 641–646.

224 Heijboer, J. (1978) *Molecular Basis of Transitions and Relaxations* (ed. D.J. Meier), Gordon and Breach, New York.

225 Aouachria, K., Belhaneche-Bensemra, N., and Massardier-Nageotte, V. (2011) *J. Vinyl Addit. Technol.*, **17**, 156–163.

226 Kremer, F. and Schönhals, A. (2003) *Broadband Dielectric Spectroscopy*, Springer, Berlin; Yin, H. and Schönhals, A. (2014) Broadband dielectric spectroscopy on polymer blends, in

*Polymer Blends Handbook* (eds L.A. Utracki and C.A. Wilkie), Springer, Dordrecht, pp. 1299–1356.

**227** van den Berg, O., Sengers, W.G.F., Jager, W.F., Pichen, S.J., and Wübbenhorst, M. (2004) *Macromolecules*, **37**, 2460–2470.

**228** Sengers, W.G.F., van den Berg, O., Wübbenhorst, M., Gotsis, A.D., and Pichen, S.J. (2005) *Polymer*, **46**, 6064–6074.

**229** Jonscher, A.K. (1983) *Dielectric Relaxation in Solids*, Chelsea Dielectrics Press, London.

**230** Havriliak, S. and Negami, S. (1967) *Polymer*, **8**, 161.

**231** Kalogeras, I.M., Roussos, M., Vassilikou-Dova, A., Spanoudaki, A., Pissis, P., Savelyev, Y.V., Shtompel, V.I., and Robota, L.P. (2005) *Eur. Phys. J. E*, **18**, 467.

**232** Nishi, T., Wang, T.T., and Kwei, T.K. (1975) *Macromolecules*, **8**, 227–234; Wagler, T., Rinaldi, P.L., Han, C.D., and Chun, H. (2000) *Macromolecules*, **33**, 1778–1789.

**233** Koningsveld, R., Kleintjens, L., and Schoffeleers, H. (1974) *Pure Appl. Chem.*, **39**, 1–32.

**234** Higgins, J.S., Lipson, J.E.G., and White, R.P. (2010) *Phil. Trans. R. Soc. A*, **368**, 1009–1025.

**235** Urakawa, O., Fuse, Y., Hori, H., Tran-Cong, Q., and Yano, O. (2001) *Polymer*, **42**, 765–773.

**236** Guo, Q. and Liu, Z. (2000) *J. Therm. Anal. Calorim.*, **59**, 101–120.

**237** Sy, J.W. and Mijovic, J. (2000) *Macromolecules*, **33**, 933–946.

**238** Kim, C.K. and Paul, D.R. (1992) *Polymer*, **33**, 1630–1639.

**239** Araki, O., Zheng, Q., Takahashi, M., Takiwaga, T., and Masuda, T. (1995) *Mater. Sci. Res. Int.*, **1**, 144–149.

**240** Chang, L.L. and Woo, E.M. (2001) *Polymer J.*, **33**, 13–17.

**241** Kim, J.K., Lee, H.H., Son, H.W., and Han, C.D. (1998) *Macromolecules*, **31**, 8566–8578.

**242** Coleman, M.M. and Painter, P.C. (1995) *Prog. Polymer Sci.*, **20**, 1–59.

**243** Painter, P.C., Veytsman, B., Kumar, S., Shenoy, S., Graf, J.F., Xu, Y., and Coleman, M.M. (1997) *Macromolecules*,

**30**, 932–942; Pruthtikul, R., Coleman, M., Painter, P.C., and Tan, N.B. (2001) *Macromolecules*, **34**, 4145–4150; Park, Y., Veytsman, B., Coleman, M., and Painter, P. (2005) *Macromolecules*, **38**, 3703–3707.

**244** Utracki, L.A. (1965) *Adv. Polymer Sci.*, **5**, 33–39.

**245** Sanchez-Cabezudo, M., Masegosa, R.M., Salom, C., and Prolongo, M.G. (2010) *J. Therm. Anal. Calorim.*, **102**, 1025–1033.

**246** Thirtha, V., Lehman, R., and Nosker, T. (2006) *Polymer*, **47**, 5392–5401.

**247** Xue, M.-L., Yu, Y.-L., Sheng, J., Chuah, H.H., and Geng, C.-H. (2005) *J. Macromol. Sci. B*, **44**, 317–329.

**248** Li, D. and Brisson, J. (1996) *Macromolecules*, **29**, 868–874; Zhang, S.H. and Runt, J. (2004) *J. Polymer Sci. B*, **42**, 3405–3415.

**249** Zhang., X., Takegoshi, K., and Hikichi, K. (1991) *Macromolecules*, **24**, 5756–5762.

**250** Savin, D.A., Larson, A.M., and Lodge, T.P. (2004) *J. Polymer Sci. B*, **42**, 1155–1163.

**251** Kratochvil, J., Sturcova, A., Sikova, A., and Dybal, J. (2011) *J. Polymer Sci. B*, **49**, 1031–1040.

**252** Chung, G.C., Kornfield, J.A., and Smith, S.D. (1994) *Macromolecules*, **27**, 5729–5741.

**253** Colmenero, J. and Arbe, A. (2007) *Soft Matter*, **3**, 1474–1485.

**254** Song, M., Hourston, D.J., Pollock, H.M., and Hammiche, A. (1999) *Polymer*, **40**, 4763–4767.

**255** Zheng, S., Huang, J., Li, Y., and Guo, Q. (1997) *J. Polymer Sci. B*, **35**, 1383–1392.

**256** Gordon, M. and Taylor, J.S. (1952) *Appl. Chem. (USSR)*, **2**, 493.

**257** Wood, L.A. (1958) *J. Polymer Sci.*, **28**, 319–330.

**258** Kelley, F.N. and Bueche, F. (1961) *J. Polymer Sci.*, **50**, 549–556.

**259** DiMarzio, E.A. and Gibbs, J.H. (1959) *J. Polymer Sci.*, **40**, 121–131.

**260** DiMarzio, E.A. (1990) *Polymer*, **31**, 2294–2298.

**261** Fox, T.G. (1956) *Bull. Am. Phys. Soc.*, **1**, 123–125.

**262** Kalogeras, I.M. and Brostow, W. (2009) *J. Polymer Sci. B*, **47**, 80–95.

263 Kalogeras, I.M. (2010) *Thermochim. Acta,* **509**, 135–146.

264 Khougaz, K. and Clas, S.-D. (2000) *J. Pharm. Sci.*, **89**, 1325–1334.

265 Konno, H. and Taylor, L.S. (2006) *J. Pharm. Sci.*, **95**, 2692–2705.

266 Kovacs, A.J. (1963) *Adv. Polymer Sci.*, **3**, 394–507.

267 Braun, G., Kovacs, A.J., and Prim, J.A. (eds) (1966) *Physics of Non-Crystalline Solids*, North-Holland, Amsterdam, p. 303.

268 Aubin, M. and Prud'homme, R.E. (1988) *Polymer Eng. Sci.*, **28**, 1355–1361; (1988) *Macromolecules*, **28**, 2945–2949.

269 Feldstein, M.M. (2001) *Polymer*, **42**, 7719–7726.

270 Chang, C.-S., Woo, E.M., and Lin, J.-H. (2006) *Macromol. Chem. Phys.*, **207**, 1404–1413.

271 Pedrosa, P., Pomposo, J.A., Calahorra, E., and Cortazar, M. (1994) *Macromolecules*, **27**, 102–109.

272 Coleman, M.M., Xu, Y., and Painter, P.C. (1994) *Macromolecules*, **27**, 127–134; Serman, C.J., Painter, P.C., and Coleman, M.M., *Polymer* (1991) **32**, 1049–1058.

273 Mpoukouvalas, K., Floudas, G., Zhang, S.H., and Runt, J. (2005) *Macromolecules*, **38**, 552–560.

274 Park, M.S. and Kim, J.K. (2002) *J. Polymer Sci. B*, **40**, 1673–1681.

275 Couchman, P.R. and Karasz, F.E. (1978) *Macromolecules*, **11**, 117–119.

276 Couchman, P.R. (1987) *Macromolecules*, **20**, 1712–1717.

277 Couchman, P.R. (1991) *Macromolecules*, **24**, 5772–5774.

278 Pochan, J.M., Beatty, C.L., and Pochan, D.F. (1979) *Polymer*, **20**, 879–886.

279 An, L., He, D., Jing, J., Wang, Z., Yu, D., and Jiang, B. (1997) *Eur. Polymer J.*, **33**, 1523–1528.

280 Pinal, R. (2008) *Entropy*, **10**, 207–223.

281 Painter, P.C., Graf, J.F., and Coleman, M.M. (1991) *Macromolecules*, **24**, 5630–5638.

282 Pomposo, J.A., Cortazar, M., and Calahorra, E. (1994) *Macromolecules*, **27**, 245–251.

283 Yang, Z. and Han, C.D. (2008) *Macromolecules*, **41**, 2104–2118.

284 ElMiloudi, K., Djadoun, S., Sbirrazzuoli, N., and Geribaldi, S. (2009) *Thermochim. Acta*, **483**, 49–54; Hamou, A.S.H., ElMiloudi, K., and Djadoun, S. (2009) *J. Polymer Sci. B*, **47**, 2074–2082; Abdellaoui-Arous, N., Hadj-Hamou, A.S., and Djadoun, S. (2012) *Thermochim. Acta*, **547**, 22–30.

285 Masser, K.A., Zhao, H.Q., Painter, P.C., and Runt, J. (2010) *Macromolecules*, **43**, 9004–9013.

286 ElMiloudi, K. and Djadoun, S. (2009) *J. Polymer Sci. B*, **47**, 923–931.

287 Kuo, S.W. (2005) *J. Polymer Res.*, **15**, 459–486.

288 Kim, J.H., Min, B.R., and Kang, Y.S. (2006) *Macromolecules*, **39**, 1297–1299.

289 Lu, X. and Weiss, R.A. (1991) *Macromolecules*, **24**, 4381–4385; Lu, X. and Weiss, R.A. (1992) *Macromolecules*, **25**, 3242–3246.

290 Chun, Y.S. and Weiss, R.A. (2004) *J. Appl. Polymer Sci.*, **94**, 1227–1235.

291 Jenckel, E. and Heusch, R. (1953) *Kolloid Z.*, **130**, 89–105.

292 Kanig, G. (1963) *Kolloid Z.*, **190**, 1–10.

293 Kwei, T.K. (1984) *J. Polymer Sci. Lett.*, **22**, 307.

294 Brekner, M.-J., Schneider, A., and Cantow, H.-J. (1988) *Makromol. Chem.*, **189**, 2085–2097.

295 Schneider, H.A. (1997) *J. Res. Natl Inst. Stand. Technol.*, **102**, 229–248.

296 Kalogeras, I.M., Stathopoulos, A., Vassilikou-Dova, A., and Brostow, W. (2007) *J. Phys. Chem. B*, **111**, 2774–2782.

297 Alegría, A., Tellería, I., and Colmenero, J. (1994) *J. Non-Cryst. Solids*, **172**, 961–965.

298 Roland, C.M. and Casalini, R. (2007) *Macromolecules*, **40**, 3631–3639.

299 Brostow, W., Chiu, R., Kalogeras, I.M., and Vassilikou-Dova, A. (2008) *Mater. Lett.*, **62**, 3152–3155.

300 Urakawa, O., Ujii, T., and Adachi, K. (2006) *J. Non-Cryst. Solids*, **352**, 5042–5049.

301 Stoelting, J., Karasz, F.E., and MacKnight, W.J. (1970) *Polymer Eng. Sci.*, **10**, 133–138; Lau, S.F., Pathak, J., and Wunderlich, B. (1982) *Macromolecules*, **15**, 1278–1283.

302 Lodge, T.P., Wood, E.R., and Haley, J.C. (2006) *J. Polymer Sci. B*, **44**, 756–763.

**303** Zhao, J., Ediger, M.D., Sun, Y., and Yu, L. (2009) *Macromolecules*, **42**, 6777–6783.

**304** Lorthioir, C., Alegría, A., and Colmenero, J. (2003) *Phys. Rev. E*, **68**, 031805; Leroy, E., Alegría, A., and Colmenero, J. (2002) *Macromolecules*, **35**, 5587. Leroy, E., Alegría, A., and Colmenero, J. (2003) *Macromolecules*, **36**, 7280–7288.

**305** Miwa, Y., Sugino, Y., Yamamoto, K., Tanabe, T., Sakaguchi, M., Sakai, M., and Shimada, S. (2004) *Macromolecules*, **37**, 6061–6068.

**306** Miwa, Y., Usami, K., Yamamoto, K., Sakaguchi, M., Sakai, M., and Shimada, S. (2005) *Macromolecules*, **38**, 2355–2361.

**307** Kumar, S.K., Shenogin, S., and Colby, R.H. (2007) *Macromolecules*, **40**, 5759–5766.

**308** Zetsche, A. and Fischer, E. (1994) *Acta Polym.*, **45**, 168–175.

**309** Kumar, S.K., Colby, R.H., Anastasiadis, S.H., and Fytas, G. (1996) *J. Chem. Phys.*, **105**, 3777–3788.

**310** Arrese-Igor, S., Alegria, A., Moreno, A.J., and Colmenero, J. (2011) *Macromolecules*, **44**, 3611–3621.

**311** Lodge, T.P. and McLeish, T.C.B. (2000) *Macromolecules*, **33**, 5278–5284.

**312** Herrera, D., Zamora, J.-C., Bello, A., Grimau, M., Laredo, E., Müller, A.J., and Lodge, T.P. (2005) *Macromolecules*, **38**, 5109–5117.

**313** Gaikwad, A.N., Wood, E.R., Ngai, T., and Lodge, T.P. (2008) *Macromolecules*, **41**, 2502–2508.

**314** Evans, C.M. and Torkelson, J.M. (2012) *Polymer*, **53**, 6118–6124.

**315** Evans, C.M., Sandoval, R.W., and Torkelson, J.M. (2011) *Macromolecules*, **44**, 6645–6648.

**316** Crispim, E.G., Rubira, A.F., and Muniz, E.C. (1999) *Polymer*, **40**, 5129–5135.

**317** Lutz, T.R., He, Y., Ediger, M.D., Pitsikalis, M., and Hadjichristidis, N. (2004) *Macromolecules*, **37**, 6440–6448.

**318** Sakaguchi, T., Taniguchi, N., Urakawa, O., and Adachi, K. (2005) *Macromolecules*, **38**, 422–428.

**319** Haley, J.C., Lodge, T.P., He, Y., Ediger, M.D., Von Meerwall, E.D., and Mijovic, J. (2003) *Macromolecules*, **36**, 6142–6151.

**320** Haley, J.C. and Lodge, T.P. (2004) *Colloid Polymer Sci.*, **282**, 793–801.

**321** He, Y., Lutz, T.R., Ediger, M.D., Pitsikalis, M., and Hadjichristidis, N. (2005) *Macromolecules*, **38**, 6216–6226.

**322** Schneider, H.A. and Breckner, M.-J. (1985) *Polymer Bull.*, **14**, 173–178.

**323** He, Y., Lutz, T.R., and Ediger, M.D. (2003) *J. Chem. Phys.*, **119**, 9956–9965.

**324** Zheng, W. and Simon, S.L. (2008) *J. Polymer Sci. B*, **46**, 418–430.

**325** Balsamo, V., Newman, D., Gouveia, L., Herrera, L., Grimau, M., and Laredo, E. (2006) *Polymer*, **47**, 5810–5820.

**326** Song, M. and Long, F. (1991) *Eur. Polymer J.*, **27**, 983–986.

**327** Hsu, W.P. (2004) *J. Appl. Polymer Sci.*, **91**, 35–39.

**328** Zawada, J.A., Ylitalo, C.M., Fuller, G.G., Colby, R.H., and Long, T.E. (1992) *Macromolecules*, **25**, 2896–2902.

**329** Li, X. and Hsu, S.L. (1984) *J. Polymer Sci. B*, **22**, 1331–1342.

**330** Zhang, S.H., Jin, X., Painter, P.C., and Runt, J. (2004) *Polymer*, **45**, 3933–3942.

**331** Zhang, S.H., Painter, P.C., and Runt, J. (2002) *Macromolecules*, **35**, 9403–9413.

**332** Zhang, S.H., Jin, X., Painter, P.C., and Runt, J. (2002) *Macromolecules*, **35**, 3636–3646.

**333** Song, M., Hammiche, A., Pollock, H.M., Hourston, D.J., and Reading, M. (1995) *Polymer*, **36**, 3313–3316.

**334** Tanaka, S., Nishida, H., and Endo, T. (2009) *Macromolecules*, **42**, 293–298.

**335** Kuo, S.W., Huang, C.F., and Chang, F.C. (2001) *J. Polymer Sci. B*, **39**, 1348–1359.

**336** Li, X.D. and Goh, S.H. (2003) *J. Polymer Sci. B*, **41**, 789–796.

**337** Chiang, W.-J. and Woo, E.M. (2007) *J. Polymer Sci. B*, **45**, 2899–2911.

**338** Benabdelghani, Z. and Etxeberria, A. (2011) *J. Appl. Polymer Sci.*, **121**, 462–468.

**339** ElMiloudi, K., Hamou, A.S.H., and Djadoun, S. (2008) *Polymer Eng. Sci.*, **48**, 458–466.

**340** Hamou, A.S.H. and Djadoun, S. (2007) *J. Appl. Polymer Sci.*, **103**, 1011–1024.

**341** M. A., deAraujo., Stadler, R., and Cantow, H.-J. (1988) *Polymer*, **29**, 2235–2243.

**342** Bouslah, N., Haddadine, N., Amrani, F., and Hammachin, R. (2008) *J. Appl. Polymer Sci.*, **108**, 3256–3261.

**343** Habi, A. and Djadoun, S. (2008) *Thermochim. Acta*, **469**, 1–7.

**344** Tanaka, S., Nishida, H., and Endo, T. (2009) *Macromol. Phys.*, **210**, 1235–1240.

**345** Lorthior, C., Alegría, A., and Colmenero, J. (2003) *Phys. Rev. B*, **68**, 031805 (1–9).

**346** Sanchis, A., Masegosa, R.M., Rubio, R.G., and Prolongo, M.G. (1994) *Eur. Polymer J.*, **30**, 781–787.

**347** Kuo, S.W. and Chang, F.C. (2001) *Macromolecules*, **34**, 5224–5228.

**348** Qin, C., Pires, A.T.N., and Belfiore, L.A. (1990) *Polymer Commun.*, **31**, 177–182.

**349** Qiu, Z., Komura, M., Ikehara, T., and Nishi, T. (2003) *Polymer*, **44**, 8111–8117.

**350** Qiu, Z., Fujinami, S., Komura, M., Nakajima, K., Ikehara, T., and Nishi, T. (2004) *Polymer*, **45**, 4515–4521.

**351** Xing, P., Dong, L., An, Y., Feng, Z., Avella, M., and Martuscelli, E. (1997) *Macromolecules*, **30**, 2726–2733.

**352** Hsu, W.-P. (2002) *J. Appl. Polymer Sci.*, **83**, 1425–1431.

**353** Zhang, L., Goh, S., and Lee, S. (1999) *J. Appl. Polymer Sci.*, **74**, 383–388.

**354** Lee, L.-T., Woo, E.M., Chen, W.-T., Chang, L., and Yen, K.-C. (2010) *Colloid Polymer Sci.*, **288**, 439–448.

**355** Meaurio, E., Zuza, E., and Sarasua, J.-R. (2005) *Macromolecules*, **38**, 1207–1215.

**356** Meaurio, E., Zuza, E., and Sarasua, J.-R. (2005) *Macromolecules*, **38**, 9221–9228.

**357** Lee, L.T. and Woo, E.M. (2004) *Polymer Int.*, **53**, 1813–1820; (2006) *J. Polymer Sci. B*, **44**, 1339, 1350.

**358** Yi, J.Z., Goh, S.H., and Wee, A.T.S. (2001) *Macromolecules*, **34**, 7411–7415.

**359** Li, X.-D. and Goh, S.H. (2003) *J. Polymer Sci. B*, **41**, 789–796.

**360** Kuo, S.-W. (2009) *J. Appl. Polymer Sci.*, **114**, 116–124.

**361** Goh, S.H. and Siow, K.S. (1987) *Polymer Bull.*, **17**, 453–458.

**362** Hill, D.J.T., Whittaker, A.K., and Wong, K.W. (1999) *Macromolecules*, **32**, 5285–5291.

**363** Arrighi, V., Cowie, J.M.G., Ferguson, R., McEwen, I.J., McGonigle, E.-A., Pethrick, R.A., and Princi, E. (2005) *Polymer Int.*, **55**, 749–756.

**364** Lee, J.Y. and Han, J.Y. (2004) *Macromol. Res.*, **12**, 94–99.

**365** Alata, H., Zhu, B., and Inoue, Y. (2007) *J. Appl. Polymer Sci.*, **106**, 2025–2030.

**366** Yang, F., Qiu, Z., and Yang, W. (2009) *Polymer*, **50**, 2328–2333.

**367** Sotele, J.J., Soldi, V., and Nunes Pires, A.T. (1997) *Polymer*, **38**, 1179–1185.

**368** Papageorgiou, G.Z., Bikiaris, D.N., and Panayiotou, C.G. (2011) *Polymer*, **52**, 4553–4561.

**369** Landry, M.P., Massa, D.J., Mandry, C.J.T., Teegarden, D.M., Colby, R.H., Long, T.E., and Henrichs, P.M. (1994) *J. Appl. Polymer Sci.*, **54**, 991.

**370** Zhang, X.Q., Takegoshi, K., and Hikichi, K. (1992) *Macromolecules*, **25**, 4871.

**371** Zhang, X.Q., Takegoshi, K., and Hikichi, K. (1991) *Macromolecules*, **24**, 5756–5762.

**372** Hirose, Y., Urakawa, O., and Adachi, K. (2004) *J. Polymer Sci. B*, **42**, 4084–4094.

**373** Huang, C.-F. and Chang, F.-C. (2003) *Polymer*, **44**, 2965–2974.

**374** Mandal, T.K. and Woo, E.M. (1999) *Polymer J.*, **31**, 226–232; Woo, E.M. and Mandal, T.K. (1999) *Macromol. Rapid Commun.*, **20**, 46–49.

**375** Bélorgey, G. and Prud'homme, R.E. (1982) *J. Polymer Sci. B*, **20**, 191–203.

**376** Chiu, F.-C. and Min, K. (2000) *Polymer Int.*, **49**, 223–234.

**377** Cheung, Y.W. and Stein, R.S. (1994) *Macromolecules*, **27**, 2512–2519.

**378** Princi, E. and Vicini, S. (2010) *J. Polymer Sci. B*, **48**, 2129–2139.

**379** Madbouly, S.A., Abdou, N.Y., and Mansour, A.A. (2006) *Macromol. Chem. Phys.*, **207**, 978–986.

**380** Rim, P.B. and Runt, J.P. (1983) *Macromolecules*, **16**, 762–768.

**381** Moskala, E.J., Varnell, D.F., and Coleman, M.M. (1985) *Polymer*, **26**, 228.

**382** Kuo, S.W. and Chang, F.C. (2001) *Macromolecules*, **34**, 7737–7743.

**383** Wang, J., Cheung, M.K., and Mi, Y. (2002) *Polymer*, **43**, 1357–1364.

**384** Prolongo, M.G., Salom, C., and Masegosa, R.M. (2002) *Polymer*, **43**, 93–102.

**385** Zhang, H., Bhagwagar, D.E., Graf, J.F., Painter, P.C., and Coleman, M.M. (1994) *Polymer*, **35**, 5378–5397.

**386** Scott, R.L. (1949) *J. Chem. Phys.*, **17**, 279.

**387** Tompa, H. (1949) *Trans. Farad. Soc.*, **45**, 1142–1152.

**388** Kwei, T.K., Frisch, H.L., Radigan, W., and Vogel, S. (1977) *Macromolecules*, **10**, 157–160; Pomposo, J.A., Cortazar, M., and Calahorra, E. (1994) *Macromolecules*, **27**, 252–259.

**389** Kuo, S.W., Lin, C.L., and Chang, F.C. (2002) *Macromolecules*, **35**, 278–285.

**390** Guo, Q. (1996) *Eur. Polymer J.*, **32**, 1409–1413.

**391** Ponoit, D. and Prud'homme, R.E. (2002) *Polymer*, **43**, 2321–2328.

**392** Min, K.E., Chiou, J.S., Barlow, J.W., and Paul, D.R. (1987) *Polymer*, **28**, 172.

**393** Goh, S.H. and Ni, X. (1999) *Polymer*, **40**, 5733–5735.

**394** Woo, E.M. and Tseng, Y.C. (2000) *Macromol. Chem. Phys.*, **201**, 1877–1886.

**395** Yau, S.N. and Woo, E.M. (1996) *Macromol. Rapid Commun.*, **17**, 615–621.

**396** Lee, S.C. and Woo, E.M. (2002) *J. Polymer Sci. B*, **40**, 747–754.

**397** Hong, B.K., Kim, J.Y., Jo, W.H., and Lee, S.C. (1997) *Polymer*, **38**, 4373–4375.

**398** Hsu, W.P. (2007) *Thermochim. Acta*, **454**, 50–56.

**399** Rabeony, M., Siano, D.B., Peiffer, D.G., Siakali-Kioulafa, E., and Hadjichristidis, M. (1994) *Polymer*, **35**, 1033.

**400** Kuo, S.W., Chan, C.S.-C., Wu, H.-D., and Chang, F.-C. (2005) *Macromolecules*, **38**, 4729–4736.

# 2

# Crystallization and Melting Behavior in Polymer Blends

*Saleh A. Jabarin, Kazem Majdzadeh-Ardakani, and Elizabeth A. Lofgren*

*Polymer Institute and Department of Chemical and Environmental Engineering, University of Toledo, Toledo, Ohio 43606-3390*

## 2.1
## Introduction

The history of polymer blends can be traced back more than a century to when the first blends of *trans-* and *cis-*1,4-polyisoprene (natural rubber with gutta-percha) were prepared in 1846 [1]. Preparing commercial blends of poly(phenylene ether) (PPE) and styrenics in 1965 led to the modern era of polymer blending. Since then, the manufacturing of polymer blends with the desired performances, by combining the unique properties of available polymers, has made such blends commercially important. The crystallization and melting characteristics of polymers and polymer blends determine many of their processing conditions and ultimate physical properties. Early reviews of polymer blends discuss the thermodynamics of mixing, phase separation, preparation, transport phenomena, chemical interactions, physical properties, and some commercial applications [1–5].

The study of polymer blends has been a very important scientific topic during the recent decades, and their applications are still growing because of the lower cost of blending processes compared to that of making new polymers. Many polymer characteristics, such as mechanical and barrier properties, chemical resistance, thermal stability, flame retardancy, and others, can be improved by blending different polymers. Volumes 4 and 5 of this encyclopedia series discuss this topic in more detail.

Important characteristics of polymer blends include the miscibility or immiscibility of their components. The properties such as crystallization and melting behavior of the blends are strongly influenced by the miscibility of the polymers. Most polymer pairs are thermodynamically immiscible; however, miscibility can be achievable with selected polymer combinations. The properties of miscible polymer blends are generally influenced by the chemical composition of each polymer. The fundamentals of miscible, immiscible, and compatible blends are included in Volume 1 of this encyclopedia series.

*Encyclopedia of Polymer Blends: Volume 3: Structure*, First Edition. Edited by Avraam I. Isayev.
© 2016 Wiley-VCH Verlag GmbH & Co. KGaA. Published 2016 by Wiley-VCH Verlag GmbH & Co. KGaA.

Many investigations have been conducted to characterize and document various preparation processes and the resultant properties of polymer blends [6–11]. Several books and review articles have also been published in this area of research [2,12–16]; however, fewer publications [12,17] comprehensively discuss the melting behavior and crystallization of polymer blends.

The following review comprises recent developments and progress related specifically to the crystallization and melting characteristics of polymer blends. It contains some basic aspects of crystallization kinetics, semicrystalline morphology, and melting behavior of miscible and immiscible polymer blends. It also includes important information from earlier books and articles and extends it to more recent achievements in this area. The miscibility of polymer blends is discussed in Section 2.2, while melting behavior and crystallization characteristics of miscible and immiscible polymer blends are included in Sections 2.3 and 2.4. Finally, compatibilized polymer blends and their crystallization behavior are discussed in Section 2.5.

## 2.2
### Miscibility of Polymer Blends

Miscibility or compatibility is achievable when two materials can be mixed in any ratio, without the separation of the two phases. One of the important issues in polymer blends is the miscibility between the polymer components. Most polymer pairs are thermodynamically immiscible because of the positive free energy change ($\Delta G$) during mixing. While this chapter primarily encompasses the crystallization and melting behavior of polymer blends, it is important to indicate that such properties are greatly influenced by the miscibility of the blend components. Poor adhesion at the interface of the polymer blends results in diminished mechanical properties such as lower impact resistance and elongation at break [17]. Miscibility is not the preferred state in all situations. In some cases, immiscibility is required to achieve two or multiple phases during blending.

From a thermodynamic point of view, the state of miscibility of any mixture is governed by the change in the free energy of mixing ($\Delta G^M$), which is defined as follows [18]:

$$\Delta G^M = \Delta H^M - T\Delta S^M < 0, \tag{2.1}$$

where $\Delta G^M$, $\Delta H^M$, $\Delta S^M$, and $T$ are, respectively, the Gibbs free energy change, enthalpy change, entropy change, and temperature of mixing.

Figure 2.1 shows the changes in the free energy of mixing with the composition of the overall mixture ($\phi_2$ is the volume fraction of component 2). Plot (a) indicates a positive free energy in entire volume fraction proportions. In case (b), the components are completely miscible in all compositions. Complete miscibility is not guaranteed by a negative free energy of mixing as shown in case (c), which shows a reversed curvature in the mid-composition range, and thus the

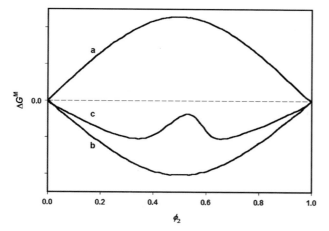

**Figure 2.1** Free energy of mixing as a function of composition for a completely immiscible (a), completely miscible (b), and partially miscible (c) binary mixtures. Reproduced from reference [20] by Paul and Barlow with permission, © 1980 Taylor & Francis.

mixture can develop an even lower free energy in this range by splitting into two phases with compositions given by the two minima. A partial miscibility is achievable in this condition.

A one-phase mixture is thermodynamically stable at fixed temperature and pressure if [18]

$$(\partial^2 \Delta G^M / \partial \phi^2)_{T,P} > 0, \tag{2.2}$$

where $\phi$ is the blend composition (volume fraction). Because of the large molecular weights (small number of moles) of components in a polymer–polymer blend, the entropy term $\Delta S^M$ in Eq. (2.1) is very small and the $T\Delta S^M$ term is negligible. The miscibility of a polymer–polymer mixture, therefore, occurs if the heat of mixing of the components is negative (exothermic mixing). Miscibility for endothermic mixtures ($\Delta H^M > 0$) only occurs at very high temperatures. To obtain nonpolar polymer blends ($\Delta H^M > 0$) at normal temperatures, the components should have similar structures or precisely matched solubility values. To lead to exothermic heats of mixing, some specific interactions such as hydrogen bonding, acid–base, charge–transfer, ion–dipole, donor–acceptor interactions should exist among the components in a polymer blend [17,19,20]. These interdependencies are not very common for many polymers and the miscibility of polymers can only exist for a group of polymers either having precisely matched solubility values or certain specific reciprocities.

Control of the phase morphology of polymer blends is another important factor that can be used to improve the miscibility. Ratios of the phase's viscosities, interfacial properties, blend composition, and processing conditions have an influence on the size, shape, and the spatial distribution of the dispersed phases in a polymer blend [17]. An empirical relationship between the capillary number and relative melt viscosity of the dispersed and continuous phases has been presented by Wu *et al.* [21] in the viscoelastic systems. If the interfacial tension is lower and the viscosity ratio is closer to unity, the dispersed drops are smaller. The interfacial tension is largely controlled by the polarities of the two phases and can be varied over several orders of magnitude by using appropriate dispersants. The reduction of the interfacial tension between two components in a polymer blend can lead to the decreased particle size of the dispersed phase and therefore improve the miscibility of polymer blends. The addition of a compatibilizer and *in situ* compatibilization are two general methods used to decrease the interfacial tension across the interface of polymer blends. Compatibilizers are high molecular weight compounds, which act as polymeric surfactants in polymer blends. Due to the existence of the compatibilizer, the resistance against coalescence of the dispersed phase is greatly increased, and the stability of the polymer blend morphology remains the same in the different processing steps. Usually, a compatibilizer consists of one constitutive block miscible with one blend component and a second block miscible with the other blend component. These compatibilizers can be premade and added to the polymer blends or can be generated during the blending process (reactive compatibilization). A good compatibilizer reduces the interfacial energy, leads to a finer dispersion during mixing, and results in an improved interfacial adhesion. Nonbonding specific interactions such as hydrogen bonding, ion–dipole, dipole–dipole, donor–acceptor, and $\pi$–electron interactions are also useful for enhancing the compatibility of polymer blends [17].

In polyester blends, as those with poly(ethylene naphthalate) (PEN) and poly (ethylene terephthalate) (PET) where the two components are inherently immiscible, the transesterification reaction between the two polyesters can lead to the production of copolymers to improve their compatibility. Some parameters such as time and temperature are key factors that control the reactions in the melt phase. These parameters should be set at the best conditions to achieve the desired compatibilization and sufficiently high levels of transesterification reactions in the polyester blends [22,23].

## 2.3
### Miscible Blends

Since crystallinity plays an important role in the properties of a polymer blend, understanding of the melting behavior, structure, and crystallization is required to control the properties of these materials. Optical microscopy (OM),small angle X-ray scattering (SAXS), differential scanning calorimetry (DSC), electron microscopy (SEM), and atomic force microscopy (AFM) are some techniques that have been used to study the melting behavior and crystallization of polymers.

Miscible polymer blends form a single-phase system and the properties of such materials are a combination of the properties of their pure components. If the free energy of crystallization is more negative than the free energy of the liquid–liquid mixture, crystallization occurs in a miscible blend. The crystallization and melting behavior of miscible crystallizable blends are complicated and affected by the structure and molecular weight of the components, the concentration of each component, and the intermolecular interactions between them [12,22].

The crystallization behaviors of miscible crystallizable blends containing one amorphous component are different from those of blends consisting of two crystallizable components.

The location of the amorphous component is very important. The amorphous component can be located in the interlamellar, interfibrillar, interspherulite, or a combination of two or more regions [24]. The type of segregation of the amorphous diluent plays a key role in the microstructure and crystallinity of a semi-crystalline polymer blend [12]. The placement of the amorphous component in a miscible blend is affected by the size of the amorphous component (radius of gyration) and intermolecular interactions between the amorphous component and the amorphous portion of the crystallizable polymer [24].

### 2.3.1
### Crystalline/Amorphous Polymer Blends

The glass transition ($T_g$) of a miscible polymer blend is generally between those of each individual component, and its crystallization occurs at a temperature between the $T_g$ and the equilibrium melting point ($T_m°$). The effect of the amorphous component on the crystallization of a crystallizable, miscible, polymer blend depends on the glass transition of the amorphous component compared to that of the crystallizable one. If the $T_g$ of the amorphous component is higher than that of the crystallizable one (as is true for the majority of polymer blends), the difference between $T_m°$ and $T_g$ of the blend decreases and, consequently, the tendency of the polymer blend to crystallize decreases. In the few cases where the $T_g$ of the amorphous component is lower than that of the crystallizable component, the tendency for crystallization is increased [12,25]. Table 2.1 shows the effects of the amorphous component on the spherulite growth rate of the crystallizable component in some polymer blends.

### 2.3.2
### Glass Transition and Melting Behavior

The glass transition temperature provides useful information on blend miscibility [42,43]. Although two $T_g$s are generally observed on the differential scanning calorimetry (DSC) scans of immiscible blends, only one $T_g$ appears on that of miscible blends or copolymers. By investigating the blends of poly(2,6-dimethyl-1,4-phenylene oxide) (PPO) with poly-(styrene-*co*-4-chlorostyrene), Fried *et al.* [44] found that the width of the glass transition region ($\Delta T_g$) in which only one $T_g$ is observed in the DSC scan can be a measure of the

**Table 2.1** Effect of the amorphous component on the spherulite growth rate of the crystallizable component polymer blends.

| Polymer blend | The effect of amorphous component on the spherulite growth rate | References |
|---|---|---|
| PEO/PMMA | decreases | [26] |
| PEO/PVAc | decreases | [27] |
| PVF$_2$/PMMA | decreases | [28] |
| PVF$_2$/PBSU | decreases | [29] |
| PCL/PVC | decreases | [30] |
| PCL/Phenoxy | decreases | [31] |
| iPS/PPO | decreases | [32] |
| PHB/PEC | decreases | [33] |
| P3HB/PVPh | decreases | [34] |
| P3HB/CE | decreases | [35] |
| P3HB/PVAc | decreases | [36] |
| iPS/PVME | increases | [37,38] |
| sPS/PVME | increases | [39] |
| PVF2/PBA | increases | [40] |
| PLLA/PBO | increases | [41] |

miscibility of the polymer blends. The $\Delta T_g$ values were less than $10\,^{\circ}\text{C}$ for homopolymers or miscible blends. On the other hand, the glass transition regions were much wider ($\Delta T_g = 30\,^{\circ}\text{C}$) for immiscible blends. Kim *et al.* [45] used the same method to evaluate the PEN/poly(ether imide) (PEI) blends and found that although only one $T_g$ was observed in the DSC scans, $\Delta T_g$ values were more than $30\,^{\circ}\text{C}$ for the blends with PEI contents of 30–70 wt%.

### 2.3.2.1 Melting Point Depression

Because of the thermodynamically favorable interactions between the crystallizable and amorphous components in a miscible polymer blend, the melting point of the crystalline component is usually less than that of the pure polymer [12,43]. Measuring the melting point depression, therefore, provides information about interactions between polymers. According to the Flory–Huggins theory, the melting point depression under thermodynamic equilibrium can be expressed as follows [46]:

$$\frac{1}{T_m} - \frac{1}{T_m^0} = -\frac{RV_{2u}}{\Delta h_u V_{1u}} \times \left[ \frac{\ln v_2}{x_2} + \left( \frac{1}{x_2} - \frac{1}{x_1} \right)(1 - v_2) + \chi_{12}(1 - v_2)^2 \right],$$

(2.3)

where $T_m^0$ and $T_m$ are the equilibrium melting points of the crystallizable polymers in the bulk and the blends, respectively. Subscript 1 corresponds with the amorphous component and subscript 2 the crystallizable polymer; $v$ represents the volume fraction; $V_{1u}$ and $V_{2u}$ are the molar volumes of the repeat units; $x$ is the degree of polymerization; $\Delta h_u$ is the heat of fusion per mole of crystalline units; $R$ is the gas constant; and $\chi_{12}$ is the Flory–Huggins interaction parameter. $x_1$ and $x_2$ are very large compared to unity in the case of polymers, and Eq. (2.3) can be reduced to the Nishi–Wang equation [46,47]:

$$\frac{1}{T_m} - \frac{1}{T_m^0} = -\frac{RV_{2u}}{\Delta h_u V_{1u}}\chi_{12}(1 - v_2)^2. \tag{2.4}$$

The authors assumed that the crystals were perfect and of finite size, and there was no recrystallization during melting. The melting point of a polymer is generally influenced not only by thermodynamic factors, but also by morphological parameters such as the crystal thickness. In miscible polymer blends, $\chi_{12}$ is usually negative and small [47–49].

The Flory–Huggins interaction parameter (neglecting the entropic contribution) can be defined by the following equation [48]:

$$\chi_{12} = \frac{BV_{1u}}{RT_m}, \tag{2.5}$$

where $B$ is the interaction energy density. By substituting Eq. (2.5) into Eq. (2.4), we have

$$\frac{1}{v_1}\left(\frac{1}{T_m} - \frac{1}{T_m^0}\right) = -\frac{BV_{2u}}{\Delta h_u}\left(\frac{v_1}{T_m}\right). \tag{2.6}$$

The interaction energy density $(B)$ can be obtained from the slope of the plot of the left-hand side of Eq. (2.6) as a function of $v_1/T_m$. From Eq. (2.4), the depression of melting point of a polymer–polymer pair is possible only when $\chi_{12}$ is negative, which is in agreement with Scott's condition for the miscibility of the two polymers [49]:

$$\chi_{12} < \frac{1}{2}\left[\frac{1}{x_1^{1/2}} + \frac{1}{x_2^{1/2}}\right]^2. \tag{2.7}$$

In some cases, Eq. (2.4) does not fit the experimental data due to the use of observed melting temperatures instead of the thermodynamic equilibrium temperatures and the concentration dependence of the interaction parameter [12]. A modified equation has therefore been defined by other authors [50,51]:

$$\frac{\Delta h_u(T_m^0 - T_m)}{v_1 RT_m^0} - \frac{T_m}{x_1} - \frac{v_1 T_m}{2x_2} = \frac{C}{R} - av_1, \tag{2.8}$$

where $C$ is a constant related to the morphological contributions, and $a$ is a constant from the relationship between interaction parameter and temperature.

Matkar and Kyu [52] have modified the Flory diluent theory that assumes the complete immiscibility of the solvent in the crystal. They have incorporated a crystal–solvent interaction parameter in addition to the amorphous–amorphous interaction parameter to develop a self-consistent theory for the determination

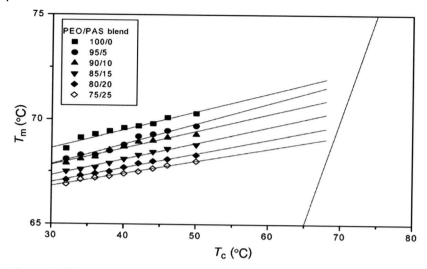

**Figure 2.2** Hoffman–Weeks plots for PAS/PEO blends. Reproduced from reference [56] by Kuo *et al.* with permission, © 2004 American Chemical Society.

of phase diagrams of a crystalline polymer solution. The changes in equilibrium melting temperatures can be explained in the context of the eutectic phase diagram approach. This has been well described by some researchers [52–55].

The equilibrium melting temperature of a polymer blend can be also determined by a plot of the experimental melting point versus the crystallization temperature (Hoffman–Weeks plot). An example of Hoffman–Weeks plots for poly (acetoxy styrene) (PAS)/poly(ethylene oxide) (PEO) blends is shown in Figure 2.2. Extrapolation from experimental data to the $T_m = T_c$ line gives the value of $T_m^0$. In some cases, the Hoffman–Weeks plots do not show a linearity because of the recrystallization and crystal defects in the sample [12,56].

## 2.3.3
## Crystallization

Primary nucleation, crystal growth, and secondary crystallization are three stages, which are involved in the crystallization of polymers. In the primary nucleation process, nuclei are formed homogeneously or heterogeneously in the melt state. Heterogeneous nucleation, which is more common in polymers, occurs in the presence of impurities. After this nucleation process, crystalline lamellae develop and form three-dimensional superstructures. In general, the crystals of a polymer are spherulites, which consist of lamellar structures, growing radially from the center of the spherulites. In most cases, the crystallization will continue with the process of secondary crystallization [12,25]. A schematic drawing of a spherulite of a semicrystalline polymer is shown in Figure 2.3.

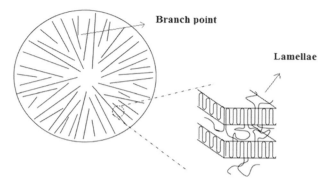

**Figure 2.3** Model of spherulitic and lamellae structure. Adapted from reference [57].

The type of segregation or rejection of the amorphous component determines the morphology of polymer blends with an amorphous phase. Three different types of segregation have been considered. In the case of an interspherulitic segregation, the spherulites are imbedded in an amorphous matrix, while in the case of intraspherulitic segregation, the amorphous component is located within the spherulites, giving interlamellar and interfibrillar segregation [58,59].

When the diameter of gyration of an amorphous component is greater than the separation between crystalline lamellae (in highly crystalline polymer blends), the amorphous component can escape from the interlamellar zones. This leads to an interfibrillar or interspherulitic segregation. Interlamellar segregation is more favorable, if there is a good intermolecular interaction between the amorphous component and the amorphous portion of the crystalline polymer [24]. If the difference between $T_g$ and $T_c$ of the crystalline polymer is low, the amorphous polymer resides in interlamellar regions, while if this difference is high, some parts of the amorphous component resides in either interfibrillar or interspherulitic regions [12,24].

Type of segregation also depends on the relative values of the diffusion rate of the amorphous component and the crystallization rate of the crystallizable component. This is discussed more extensively in Section 2.3.4. The type of segregation of the amorphous component for some polymers is listed in Table 2.2 (Some data have been selected from the *Polymer Blends Handbook* [12] and some polymer blends are added from other articles).

In partially miscible polymer blends (blends with miscibility gaps in their composition ranges), two crystallization and demixing phenomena can affect the phase transitions and the morphology, depending on the amount of the crystallizable component. In crystallizable/amorphous polymer blends, if the phase separation region intersects the crystallization range, a competition between the kinetics of these two phenomena can be observed. Another phenomenon, observed in partially miscible blends, is interface crossing crystallization. In this case, there are two phases with different amounts of crystallizable component

**Table 2.2** The type of segregation of the amorphous component in miscible polymer blends.

| Polymer Blend | Amorphous Comp. (wt%) | Technique | Type of Segregation | References |
|---|---|---|---|---|
| PCL/CPE | 0–30 CPE | SAXS, OM | interfibrillar | [60] |
| PCL/ Phenoxy | 0–50 Phenoxy | SAXS, OM | interlamellar/ interfibrillar | [60] |
| | 0–10 Phenoxy | SAXS | interlamellar | [61] |
| PEG/PEMA | 0–20 PEMA | OM | intraspherulitic | [62] |
| aPMMA/ iPHB | 0–50 aPMMA | DSC, SAXS | interlamellar | [63] |
| PVDF/ PMMA | 0 and 40 PMMA | SAXS, OM | interlamellar/ interfibrillar | [64] |
| PVF$_2$/ PMMA | PMMA | SAXS, OM | interlamellar | [65] |
| PEO/PMMA | 0–50 PMMA | SAXS | interlamellar | [66] |
| *Cis*-PI/PVE | 5–50 PVE | DSC | interlamellar | [67] |
| PBT/PAr | 10–70 PAr | DSC, SAXS | interlamellar | [68] |
| PEEK/PEI | 0–75 PEI | SAXS | interfibrillar/ intraspherulitic | [69] |
| iPP/aPP | 10–75 aPP | SAXS, SEM | interfibrillar/ intraspherulitic | [70] |
| PEEK/PEI | 25 and 75 PEI | TEM, OM | interfibrillar/ intraspherulitic | [71] |
| PBT/PAr | 0–80 PAr | SAXS | interfibrillar/ intraspherulitic | [72] |

and different number density of heterogeneous nuclei in the phase separation region of partially miscible blends. The phase with a higher content of crystallizable component will crystallize first, and its crystallization will continue until the crystals reach the interface with the second phase. At this point, crystallization of the second phase will be induced, by a secondary nucleation process, on the crystals in contact with the melt [16].

2.3.4
**Spherulite Growth Rate of the Crystallizable Component**

An equation for the growth rate kinetics of a crystal in a homopolymer has been developed as follows by Hoffman and Lauritzen [73,74]:

$$G = G^\circ \, \exp\left(-\frac{\Delta E}{R(T_c - T_0)}\right) \exp(-\Delta F^*/k_B T_c) \tag{2.9}$$

where $G$ is the growth rate of a crystal, $G°$ is a constant that depends on the regime of crystallization, $T_0$ is the temperature at which the required motions for the transport of molecules into the liquid–solid boundary occurs, $T_c$ is the crystallization temperature, $\Delta E$ is the activation energy required for transferring crystals through the solid–liquid interface, and $\Delta F^*$ is the energy required to form a nucleus. Two terms are involved in the growth rate kinetics of a crystal in a homopolymer: its ability to form surface nuclei and diffusion of crystalline molecules toward the crystal growth front. Based on these two opposing terms, a plot of the growth rate as a function of crystallization temperature in a homopolymer has a maximum (a bell-shape curve). The growth rate is nucleation controlled at low undercooling temperature and diffusion controlled at high undercooling temperature [12,26].

In most cases, adding an amorphous component to the polymer blend decreases the equilibrium melting temperature and, consequently, the spherulite growth rate of the crystallizable component [25]. This reduction in the equilibrium melting point occurs because of interactions between the two components [12].

Alfonso and Russel [26] have proposed a theoretical treatment for the growth rates in the miscible polymer blends containing an amorphous component. The interactions between the two components change the chemical potential of the liquid phase. These interactions affect the free energy required for forming a nucleus on the crystal surface and the mobility of both components. In addition, the amorphous component must diffuse from the growth front to the interlamellar area (segregation). There is a competition between the segregation of the amorphous component and the capability of the crystal to grow, with the crystal growth rate controlled by the slower process. One conclusion is that the crystallization growth rate depends on the molecular weight of both components. The crystallization of crystallizable component is also affected by the volume fraction of this component. Finally, the $T_g$ of the amorphous phase has an influence on the segregation of the amorphous component at the solid–liquid interface. If the $T_g$ of the amorphous component is higher than that of the crystalline component, the temperature range over which crystallization can occur is narrowed. Alternatively, the crystallization temperature range is widened if the $T_g$ of the crystalline component is higher than that of the amorphous component.

Based on the above parameters and their effects, the crystallization growth rate of a polymer blend can be defined according to the following equation:

$$G_m = \frac{\phi_2 k_1 k_2}{k_1 + k_2} e^{-\Delta F_m^*/k_B T_c}, \tag{2.10}$$

where $\phi_2$ is the volume fraction of the crystallizable component, $T_c$ is the crystallization temperature, $k_1$ is the rate of transport of the crystallizable molecule across the liquid–solid interface, and $k_2$ is the rate of segregation of the amorphous component from the growth front. $\Delta F_m^*$ is the work required to form a nucleus on the crystal surface modified by the presence of the amorphous component. If the transport of the crystallizable component across the interface is faster than the rate of segregation ($k_1 \gg k_2$), then

$$G_m \approx \phi k_2 e^{-\Delta F_m^*/k_B T_c}. \tag{2.11}$$

Alternatively, if $k_2 \gg k_1$, then

$$G_m \approx \phi k_1 e^{-\Delta F_m^*/k_B T_c}. \tag{2.12}$$

To calculate the crystallization growth rate of a polymer blend, the values for $\Delta F*$, $k_1$, and $k_2$ should be defined. For semicrystalline polymers in the presence of a low molecular weight diluent, Flory [75] has shown that

$$\Delta F_m^* = \frac{2b\sigma\sigma_e}{\Delta h_u f \left( 1 - \dfrac{T}{T_m^0} - \dfrac{RT\chi_{12}}{\Delta h_u f} \dfrac{V_{2u}}{V_{1u}} (1 - \phi_2)^2 \right)}, \tag{2.13}$$

where $V_{1u}$ and $V_{2u}$ are the molar volume of the repeating unit of the amorphous component and crystallizable one; $\chi_{12}$ is the Flory–Huggins interaction parameter; $\Delta h_u$ is the enthalpy of fusion per mole of repeating unit of crystallizable component; $b$ is the thickness of the critical nucleus; $\sigma\sigma_e$ is the product of the lateral and fold surface free energies; $\phi_2$ is the volume fraction of the crystallizable component; and parameter $f$ is defined as follows by Hoffman and Weeks [76,77]:

$$f = \left( \frac{2T}{(T + T_m^0)} \right). \tag{2.14}$$

In addition to the assumptions applied for the derivation of the Flory equation, $\chi_{12}$ and $\sigma\sigma_e$ are assumed to be independent of temperature and composition. $k_1$ is also defined as

$$k_1 = G^0 \exp\left[-\Delta E/R(T_c - T_0')\right] \sqrt{b^2 - 4ac}, \tag{2.15}$$

where $G^0$ is a constant dependent on the regime of the crystallization, $\Delta E$ is the energy of the transport, and $T_0'$ is the value of $T_0$ in the polymer blend.

$$k_2 = D/d = 2\overline{D}/L, \tag{2.16}$$

where $d$ is the maximum distance over which the amorphous component must diffuse away from the growth front, and $D$ is the diffusion coefficient. $D$ can be replaced by $\overline{D}$ which is the mutual diffusion coefficient (a parameter indicating the cooperative diffusion of the amorphous and crystallizable components). $\overline{D}$ can be determined experimentally or theoretically; however, the theoretical predictions by Kramer *et al.* [78] and Sillescu [79] are not accurate enough. Since the amorphous component diffuses in a direction normal to the growth direction, the maximum distance ($d$) would be half of the crystal thickness ($L/2$).

With the substitution of all the parameters into Eq. (2.10), the following equation can be derived:

$$
G_m = \left\{ \frac{\phi_2 G^0 e^{-\frac{\Delta E}{R(T-T'_0)}} 2\overline{D}/L}{G^0 e^{-\frac{\Delta E}{R(T-T'_0)}} + 2\overline{D}/L} \right\}
$$

$$
\exp\left\{ \frac{-2b\sigma\sigma_e}{k_B T \Delta h_u f (1 - \frac{T}{T_m^0} - \frac{RT}{\Delta h_u f}\frac{V_{2u}}{V_{1u}}\chi(1-\phi_2)^2} \right\}. \tag{2.17}
$$

This can be rearranged to separate the kinetic term from the thermodynamic one:

$$
\alpha = -\sigma\sigma_e \vartheta, \tag{2.18}
$$

where $\alpha$ is the kinetic term and $\vartheta$ contains thermodynamic parameters.

$$
\alpha = \ln G_m - \ln \phi_2 - \ln G^0 + (\Delta E/R(T-T'_0)) + \ln\left(1 + \frac{G^0 L e^{-\Delta E/R(T-T'_0)}}{2\overline{D}}\right) \tag{2.19}
$$

and

$$
\vartheta = \left(\frac{2b}{k_B T}\right)\left\{ \left(\frac{\Delta h_u f \Delta T}{T_m^0}\right) - \frac{RT V_{2u}}{V_{1u}}\chi(1-\phi_2)^2 \right\}^{-1}. \tag{2.20}
$$

Since $\sigma\sigma_e$ are assumed to be independent of temperature and composition, a plot of $\alpha$ as a function of $\vartheta$ is a straight line, and $\sigma\sigma_e$ can be determined by the slope of this plot. Alfonso and Russell [26] have shown that deviation from a straight line can be observed when the molecular weight of the amorphous component is approximately equal to the critical molecular weight for entanglement coupling. This deviation corresponds to increasing $\sigma\sigma_e$ with decreasing temperature.

In Eqs. (2.17) and (2.19), the ratio of $\overline{D}/G$ is a modified version of the $\delta$ parameter defined by Keith and Padden [80,81]. Equation (2.17) is derived based on the incorporation of the amorphous component in the interlamellar regions and the consistency of the growth rate on a microscopic level. It is important that the theoretical developments are more accurate in the case of low concentrations of diluent and in a temperature range near the melting point [26].

## 2.3.5
### Overall Crystallization Kinetics

#### 2.3.5.1 Isothermal Kinetics
The overall crystallization kinetics of polymer blends have been analyzed by the Avrami equation [82]:

$$
\theta_a = \exp(-kt^n), \tag{2.21}
$$

where $\theta_a$ is the fraction of an uncrystallized material at time t, $k$ is the Avrami constant, and $n$ is the Avrami exponent describing the type of nucleation and the crystal growth geometry. The kinetic rate constant ($k$) is a function of the type of nucleation and spherulite growth rate. The kinetic parameters ($n$ and $k$) can be obtained by plotting $\ln(-\ln(\theta_a))$ as a function of $\ln(t)$ according to Eq. (2.22):

$$\ln(-\ln(\theta_a)) = \ln(k) + n\ln(t). \tag{2.22}$$

This plot gives a straight line with the slope $n$ and an intercept of $\ln(k)$. A typical example of an Avrami plot for the crystallization behavior of a 40/60 PEN/PET blend is illustrated in Figure 2.4. Summaries of $n$ values and the rate constants for the crystallization behaviors of the various blends are given in Table 2.3.

The temperature dependence of the rate constant ($k$) can be obtained from the Arrhenius equation [83]:

$$k = A\exp(-E_a/RT_c), \tag{2.23}$$

where $E_a$ is the apparent activation energy of rate constant ($k$), $A$ is the frequency factor, $T_c$ is the crystallization temperature, and $R$ is the gas constant. $E_a$ can be calculated from the slope of the plot of $\ln(k)$ as a function of $1/T_c$.

In most cases, the $n$ value is a noninteger because of the assumptions applied in the Avrami model, which assumed that the shape of the growing crystal, radial density, and the rate of radial growth are constant. Other assumptions are also applied in his model such as the absence of the secondary nucleation, no overlap between the growth fronts, lack of induction time, complete crystallinity of the sample, uniqueness of the nucleation mode, and random distribution of nuclei [12,82].

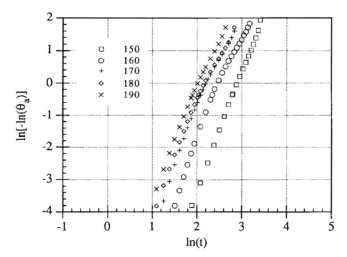

**Figure 2.4** $\ln[-\ln(\theta_a)]$ as a function of $\ln(t)$ for 40/60 PEN/PET blends. Reproduced from reference [83] by Shi and Jabarin with permission, © 2001 John Wiley & Sons, Inc.

**Table 2.3** Rate constant, Avrami index as function of temperature and blend composition for some miscible blends.

| Polymer blend | Composition | $T_c$ (°C) | $k$ (sec$^{-n}$) | $n$ | Reference |
|---|---|---|---|---|---|
| PCL/PVC | 100/0 | 29.6 | $19.95 \times 10^{-3}$ | 3.5 | [84] |
| | | 38.0 | $3.88 \times 10^{-3}$ | 2.2 | |
| | | 42.1 | $1.01 \times 10^{-3}$ | 2.4 | |
| | 90/10 | 28.4 | $13.8 \times 10^{-3}$ | 2.1 | |
| | | 38.4 | $2.16 \times 10^{-3}$ | 1.7 | |
| | | 40.4 | $0.87 \times 10^{-3}$ | 1.9 | |
| PEG/PMMA | 100/0 | 47 | 1.09 | 2.53 | [85] |
| | | 54 | $3.52 \times 10^{-5}$ | 2.63 | |
| | 80/20 | 42 | $3.57 \times 10^{-2}$ | 2.45 | |
| | | 47 | $4.94 \times 10^{-4}$ | 2.46 | |
| | | 50 | $5.40 \times 10^{-5}$ | 2.50 | |
| PET/PEN | 100/0 | 120 | $1.12 \times 10^{-5}$ | 2.7 | [43] |
| | | 140 | $3.08 \times 10^{-4}$ | 2.9 | |
| | 80/20 | 150 | $2.92 \times 10^{-6}$ | 3.8 | |
| | | 190 | $7.31 \times 10^{-3}$ | 3.2 | |
| | 60/40 | 150 | $1.17 \times 10^{-6}$ | 3.9 | |
| | | 170 | $1.99 \times 10^{-4}$ | 3.8 | |
| | 20/80 | 130 | $4.47 \times 10^{-4}$ | 3.9 | |
| | | 140 | $6.28 \times 10^{-3}$ | 3.9 | |
| | 0/100 | 170 | $3.67 \times 10^{-7}$ | 3.7 | |
| | | 200 | $5.36 \times 10^{-4}$ | 3.4 | |

A modified Avrami equation was developed by Perez-Cardenas *et al.* [86]. In their model, the effects of the secondary crystallization are taken into account.

$$\alpha = \alpha_p + \alpha_s, \tag{2.24}$$

where $\alpha$ is the weight fraction of crystallinity, and $\alpha_p$ and $\alpha_s$ are the weight fraction of the primary and secondary crystallization, respectively. The crystallization process can be divided into three regions that include: initial primary crystallization (region 1), primary and secondary crystallization (region 2), and secondary crystallization (region 3). The total crystallization process is described by the following equations:

$$1-\alpha = \exp(-kt^n - k't^{n'}) \left[ kn(1-\xi) \int_0^t \exp(k\tau^n + k'\tau^{n'})\tau^{n-1}d\tau + 1 \right]; \quad \alpha \leq \xi, \tag{2.25}$$

$$1 - \alpha = (1 - \xi)\exp(k't^{*n'})\exp\left(-k't^{n'}\right); \quad \alpha > \xi, \tag{2.26}$$

where $\xi$ is the weight fraction of the polymer crystallized by primary and secondary crystallization when the primary crystallization is finished (end of region 2); $k$ and $n$ are the primary crystallization parameters; $k'$ and $n'$ are secondary crystallization parameters; $t^*$ is the time in which pure secondary crystallization occurs (region 3) [12,86]. Experimental isothermal crystallization data can be fitted to Eqs. (2.25) and (2.26) by using different values for the parameters in order to find the most accurate ones.

### 2.3.5.2 Nonisothermal Kinetics

The theory of Avrami is used for isothermal conditions. This theory has been extended by Ozawa [87] for nonisothermal conditions, assuming that the non-isothermal condition is the result of infinite isothermal processes:

$$\ln(-\ln(\theta_a)) = \ln k(T) - n \ln(\beta), \tag{2.27}$$

where $\theta_a$ is the fraction of uncrystallized material at temperature $T$, $k(T)$ is the cooling crystallization function, $\beta$ is the cooling rate, and $n$ is the Avrami exponent. The kinetic parameters ($n$ and $k(T)$) can be obtained by plotting $\ln(-\ln(\theta_a))$ as a function of $\ln(\beta)$ at a given temperature for various cooling rates.

Another approach used to describe a nonisothermal crystallization process is to apply the theory of Avrami to the results of nonisothermal crystallization thermograms by plotting $\ln(-\ln(\theta_a))$ as a function of $\ln(t)$ (Eq. (2.22)) at each cooling rate. Since the temperature changes constantly, $n$ and $k$ cannot be related to the isothermal case (the spherulite growth rate and nucleation are temperature dependent) and do not have the same physical significance as in the isothermal crystallization [88–90].

A third approach was developed by Ziabicki [91,92], describing the noniso-thermal processes:

$$\theta_a = \exp(-E(t)). \tag{2.28}$$

Under quasistatic conditions

$$E(t) = \ln 2 \left[ \int_0^t \left( \frac{ds}{t_{1/2}(T(s))} \right) \right]^n, \tag{2.29}$$

where $T(s)$ is the thermal history, $s$ is the time required for the nucleation of crystals, $t_{1/2}$ is the half time of crystallization, and $n$ is the Avrami exponent.

Liu et al. [93,94] have presented a different kinetic equation by combining Ozawa and Avrami equations:

$$\log k + n \log t = \log k(T) - n \log \beta, \tag{2.30}$$

which can be rewritten as

$$\log \beta = \log F(T) - b \log t, \tag{2.31}$$

where $\beta$ is the cooling rate, $k$ is the Avrami constant, $k(T)$ is the cooling crystallization function, $F(T) = [k(T)/k]^{1/n}$, with $n$ the Avrami exponent calculated with the Ozawa method, and $b$ is the ratio between the Avrami and Ozawa exponents.

$F(T)$ refers to the value of the cooling rate selected at unit crystallization time, when the system has a defined degree of crystallinity. A plot of log $\beta$ as a function of log $t$ at a given degree of crystallinity gives a straight line with log $F(T)$ as the intercept and $-b$ as the slope.

The nonisothermal crystallization behavior of a miscible poly(hydroxy ether of bisphenol-A) (PHE)/poly-$\varepsilon$-caprolactone (PCL) polymer blend has been investigated by de Juana et al. [88] at various cooling rates. The presence of PH reduced the overall PCL crystallization rate at a given cooling rate. Their experimental data analysis was in agreement with treatments by both Ozawa and Avrami, and agreed quite well with the theoretical results obtained using the Ziabicki method. Ozawa's treatment (the plot of log$(-\ln(1 - X(t)))$) as a function of log($\beta$) for PHE/PCL 20/80 blend at several temperatures is shown in Figure 2.5. For all temperatures, straight lines confirmed the validity of applying Ozawa's method to PHE/PCL blends. The Avrami exponents were approximately 3, confirming a three dimensional growth process and heterogeneous nucleation.

Nonisothermal crystallization behaviors of poly(trimethylene terephthalate) (PTT)/PEN blends were studied by Run et al. [95]. Plots of log $\beta$ as a function of log $t$ for 20/80 and 40/60 PTT/PEN blends are shown in Figure 2.6. Based on the log $F(T)$ and $b$ calculated by the authors, log $F(T)$ values were increased with the relative crystallinity from 3.81 to 4.04 (for 20/80 PTT/PEN blend) and 3.08 to 3.45 (for 40/60 PTT/PEN blend). This indicated that a lower crystallization rate was needed to reach the given crystallinity within unit time. A small increase in parameter $b$ was observed with increasing the crystallinity, ranging from 1.31 to 1.35 (for 20/80 PTT/PEN blend) and 1.01 to 1.09 (for 40/60 PTT/PEN blend), respectively. The results showed that with increasing PTT content, higher crystallization rate was achieved. The authors also confirmed their results with

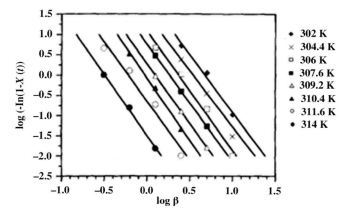

**Figure 2.5** Ozawa plot of log$[-\ln(1 - X(t))]$ as a function of log ($\beta$) for 20/80 PHE/PCL blend at various temperatures. Reproduced from reference [88] by de Juana et al. with permission, © 1996 Elsevier.

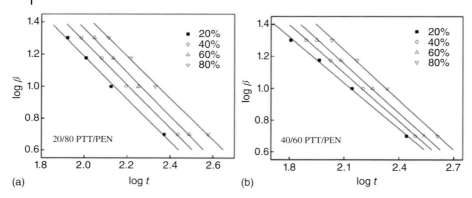

**Figure 2.6** $\log \beta$ as a function of $\log t$ from the equation defined by Liu *et al.* [93] for (a) 20/80 and (b) 40/60 PTT/PEN blends. Reproduced from reference [95] by Run *et al.* with permission, © 2006 Elsevier.

treatments by Avrami and Ozawa. They concluded that the equation proposed by Liu *et al.* [93] successfully describes the nonisothermal crystallization process of the binary blends for the whole crystallization process.

### 2.3.6
### Crystalline/Crystalline Polymer Blends

Crystalline/crystalline polymer blends have received less attention compared to the amorphous/amorphous or amorphous/crystalline systems. Since both components present in the mixture are crystallizable, the melting and crystallization behaviors of such systems are more complicated. The crystallization and morphology of these systems are related to differences in the melting points of each of component [96]. In such polymer blends, the morphology of the final mixture is also affected by thermodynamic and kinetic factors during crystallization. Eutectic or concurrent crystallization (simultaneous crystallization of the components along with fine dispersion of the crystals) is very rare in polymers. However, it may occur in low molecular weight materials. In most polymer cases, the components crystallize successively (separate crystallization), and the crystallization behavior of the first crystallized component is influenced by the other component. Since the component that crystallizes later is still in the melt state, it acts as a diluent for its crystallized partner [97]. In this case, spherulites of the component that crystallizes later will contact with the spherulites of the other component and start to grow within its spherulites. Generally, when the difference between the $T_m$s of two components is large, the two components will crystallize separately. However, when the difference between the $T_m$s of two components is small, they can be crystallized simultaneously [96]. Table 2.4 shows some examples of polymer blends with different types of crystallization.

Qiu *et al.* [96] have studied a miscible poly(vinylidene fluoride) (PVDF)/poly (butylene succinate-*co*-butylene adipate) (PBSA) polymer blend. Since the

**Table 2.4** Types of crystallization in some crystalline/crystalline miscible blends.

| Polymer Blend | Comp. (wt%) | Technique | Type of Crystallization | References |
|---|---|---|---|---|
| PHB/PEO | 0–80 PEO | OM, DSC | Separate | [98] |
| PHB/POM | 0–100 POM | DSC, SAXS | Separate | [99] |
| PEO/PES | 0–100 PEO | OM, SAXS | Separate | [100] |
| PBT/PAr | 0–100 PAr | DSC, WAXD | Separate | [101] |
| PVF$_2$/PBA | 0–100 PBA | OM, WAXD | Separate | [102] |
| PBSU/PEO | 0–100 PEO | OM, DSC | Separate | [103] |
| PES/PVPh | 0–100 PVPh | OM, DSC | Separate | [104] |
| PHB/PLLA | 0–100 PLLA | OM | Simultaneous | [105] |
| PEC/PLLA | 0–100 PLLA | DSC, SAXS | Simultaneous | [106] |
| PEC/PLLA | 20 PLLA | AFM | Simultaneous | [107] |
| PES/PEO | 0–100 PEO | OM | Simultaneous | [108] |
| PEO/PBAS | 0–100 PBAS | DSC, OM | Simultaneous | [109] |
| PED/EVAc | 0–100 EVAc | DSC, WAXD | Simultaneous | [110] |
| LLDPE/VLDPE | 0–100 VLDPE | DSC, SAXS | Simultaneous | [111] |
| PEEK/PEK | 0–100 PEK | DSC | Simultaneous | [112] |

difference in the melting points of the two components was large (165 °C and 95 °C for PVDF and PBSA, respectively), they crystallized separately. PVDF crystallized first, so PBSA which was still in the melt state acted as a diluent during the crystallization of PVDF. The crystallization of the component that crystallizes first (high temperature crystallizing component [HTC]) is a transition from the amorphous/amorphous to the amorphous/crystalline state, while the crystallization of the second component (low temperature crystallizing component [LTC]) is a transition from the amorphous/crystalline to crystalline/crystalline state.

Figure 2.7 illustrates the effect of crystallization temperature and composition on the spherulitic growth rates of PVDF (the component that is first crystallized).

The spherulitic growth rates decrease with increasing crystallization temperature regardless of blend composition. The spherulitic growth rates also decrease with the increasing LTC content at a constant temperature. The change in this growth rate is less significant at high temperatures. This indicates that supercooling plays a key role in the morphology and growth rates of such systems. It can be concluded that factors causing reduction in the spherulitic growth rates of HTC include the depression of its $T_m$ after blending, and the diluent role of LTC during the crystallization of HTC [96].

The spherulitic morphologies of LTC are shown in Figure 2.8. Figure 2.8a shows the spherulites of the neat PBSA crystallized at 70 °C for 20 min. The spherulites are compact, and the lamellae bundles are thick. In the case of the 20/80 blend (Figure 2.8b), PVDF is crystallized first at 120 °C for 5 min. With decreasing the crystallization temperature to 70 °C, the PBSA spherulites start to grow until impinging on other PBSA spherulites. A comparison of these two

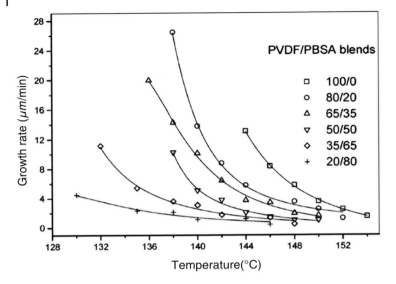

**Figure 2.7** Spherulitic growth rate of PVDF as a function of temperature for the PVDF/PBSA blends. Reproduced from reference [96] by Oiu *et al.* with permission, © 2007 American Chemical Society.

figures indicates that the nucleation of PBSA has increased in the presence of the PVDF, because the pre-existing PVDF crystals increase the nucleation of PBSA. The overall crystallization rate of the blends is larger than that of neat PBSA in this system; however, with increasing PVDF content, the overall crystallization rate decreases.

We can conclude that, in the crystalline/crystalline polymer blends, the overall crystallization of LTC is influenced by the presence of the crystals of HTC, the dilution effect, and physical hindrances in the matrix of HTC. The first

**Figure 2.8** Spherulitic structures of (a) neat PBSA at 70 °C for 20 min, and (b) 20/80 PVDF/PBSA blend at 70 °C for 6 min after crystallization of PVDF at 120 °C for 5 min. Reproduced from reference [96] by Oiu *et al.* with permission, © 2007 American Chemical Society.

parameter accelerates the overall crystallization rate, while the last two parameters slow down the rate.

Isotactic polypropylene (PP)/maleic anhydride grafted polypropylene (mPP) blends have been studied by Cho *et al.* [113]. Based on their results, the cooling rate has a significant effect on the DSC fusion endotherms of the blends. For a blend sample prepared at a low cooling rate such as 1 °C/min, a separate crystallization (phase separation) occurs and the DSC thermographs consist of two melting peaks (Figure 2.9). For the samples prepared at high cooling rates (40 °C/min), crystallization occurs simultaneously and a single melting peak is observed in DSC thermographs. Isothermal DSC measurements show that at a low degree of supercooling (crystallization at higher temperatures), the components crystallize separately, while at a high degree of supercooling, co-crystallization occurs and a single endotherm is observed from DSC data. This is in agreement with results from nonisothermal experiments.

Two characteristic features of eutectic crystallization are simultaneous crystallization of two components and finely dispersed structure of the crystals composed of mixed lamellae or mixed lamellar stacks. The eutectic crystallization of polymer blends is rare, and in the most cases, only one of these two features is fulfilled. In

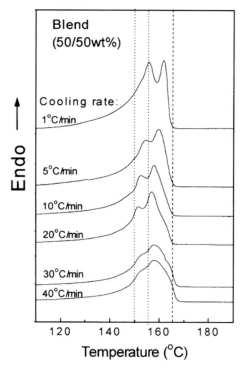

**Figure 2.9** DSC endotherms for 50/50 blend by nonisothermal crystallization at different cooling rates (the heating rate is 10 °C/min). Reproduced from reference [113] by Cho *et al.* with permission, © 1999 Elsevier.

some cases, simultaneous crystallization occurs without achieving the finely dispersed structures of the crystals (pseudoeutectic type I), and in some cases, the second feature (fine dispersion) is fulfilled alone (pseudoeutectic type II) [12,97].

Polymer blends with different crystallization processes are listed by Liu and Jungnickel [97]. A few polymer blends containing two crystallizable components fulfill the requirements of a homogeneously mixed melt for the whole composition range (at different temperatures). These systems can be found in the literature: PVDF/poly(hydroxy butyrate) (PHB) [99,114,115], bisphenol A polycarbonate (PC)/PCL [116], PVDF/PBA [102,117], and PHB/PEO [98]. Some blends such as PHB/PEO [98] and PVDF/PBA [102] show step crystallization when they are cooled from their melt. The melting point difference of the two components is large in these systems. Step crystallization also occurs in polymer blends when the melting points of the components are closer but the crystallizabilities are different (PHB/poly(oxymethylene) (POM) [99]). Simultaneous crystallization (pseudoeutectic II) only occurs in a few blend systems such as poly (butylene succinate) (PBSU)/poly(vinylidene choloride-co-vinyl chloride) (PVDCVC) [118] and poly(epichlorohydrin) (also poly(butylene succinate-co-butylene carbonate)) (PEC)/polylactic acid (PLLA) [106].

Liu and Jungnickel [97] have studied the crystallization of PVDF/PHB blends in which PVDF and PHB are the HTC and LTC components, respectively. As the polymer blend was quenched from the melt state, the PVDF first crystallized and then the crystallization of PHB occurred. Based on the crystallization temperature and the composition, a large variety of spherulitic structures could be obtained for PVDF. These structures change between dendritic (open and irregular) and compact (close and dense) spherulites. At lower crystallization temperatures (135 °C), dendrites with nonresolvable banding were achievable, while at higher crystallization temperatures (150 °C), banded compact spherulites were obtained. Since there are different structures for HTC, the crystallization of LTC depends on the spherulitic type of HTC. The spherulites of PHB (LTC component) can only grow around the banded spherulites of PVDF (HTC component), and there is no change in the PVDF spherulites (Figure 2.10).

Different morphologies are achievable in the presence of dendrite spherulites of PVDF. In compositions lower than 70 wt% of PHB, the spherulites form within the PVDF dendrites (Figure 2.11). There are no separate spherulites for PHB, but it is crystallized with a defined growth front and controlled by the orientation of the fibrils and the interlamellar packing density of the PVDF dendrites (Figure 2.11b). Figure 2.11(c) shows the completion of the crystallization of PHB.

In the case of higher compositions of PHB, such as 20/80 (w/w) of PVDF/PHB inwhich the HTC dendrites are not volume-filling, the PHB spherulites will form in the space between the PVDF dendrites. Figure 2.12a shows the spherulitic morphology of PVDF crystallized at 132 °C. Figure 2.12b indicates that PHB spherulites start to grow and interpenetrate those of PVDF. After reheating of the blend, the PHB spherulites melt in two stages. First, PVDF dendrites start to melt and then the PHB outside the dendrites will be melted. Finally, the PVDF morphology that was present before the PHB crystallization is recovered

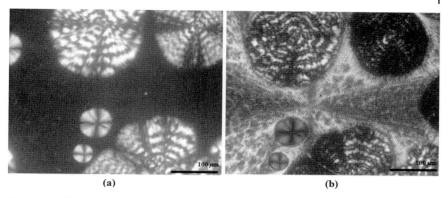

**Figure 2.10** (a) Banded PVDF spherulites crystallized at 150 °C and (b) PHB spherulites grown in the same area at 60 °C in 40/60 PVDF/PHB blends. Reproduced from reference [97] by Liu and Jungnickel with permission, © 2003 John Wily & Sons, Inc.

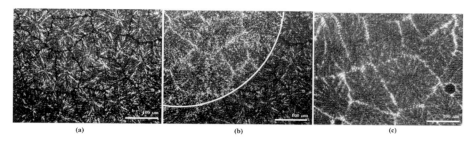

**Figure 2.11** (a) PVDF dendrites crystallized at 135 °C (uncrystallized material between the spherulites consists of almost pure PHB), (b) PHB crystals formed because of further cooling to 60 °C after 10 min (growth front has been indicated by the white line), and (c) completion of PHB crystallization at 60 °C in 40/60 PVDF/PHB blends. Reproduced from reference [97] by Liu and Jungnickel with permission, © 2003 John Wily & Sons, Inc.

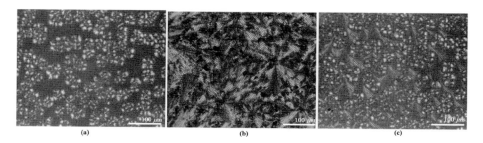

**Figure 2.12** (a) PVDF spherulitic structure crystallized at 132 °C, (b) PHB spherulitic structure crystallized at 60 °C, and (c) PHB spherulitic structure reheated to 120 °C in a 20/80 PVDF/PHB blend. Further heating to 130 °C leads to the recovery of the initial structure. Reproduced from reference [97] by Liu and Jungnickel with permission, © 2003 John Wily & Sons, Inc.

(Figure 2.12c). There is a significant difference between the PHB lamellae inside and outside the PVDF dendrites. The morphology of PHB lamellae inside the PVDF dendrites is dependent on the morphology of PVDF, while the morphology of PHB lamellae outside the interfibrillar regions of PVDF is affected by the crystallization temperature.

## 2.4
## Immiscible Blends

Immiscible blends can also contain two amorphous components, two crystallizable components or one crystallizable and one amorphous component. Many authors have investigated the crystallization of immiscible blends [119–130]. A comprehensive overview of the crystallization behavior of these systems has also been reported in some handbooks [12,16,131].

Although the matrix and dispersed phases are separated, they have a significant influence on each other. Bartczak and Galeski [132] showed that the presence of a second crystallizable phase or an amorphous phase (either as the matrix or the dispersed phase) affects the primary nucleation, crystal growth, and morphology of immiscible polymer blends.

The superstructures of immiscible polymer blends are influenced by the composition, the interfacial tension, rheological characteristics, and processing conditions. Finally, the properties of such blends are influenced by the phase morphology and the crystallinity of the system [12].

The subject of immiscible blends can be divided into two major parts: blends with a crystallizable matrix and an amorphous dispersed phase, and blends with an amorphous matrix and a crystallizable dispersed phase. These topics are discussed in Sections 2.4.1 and 2.4.2. The effect of fillers on the crystallization of immiscible blends is briefly discussed in Section 2.4.3.

### 2.4.1
### Blends with an Amorphous Dispersed Phase in a Crystallizable Matrix

During the melt mixing of two immiscible components, threads of the dispersed phase will form in the matrix; these threads start to break into the droplet-like particles that then coalesce into larger particles. This phenomenon reaches an equilibrium between the breakup and the coalescence of the particles [16,25,133]. The rate of coalescence decreases with cooling of an immiscible polymer blend from its melt state because of an increase in the viscosity of the blend components. The coalescence is completed by the solidification of one of the components [126].

#### 2.4.1.1   Nucleation of the Crystalline Phase
As with nucleation in a crystallizable pure component, nuclei are formed by heterogeneities such as impurities, residual catalysts, and crystalline residues due to incomplete melting in a crystallizable matrix of a polymer blend. According to

the activation energy for the formation of a critical sized nucleus at each value of undercooling $(T_m{}^\circ - T_c)$, different heterogeneities become active and nucleation is induced. Melt annealing, thermal history of the sample, and crystallization temperature change the number of active nuclei for the crystallizable component in an immiscible polymer blend. If the blend is kept at higher temperatures during melting, the residual crystalline parts and the number of nuclei would be less, and consequently, there would be fewer and larger spherulites in the matrix. Since the processes of primary nucleation in both a pure crystallizable polymer and a crystallizable matrix in an immiscible blend are the same, crystallization temperatures of the pure polymer and the blend component are similar. Compared to nucleation of homopolymers, however, the nucleation of a crystallizable polymer in a blend is influenced by migration of impurities during melt mixing and nucleation at the interface between the two phases [134,135].

If the interfacial free energy of the impurity within its melt phase is higher than the interfacial free energy within the second melt phase, it can migrate across the interface between two phases [12,16,136,137]. Mechanical mixing, the phase morphology during mixing, and the interfacial free energy affect this migration of impurities within the two phases [12,136]. If the amount of the amorphous phase is increased, coalescence occurs and the interfacial area decreases. The blend composition, therefore, also affects the migration of nuclei from one phase to another. Nucleation at the interface between the two phases is influenced by the physical state of the amorphous phase, the total amount of interface, and the crystallization temperature. Smaller and finer dispersions of the amorphous droplets lead to higher total amounts of interface and thus higher possibilities of nucleation at the interface. Lower crystallization temperatures also provide higher rates of nucleation at these interfaces [134,138,139].

### 2.4.1.2 Spherulite Growth Rate of the Crystalline Phase

The growth rate of a crystal in a homopolymer is measured according to Eq. (2.9). The spherulitic growth rate for immiscible blends, however, is influenced by the amorphous component and does not follow this equation. In fact, the droplet-like domains of the amorphous phase can be either rejected to the interspherulitic region by the spherulitic growth front or physically trapped within a growing spherulite (occluded). This leads to deformation of the droplet-like domains. A combination of both phenomena is generally observed, and small droplets are rejected over some distance and then deformed by the stacks of lamellae. This rejection and deformation of the droplets disturbs the spherulite growth and creates new energy barriers to be overcome. Equation (2.9) must be modified as shown in Eq. (2.32) in order for it to be applicable to these kinds of blends [12,120,140]:

$$G = G_1 \exp\left[-(E_1 + E_2 + E_3 + E_4)/k_B T_c\right], \tag{2.32}$$

where $G_1$ is the growth rate of the plain crystallizable polymer (Eq. (2.9)), $E_1$ is the energy required for the rejection of the droplets (proportional to the melt viscosity), $E_2$ is the energy required to overcome the inertia of the droplets, $E_3$ is the energy required to form a new interface when the droplets are engulfed, and $E_4$ is the energy dissipated for the deformation of the occluded particles. The equations needed to measure these energies can be found in the literature [120,140]. Measurements by these authors have shown that the rejection of small droplets and the deformation of large engulfed droplets have major contributions to depression of the spherulite growth rate. The effect of an immiscible amorphous component on the spherulite growth rate has been experimentally measured for some polymer pairs and summarized in the literature [25]. In some polymer blends, the addition of a second component produces a measurable contribution to the spherulite growth rate ($G$) and reduces it. Examples of such systems include the blends of isotactic polypropylene (iPP) with poly(dicyclohexylitaconate) (PDCHI) [141], POM with P3HB [142], and iPP with ethylene propylene rubber (EPR) [140,143]. In some cases, however, the energy barriers introduced by the presence of dispersed domains are small and have no effect on the spherulite growth rate as with blends of PHB with EPR [36], PEO with PDCHI [144], PEO with poly(vinyl chloride) PVC [145], and iPP with poly(vinyl butyral) (PVB) [146].

The difference in the interfacial free energy between the crystallizing solid and droplets ($\gamma_{PS}$) and the interfacial free energy between the melt and droplets ($\gamma_{PL}$) is the driving force for rejection, occlusion, or deformation processes [140]. If this difference is positive, the droplet is rejected, and if it is negative, the droplet is occluded [139].

Crystallization temperature and blend composition have significant effects on the spherulite growth rate. The temperature dependence of the spherulitic growth rate can be calculated by the following equation [147–149]:

$$G(T) = G^0 \exp\left[-C_1 C_2 / (C_2 + T - T_g)\right] \exp\left[-C_3 / T\left(T_m^0 - T\right)\right]; \quad T_g - C_2 < T < T_m^0,$$

(2.33)

where $T_g$ is the glass transition temperature, $T_m^0$ is the equilibrium melting temperature, and $G^0$, $C_1$, $C_2$, $C_3$ are the parameters describing the growth rate behavior. For example for polypropylene, $C_1$ and $C_2$ (WLF constants) are 25 and 30 K from the literature [150]. The value for $G(T)$ will be zero when $T \le T_g - C_2$ or $T \ge T_m^0$. The values of $T_g$ and $T_m^0$ can be measured for a pure crystallizable polymer. $G^0$ and $C_3$, which are strongly dependent on the melting temperature, can be measured by plotting the value $\ln G + C_1 C_2 / (C_2 + T - T_g)$ versus $1/[T (T_m^0 - T)]$. With decreasing blend composition (lower content of amorphous phase), finer droplets can be achieved. This leads to more rejection of droplets and higher energy consumption, which reduces the spherulite growth rate. A theoretical prediction of the dependence of growth rate on particle size, crystallization temperature, and the content of the amorphous component is shown in Figure 2.13. The growth rate is considerably depressed at higher $T_c$. At high crystallization temperatures, the influence of rejection of the dispersed domains is limited because at higher temperatures, the melt viscosity is lower, leading to

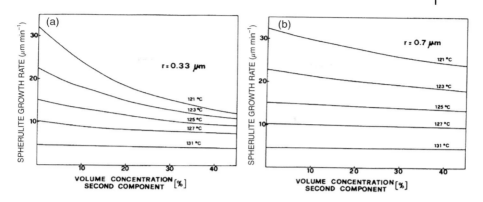

**Figure 2.13** Effect of particle size, crystallization temperature and the content of the second component on the spherulite growth rate depression of isotactic polypropylene: (a) r = 0.33 μm; (b) r = 0.70 μm. Reproduced from reference [140] by Martuscelli with permission, © 1984 John Wily & Sons, Inc.

easier rejection of small droplets and less energy consumption. Moreover, the size of the domains has a significant effect on G. A finer dispersion of the second component produces a stronger decrease of the growth rate [140].

Galeski *et al.* [151] investigated the spherulite growth rate of an iPP/low density polyethylene (LDPE) blend, at crystallization temperatures where the LDPE did not crystallize. They found that after passing the crystallization growth front, the LDPE droplets that had been occluded within the growing spherulites were not deformed and remained undisturbed. The LDPE droplets acted as obstacles to the spherulite growth front and facilitated the formation of concavities as shown in Figure 2.14. The shape of the concavities resulted from creation of a boundary beyond the occlusion.

### 2.4.1.3 Overall Crystallization Kinetics

The overall crystallization rate of an immiscible polymer, which is a combination of nucleation and spherulite growth rate, can be either depressed or increased with increasing content of the amorphous phase. In the work reported by Martuscelli [140], the spherulite growth rate of the PP is not affected with increasing LDPE amorphous polymer content. The nuclei density, however, decreases with increasing LDPE content because of the migration of impurity from PP to the LDPE amorphous phase [12]. The overall rate of crystallization of the PP matrix was, therefore, reduced with increasing content of LDPE. The reciprocal of $t_{1/2}$ (half-time of PHBV crystallization) can be plotted as a function of $T_c$ for different concentrations of ENR amorphous phase (Figure 2.15). This illustrates that the overall crystallization rate was increased with increasing content of the amorphous phase [152].

Nucleation density, spherulite growth rate, and blend composition are three important parameters that determine the final crystalline morphology of

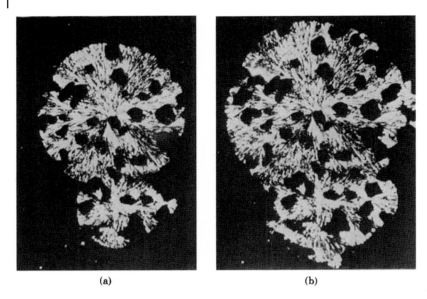

**(a)**     **(b)**

**Figure 2.14** iPP spherulites in a 50/50 iPP/ LDPE blend at $T_c = 135\,°C$ after: (a) $t = 135\,min$, (b) $t = 165\,min$. The dark area is the molten LDPE occlusions surrounded by iPP crystalline lamellae. Reproduced from reference [151] by Galeski *et al.* with permission, © 1984 John Wily & Sons, Inc.

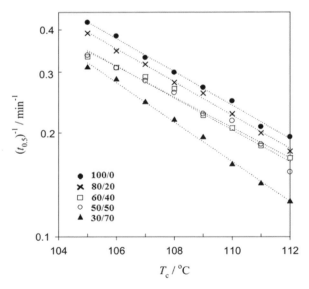

**Figure 2.15** Rate of crystallization of PHBV in PHBV/ENR blends as a function of crystallization temperature at different blend compositions. Reproduced from reference [152] by Han *et al.* with permission, © 2011 International Journal of Pharmacy and Pharmaceutical Sciences.

immiscible blends. The migration of nuclei from one phase to another or nucleating activity at the interphase between the two phases can cause increases in the nucleation density and the formation of more and smaller spherulites. The spherulite growth rate also affects the final morphology. At high spherulite growth rates, the droplets are rejected into new boundaries behind occluded particles and deformed, while at low growth rates, the droplets are rejected very fast and most of them reside in the interspherulitic regions. Higher amorphous phase content results in the coalescence and formation of a coarse melt with large droplets. These large droplets will be deformed by the growth front.

### 2.4.1.4 Glass Transition of the Amorphous Component and Melting Behavior of the Crystalline Matrix in Immiscible Polymer Blends

Although the $T_g$s of the components move toward each other in miscible blends, the composition of an immiscible blend does not have an important effect on the $T_g$ of the components. They are, therefore, expected to be close to their bulk value. The $T_g$ values of components shift in a blend when the thermal expansion coefficients of the soft and the hard phases of the blend are very different. This phenomena has been observed for polystyrene (PS)/polybutadiene (PB) blends [153] and PET blends containing a rigid PC phase [154].

Figure 2.16 illustrates the change in $T_g$ of PS as a function of its composition in the PS/PP blends. The $T_g$ values increase with decreasing composition of PS in the blend. Similar behaviors were observed for poly(methyl methacrylate) (PMMA)/PP and PS/polyethylene (PE) blends [122,155]. Because of the lack of

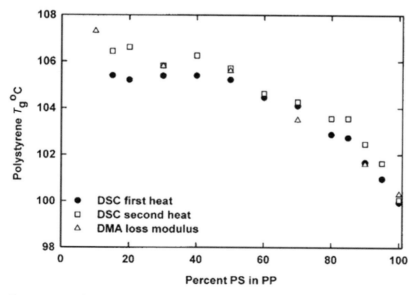

**Figure 2.16** $T_g$ of PS as a function of composition in PS/PP blends. Reproduced from reference [155] by Thirtha *et al.* with permission, © 2005 John Wily & Sons, Inc.

miscibility between PS and PP, the $T_g$ of PS moves to higher values and not to the $T_g$ value of PP, which is lower than that of PS. The difference between the thermal expansion coefficients of the components results in the high $T_g$ values of PS at lower percentages. At low PS concentrations, compressive stress generated by the contracting PS matrix is the significant mechanism in raising the $T_g$ value of PS [155].

Since the droplets are separated and dispersed in the molten crystallizable matrix, its melting temperature is almost constant [140]. If blend composition has no effect on the crystallization behavior (nucleating activity, spherulite growth rate, etc.) of the matrix, melting temperature does not change significantly [12]. A slight reduction of melting point may, however, occur if there is an interaction between the amorphous phase and the crystalline matrix [156].

### 2.4.2
### Blends with a Crystallizable Dispersed Phase in an Amorphous Matrix

If the content of crystalline polymer is lower than that of the amorphous component in an immiscible polymer blend, the crystallizable component usually forms the dispersed phase. Some authors have shown that crystallization of the dispersed phase in these systems occurs in several steps (fractionated crystallization) when the number density of dispersed particles is greater than that of heterogeneities ($N*$), which act as nucleation agents for the pure crystallizable component [121,123,130,157–160]. If $N* \leq 1/V_d$ ($V_d$ is the average volume of dispersed droplets), fractionated crystallization occurs. Alternatively, if $N* > 1/V_d$, the crystallization behavior of the crystallizable phase in the blend is like that of the pure component [16].

#### 2.4.2.1 Fractionated Crystallization
In homopolymers, crystallization starts with primary heterogeneous nucleation (impurities) and then it continues to grow with secondary nucleation. The crystallization will generally be completed before activation of other heterogeneities, which require higher degrees of undercooling. During the cooling, different types of heterogeneities become active for nucleation at each value of undercooling ($T_m^0 - T_c$). For this reason, DSC cooling thermograms of such systems exhibit single crystallization exotherms. In the case of polymer blends containing fine droplets of the crystallizable phase, crystallization can take place in several steps, which are initiated at different undercoolings. Multiple crystallization exotherms are observed in the DSC thermograms of these polymer blends. The lowest value of crystallization exotherm that appears at the homogeneous crystallization temperature is related to homogeneous nucleation [16,157–160]. The crystallizable phase droplets in a polymer blend crystallize at larger undercoolings than the pure crystallizable component [12]. It has been reported that fractionated crystallization behavior depends on the blend morphology (size of the droplets) and the thermal history of the samples (heterogeneous nucleation density) [160,161].

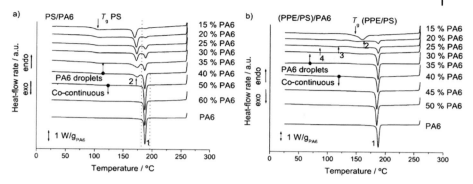

**Figure 2.17** DSC thermograms for (a) PS/PA6; and (b) (PPE/PS)/PA6 blends at various compositions (cooling at 10 K/min). Reproduced from reference [160] by Tol et al. with permission, © 2005 Elsevier.

There is a shift in the heterogeneous nucleation spectrum to larger undercooling for premelted crystallizable dispersed phase at higher temperatures after longer annealing times. Having smaller dispersed droplets also leads to crystallization at larger undercooling. The significant effect of the size distribution of droplets on the fractionated crystallization behavior of polymer blends makes the role of compatibilizers more important. Since the droplet size is reduced in the presence of compatibilizers, this can increase the fractionated crystallization and shift the crystallization temperature to lower temperatures. The DSC cooling curves of PS/polyamide 6 (PA6) and (PPE/PS)/PA6 blends with a (PPE/PS) ratio of 50/50 wt/wt as a function of the blend composition are shown in Figure 2.17. According to the thermograms, the PA6 crystallization is strongly dependent on the blend composition in both blend series. At high PA6 concentrations where PA6 constitutes the matrix, there are no significant changes in the crystallization behavior. At lower PA6 concentrations, where PA6 droplets are dispersed in the amorphous matrix, a fractionated crystallization behavior is observed in both blend series (multiple crystallization peaks at different degrees of undercooling). Tol et al. [160] showed that when PA6 droplet size is relatively large, at higher concentrations of PA6, most of the droplets crystallize around the PA6 bulk temperature, identified by "1" in the figures. With increasing numbers of PA6 droplets per unit volume, at lower concentrations, most of the droplets crystallize at a higher degree of undercooling. This is because the activation of some heterogeneous nucleation occurs at higher undercooling. The fractionated crystallization behavior, however, is different for the two blend series. The crystallization peak at a very large degree of undercooling corresponds to homogeneous nucleation of PA6 due to the exhaustion of all heterogeneous nuclei.

### 2.4.2.2 Determination of the Number Density of Heterogeneities
The number density of heterogeneities can be determined from the fractionated crystallization behavior. If the distribution of heterogeneities over the droplet

population follows a Poisson distribution (based on the approximation of Pound and LaMer [162] for the distribution of heterogeneities for monodisperse tin droplets), the fraction of droplets that contain exactly $z$ heterogeneities of type $i$ (inducing normal crystallization in the bulk polymer at $T_c^0$) can be defined as [162]

$$f_z^i = [(M^i V_d)^z / z!] \exp(-M^i V_d), \tag{2.34}$$

where $M^i$ is the concentration of heterogeneities of type $i$, and $M^i V_d$ is the mean number of heterogeneities of type $i$ per droplet with volume $V_d$. In Eq. (2.34), a large number of small polymer droplets, each having a volume $V_d$ is considered. The fraction of droplets that contain at least one heterogeneity of type $i$ is defined by

$$f_{z>0}^i = 1 - \exp(-M^i V_d) \tag{2.35}$$

or

$$M^i = -[\ln(1 - f_{z>0}^i)] / V_d. \tag{2.36}$$

This fraction can be calculated from the relative partial area of each crystallization exotherm during cooling in the DSC. If the mean size of the droplets is known, calculations can be done with respect to the concentration of the respective heterogeneities based on the assumption that one nucleus is sufficient to crystallize the whole droplet [157,160]. When the usual crystallization from the heterogeneities of type 1, which have the lowest value of activation energy, is totally suppressed, the equation will be simplified to

$$M^1 V_d \ll 1. \tag{2.37}$$

### 2.4.3
### Effect of the Fillers on the Crystallization of Immiscible Polymer Blends

As an example of an immiscible polymer blend, iPP/high density polyethylene (HDPE) has been studied by Zhang *et al.* [163]. These two components crystallized separately into a two-phase system. Short carbon fiber fillers, added to the polymer blend, were located in their favorable phase, based on the affinity of the filler to each component. In the iPP/HDPE blend, vapor grown carbon fibers (VGCF) preferred to reside in the HDPE phase. Since fillers change the viscosity and perturb the kinetics of the phase coalescence during mixing, the addition of filler affects the morphology of the blends. The driving force for the migration of filler between the components in the blends is the difference between the interfacial free energy of filler surrounded by one component of the melt and that of filler surrounded by the second phase [137,164,165].

Nonisothermal crystallization behaviors of such systems show that at very low cooling rates (0.5 and 1 °C/min), two crystallization peaks appear on the DSC thermograms, while with increasing cooling rate, overlapping of two crystallization peaks occurs. Comparing the DSC crystallization exotherms of unfilled blends and those of filled blends, at a constant cooling rate, indicates that with increasing composition of the favorable phase, fillers show two opposite effects on the crystallization of such systems. In the presence of a favorable phase for the fillers, the number of heterogeneous nuclei in the unfavorable phase decreases. This reduces the crystallization of the unfavorable phase (a decrease in the crystallization temperature occurs). Alternatively, with increasing the content of the favorable phase, the probability of the migration of the fillers from the favorable phase to the unfavorable one increases. Based on the size of the filler and the domain size of the favorable phase, some fillers near the boundaries of two phases, however, penetrate into the unfavorable phase. The favorable phase crystals growing close to the surface of the nucleating agents (fillers), therefore, could act as nucleating agents for the unfavorable phase. This, alternatively, increases the crystallization of the unfavorable phase (an increase in the crystallization temperature occurs). The overall crystallization behavior of the unfavorable phase is, therefore, affected by these two opposite tendencies in blends containing fillers. In such systems, the interface of the two phases has a key role in the crystallization behavior of the blends.

## 2.5
## Compatibilized Polymer Blends

The process of interface modification by the formation of chemical or physical bonds between the polymers to suppress phase separation and improve adhesion is called compatibilization. The components, which create these bonds, are identified as compatibilizers [2,14,166,167]. A strict definition of "compatibility" in a thermodynamic sense would mean that the polymer blends are miscible on a molecular scale. Another definition for compatibility is "technological compatibility," which means the blend would not be separated into its components and results in a desirable set of properties [168,169]. The three principle aspects of compatibilization are: reduction of the interfacial tension to improve the phase dispersion, stabilization of morphology during the high stress and strain processing, and improvement of adhesion between phases in the solid state [168]. Two general methods have been reported for compatibilization of immiscible polymers: addition of blocks or graft copolymers and reactive blending or reactive compatibilization [167].

### 2.5.1
### Addition of Blocks or Graft Copolymers

The addition of (nonreactive) block or graft copolymers is one method found to make the individual components compatible with each other in polymer blends. Each segment of a block or a graft copolymer is miscible with its corresponding polymer component and shows a tendency to be localized at the interface between immiscible blend phases. This reduces interfacial tension and stabilizes dispersion against coalescence [167–175]. The achievement of a preferred local- ization at the interface between the components is related to the molecular weight and chain structure (type, number, and molecular parameters) of the copolymer segments [2,16,170,173,176]. Although random copolymers also have been used as compatibilizers, they are not able to stabilize the dispersion very well [167,177].

Smith *et al.* [178] have used a cryogenic mechanical alloying to incorporate a poly(methyl methacrylate-b-isoprene) (MI) diblock copolymer into blends of PMMA and polyisoprene (PI). They investigated the effects of mechanical mill- ing time and copolymer concentration on the morphology and impact strength of the blends. They found that the characteristic size scale of the minor phase decreases with the increasing copolymer content, as well as milling time. The compatibilization of LDPE/PS blends by adding graft (PS-g-LDPE) and block copolymers has been reported by various authors, indicating the improvement of interfacial adhesion and mechanical properties [179,180]. Cao *et al.* [181] recently showed that graphene oxide sheets (GOSs) can be used as compatibil- izers for immiscible PA/PPO blends. With the addition of GOSs, the droplet diameter of the dispersed minor phase (PPO) is reduced by more than 1 order of magnitude. Since PPO can be adsorbed on GOSs basal planes and PA can be strongly hydrogen bonded through their edge-located oxygen functional groups, they act as coupling agents for the two components.

Some commercial copolymers that have been used in this area of research include styrene butadiene block copolymers and their styrene-hydrogenated butadiene analogues for compatibilization of styrene polymers such as PS, high impact polystyrene (HIPS), styrene-acrylonitrile (SAN), acrylonitrile butadiene styrene (ABS) with polyolefins [182], ethylene–propylene copolymers for compa- tibilization of different polyolefins [167,183], polyester amide elastomer, and maleated PP or EPR (Dexcarb®) for the compatibilization of PA with PC [184,185], and hydrogenated styrene-b-butadiene-b-styrene triblock copolymer (SEBS) for the compatibilization of PP/PC blends [186,187].

This method has some practical limitations. Obtaining desired copolymer for the compatibilization of each immiscible polymer pair requires a specific syn- thetic procedure; therefore, this method is usually expensive, and sometimes no feasible technology can be readily found. In addition, an excess amount of copolymer is usually required because of the possibility of it being trapped in the bulk phase during blending and remaining far from the interface [167].

2.5.2
**Reactive Compatibilization**

The most commonly used method of compatibilization is reactive compatibiliza- tion in which a specific heterogeneous chemical reaction occurs between two polymeric components during melt blending. This results in an *in situ* formation of (graft, block, or random) interchain copolymers acting as compatibilizers, without addition of other polymer components. These compatibilizers are sometimes produced through chain cleavage of polymers and recombination during reactive processing in the melt. They also can be formed through mechanical scission and recombination under high shear processing [16,186]. Copolymers, formed at the interfaces, link the immiscible phases by covalent or ionic bonds and consequently improve adhesion and reduce the size of the dispersed phase [167]. Generally, functional groups such as amino, carboxyl, and hydroxyl end groups can be connected to the functional groups (cyclic anhydrides, carboxyl acids, epoxides, oxazolines, and isocyanates) of polyolefins, elastomers, and related copolymers by using covalent bonds. Block and random interchain copolymers are usually produced during interchange reactions in the melt between polycarbonate, polyesters, and polyamides. In some cases, low molecular weight compounds have been used to facilitate the reactive compatibilization of polymers. Combinations of peroxide with oligomer coagents have been used for preparation of PE/PP blends. Bis-maleic imide has been applied to prepare PE/PBT blend. Radical-initiated reactions of monomers, forming homopolymers and grafts on the chains of dissolved polymers, have been used for manufacture of some important polymers such as HIPS and ABS [16,167,188–192].

Reactive compatibilization has some advantages compared to the previous methods since the copolymer is formed at the interface between the two immiscible polymers. In the method of addition of copolymers, compatibilizers must diffuse to the interface to improve morphology stabilization. Another advantage is that the molecular weight of the formed copolymer is similar to that of the bulk phases in which the segments need to be dissolved [169,193,194].

### 2.5.2.1 Reactive Compatibilization in Bio-based Polymer Blends

In recent years, the production and use of sustainable polymer systems has continuously increased because of concerns raised by global warming and the depletion of oil reserves. Most biopolymers, however, have less desirable properties compared to the commodity polymers. Blending is one modification method used to improve the properties (especially resistance to brittleness) of such materials. Various processes, such as copolymerization, grafting, transesterification, and use of reactive coupling agents, have been applied to improve the properties of bio-based polymer blends [195,196]. Since simultaneous improvements of deformability and strength are seldom achieved by physical compatibilization [195,197,198], and the chemical structures of biopolymers make the reactive modification of these materials easier, reactive compatibilization methods are often used for such systems [195,199–202]. Hassouna *et al.* [199] improved the compatibility of PEG/PLA blends by using maleic anhydride (MA) grafted PLA and hydroxyl terminated PEG. Epoxy [203] and isocyanate [204] functional groups have been used in natural rubber/PLA and starch/PBSU bio-based polymer blend systems, respectively. Radical reaction routes using peroxide initiators have been applied to attach unsaturated anhydrides, such as maleic anhydride to biopolymers. This method is suitable for the modification of different polyesters [195].

2.5.3
**Crystallization of Compatibilized Blends**

The addition of a copolymer as a compatibilizer may either decrease or increase the tendency for crystallization. Although the addition of a compatibilizer increases the interfacial area, and consequently, the nucleation rate, it also increases the thickness of the interphase and hinders the diffusion of nucleating agent to the crystallizable phase. If the size of crystallizable polymer domains is below a certain limit, the crystallization rate will be reduced. This leads to fractionated crystallization in the case of blends containing wide variations of droplet sizes [157,168]. Pracella *et al.* [205] have studied the crystallization behavior of PET/HDPE blends compatibilized by the addition of various functionalized polyolefins (EPR-*g*-MA, HDPE-*g*-MA, and E-AA, E-GMA). They showed that crystallization temperature of PET in the blend shifted to temperatures lower than those of unblended PET as well as the noncompatibilized blends. In addition, the crystallinity levels of the blends are dependent on the type and concentration of compatibilizer. Because of the chemical reactions of functional groups with PET and compatibilization of two phases, the crystallinities of both PET and HDPE components were decreased upon addition of compatibilizers. Reduction in the mobility of the chains because of interchain grafting reactions affected the crystallization of PET molecules close to the interface and decreased the crystallization temperature [16]. Polypropylene and polyamide blends (composition 70/30) have been studied with and without the addition of maleic anhydride functionalized polypropylene at different loading. The isothermal crystallization rate of the PA component was reduced in the presence of PP due to the diluent effect of PP. The crystallization rate of PP in the blends is, however, higher than that of pure PP. This effect is higher in the blends without compatibilizer because of the nucleating activity of the PA component [206]. Blends of polyamides and polyolefins modified with different compatibilizers such as maleic anhydride, acrylic acid, and diethylmaleate have also been investigated by other authors [207–209]. The effects of interfacial compatibilization of maleated thermoplastic elastomer (TPEg) on the crystallization behaviors of the PP/PA6 blends were investigated by Liu *et al.* [210]. Crystallization behaviors of the individual components (especially those of PA6) in the blends were significantly influenced by the addition of TPEg. The presence of TPEg reduced the nucleating effect of the PA6 on the PP matrix. Crystallization temperature and crystallization enthalpy associated with PA6 decreased with increasing concentration of the compatibilizer. For TPEg amounts, up to 24 wt%, concurrent crystallization of PA6 and PP occurred. The addition of compatibilizer reduced the PA6 particle sizes and led to the lack of active heterogeneities. The crystallization behavior of 20/80 blends of polyamide 6 (PA6) and acrylonitrile butadiene styrene copolymer (ABS) in the presence of styrene maleic anhydride copolymer (SMA) and multiwall carbon nanotubes (MWNT) were investigated by Bose *et al.* [211]. The fractionated crystallization of the PA6 phase was observed in this case. It was also found that the fractionated crystallization was influenced by the presence of both SMA and MWNT. Wide domain size distribution of PA6 droplets in the amorphous ABS matrix was observed. The formation of less perfect crystallites formed at higher degree of supercooling during cooling from the melt led to the formation of multiple lamellae stacking of the PA6 phase in the presence or absence of SMA and MWNT in 20/80 PA6/ABS blends.

Blends of Nylon 6 (Ny6) with acrylic acid-modified isotactic polypropylene (PP-AA) and modified polyethylene (PE-AA) have been compared to blends of Ny6 with PP and LDPE by Psarski *et al.* [212]. Separated crystallization peaks were observed in the DSC thermograms of incompatible Ny6/PP and Ny6/LDPE blends, whereas the compatibilized blends exhibited fractionated and/or coincident crystallization. The authors have reported that the weight fraction of $\gamma$ form crystals monotonically increased with decreasing concentration of Ny6 in the dispersed phase, while the fraction of $\alpha$ form crystals decreased. No significant change in the fraction of the two crystalline modifications were, however, observed for blends in the same composition range without compatibilizer. Fractionated crystallization of the PA led to the shift of the crystallization of Ny6 to lower temperatures. Since the $\gamma$ phase is more stable at lower temperatures, it is more favored than $\alpha$ phase [16,212]. Bio-based PTT/PEEA blends with different weight ratios of ionomers, such as lithium-neutralized poly(ethylene-*co*-methacrylic acid) copolymer (EMAA-Li) and sodium-neutralized poly(ethylene-*co*-methacrylic acid) copolymer (EMAA-Na), were prepared by Kobayashi *et al.* [213]. They showed that the crystallization rate for PTT increased, significantly, by using EMAA-Na. The crystallization rate, however, did not increase by using EMAA-Li. Interfacial segregation of PEEA to domains of EMAA-Li and EMAA-Na was observed at the center of the molded samples. Ionomer domains showed core–shell morphology. They were partially or completely encapsulated by PEEA.

### 2.5.3.1 Differences Between the Crystallization Behaviors of Polymer Blends and Copolymers

There are two distinct differences between the phase diagram of a block copolymer and the one obtained for a mixture of two homopolymers. In block copolymers, owing to the chemical links between blocks, micro-domains instead of macroscopic phases are observed. The size of micro-domains can be controlled by varying the molecular weight and composition. Furthermore, since the type of morphology depends on the concentration as well as on the transition temperatures of the individual phases, the phase diagram of block copolymers shows comparable complexity to those of metallic alloys [168].

There are significant differences between the crystallization behaviors of polymer blends and copolymers. To illustrate this, a PET/PEN system can be considered. Blends, prepared from PET and PEN, are able to crystallize over a broader compositional range than copolymers with equivalent compositions, as shown in Figure 2.18 [22,214]. It must be noted that blends such as these, prepared through reactive extrusion, have undergone *in situ* ester-interchange reactions [215]. Random PET/PEN copolymers have been formed to act as

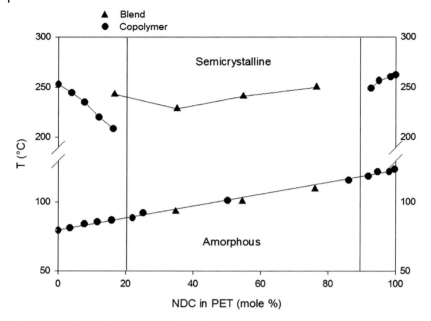

**Figure 2.18** Thermal Properties of PET/PEN blends and copolymers. Adapted from reference [214].

compatibilizers, improve miscibility, and blend clarity. The blend components and their molecular weights, as well as time, temperature and level of mixing all control the extent of the ester-interchange reactions and thus blend crystallization behavior. For applications that require some level of crystallinity in order to have the desired mechanical and barrier properties [214], PET/PEN blends would be superior to similar copolymer compositions. As shown in Figure 2.18 (in the composition range of 15–85% of monomer NDC), PET and PEN blends are crystallizable, while the copolymer is amorphous in this region.

2.5.4
**The role of Transesterification on the Miscibility and Morphology of Polyester Blends**

When polyesters are melt blended together, the end groups react with each other through a reaction called transesterification. This reaction creates block copolymers that act as a compatibilizer between the two phases, improving miscibility of the blends [57,216–219].

Transesterification can increase the miscibility of crystalline/crystalline polyester blends and, consequently, bring down the particle size of the dispersed phase to obtain a transparent polymer.

Kimura and Porter [220], and Devaux *et al.* [221] studied the transesterification reaction in PET/PC blends and found that the blends were first converted to the block copolymers. With increasing reaction time, completely random copolymers were formed.

In PET/PEN blends, as the transesterification continues, the blends become more miscible [222]. Patcheak and Jabarin [57,223] have applied solid state polymerization (SSP; a method widely used for increasing the relatively low molecular weight melt polymerized PET to a higher value required for bottle-grade PET resin) to the PEN/PET blends to obtain blends with both higher molecular weights and improved miscibility between these two polymers. Their results also showed that quiescent crystallization of blends led to independent crystallization of the major component while the minor component was rejected into the amorphous phase. When the two components had similar fractions (such as 40% PEN/PET blends) both phases crystallized independently. They also showed that PET and PEN co-crystallized in the case of strain induced crystallization of blends. In this case, it was assumed that the dominant morphologies for PET, PEN, and the blends were interlamellar. PET and PEN both crystallized in a spherulitic superstructure, shown by optical microscopy to be volume filling, thus indicating that little if any interspherulitic morphology existed. In addition, the benzene and naphthalene rings of the PET and PEN, respectively, stiffen the polymer backbone, thus slowing diffusion. Finally, the transesterification reaction makes it less likely that the minor component could segregate out and form clusters around the spherulites [57].

## 2.6
## Summary and Conclusions

Physical and mechanical properties of polymer blends are influenced by the crystallization and melting characteristics of these materials. An understanding of the crystallization process and kinetics is, therefore, important in order to analyze the relationship between structure and properties of such systems. The addition of a second component to the pure polymer affects the crystallization and melting behavior of both polymer phases. These processes are complicated and influenced by parameters, such as the structures and molecular weights of both components, intermolecular interactions between them, blend compositions, equilibrium thermodynamics, and melt rheology. Miscibility of polymer blends and crystallization and melting behavior of miscible and immiscible polymer blends, in which either one or both components are crystallizable, have been presented. The addition of compatibilizers and reactive compatibilization of polymer blends were also reviewed.

The crystallization and melting behavior of polymer blends are significantly influenced by the miscibility of the components. A depression of the equilibrium melting point is often observed in the case of miscible polymer blends because of favorable interactions between the components. Different melting behavior, however, can be observed because of the occurrence of recrystallization or demixing phenomena. In the case of crystallizable/amorphous polymer blends, the amorphous component is rejected during crystallization of the crystallizable phase. This segregation can occur into an interspherulitic or intraspherulitic region. In the latter case, the segregation can be either interlamellar or interfibrillar. The specific type of segregation depends on the relative values of the diffusion rate of the amorphous component and the crystallization rate of the crystallizable one. In partially miscible polymer blends, two crystallization and demixing phenomena affect the phase transitions and the morphology, depending on the composition of the crystallizable component. The Spherulite growth rate consists of both kinetic and the thermodynamic terms. It is related to the concentration of crystallizable component, changes in $T_g$ and melting point, interactions between the two components, and the diffusion of both components. Overall crystallization kinetics (considering both the effect of primary nucleation and growth) can be analyzed by the Avrami model. Theories, derived using simplified assumptions, have been developed for real processing conditions, such as nonisothermal processes. Different approaches, such as those of the Ozawa and Lu models, were introduced to analyze the overall nonisothermal crystallization kinetics of polymer blends. In crystallizable/crystallizable polymer blends, separate crystallization, concurrent crystallization, or co-crystallization can occur.

Although the matrix and dispersed phases are separated in immiscible polymer blends, the presence of a second component affects the primary nucleation, crystal growth, and morphology of these systems. In blends with an amorphous dispersed phase in a crystallizable matrix, nucleation of the crystallizable polymer is affected by migration of impurities during melt mixing, and nucleation at the interface between the two phases. The rejection of droplet domains to the interspherulitic region by the spherulitic growth front can occur along with occlusion by the growing lamellar stacks in the intraspherulitic regions. The rejection of small droplets and the deformation of large engulfed droplets show major contributions to depression of the spherulite growth rate in these systems. The overall crystallization rate of an immiscible polymer, which is a combination of nucleation and spherulite growth rate, can be either depressed or increased with increasing content of the amorphous phase. Fractionated crystallization occurs in the case of blends with a crystallizable dispersed phase in an amorphous matrix, when the number density of dispersed particles is greater than that of heterogeneities. Fractionated crystallization behavior depends on the blend morphology and the thermal history of the samples.

The addition of a block or graft copolymers and reactive compatibilization are two methods for compatibilization of immiscible polymer blends. Compatibilizers may either decrease or increase the tendency for crystallization of polymer blends. Finely dispersed crystallizable components in the matrix, in the presence of a compatibilizer, result in the fractionated or coincident crystallization, depending on the primary nucleation processes of the dispersed polymer phase. Most biopolymers show less desirable properties compared to the commodity polymers. Blending is one of the modification methods used to improve the properties of these materials. Although biopolymers and their blends have recently been used to some extent, further study is necessary to understand the relationships between crystallization behaviors and properties of these materials, in order to obtain the greatest benefit of blending.

Recently, polymer blends containing liquid-crystalline components have received increasing attention. The addition of both low and high molar weight liquid crystal compounds to the conventional polymer blends creates new materials that can be used in advanced technological applications. The comprehensive understanding and development of theoretical aspects and properties of these polymer blends constitutes an interesting research area.

## 2.7
## Nomenclature

### 2.7.1
### Abbreviations

| | |
|---|---|
| ABS | acrylonitrile butadiene styrene |
| aPP | atactic polypropylene |
| aPMMA | atactic poly(methyl methacrylate) |
| CE | cellulose ester |
| CPE | chlorinated polyethylene |
| EAA | ethylene-*co*-acrylic acid copolymer |
| EGMA | ethylene-*co*-glycidyl methacrylate copolymer |
| EMAA-Li | lithium-neutralized poly(ethylene-*co*-methacrylic acid) copolymer |
| EMAA-Na | sodium-neutralized poly(ethylene-*co*-methacrylic acid) copolymer |
| ENR | epoxidized natural rubber |
| EPR | ethylene propylene rubber |
| EPR-*g*-MA | EPR grafted with maleic anhydride |
| EVAc | poly(ethylene-co-vinyl acetate) |
| GOSs | graphene oxide sheets |
| HDPE | high density polyethylene |
| HDPE-*g*-GMA | HDPE grafted with glycidyl methacrylate |
| HIPS | high impact polystyrene |
| iPHB | isotactic poly(hydroxy butyrate) |
| iPS | isotactic polystyrene |
| LDPE | low density polyethylene |
| LLDPE | liner low density polyethylene |
| MA | maleic anhydride |
| mPP | maleic anhydride grafted polypropylene |
| MWNT | multiwall carbon nanotubes |
| NDC | 2,6-napthalene dicarboxylate |
| Ny6 | Nylon 6 |

| | |
|---|---|
| PA | polyamide |
| Par | polyarylate |
| PAS | poly(acetoxy styrene) |
| PB | polybutadiene |
| PBA | poly(butyl acrylate) (also poly(butylene adipate) |
| PBAS | poly(butylene adipate-*co*-butylene succinate) |
| PBO | poly(*p*-phenylene benzobisoxazole) |
| PBSA | poly(butylene succinate-*co*-butylene adipate) |
| PBSU | poly(butylene succinate) |
| PBT | poly(butylene terephthalate) |
| PC | bisphenol A polycarbonate |
| PCL | poly-$\varepsilon$-caprolactone |
| PDCHI | poly(dicyclohexylitaconate) |
| PE | polyethylene |
| PE-AA | acid-modified polyethylene |
| PEC | poly(epichlorohydrin) (also poly(butylene succinate-co-butylene carbonate)) |
| PED | n-docosyl ester terminated poly(ethylene glycol) |
| PEEA | poly(ether esteramide) |
| PEEK | poly(ether ketone) |
| PEG | poly(ethylene glycol) |
| PEI | poly(ether imide) |
| PEK | poly(ether ketone) |
| PEMA | poly(ethyl methacrylate) |
| PEN | poly(ethylene naphthalate) |
| PEO | poly(ethylene oxide) |
| PES | poly(ethylene succinate) |
| PET | poly(ethylene terephthalate) |
| PHE | poly(hydroxy ether of bisphenol-A) |
| PHB | poly(hydroxy butyrate) |
| PHBV | poly(*R*-3-hydroxybutyrate-*co*-*R*-3-hydroxyvalerate) |
| Phenoxy | poly(hydroxyl ether of bisphenol A) |
| PI | polyisoprene |
| Cis-PI | *cis*-l,4-polyisoprene |
| PLLA | polylactic acid |
| PMMA | poly(methyl methacrylate) |
| POM | poly(oxymethylene) |
| PP | isotactic polypropylene (also iPP) |
| PP-AA | acrylic acid-modified isotactic polypropylene |
| PPE | poly(phenylene ether) |
| PPO | poly(phenylene oxide) |
| PS | polystyrene |
| PTT | poly(trimethylene terephthalate) |
| PVAc | polyvinylacetate |
| PVB | poly(vinyl butyral) |

| PVC | poly(vinyl chloride) |
|---|---|
| PVDCVC | poly(vinylidene choloride-*co*-vinyl chloride) |
| PVDF | poly(vinylidene fluoride) |
| PVE | poly(vinylethylene) |
| PVF | poly(vinyl fluoride) |
| PVME | poly(vinyl methyl ether) |
| PVPh | poly(vinyl phenol) |
| SAN | styrene-acrylonitrile |
| SEBS | hydrogenated styrene-b-butadiene-b-styrene triblock copolymer |
| SMA | styrene maleic anhydride copolymer |
| sPS | syndiotactic polystyrene |
| TPEg | maleated thermoplastic elastomer |
| VGCF | vapor grown carbon fiber |
| VLDPE | very low density polyethelene |

## 2.7.2
## Notations

| AFM | Atomic Force Microscopy |
|---|---|
| DSC | Differential Scanning Calorimetry |
| HTC | High Temperature Crystallizing Component |
| LTC | Low Temperature Crystallizing Component |
| OM | Optical Microscopy |
| SANS | Small Angle Neutron Scattering |
| SAXS | Small Angle X-Ray Scattering |
| SEM | Scanning Electron Microscopy |
| SSP | Solid State Polymerization |
| TEM | Transmission Electron Microscopy |
| WLF | Williams Landel and Ferry |

## 2.7.3
## Symbols

### 2.7.3.1  Roman Letters

| $a$ | constant from the relationship between interaction parameter and temperature |
|---|---|
| $A$ | frequency factor |
| $b$ | thickness of the critical nucleus |
| $B$ | interaction energy density |
| $C$ | constant related to the morphological contributions |
| $C_1, C_2, C_3$ | WLF constants |
| $d$ | maximum distance over which the amorphous component must diffuse away from the growth front |

| | |
|---|---|
| $D$ | diffusion coefficient |
| $\overline{D}$ | mutual diffusion coefficient |
| $E_1$ | energy required for the rejection of the droplets |
| $E_2$ | energy required to overcome the inertia of the droplets |
| $E_3$ | energy required to form a new interface when the droplets are engulfed |
| $E_4$ | energy dissipated for the deformation of the occluded particles |
| $E_a$ | apparent activation energy of rate constant |
| $\Delta E$ | activation free energy for the transport of chains through the solid-liquid interface |
| $f_z^i$ | fraction of droplets which contain $z$ heterogeneities of type $i$ |
| $\Delta F^*$ | free energy for the formation of a nucleus of critical size |
| $\Delta F_m^*$ | free energy for the formation of a nucleus of critical size in the presence of the amorphous component |
| $G$ | isothermal spherulite growth rate (also $G_1$) |
| $G^\circ$ | theoretical spherulite growth rate |
| $G_m$ | spherulite growth rate of crystalline phase in a polymer blend |
| $\Delta G$ | change in free energy of melting |
| $\Delta G^M$ | change in free energy of mixing (also $\Delta G_{mix}$) |
| $\Delta h_u$ | heat of fusion per mole of crystalline units |
| $\Delta H^M$ | change in enthalpy of mixing |
| $k_1$ | rate of transport of the crystalline molecule across the liquid-solid interface |
| $k$ | Avrami constant for primary crystallization |
| $k'$ | Avrami constant for secondary crystallization |
| $k_2$ | rate of segregation of the amorphous component |
| $k_B$ | Boltzman constant |
| $k(T)$ | cooling crystallization function |
| $L$ | crystal thickness |
| $M^i$ | concentration of heterogeneities of type $i$ |
| $n$ | Avrami exponent for primary crystallization |
| $n'$ | Avrami exponent for secondary crystallization |
| $N*$ | number density of dispersed particles |
| $R$ | gas constant |
| $s$ | time required for the nucleation of crystals |
| $\Delta S^M$ | change in entropy of mixing |
| $t^*$ | time at which pure secondary crystallization occurs |
| $t_{1/2}$ | half time of crystallization |
| $T_0$ | temperature required for the transport of molecules into the liquid-solid boundary |
| $T'_0$ | value of $T_0$ in the polymer blend |
| $T_c$ | crystallization temperature |
| $T_c^0$ | crystallization temperature of the bulk homopolymer |
| $T_g$ | glass transition temperature |
| $\Delta T_g$ | change in glass transition temperature |

| $T_m$ | equilibrium melting point of the crystallizable polymer in the blend |
|---|---|
| $T_m{}^\circ$ | equilibrium melting point of the crystallizable polymer in the bulk |
| $T(s)$ | thermal history |
| $v$ | volume fraction (also $\phi$) |
| $V_d$ | average volume of dispersed droplets |
| $V_{iu}$ | molar volume of the repeating unit of component $i$ |
| $x$ | degree of polymerization |
| $z$ | number of heterogeneities of type $i$ inducing normally crystallization in the bulk polymer at $T_c^0$ |

## 2.7.3.2 Greek Letters

| $\alpha$ | weight fraction of crystallinity |
|---|---|
| $\alpha_p$ | weight fraction of the primary crystallization |
| $\alpha_s$ | weight fraction of the secondary crystallization |
| $\beta$ | cooling rate |
| $\gamma_{PL}$ | interfacial free energy between the melt and droplets |
| $\gamma_{PS}$ | interfacial free energy between crystallizing solid and droplets |
| $\delta$ | Keith and Padden parameter |
| $\theta_a$ | fraction of uncrystallized material at time t |
| $\xi$ | weight fraction of the polymer crystallized by primary and secondary crystallization when the primary crystallization is finished |
| $\sigma\sigma_e$ | product of the lateral and fold surface free energies |
| $\phi_2$ | volume fraction of component 2 (also $v$) |
| $\chi_{12}$ | Flory– Huggins interaction parameter |

## References

1 Utracki, L.A. (2002) Preface, in *Polymer Blends Handbook* (ed. L.A. Utracki), Kluwer Academic Publishers, Dordrecht, Netherlands.

2 Paul, D.R. and Newman, S. (1978) *Polymer Blends*, vol. **I**, Academic Press, NY.

3 Gesner, B.D. (1969) Polyblends, in *Encyclopedia of Polymer Science and Technology*, vol. **10** (ed. N.M. Bikales), Interscience Publishers, John Wiley & Sons, Inc., NY, pp. 694–709.

4 Denchev, Z. (2002) Chemical interactions in blends of condensation polymersinvolving polyesters, in *Handbook of Thermoplastic Polyesters*, vol. **2** (ed. S. Fakirov), Wiley-VCH Verleg GmbH, Weinheim, Germany, Chapter 17.

5 Weber, M. (2001) *Macromol. Symp*, vol. **163**, Wiley-VCH Verleg GmbH, Weinheim, Germany, p. 235.

6 Isayev, A.I. and Modic, M.J. (1987) Self-reinforced melt processible polymer composites: Extrusion, compression, and injection molding. *Polym. Comp.*, **8**, 158.

7 Charrier, J.-M. and Ranchouse, R.J.P. (1971) Certain effects of melt-shearing and cooling on the stress-strain behavior of thermoplastic elastomers. *Polym. Eng. Sci.*, **11**, 381.

8 Mehta, A. and Isayev, A.I. (1991) Rheology, morphology, and mechanical characteristics of poly(etherether ketone)-liquid crystal polymer blends. *Polym. Eng. Sci.*, **31**, 971.

9 Pukanszky, B., Tudos, F., Kallo, A., and Bodor, G. (1989) Multiple morphology in

polypropylene/ethylene-propylene-diene terpolymer blends. *Polymer*, **30**, 1399.

10 Stehling, F.C., Huff, T., Speed, C.S., and Wissler, G. (1981) Structure and properties of rubber-modified polypropylene impact blends. *J. Appl. Polym. Sci.*, **26**, 2693.

11 Jang, B.Z., Uhlmann, D.R., and Van der Sande, J.B. (1985) Rubber-toughening in polypropylene. *J. Appl. Polym. Sci.*, **30**, 2485.

12 Groeninckx, G., Vanneste, M., and Everaert, V. (2002) Crystallization, morphological structure, and melting of polymer blends, in *Polymer Blends Handbook* (ed. L.A. Utracki), Kluwer Academic Publishers, Dordrecht, Netherlands, Chapter 3.

13 Vasile, C., Kulshreshtha, A.K., and Bumbu, G.G. (2003) *Handbook of Polymer Blends and Composites*, Rapra Technology Ltd., Shawbury.

14 Utracki, L.A. (1989) *Polymer Alloys and Blends*, Hanser Publishers, Munich.

15 Olabisi, O., Robeson, L.M., and Shaw, T. (1979) *Polymer-Polymer Miscibility*, Academic Press, NY.

16 Pracella, M. (2013) Crystallization of polymer blends, in *Handbook of Polymer Crystallization* (eds E. Piorkowska and G.C. Rutledge), John Wiley & Sons, Inc., Hoboken, NJ, Chapter 10.

17 Xie, F. (2006) The study of PET/MXD6 blends: interchange reaction, melting, crystallization, morphology, barrier property and solid state polymerization. Ph.D. dissertation, The University of Toledo, OH.

18 Ougizawa, T. and Inoue, T. (1999) Miscibility and interfacial behavior in polymer–polymer mixtures, in *Polymer Blends and Alloys* (eds G.O. Shonaike and G.P. Simon), Marcel Dekker, Inc, NY, Chapter 3.

19 Deanin, R.D. and Manion, M.A. (1999) Compatibilization of polymer blends, in *Polymer Blends and Alloys* (eds G.O. Shonaike and G.P. Simon), Marcel Dekker, Inc, NY, Chapter 1.

20 Paul, D.R. and Barlow, J.W. (1980) Polymer blends. *J. Macromol. Sci C.*, **18**, 109.

21 Wu, S.H. (1987) Formation of dispersed phase in incompatible polymer blends: Interfacial and rheological effects. *Polym. Eng. Sci.*, **27**, 335.

22 Fu, Q. (2005) Solid state polymerization, processing and properties of PEN/PET blends. Ph.D. dissertation, The University of Toledo, OH.

23 Stewart, M.E., Cox, A.J., and Taylor, D.M. (1993) Reactive processing of poly (ethylene 2,6-naphthalene dicarboxylate)/ poly(ethylene terephthalate) blends. *Polymer*, **34**, 4060.

24 Talibuddin, S., Wu, L., and Runt, J. (1996) Microstructure of melt-miscible, semicrystalline polymer blends. *Macromolecules*, **29**, 7527.

25 Lorenzo, M.L.D. (2003) Spherulite growth rates in binary polymer blends. *Prog. Polym. Sci.*, **28**, 663.

26 Alfonso, G.C. and Russel, T.P. (1986) Kinetics of crystallization in semicrystalline/amorphous polymer mixtures. *Macromolecules*, **19**, 1143.

27 Martuscelli, E., Silvestre, C., and Gismondi, C. (1985) Morphology, crystallization and thermal behaviour of poly(ethylene oxide)/poly(vinyl acetate) blends. *Macromol. Chem.*, **186**, 2161.

28 Wang, T.T. and Nishi, T. (1977) Spherulitic crystallization in compatible blends of poly(vinylidene fluoride) and poly(methyl methacrylate). *Macromolecules*, **10**, 421.

29 Lee, J.C., Tazawa, H., Ikehara, T., and Nishi, T. (1998) Miscibility and crystallization behavior of poly(butylene succinate) and poly(vinylidene fluoride) blends. *Polym. J.*, **30**, 327.

30 Robeson, L.M. (1973) Crystallization kinetics of poly-ε-caprolactone from poly-ε-caprolactone/poly(vinyl chloride) solutions. *J. Appl. Polym. Sci.*, **17**, 3607.

31 Defieuw, G., Groeninckx, G., and Reynaers, H. (1989) Miscibility, crystallization and melting behaviour, and semicrystalline morphology of binary blends of polycaprolactone with poly (hydroxy ether of bisphenol A). *Polymer*, **30**, 2164.

32 Berghmans, H. and Ovembergh, N. (1977) Crystallization and melting of the system isotactic polystyrene + poly(2,6-dimethyl phenylene oxide). *J. Polym. Sci. Pol. Phys.*, **15**, 1757.

33 Finelli, L., Sarti, B., and Scandola, M. (1997) Miscibility and biodegradation behavior of melt-processed blends of bacterial poly(3-hydroxybutyrate) with poly(epichlorohydrin). *J. Macromol. Sci. A.*, **34**, 13.

34 Xing, P.X., Dong, L.S., An, Y.X., Feng, Z.L., Avella, M., and Martuscelli, E. (1997) Miscibility and crystallization of poly(β-hydroxybutyrate) and poly(p-vinylphenol) blends. *Macromolecules*, **30**, 2726.

35 Pizzoli, M., Scandola, S., and Ceccurulli, G. (1994) Crystallization kinetics and morphology of poly(3-hydroxybutyrate)/cellulose ester blends. *Macromolecules*, **27**, 4755.

36 Greco, P. and Martuscelli, E. (1989) Crystallization and thermal behaviour of poly(D(—)-3-hydroxybutyrate)-based blends. *Polymer*, **30**, 1475.

37 Amelino, L., Martuscelli, E., Sellitti, C., and Silvestre, C. (1990) Isotactic polystyrene/poly(vinyl methyl ether) blends: miscibility, crystallization and phase structure. *Polymer*, **31**, 1051.

38 Martuscelli, E., Sellitti, C., and Silvestre, C. (1985) Enhancement of spherulite radial growth rate in isotactic polystyrene/poly(methyl vinyl ether) blends. *Macromol. Rapid Comm.*, **6**, 125.

39 Cimmino, S., Di Pace, E., and Martuscelli, E., and Silvestre, C. (1993) Syndiotactic polystyrene-based blends: crystallization and phase structure. *Polymer*, **34**, 2799.

40 Liu, L.Z., Chu, B., Penning, J.P., and Manley, R.S. (1998) A time-resolved synchrotron SAXS study of crystallization in a miscible blend of poly(vinylidenefluoride) and poly(1,4-butylene adipate). *J. Macromol. Sci. B*, **37**, 485.

41 Yang, J.M., Chen, H.L., You, J.W., and Hwang, J.C. (1997) Miscibility and crystallization of poly(l-lactide)/poly(ethylene glycol) and poly(l-lactide)/poly(ε-caprolactone) blends. *Polym J.*, **29**, 657.

42 Porter, R.S. (1992) Compatibility and transesterification in binary polymer blends. *Polymer*, **33**, 2019.

43 Shi, Y. and Jabarin, S.A. (2001) Glass-transition and melting behavior of poly(ethylene terephthalate)/poly(ethylene 2,6-naphthalate) blends. *J. Appl. Polym. Sci.*, **81**, 11.

44 Fried, J.R., Karasz, F.E., and MacKnight, W.J. (1978) Compatibility of poly(2,6-dimethyl-1,4-phenylene oxide) (ppo)/poly(styrene-co-4-chlorostyrene) blends. 1. Differential scanning calorimetry and density studies. *Macromolecules*, **11**, 150.

45 Kim, J.C., Cakmak, M., and Geil, P.H. (1997) Miscibility and crystallization behavior of PEN and PEN/PEI blends. SPE/ANTEC 1997 Proceedings, Toronto, Canada, 1572.

46 Flory, P.J. (1953) *Principles of Polymer Chemistry*, Cornell University Press, Ithaca, NY.

47 Nishi, T. and Wang, T.T. (1975) Melting point depression and kinetic effects of cooling on crystallization in poly(vinylidene fluoride)-poly(methyl methacrylate) mixtures. *Macromolecules*, **8**, 909.

48 Hsiao, B.S. and Sauer, B.B. (1993) Glass transition, crystallization, and morphology relationships in miscible poly(aryl ether ketones) and poly(ether imide) blends. *J. Polym. Sci. Poly. Phys.*, **31**, 901.

49 Scott, R.L.J. (1949) The thermodynamics of high polymer solutions. V. phase equilibria in the ternary system: polymer 1—polymer 2—solvent. *Chem. Phys.*, **17**, 279.

50 Kwei, T.K. and Frisch, H.L. (1978) Interaction parameter in polymer mixtures. *Macromolecules*, **11**, 1267.

51 Walsh, D.J., Rostami, S., and Singh, V.B. (1985) The thermodynamics of polyether sulfone-poly(ethylene oxide) mixtures. *Macromol. Chem.*, **186**, 145.

52 Matkar, R.A. and Kyu, T. (2006) Role of crystal–amorphous interaction in phase equilibria of crystal–amorphous polymer blends. *J. Phys. Chem. B*, **110**, 12728.

53 Xie, F., Lofgren, E.A., and Jabarin, S.A. (2010) Melting and crystallization behavior of poly(ethylene terephthalate) and poly(m-xylylene adipamide) blends. *J. Appl. Polym. Sci.*, **118**, 2153.

54 Hirami, M. and Matsda, T. (1999) Phase behavior of binary mixture systems of two crystallizable components.

Semicrystalline polymers and wholly-crystalline substances. *Polym. J.*, **31**, 801.

55 Matkar, R.A. and Kyu, T. (2006) Phase diagrams of binary crystalline–crystalline polymer blends. *J. Phys. Chem. B*, **110**, 16059.

56 Kuo, S.W., Huang, W.J., Huang, C.F., Chan, S.C., and Chang, F.C. (2004) Miscibility, specific interactions, and spherulite growth rates of binary poly (acetoxystyrene)/poly(ethylene oxide) blends. *Macromolecules*, **37**, 4164.

57 Patcheak, T.D. (2001) A study of the structures and morphologies of PET/PEN blends. Ph.D. dissertation, The University of Toledo, OH..

58 Russell, T.P. and Stein, R.S.J. (1983) An investigation of the compatibility and morphology of semicrystalline poly (ε-caprolactone)–poly(vinyl chloride) blends. *Polym. Sci. Pol. Phys.*, **21**, 999.

59 Stein, R.S., Khambatta, F.B., Warner, F.P., Russel, T., Escala, A., and Balizer, E. (1978) X-ray and optical studies of the morphology of polymer blends. *J. Polym. Symp.*, **63**, 313.

60 Defieuw, G., Groeninckx, G., and Reynaers, H. (1989) Miscibility and morphology of binary polymer blends of polycaprolactone with solution-chlorinated polyethylenes. *Polymer*, **30**, 595.

61 Vanneste, M. (1993) Ph.D. dissertation, University of Leuven, Belgium.

62 Cimmino, S., Martuscelli, E., Silvestre, C., Canetti, M., De Lalla, C., and Seves, A. (1989) Poly(ethylene oxide)/poly(ethyl methacrylate) blends: Crystallization, melting behavior, and miscibility. *J. Polym. Sci. Pol. Phys.*, **27**, 1781.

63 Canetti, M., Sadocco, P., Siciliano, A., and Seves, A. (1994) Investigation of the phase structure of poly(D(−)3-hydroxybutyrate)/atactic poly(methyl methacrylate) blends by small-angle X-ray scattering. *Polymer*, **35**, 2884.

64 Saito, H. and Stuhn, B. (1994) Exclusion of noncrystalline polymer from the interlamellar region in poly(vinylidene fluoride)/poly(methyl methacrylate) blends. *Macromolecules*, **27**, 216.

65 Morra, B.S. and Stein, R.S. (1984) The crystalline morphology of poly(vinylidene

fluoride)/poly(methylmethacrylate) blends. *Polym. Eng. Sci.*, **24**, 311.

66 Russell, T.P., Ito, H., and Wignall, G.D. (1988) Neutron and x-ray scattering studies on semicrystalline polymer blends. *Macromolecules*, **21**, 1703.

67 Zemel, I.S. and Roland, C.M. (1992) Anomalies in the crystallization of cis-1,4-polyisoprene in blends with poly (vinylethylene). *Polymer*, **33**, 3427.

68 Runt, J.P., Zhang, X., Miley, D.M., Gallagher, K.P., and Zhang, A. (1992) Phase behavior of poly(butylene terephthalate)/polyarylate blends. *Macromolecules*, **25**, 3902.

69 Crevecoeur, G. and Groeninckx, G. (1991) Binary blends of poly(ether ether ketone) and poly(ether imide): Miscibility, crystallization behavior and semicrystalline morphology. *Macromolecules*, **24**, 1190.

70 Lohse, D.J. and Wissler, G.E.J. (1991) Compatibility and morphology of blends of isotactic and atactic polypropylene. *J. Mater. Sci.*, **26**, 743.

71 Hudson, S.D., Davis, D.D., and Lovinger, A.J. (1992) Semicrystalline morphology of poly(aryl ether ether ketone)/poly(ether imide) blends. *Macromolecules*, **25**, 1759.

72 Huo, P.P., Cebe, P., and Capel, M. (1993) Dynamic mechanical relaxation and x-ray scattering study of poly(butylene terephthalate)/polyarylate blends. *Macromolecules*, **26**, 4275.

73 Lauritzen, J.I. and Hoffman, J.D. (1973) Extension of theory of growth of chain-folded polymer crystals to large undercoolings. *J. Appl. Phys.*, **44**, 4340.

74 Hoffman, J.D., Davis, G.T., and Lauritzen, J.I. (1976) The rate of crystallization of linear polymers with chain folding, in *Treatise on Solid State Chemistry*, vol. **3** (ed. N.B. Hannay), Plenum Press, NY.

75 Flory, P.J.J. (1949) Thermodynamics of crystallization in high polymers. IV. A theory of crystalline states and fusion in polymers, copolymers, and their mixtures with diluents. *Chem. Phys.*, **17**, 223.

76 Hoffman, J.D. and Weeks, J.J. (1962) Melting process and the equilibrium melting temperature of polychlorotrifluoroethylene. *J. Res. Nbs. A Phys. Ch.*, **66**, 13.

77 Hoffman, J.D. and Weeks, J.J. (1962) Rate of spherulitic crystallization with chain folds in polychlorotrifluoroethylene. *J. Chem. Phys.*, **37**, 1723.

78 Kramer, E.J., Green, P., and Palmstrom, C.J. (1984) Interdiffusion and marker movements in concentrated polymer-polymer diffusion couples. *Polymer*, **25**, 473.

79 Sillescu, H. (1984) Relation of interdiffusion and self-diffusion in polymer mixtures. *Makromol. Chem., Rapid Commun.*, **5**, 519.

80 Keith, H.D. and Padden, F.J.J. (1964) Spherulitic crystallization from the melt. I. Fractionation and impurity segregation and their influence on crystalline morphology. *Appl. Phys.*, **35**, 1270.

81 Keith, D.H. and Padden, F.J.J. (1963) A phenomenological theory of spherulitic crystallization. *Appl. Phys.*, **34**, 2409.

82 Avrami, M.J. (1939) Kinetics of phase change. I General theory. *Chem. Phys.*, **7**, 1103.

83 Shi, Y. and Jabarin, S.A. (2001) Crystallization kinetics of poly(ethylene terephthalate)/poly(ethylene 2,6-naphthalate) blends. *J. Appl. Polym. Sci.*, **81**, 23.

84 Nojima, S., Tsutsui, H., Urushihara, M., Kosaka, W., Kato, N., and Ashida, T. (1986) A Dynamic study of crystallization of poly(ε-caprolactone) and poly (ε-caprolactone)/poly(vinyl chloride) blend. *Polym. J.*, **18**, 451.

85 Martuscelli, E., Pracella, M., and Wang, P.Y. (1984) Influence of composition and molecular mass on the morphology, crystallization and melting behaviour of poly(ethylene oxide)/poly(methyl methacrylate) blends. *Polymer*, **25**, 1097.

86 Perez-Cardenas, F.C., Felipe Del Castillo, L., and Veragraziano, R. (1991) Modified avrami expression for polymer crystallization kinetics. *J. Appl. Polym. Sci.*, **43**, 779.

87 Ozawa, T. (1971) Kinetics of non-isothermal crystallization. *Polymer*, **12**, 150.

88 de Juana, R., Jauregui, A., Calahorra, E., and Cortfizar, M. (1996) Non-isothermal crystallization of poly(hydroxy ether of bisphenol-A)/poly(ε-caprolactone), PH/PCL blends. *Polymer*, **37**, 3339.

89 Herrero, C.H. and Acosta, J.L. (1994) Effect of poly(epichlorhydrin) on the crystallization and compatibility behavior of poly(ethylene oxide)/polyphosphazene blends. *Polymer J.*, **26**, 786.

90 Cebe, P. (1988) Non-isothermal crystallization of poly(etheretherketone) aromatic polymer composite. *Polym. Comp.*, **9**, 271.

91 Ziabicki, A. (1967) Kinetics of polymer crystallization and molecular orientation in the course of melt spinning. *Appl. Polym. Symp.*, **6**, l.

92 Ziabicki, A. (1976) *Fundamentals of Fiber Formation*, John Wiley & Sons Ltd, London.

93 Liu, S.Y., Yu, Y.N., Cui, Y., Zhang, H.F., and Mo, Z.S. (1998) Isothermal and nonisothermal crystallization kinetics of nylon-11. *J. Appl. Polym. Sci.*, **70**, 2371.

94 Liu, M.Y., Zhao, Q.X., Wang, Y.D., Zhang, C.G., Mo, Z.S., and Cao, S.K. (2003) Effect of poly(epichlorhydrin) on the crystallization and compatibility behavior of poly(ethylene oxide)/polyphosphazene blends. *Polymer*, **44**, 2537.

95 Run, M., Wang, Y., Yao, C., and Gao, J. (2006) Non-isothermal crystallization kinetics of poly(trimethylene terephthalate)/poly(ethylene 2,6-naphthalate) blends. *Thermochim. Acta*, **447**, 13.

96 Qiu, Z., Yan, C., Lu, J., and Yang, W. (2007) Miscible crystalline/crystalline polymer blends of poly(vinylidene fluoride) and poly(butylene succinate-co-butylene adipate): Spherulitic morphologies and crystallization kinetics. *Macromolecules*, **40**, 5047.

97 Liu, J. and Jungnickel, B.J. (2003) Crystallization and morphology of poly (vinylidene fluoride)/poly(3-hydroxybutyrate) blends. I. Spherulitic morphology and growth by polarized microscopy. *J. Polym. Sci. Pol. Phys.*, **41**, 873.

98 Avella, M. and Martuscelli, E. (1988) Poly-D-(−)(3-hydroxybutyrate)/poly (ethylene oxide) blends: phase diagram, thermal and crystallization behaviour. *Polymer*, **29**, 1731.

99 Marand, H. and Collins, M. (1990) Crystallization and morphology of poly

(vinylidene fluoride) poly (3-hydroxybutyrate) blends. *Am. Chem. S. (Div. Polym. Sci.)*, **31**, 552.

100 Chen, H.L. and Wang, S.F. (2000) Crystallization induced microstructure of polymer blends consisting of two crystalline constituents. *Polymer*, **41**, 5157.

101 Liu, A.S., Liau, W.B., and Chiu, W.Y. (1998) Studies on blends of binary crystalline polymers. 1. Miscibility and crystallization behavior in poly(butylene terephthalate)/polyarylates based on bisphenol a isophthalate. *Macromolecules*, **31**, 6593.

102 Penning, J.P. and Manley, R.S.J. (1996) Miscible blends of two crystalline polymers. 2. Crystallization kinetics and morphology in blends of poly(vinylidene fluoride) and poly(1,4-butylene adipate). *Macromolecules*, **29**, 84.

103 Qiu, Z., Ikehara, T., and Nishi, T. (2003) Miscibility and crystallization in crystalline/crystalline blends of poly (butylene succinate)/poly(ethylene oxide). *Polymer* (2003) **44**, 2799.

104 Qiu, Z., Fujinami, S., Komura, M., Nakajima, K., Ikehara, T., and Nishi, T. (2004) Miscibility and crystallization of poly(ethylene succinate)/poly(vinyl phenol) blends. *Polymer*, **45**, 4515.

105 Blumm, E. and Owen, A. (1995) Miscibility, crystallization and melting of poly(3-hydroxybutyrate)/poly(L-lactide) blends. *Polymer*, **36**, 4077.

106 Lee, J.C., Tazawa, H., Ikehara, T., and Nishi, T. (1998) Crystallization kinetics and morphology in miscible blends of two crystalline polymers. *Polym. J.*, **30**, 780.

107 Ikehara, T., Nishikawa, Y., and Nishi, T. (2003) Evidence for the formation of interpenetrated spherulites in poly (butylene succinate-co-butylene carbonate)/poly(L-lactic acid) blends investigated by atomic force microscopy. *Polymer*, **44**, 6657.

108 Qiu, Z., Ikehara, T., and Nishi, T. (2002) Unique morphology of poly(ethylene succinate)/poly(ethylene oxide) blends. *Macromolecules*, **35**, 8251.

109 Ikehara, T., Kimura, H., and Qiu, Z. (2005) Penetrating spherulitic growth in poly(butylene adipate-co-butylene succinate)/poly(ethylene oxide) blends. *Macromolecules*, **38**, 5104.

110 Clough, N.E., Richards, R.W., and Ibrahim, T. (1994) Melting point depression in ethylene-vinyl acetate copolymer mixtures. *Polymer*, **35**, 1044.

111 Huang, B., Wang, J., and Pang, D. (1990) *Third European Symposium on Polymer Blends (PRI)*, Cambridge, UK.

112 Harris, J.E. and Robeson, L.M. (1987) Isomorphic behavior of poly(aryl ether ketone) blends. *J. Polym. Sci. Pol. Phys.*, **25**, 311.

113 Cho, K., Li, F., and Choi, J. (1999) Crystallization and melting behavior of polypropylene and maleated polypropylene blends. *Polymer*, **40**, 1719.

114 Edie, S.L. and Marand, H. (1991) Study of miscible blends of poly (vinylidene fluoride) and poly (3-hydroxybutyrate). *Am. Chem. S. (Div. Polym. Sci.)*, **32**, 329.

115 Chiu, H.J., Chen, H.L., and Lin, J.S. (2001) Crystallization induced microstructure of crystalline/crystalline poly (vinylidenefluoride)/poly(3-hydroxybutyrate) blends probed by small angle X-ray scattering. *Polymer*, **42**, 5749.

116 Cheung, Y.W. and Stein, R.S. (1994) Critical analysis of the phase behavior of poly(.epsilon.-caprolactone) (PCL)/polycarbonate (PC) blends. *Macromolecules*, **27**, 2512.

117 Fujita, K. and Kyu, T. (1996) Miscible blends of two crystalline polymers. 3. Liquid–liquid phase separation in blends of poly(vinylidene fluoride)/poly(butylene adipate). *Macromolecules*, **29**, 91.

118 Terada, Y., Hirano, S., Ikehara, T., and Nishi, T. (2001) Formation of interpenetrated spherulites based on miscible crystalline polymer blends. *Macromol. Symp.*, **175**, 209.

119 Liang, B. and Pan, L. (1994) Compatibility and morphology studies of PPO multicomponent blends. *J. Appl. Polym. Sci.*, **54**, 1945.

120 Bartczak, Z., Galeski, A., and Martuscelli, E. (1984) Spherulite growth in isotactic polypropylene-based blends: Energy and morphological considerations. *Polym. Eng. Sci.*, **24**, 1155.

121 Santana, O.O. and Muller, A.J. (1994) Homogeneous nucleation of the dispersed crystallisable component of immiscible polymer blends. *Polym. Bull.*, **32**, 471.

122 Thirtha, V., Lehman, R., and Nosker, T. (2005) Glass transition effects in immiscible polymer blends. *ANTEC 2005 Proceedings, Boston, MA*, 2380.

123 Morales, R.A., Arnal, M.L., and Muller, A.J. (1995) The evaluation of the state of dispersion in immiscible blends where the minor phase exhibits fractionated crystallization. *Polym. Bull.*, **35**, 379.

124 Coppola, F., Greco, R., Martuscelli, E., Kammer, H.W., and Kummerlowe, C. (1987) Mechanical properties and morphology of isotactic polypropylene/ethylene-propylene copolymer blends. *Polymer*, **28**, 47.

125 Karger-Kocsis, J., Kallo, A., Szafner, A., Bodor, G., and Senyei, Z.S. (1979) Morphological study on the effect of elastomeric impact modifiers in polypropylene systems. *Polymer*, **20**, 37.

126 Dimzoski, B., Fortelny, I., Slouf, M., Sikora, A., and Michalkova, D. (2013) Morphology evolution during cooling of quiescent immiscible polymer blends: matrix crystallization effect on the dispersed phase coalescence. *Polym. Bull.*, **70**, 263.

127 Kalfoglou, N.K. (1985) Physical characterization of isotactic polypropylene/ethylene-propylene copolymer blends. *Angew. Makromol. Chem.*, **129**, 103.

128 Nadkarni, V.M. and Jog, J.P. (1987) Interrelationship between the crystallization behavior, injection molding conditions, and morphology of poly(ethylene terephthalate)/poly(methyl methacrylate) alloys. *Polym. Eng. Sci.*, **27**, 451.

129 Aref-Azar, A., Hay, J., Marsden, B., and Walker, J. (1980) Crystallization characteristics of polymer blends. I. Polyethylene and polystyrene. *J. Polym. Sci. Pol. Phys.*, **B18**, 637.

130 Ghijsels, A., Groesbeek, N., and Yip, C.W. (1982) Multiple crystallization behaviour of polypropylene/thermoplastic rubber blends and its use in assessing blend morphology. *Polymer*, **23**, 1913.

131 Quirk, R.P., Ma, J.J., Chen, C.C., Min, K., and White, J.L. (1989) Multiphase macromolecular systems, in *Contemporary Topics in Polymer Science* (ed. B.M. Culbertson), Plenum Press, New York and London.

132 Bartczak, Z. and Galeski, A. (1990) Homogeneous nucleation in polypropylene and its blends by small-angle light scattering. *Polymer*, **31**, 2027.

133 Scott, C.E. and Macosko, C.W. (1995) Morphology development during the initial stages of polymer-polymer blending. *Polymer*, **36**, 461.

134 Bartczak, Z., Galeski, A., and Krasnikova, N.P. (1987) Primary nucleation and spherulite growth rate in isotactic polypropylene-polystyrene blends. *Polymer*, **28**, 1627.

135 Wenig, W., Fiedel, H.-W., and Scholl, A. (1990) Crystallization kinetics of isotactic polypropylene blended with atactic polystyrene. *Colloid Polym. Sci.*, **268**, 528.

136 Galeski, A., Bartczak, Z., and Pracella, M. (1984) Spherulite nucleation in polypropylene blends with low density polyethylene. *Polymer*, **25**, 1323.

137 Bartczak, Z., Galeski, A., and Pracella, M. (1986) Spherulite nucleation in blends of isotactic polypropylene with high-density polyethylene. *Polymer*, **27**, 537.

138 Nadkarni, V.M. and Jog, J.P. (1991) Crystallization behavior in polymer blends, in *Two-phase Polymer Systems* (ed. L.A. Utracki), Hanser Publishing, Munich and New York.

139 Wei-Berk, C. (1993) Crystallization behaviors of PP/PS blends and alloys. *ACS*, **68**, 299.

140 Martuscelli, E. (1984) Influence of composition, crystallization conditions and melt phase structure on solid morphology, kinetics of crystallization and thermal behavior of binary polymer/polymer blends. *Polym. Eng. Sci.*, **24**, 563.

141 Silvestre, C., Cimmino, S., Di Pace, E., Martuscelli, E., Monaco, M., and Buzarovska, A., and Koseva, S. (1996) Morphology, phase structure, and properties of polyolefin/hydrogenated oligocyclopentadiene blends. *Polym. Network Blends*, **6**, 73.

142 Avella, M., Martuscelli, E., Orsello, G., Raimo, M., and Pascucci, B. (1997) Poly (3-hydroxybutyrate)/poly (methyleneoxide) blends: thermal, crystallization and mechanical behaviour. *Polymer*, **38**, 6135.

143 Yokoyama, Y. and Ricco, T. (1997) Crystallization and morphology of reactor-made blends of isotactic polypropylene and ethylene-propylene rubber. *J. Appl. Polym. Sci.*, **66**, 1007.

144 Buzarovska, A., Cimmino, S., Di Pace, E., Martuscelli, E., and Silvestre, C., and Slobodanka, K. (1995) High density polyethylene/hydrogenated oligo (cyclopentadiene) blends: Tensile stress-strain behavior. *Polym. Network Blends*, **5**, 63.

145 Marentette, J.M. and Brown, G.R. (1998) The crystallization of poly(ethylene oxide) in blends with neat and plasticized poly(vinyl chloride). *Polymer*, **39**, 1415.

146 Cascone, E., David, D.J., Di Lorenzo, M.L., Karasz, F.E., Macknight, W.J., and Martuscelli, E., and Raimo, M. (2001) Blends of polypropylene with poly(vinyl butyral). *J. Appl. Polym. Sci.*, **82**, 2934.

147 Hoffmann, J.D., Frolen, L.J., Ross, G.S., and Lauritzen, J.I. (1975) On the growth rate of spherulites and axialites from the melt in polyethylene fractions: Regime I and regime II crystallization. *J. Res. Natl. Bur. Stand.*, **79**, 671.

148 Hoffmann, J.D. (1983) Regime III crystallization in melt-crystallized polymers: The variable cluster model of chain folding. *Polymer*, **24**, 3.

149 Suzuki, T. and Kovacs, A. (1970) Temperature dependence of spherulitic growth rate of isotactic polystyrene. A critical comparison with the kinetic theory of surface nucleation. *Polymer J.*, **1**, 82.

150 Icenogle, R.D. (1985) Temperature-dependent melt crystallization kinetics of poly(butene-1): A new approach to the characterization of the crystallization kinetics of semicrystalline polymers. *J. Polym. Sci. Pol. Phys.*, **23**, 1369.

151 Galeski, A., Pracella, M., and Martuscelli, E. (1984) Polypropylene spherulite morphology and growth rate changes in blends with low-density polyethylene. *J. Polym. Sci. Pol. Phys.*, **22**, 739.

152 Han, C.C., Kammer, H.W., Har, S.L., and Winie, T. (2011) Morphologies and kinetics of isothermal crystallization for green polymer blends comprising PHBV and ENR: Influence of rubbery phase. *Int. J. Pharm. Pharm. Sci.*, **3**, 10.

153 Bates, F.S., Cohen, R.E., and Argon, A.S. (1983) Dynamic mechanical properties of polystyrene containing microspherical inclusions of polybutadiene: influence of domain boundaries and rubber molecular weight. *Macromolecules*, **16**, 1108.

154 Reinsch, V.E. and Rebenfeld, L. (1996) Crystallization of poly(ethylene terephthalate)/polycarbonate blends. I. Unreinforced systems. *J. Appl. Polym. Sci.*, **59**, 1913.

155 Thirtha, V.M., Lehman, R.L., and Nosker, T.J. (2005) Glass transition phenomena in melt-processed polystyrene/ polypropylene blends. *Polym. Eng. Sci.*, **45**, 1187.

156 Mucha, M. (1986) Miscibility of isotactic polypropylene with atactic polystyrene. *Coll. Polym. Sci.*, **264**, 859.

157 Frensch, H., Harnischfeger, P., and Jungnickel, B.J. (1989) Multiphase polymers, blends: Ionomers and interpenetrating networks. (eds L.A. Utracki and R.A. Weiss), *ACS Symp. Ser.*, **395**, 101.

158 Robitaille, C. and Prud'homme, J. (1983) Thermal and mechanical properties of a poly(ethylene oxide-b-isoprene-b-ethylene oxide) block polymer complexed with sodium thiocyanate. *Macromolecules*, **16**, 665.

159 Bailtoul, M., Saint-Guirons, H., Xans, P., and Monge, P. (1981) Etude par analyse thermique differentielle de melanges diphasiques de polyethylene basse densite et de polystyrene atactique. *Eur. Polym. J.*, **17**, 1281.

160 Tol, R.T., Mathot, V.B.F., and Groeninckx, G. (2005) Confined crystallization phenomena in immiscible polymer blends with dispersed micro – and nanometer sized PA6 droplets, part 1: uncompatibilized PS/PA6, (PPE/PS)/PA6 and PPE/PA6 blends. *Polymer*, **46**, 369.

161 Koutsky, J.A., Walton, A.G., and Baer, E. (1967) Nucleation of polymer droplets. *J. Appl. Phys.*, **38**, 1832.

162 Pound, G.M. and LaMer, V.K. (1952) Kinetics of crystalline nucleus formation in supercooled liquid tin. *J. Am. Chem. Soc.*, **74**, 2323.

163 Zhang, C., Yi, X.-S., Asai, S., and Sumita, M. (2000) Morphology, crystallization and melting behaviors of isotactic polypropylene/high density polyethylene blend: effect of the addition of short carbon fiber. *J. Mater. Sci.*, **35**, 673.

164 Sumita, M., Sakata, K., Asai, S., Miyasaka, K., and Nakagawa, H. (1991) Dispersion of fillers and the electrical conductivity of polymer blends filled with carbon black. *Polym. Bull.*, **25**, 265.

165 Asai, S., Sakata, K., Sumita, M., and Miyasaka, K. (1992) Effect of interfacial free energy on the heterogeneous distribution of oxidized carbon black in polymer blends. *Polym. J.*, **24**, 415.

166 Paul, D.R., Barlow, J.W., and Keskula, H. (1985) *Encyclopedia of Polymer Science and Engineering*, 2nd edn (ed. J.I. Kroschwitz), Wiley-Interscience, NY.

167 Horák, Z., Fortelný, I., Kolařík, J., Hlavatá, D., and Sikora, A. (2005) Polymer blends, in *Encyclopedia of Polymer Science and Technology* (ed. H.F. Mark), John Wiley & Sons, Inc., Hoboken, NJ.

168 Ajji, A. (2002) Interphase and compatibilization by addition of a compatibilizer, in *Polymer Blends Handbook* (ed. L.A. Utracki), Kluwer Academic Publishers, Dordrecht, Netherlands, Chapter 4.

169 Bonner, J.G. and Hope, P.S. (1993) Compatibilisation and reactive blending, in *Polymer Blends and Alloys* (eds M.J. Folkes and P.S. Hope), Blackie Academic & Professional, London, Chapter 3.

170 Wu, S. (1982) *Polymer Interface and Adhesion*, Dekker, Inc., NY.

171 Leibler, L. (1981) Theory of phase equilibria in mixtures of copolymers and homopolymers, 1. Phase diagram. *Makromol. Chem., Rapid Commun.*, **2**, 393.

172 Eastwood, E.A. and Dadmun, M.D. (2002) Multiblock copolymers in the compatibilization of polystyrene and poly (methyl methacrylate) blends: Role of polymer architecture. *Macromolecules*, **35**, 5069.

173 Li, J. and Favis, D. (2002) Strategies to measure and optimize the migration of the interfacial modifier to the interface in immiscible polymer blends. *Polymer*, **43**, 4935.

174 Datta, S. and Lohse, D.J. (1996) *Polymeric Compatibilizers*, Carl Hanser Verlag., Munich.

175 Fayt, R., Jerome, R., and Teyssie, P. (1990) Molecular design of multicomponent polymer systems. XVIII: Emulsification of high-impact polystyrene and low density polyethylene blends into high-impact alloys. *Polym. Eng. Sci.*, **30**, 937.

176 Noolandi, J. (1984) Recent advances in the theory of polymeric alloys. *Polym. Eng. Sci.*, **24**, 70.

177 Hellmann, G.P. and Dietz, M. (2001) Random and block copolymers as compatibilizers, directly compared. *Macromol. Symp.*, **170**, 1.

178 Smith, A.P., Ade, H., Koch, C.C., Smith, S.D., and Spontak, R.A. (2000) Addition of a block copolymer to polymer blends produced by cryogenic mechanical alloying. *Macromolecules*, **33**, 1163.

179 Heikens, D., Hoen, N., Barentsen, W., Piet, P., and Ladan, H.J. (1978) *Polym. Sci. Polym. Symp.*, **62**, 309.

180 Fayt, R., Jerome, R., and Teyssie, P. (1989) Molecular design of multicomponent polymer systems. XIV. Control of the mechanical properties of polyethylene–polystyrene blends by block copolymers. *J. Polym. Sci. B Polym. Phys.*, **27**, 775.

181 Cao, Y., Zhang, J., Feng, J., and Wu, P. (2011) Compatibilization of immiscible polymer blends using graphene oxide sheets. *AM. Chem. S.*, **5**, 5920.

182 Fayt, R., Jerome, R., and Teyssie, P. (1982) Molecular design of multicomponent polymer systems. III. Comparative behavior of pure and tapered block copolymers in emulsification of blends of low-density polyethylene and polystyrene. *J. Polym. Sci., Polym. Phys.*, **20**, 2209.

183 Fayt, R., Jerome, R., and Teyssie, P. (1987) Characterization and control of interfaces in emulsified incompatible polymer blends. *Polym. Eng. Sci.*, **27**, 328.

184 Perron, P.J. (1984) New developments in thermoplastic alloys. *Adv. Polym. Technol.*, **6**, 79.

185 Perron, P.J. and Bourbonais, E.A. (1988)
Compatibilizing agent for polycarbonate
and polyamide polymer blends. US
Patent 4,782,114.

186 Utracki, L.A. (2002) Compatibilization of
polymer blends. *Can. J. Chem. Eng.*, **80**,
1008.

187 Srinivasan, K.R. and Gupta, A.K. (1994)
Mechanical properties and morphology
of PP/SEBS/PC blends. *J. Appl. Polym.
Sci.*, **53**, 1.

188 Robeson, L.M. (1985) Phase behavior of
polyarylate blends. *J. Appl. Polym. Sci.*,
**30**, 4081.

189 Baker, W.E., Scott, C., and Hu, G.H.
(2001) *Reactive Polymer Blending*, Carl
Hanser Verlag., Munich.

190 Krulis, Z., Horak, Z., Lednicky, F.,
Pospısil, J., and Sufcak, M. (1998)
Reactive compatibilization of polyolefins
using low molecular weight
polybutadiene. *Angew. Makromol. Chem.*,
**258**, 63.

191 Zhang, H.X. and Hourston, D.J. (1999)
Reactive compatibilization of poly
(butylene terephthalate)/low-density
polyethylene and poly(butylene
terephthalate)/ethylene propylene diene
rubber blends with a bismaleimide. *J.
Appl. Polym. Sci.*, **71**, 2049.

192 Schies, J. and Priddy, D.B. (2003) *Modern
Styrenic Polymers: Polystyrene and
Styrenic Copolymers*, John Wiley & Sons
Ltd., Chichester, UK

193 Brown, S.B. (2002) Reactive
Compatibilization of Polymer Blends, in
*Polymer Blends Handbook* (ed. L.A.
Utracki), Kluwer Academic Publishers,
Dordrecht, Netherlands, Chapter 5.

194 Ajji, A. and Utracki, L.A. (1996)
Interphase and compatibilization of
polymer blends. *Polym. Eng. Sci.*, **36**,
1574.

195 Imre, B. and Pukánszky, B. (2013)
Compatibilization in bio-based and
biodegradable polymer blends. *Eur.
Polym. J.*, **49**, 1215.

196 Flores, A.M. (2008) Ph.D. dissertation,
Northwestern University, Evanston, IL.

197 Na, Y.-H., He, Y., Shuai, X., Kikkawa, Y.,
Doi, Y., and Inoue, Y. (2002)
Compatibilization effect of poly
(ε-caprolactone)-b-poly(ethylene glycol)

block copolymers and phase morphology
analysis in immiscible poly(lactide)/poly
(ε-caprolactone) blends.
*Biomacromolecules*, **3**, 1179.

198 Wang, Y. and Hillmyer, M.A. (2001)
Polyethylene-poly(L-lactide) diblock
copolymers: Synthesis and
compatibilization of poly(L-lactide)/
polyethylene blends. *J. Polym. Sci. Polym.
Chem.*, **39**, 2755.

199 Hassouna, F., Raquez, J.M., Addiego, F.,
Dubois, P., Toniazzo, V., and Ruch, D.
(2011) New approach on the
development of plasticized polylactide
(PLA): Grafting of poly(ethylene glycol)
(PEG) via reactive extrusion. *Eur. Polym.
J.*, **47**, 2134.

200 Zhang, J.-F. and Sun, X. (2004)
Mechanical properties of poly(lactic
acid)/starch composites compatibilized
by maleic anhydride. *Biomacromolecules*,
**5**, 1446.

201 Yao, M., Deng, H., Mai, F., Wang, K.,
Zhang, Q., Chen, F., and Fu, Q. (2011)
Modification of poly(lactic acid)/poly
(propylene carbonate) blends through
melt compounding with maleic
anhydride. *Exp. Polym. Lett.*, **5**, 937.

202 Li, H. and Huneault, M.A. (2011) Effect
of chain extension on the properties of
PLA/TPS blends. *J. Appl. Polym. Sci.*,
**122**, 134.

203 Juntuek, P., Ruksakulpiwat, C.,
Chumsamrong, P., and Ruksakulpiwat, Y.
(2012) Effect of glycidyl methacrylate-
grafted natural rubber on physical
properties of polylactic acid and natural
rubber blends. *J. Appl. Polym. Sci.*, **125**,
745.

204 Zeng, J.-B., Jiao, L., Li, Y.-D., Srinivasan,
M., Li, T., and Wang, Y.-Z. (2011) Bio-
based blends of starch and poly(butylene
succinate) with improved miscibility,
mechanical properties, and reduced water
absorption. *Carbohyd. Polym.*, **83**, 762.

205 Pracella, M., Rolla, L., Chionna, D., and
Galeski, A. (2002) Compatibilization and
properties of poly(ethylene
terephthalate)/polyethylene blends based
on recycled materials. *Macromol. Chem.
Phys.*, **203**, 1473.

206 Campoy, I., Arribas, J.M., Zaporta,
M.A.M., Marco, C., Gómez, M.A., and

Fatou, J.G. (1995) Crystallization kinetics of polypropylene–polyamide compatibilized blends. *Eur. Polym. J.*, **31**, 475.

207 Keskkula, H. and Paul, D.R. (1994) Toughened nylons, in *Nylon Plastics Handbook* (ed. M. Kohan), Carl Hanser, Munich, Chapter 11.

208 Lazzeri, A., Malanima, M., and Pracella, M.J. (1999) Reactive compatibilization and fracture behavior in nylon 6/VLDPE blends. *J. Appl. Polym. Sci.*, **74**, 3455.

209 Sanchez, A., Rosales, C., Laredo, E., Muller, A.J., and Pracella, M. (2001) Compatibility Studies in Binary Blends of PA6 and ULDPE-graft-DEM. *Macromol. Chem. Phys.*, **202**, 2461.

210 Liu, H., Xie, T., Zhang, Y., Ou, Y., and Yang, G. (2006) Crystallization behaviors of polypropylene/polyamide-6 blends modified by a maleated thermoplastic elastomer. *Polym. J.*, **38**, 21.

211 Bose, S., Bhattacharyya, A.R., Kodgire, P.V., and Misra, A. (2007) Fractionated crystallization in PA6/ABS blends: Influence of a reactive compatibilizer and multiwall carbon nanotubes. *Polymer*, **48**, 356.

212 Psarski, M., Pracella, M., and Galeski, A. (2000) Crystal phase and crystallinity of polyamide 6/functionalized polyolefin blends. *Polymer*, **41**, 4923.

213 Kobayashi, T., Wood, B.A., and Takemura, A. (2011) Crystallization, morphology, and electrical properties of bio-based poly(trimethylene terephthalate)/poly(ether esteramide)/ionomer blends. *J. Appl. Polym. Sci.*, **119**, 2714.

214 Hoffman, D.C. and Caldwell, J.K. (1995) Specialty Polyester '95 presentation.

215 Callander, D.D. (2003) in *Modern Polyesters: Chemistry and Technology of Polyesters and Copolyesters* (eds J. Scheirs and T.E. Long), John Wiley & Sons Ltd., West Sussex, England, Chapter 9.

216 Jensen, P.W. (1958) The determination of apparent second-order transition temperatures of polymers. *J. Polym. Sci. Pol. Chem.*, **28**, 635.

217 Kolb, H.J. and Izard, E.F. (1949) Dilatometric studies of high polymers. I. second-order transition temperature. *J. Appl. Phys.*, **20**, 564.

218 Ghanem, A.M. and Porter, R.S. (1989) Cold crystallization and thermal shrinkage of uniaxially drawn poly (ethylene 2,6-naphthalate) by solid-state coextrusion. *J. Polym. Sci. Pol. Phys.*, **27**, 2587.

219 Cakmak, M., Wang, Y.D., and Simhambhatla, M. (1990) Processing characteristics, structure development, and properties of uni and biaxially stretched poly(ethylene 2,6 naphthalate) (PEN) films. *Polym. Eng. Sci.*, **30**, 721.

220 Kimura, M. and Porter, R.S. (1983) Blends of poly(butylene terephthalate) and a polyarylate before and after transesterification. *J. Polym. Sci. Pol. Phys.*, **21**, 367.

221 Devaux, J., Godard, P., and Mercier, J.P. (1982) Bisphenol-A polycarbonate–poly (butylene terephthalate) transesterification. I. Theoretical study of the structure and of the degree of randomness in four-component copolycondensates. *J. Polym. Sci. Pol. Phys.*, **20**, 1875.

222 Shi, Y. and Jabarin, S.A. (2001) Transesterification reaction kinetics of poly(ethylene terephthalate/poly(ethylene 2,6-naphthalate) blends. *J. Appl. Polym. Sci.*, **80**, 2422.

223 Patcheak, T.D. and Jabarin, S.A. (2001) Structure and morphology of PET/PEN blends. *Polymer*, **42**, 8975.

# 3
# Morphology and Structure of Crystalline/Crystalline Polymer Blends

*Zhaobin Qiu and Shouke Yan*

*State Key Laboratory of Chemical Resource Engineering, Beijing University of Chemical Technology, Beijing 100029, China*

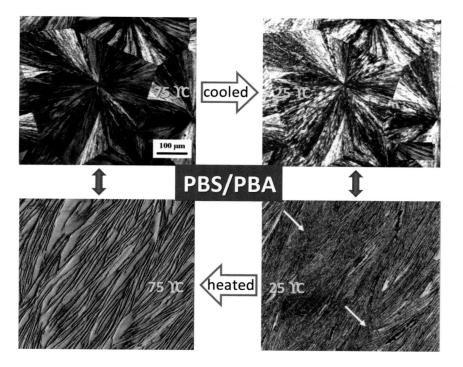

Optical micrographs of PBS/PBA blend crystallized from the melt first at 75 °C (upper left) and then at 25 °C (upper right). The corresponding AFM images of the PBS/PBA blend scanned at room temperature (lower right) and at 75 °C (lower left).

*Encyclopedia of Polymer Blends: Volume 3: Structure*, First Edition. Edited by Avraam I. Isayev.
© 2016 Wiley-VCH Verlag GmbH & Co. KGaA. Published 2016 by Wiley-VCH Verlag GmbH & Co. KGaA.

## 3.1
## Introduction

Tailoring high-value polymers to produce optimal properties for specific applications has become increasingly important in polymer science and engineering. To this end, attention has been focused significantly on polymer blends, which provides an efficient way for developing new polymeric materials with desirable property combinations, to achieve a large inventory of morphologies and associating these with end uses. The complicated phase structure of polymer blend influences the intrinsic multiscale structures of each component and shows pronounced influence on the property of the materials [1–43]. For the polymer blends, increasing interest has been seen for those containing at least one component capable of crystallizing, since a certain degree of crystallinity is normally important in order to ensure satisfactory high-temperature strength and environmental resistance. In these systems, liquid–solid phase separation during the crystallization of one component can lead to a large variety of morphologies [44–49]. Especially in crystalline/crystalline polymer blends, the difference in the melting point $(T_m)$ and crystallization temperature $(T_c)$ of the components can lead to the systems existing in amorphous/amorphous, crystalline/amorphous, and crystalline/crystalline states depending on the temperature. The occurrence of liquid–solid phase separation offers an effective route to produce a wide variety of morphological patterns. For such systems, an even wider range of morphologies has been opened. Therefore, a number of works have been reported on melt-miscible blends of two crystalline polymers with different chemical structures, which provide various conditions to study the crystallization behavior, morphology, and structure of polymer blends [50–88]. For the crystalline/crystalline polymer blends, the crystallization processes of the two components have significant influence on the resultant phase structure of the blend and morphologies of each component in the blends. When the systems have a large $T_m$ difference, the high-$T_m$ component usually crystallizes first at higher temperature [50–77]. The low-$T_m$ component acts as temporary amorphous diluent. The morphological pattern of this kind of blend system is determined by the distance over which the low-$T_m$ component is expelled. Generally, three basic types of separation, that is, interlamellar, interfibrillar, or interspherulitic segregations, can be generated. These morphological patterns represent the diluent dispersion from the nanometer scale (interlamellar segregation) to the micrometer scale (interspherulitic segregation) [78]. After the sample was cooled to a temperature below $T_m$ of the low-$T_m$ component, the crystallization of the low-$T_m$ component is often confined to, or initiated in, relatively small spaces. It can crystallize in any of the following modes: as another set of spherulites, in the spaces between first-formed spherulites, between growth arms in the first-formed spherulites, or between crystalline lamellae of the first-formed spherulites. There is another possibility of the two species to cocrystallize, if the components are sufficiently similar. For the systems with a small $T_m$ difference, two components can crystallize separately or simultaneously depending on blend composition and crystallization temperature [78–86]. If the crystallization rate of one component is significantly faster than that of the other, the rapidly growing component crystallizes first with its spherulites filling the entire volume, then the slowly growing component must crystallize subsequently as tiny crystals at the same orientation as the crystallites of the host spherulites of rapidly growing component, when they are crystallized at the same $T_c$ [79–86]. When their crystallization rates are comparable, both components are able to crystallize simultaneously at the same $T_c$, forming two distinct spherulite types; moreover, interpenetrating spherulites may usually be formed [79–86]. The feature of interpenetrating spherulites is that spherulites of one component continue to grow in the spherulites of the other component after they contact with each other. In the following sections, we will introduce the morphology and structure of melt-miscible crystalline/crystalline polymer blends with small and large melting point differences, respectively.

## 3.2
## Systems with Small Melting Point Difference

### 3.2.1
### Preliminary Study on Morphology and Structure of PES/PEO Blends

Depending on blend composition and crystallization temperature, two components can crystallize separately or simultaneously for the systems with small $T_m$ difference. The interpenetrating spherulites formation processes have been found in poly(butylene succinate) (PBS)/poly(vinylidene chloride-*co*-vinyl chloride) (PVDCVC), poly(ester carbonate) (PEC)/poly(L-lactide) (PLLA), poly(ethylene succinate) (PES)/poly(ethylene oxide) (PEO), poly(butylene succinate-co-butylene adipate) (PBSA)/PEO blends, and so on [79–86]. In this section, we choose PES/PEO blends as a model system to show the morphology and structure of miscible crystalline/crystalline polymer blends with small melting point difference. PES/PEO blends are typical miscible crystalline/crystalline polymer blends [81,82]. The melting point difference of the two components is small ($\sim$35 °C); therefore, PES/PEO blends are an ideal system to study the morphology and structure of crystalline/crystalline polymer blends, which may show simultaneous crystallization and the formation of interpenetrating spherulites.

Figure 3.1 illustrates a series of POM images showing the spherulitic morphology of neat PEO, neat PES, and their blends (80/20, 60/40, 40/60), which were crystallized at 50 °C to study the possibility of simultaneous crystallization of the two components. Both neat PEO and neat PES show negative birefringence. In the case of PES/PEO blends, the nucleation of PEO spherulites cannot be found down to the composition of PES/PEO = 40/60 (see Figure 3.1c–e), because PES crystallized quickly and filled the whole space. The PES spherulites become larger after blending with PEO, indicative of a decrease in nucleation density; moreover, the coarseness of PES spherulites increases with an increase in the PEO content in the blends.

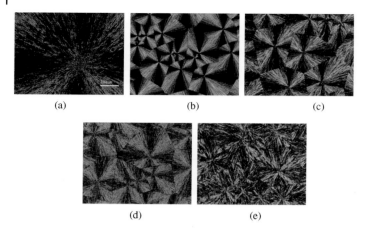

(a)     (b)     (c)

(d)     (e)

**Figure 3.1** Polarized optical micrographs of the spherulitic morphology of PES/PEO blends after complete crystallization at 50 °C: (a) 0/100, (b) 100/0, (c) 80/20, (d) 60/40, and (e) 40/60 (Same magnification, bar = 100 μm). Both PES ($M_w$ = 10,000 g/mol, $T_g$ = −11 °C, $T_m$ = 103 °C) and PEO ($M_w$ = 100,000 g/mol, $T_g$ = −55 °C, $T_m$ = 67 °C) samples were purchased from Scientific Polymer Products, Inc. (Reproduced from Ref. [81] with permission, © 2002, American Chemical Society.)

The most interesting crystalline morphology, that is, the formation of interpenetrating spherulites was observed for a PES/PEO 20/80 sample. Figure 3.2 shows two different types of spherulites growing during the crystallization. The large and compact spherulites are attributed to PEO-type spherulites, and the small and open ones are PES-type spherulite. By heating the sample to a temperature just above $T_m$ of PEO while below that of PES, the PEO-type spherulites disappeared, but PES-type spherulites still existed. Figure 3.2a displays that PES and PEO crystallized simultaneously with PEO spherulite generally growing faster than PES spherulite. Instead of growth being arrested, the PEO spherulite continued to crystallize inside the PES spherulite; moreover, the growth front of the PEO-type spherulite became distorted when it reached the PES-type spherulite. The brightness increases in the part of PES spherulite where the crystallization of PEO has occurred, and does not change where the crystallization of PEO

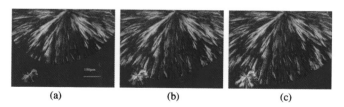

(a)     (b)     (c)

**Figure 3.2** The process of interpenetration in a 20/80 PES/PEO blend at 50 °C: (a) 3 min, (b) 5 min, and (c) 6.5 min (Same magnification, bar = 100 μm). (Reproduced from Ref. [81] with permission, © 2002, American Chemical Society.)

has not occurred (see Figure 3.2b). These results indicate that PES spherulite was penetrated by PEO spherulite. The outline of the brighter area is not consistent with the shape of the rest of the PEO spherulite; moreover, the brighter area seems too large. It can be reasoned that the PEO fibrils grow faster inside the PES spherulite because the PEO content in the amorphous regions of the PES spherulite should be expected to be higher than the nominal melt concentration due to rejection of PEO from PES crystals.

### 3.2.2
### Effect of Blend Composition on the Formation of Interpenetrating Spherulites of PES/PEO Blends

It is expected that such an interesting crystalline morphology must be related to blend composition and crystallization conditions. The effects of blend composition and crystallization temperature on the crystalline morphologies of PES/PEO blends were further studied in detail. For example, for a 30/70 blend, PES and PEO could also crystallize simultaneously at 47.5 °C and form interpenetrated spherulites [82]. It should be noted that the penetration of PES-type spherulites by PEO-type spherulites can be found in a wide crystallization temperature range for a variety of blend compositions, by adjusting blend composition and crystallization temperature. Depending on blend composition and crystallization temperature, the spherulites of both components may interpenetrate each other in some special cases, forming a novel morphological pattern. Figure 3.3 illustrates the crystallization process of a 50/50 blend at 47.5 °C. Figure 3.3a shows that both PES and PEO spherulites grow simultaneously with the growth rate of PEO being higher than that of PES. When the two types of spherulites developed freely in the undercooled melt, the growth rates were around 36 and 190 μm/min for the PES-type spherulite and PEO-type spherulite, respectively. As shown in Figure 3.3b, the two spherulites continued to crystallize inside the spherulite of the other component when they contacted, forming interpenetrated spherulites.

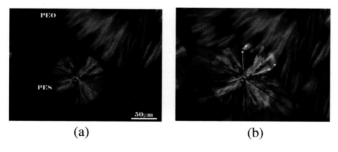

(a)                                        (b)

**Figure 3.3** The simultaneous crystallization and formation process of interpenetrated spherulites in a 50/50 blend at $T_c = 47.5$ °C at different crystallization time: (a) 221 s and (b) 245 s. Black arrows indicate the growth direction of PEO spherulite, and white arrows indicate the growth direction of PES spherulite. (Reproduced from Ref. [82] with permission, © 2008, American Chemical Society.)

It is interesting to note that the PEO fibrils traveled at a growth rate of around $6.6 \times 10^2$ μm/min inside the PES spherulite during this penetration process; however, the growth rate value was only around 190 μm/min for the same PEO spherulite developing along other directions in the undercooled melt. Such growth rates difference results in the distortion phenomenon of the PEO spherulite growth front. The growth rate of PEO recovered to be around 190 μm/min again after passing through the PES spherulite. As also indicated by the arrows in Figure 3.3b, the PES spherulite becomes a new growth site for the PEO component, indicative of the penetration of PES spherulite by PEO spherulite. PEO must continue to grow along the fibrils of PES spherulites; as a result, the growth direction of PEO must change, forming new boundaries with original PEO spherulite. As is well known, the boundary can only appear when neighboring spherulites contact each other; therefore, it is of great interest that new boundaries are found for a single PEO spherulite after penetrating the PES spherulite. Unlike the crystalline morphology pattern that PES stopped growing inside the PEO spherulite in the 20/80 and 30/70 samples, PES may continue to crystallize inside the PEO spherulite in a 50/50 blend, as indicated by the increase of brightness and the growth front shown in Figure 3.3b. Such a result indicated that PES spherulite also penetrated into PEO spherulite. Thus, PES and PEO formed interpenetrating spherulites. The growth rates was estimated to be around 50 μm/min for the PES fibrils, which continued to develop inside the PEO spherulite as indicated by the white arrow in Figure 3.3b. Such an increase in the growth rate of PES fibrils can also be attributed to the melt concentration change during the formation of penetrated spherulites.

Blend composition plays a dominant role on the development of PES spherulite inside PEO spherulite. For $\phi_{PEO} \geq 70\%$, the advancement of PES spherulites terminated inside PEO spherulites. For $50\% \leq \phi_{PEO} < 70\%$, PES continued to grow inside PEO spherulite during the interpenetration. Such results may be explained as follows. For $\phi_{PEO} \geq 80\%$, the structure of PEO spherulite is so compact that there is not enough space for PES to continue to crystallize inside. For $50\% \leq \phi_{PEO} < 70\%$, the compactness of the structure of PEO spherulite reduced, providing the space for PES to continue to crystallize inside further. The important factors in realizing interpenetrating spherulites are the difference in the lamella population density in the different types of spherulites of the two components and the sufficient amount of the melt of one component inside the spherulites of the other component. As a result, interpenetrating spherulites appeared for a wide crystallization temperature range and blend composition in PES/PEO blends. The blends with $\phi_{PEO}$ between 50 and 90% present interpenetrating spherulites morphology in a wide $T_c$ range of 37.5–57.5 °C.

Both components crystallized as separate spherulites and formed interpenetrating spherulites, when they contacted each other in most cases. A novel crystalline morphology pattern was also found in some cases for the samples with $50 \leq \phi_{PEO} \leq 70\%$ at $37.5 \leq T_c \leq 50$ °C. PES and PEO usually crystallize as separate spherulites, forming interpenetrating spherulites in some area of the sample; however, different crystalline morphology may also be observed in some area of

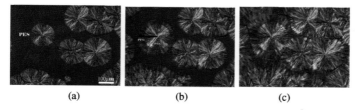

(a)  (b)  (c)

**Figure 3.4** The formation of blended spherulites showing two growth fronts in a 50/50 blend at $T_c$ = 42.5 °C with the nucleation of PEO spherulite inside PES spherulite at different crystallization time: (a) 310 s, (b) 335 s, and (c) 390 s. Black arrow indicates the growth front of PEO spherulite, and white arrow indicates the growth front of PES spherulite. (Reproduced from Ref. [82] with permission, © 2008, American Chemical Society.)

the sample. Figure 3.4 displays the crystallization process of a 50/50 blend at 42.5 °C. Figure 3.4a shows that PES crystallized according to the spherulitic morphology with a growth rate of around 22 μm/min. With prolonging crystallization time, PEO started to nucleate at the center part of PES spherulites as shown in Figure 3.4b, which further grew through a spherulitic shape inside the PES spherulites and surpassed the growth front of original PES spherulites. Finally, PEO spherulites stopped growing and formed straight boundaries when they contacted each other as shown in Figure 3.4c. From Figure 3.4, PEO spherulite started to crystallize inside PES spherulite, showing two growth fronts, corresponding to the growth of PES and PEO components, respectively, in one spherulite. By heating the sample to a temperature between $T_m$ of two components, the outer front part disappeared, while the inner part still existed, confirming that the inner and outer fronts of the blended spherulite are from PES and PEO, respectively. The growth rate of PES corresponding to the inner growth front was determined to be around 28 μm/min, which is higher than that developed freely in the undercooled melt at the same $T_c$. Such an increase in the growth rate may be related to the melt concentration change during the crystallization process of PES surpassed by PEO. The growth rate corresponding to the growth front of PEO was determined to be 190 μm/min. The growth rate of PEO spherulite is significantly faster than that of PES spherulite. Similar results were also found a 40/60 blend at 42.5 °C. Such a novel crystalline morphology is named as blended spherulites showing two growth fronts.

PEO would form separate spherulites and form interpenetrating spherulites, when they contact PES spherulites, if PEO nucleated from the homogeneous melt. However, PEO would form blended spherulites with two growth fronts, if PEO nucleated inside the preexisting PES spherulites. Where PEO started to nucleate is the key factor influencing the crystalline morphologies. Whether the formation of PES spherulites penetrated by PEO spherulites or the blended spherulites with two growth fronts occurred mainly depended on the nucleation sites of PEO in binary miscible crystalline/crystalline polymer blends. The fact that blended spherulites showing two growth fronts were only observed with $50 \leq \phi_{PEO} \leq 70\%$ may be explained as follows. In the case of the PEO content above 70%, the PES content is low. Therefore, PES spherulites grow slowly with open spherulitic structure and leave sufficient undercooled melt in the blends. In this case, it will be easier for PEO to nucleate in the undercooled melt rather than inside the PES spherulites. Thus, it will be difficult to form blended spherulite showing two growth fronts.

### 3.2.3
### Effect of Crystallization Temperature on the Crystalline Morphologies of PES/PEO Blends

In addition to blend composition, crystallization temperature is another important factor, which influences the structure and morphology of crystalline/crystalline polymer blends. For the miscible PES/PEO blends, a 20/80 blend is chosen as a model blend to investigate the effect of $T_c$ on the crystalline morphology. Depending on $T_c$, a 20/80 blend may present various crystalline morphologies. When $T_c$ is above 60 °C, only PES can crystallize through spherulitic growth, because PEO is still in the melt due to the different supercooling required by the two components. On further lowering $T_c$, both components can crystallize. Figure 3.5 shows the crystalline morphology of a 20/80 blend at 57.5 °C. At this high $T_c$, PES crystallized first through spherulitic growth with growth rate of around 6.1 μm/min and filled the whole sample; moreover, PEO is still in the melt due to the low supercooling. The crystallization of PES corresponds to the phase transition from amorphous/amorphous to crystalline/amorphous state. By prolonging crystallization time, PEO also crystallized in the matrix of PES spherulites. The crystallization of PEO corresponds to the phase transition from crystalline/amorphous to crystalline/crystalline state. Figure 3.5 shows the growth process of PEO component. As shown in Figure 3.5a, one PEO spherulite with strong birefringence was found to nucleate and continue to grow within the preexisting PES spherulites. As shown in Figure 3.5b, the growth of PEO spherulites proceeded with keeping almost the spherulitic shape until they impinged on other PEO spherulites. The growth rate of PEO nucleating and growing inside the preexisting PES spherulites was determined to be around 2.7 μm/min.

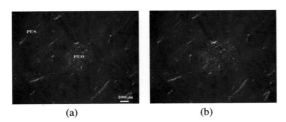

(a)                              (b)

**Figure 3.5** Crystalline morphology in a 20/80 blend at 57.5 °C: (a) spherulitic morphology and growth of PEO spherulites after the PES spherulites filled the whole space sample at 150 min and (b) 202 min. (Reproduced from Ref. [82] with permission, © 2008, American Chemical Society.)

It is usually difficult for the low-$T_m$ component to crystallize through the spherulitic growth in crystalline/crystalline polymer blends, because it must crystallize in the matrix of the preexisting crystals of the high-$T_m$ component. Only tiny crystals of the low-$T_m$ component are found to grow within the spherulites or at the boundaries of the spherulites of the high-$T_m$ component for most crystalline/crystalline polymer blends. The possibility of the observation of the nucleation and spherulitic growth of the low-$T_m$ component within the matrix of the preexisting crystals of the high-$T_m$ component may be related to the following two factors. One is the nucleation ability of the low-$T_m$ component, and the other is the amount of its melt in the matrix of the preexisting crystals of the high-$T_m$ component. The lower nucleation ability and the higher amount of the melt of the low-$T_m$ component favor the observation of its nucleation and spherulitic growth. On the basis of the aforementioned factors, PEO nucleated and grew within the preexisting PES spherulites for the blend with high PEO content at higher $T_c$ values.

On further lowering $T_c$ to 55 °C and below, the nucleation ability of PEO was improved because of relatively high supercooling. In this case, PEO can also nucleate and grow through spherulitic growth before PES filled the whole sample. In this case, two types of spherulites, corresponding to PES and PEO spherulites, respectively, can grow simultaneously. The penetration of PES spherulites with slow growth rate by PEO spherulites with fast growth rate can be found in the 20/80 blend, as shown in Figure 3.2.

Upon further lowering $T_c$ to 45 °C and below, the crystallizability of PEO was improved significantly. PEO nucleated first and continued to fill the whole sample before the nucleation of PES. When a 20/80 blend was quenched to 45 °C and below, PEO spherulites contacted each other and showed clear boundaries between neighboring spherulites first. The nucleation of PES component was impossible to be observed, before PEO-type spherulites filled the whole sample. However, the brightness of PEO spherulites increased apparently with further prolonging crystallization time, indicative of the occurrence of crystallization of PES inside the PEO spherulites. Unlike PEO that could be observed to crystallize through the so-called spherulitic growth, PES spherulites cannot be observed, when they crystallized inside the PEO spherulites in the interlamellar and interfibrillar regions. The different crystalline morphologies may arise from the difference in the nucleation ability of the two components and the supercooling. At $T_c$ as high as 57.5 °C the nucleation ability of PES was stronger than that of PEO; therefore, PES crystallized first and filled the whole sample, followed by the nucleation and growth of PEO spherulites inside the preexisting PES spherulites. At $T_c$ as low as 40 °C, the nucleation ability of PEO was stronger than that of PES; therefore, PEO crystallized first and filled the whole sample, followed by the crystallization of PES crystals inside the preexisting PEO spherulites. In brief, crystallization temperature plays a dominant role on the crystalline morphology for a given blend composition sample in miscible crystalline/crystalline polymer blends. Various crystalline morphologies can be obtained by just controlling crystallization temperature.

The crystalline morphologies of miscible crystalline/crystalline polymer blends depend not only on blend composition but also crystallization temperature. The difference in the crystallization rates (or spherulitic growth rates) of both components is a key factor of influencing the crystalline morphology patterns. By adjusting crystallization temperature and blend composition, three different kinds of crystalline morphologies may be formed. The first is that the growth rate of PEO is significantly faster than that of PES. In this case, the low-$T_m$ component PEO crystallized first with compact structure and filled the whole space, while the high-$T_m$ component PES crystallized subsequently as microcrystals inside the preexisting PEO spherulites with prolonging crystallization time. The second is that the growth rates of both components are comparable. Both components can crystallize simultaneously. Depending on the nucleation site of the low-$T_m$ component PEO, both components can form interpenetrated spherulites or blended spherulite containing two growth fronts from the same growth site with PEO growing faster than PES. The third is that the growth rate of PES is significantly faster than that of PEO. In this case, the high-$T_m$ component PES crystallized first with very open structure, while the low-$T_m$ component PEO nucleated and continued to grow subsequently in the preexisting PES spherulites with prolonging crystallization time.

Figure 3.6 summarizes the crystallization temperature and blend composition dependence of morphological patterns of PES/PEO blends in detail. Only PES component crystallized for $\Phi_{PEO} \leq 40\%$ in the blends. Both components crystallized at the same crystallization temperature and formed interpenetrated

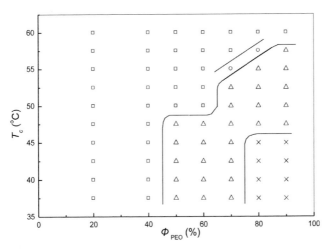

**Figure 3.6** Dependence of the spherulitic morphologies in PES/PEO blends on the crystallization temperature $T_c$ and the PEO content $\phi_{PEO}$ (%): square, only PES crystallized; circle, PES first filled the whole space while PEO crystallized in the matrix of PES spherulites later; triangle, simultaneous crystallization and penetration of spherulites took place; cross, PEO first filled the whole space while PES crystallized inside the PEO spherulites later. (Reproduced from Ref. [82] with permission, © 2008, American Chemical Society.)

spherulites for the samples with $50 \leq \phi_{\text{PEO}} \leq 90\%$ at $37.5 \leq T_c \leq 57.5\,°C$. Furthermore, a new crystalline morphology pattern of blended spherulite with two growth fronts could be observed for the samples with $50 \leq \phi_{\text{PEO}} \leq 70\%$ at $37.5 \leq T_c \leq 50\,°C$. For the samples with high $\phi_{\text{PEO}}$ at high $T_c$, PES crystallized first and filled the whole space followed by the crystallization of PEO inside the preexisting spherulites through spherulitic growth. For the samples with $80 \leq \phi_{\text{PEO}} \leq 90\%$ at $T_c \leq 45\,°C$, PEO crystallized first and filled the whole space followed by the crystallization of PES inside the preexisting PEO spherulites as microcrystals. Although such summary is on the basis of PES/PEO blends, it will also be useful for a better understanding and prediction of morphological patterns for other miscible crystalline/crystalline polymer blends.

## 3.3
## Systems with Large Melting Point Difference

Unlike the systems with small melting point difference, the crystallization of each component in the systems with a large $T_m$ difference can only take place separately. The high-$T_m$ component usually crystallizes first at higher temperature. During this process, the low-$T_m$ component acts as temporary amorphous diluent. The amorphous diluent influences the crystallization behavior of the high-$T_m$ component, for example, the change in crystallization kinetics. Also the morphological pattern of this kind of blend system is determined by the distance over which the low-$T_m$ component is expelled. Generally, three basic types of separation, that is, interlamellar segregation, interfibrillar segregation or interspherulitic segregation, can be generated. These morphological patterns represent the diluent dispersion from the nanometer scale (interlamellar segregation) to the micrometer scale (interspherulitic segregation) [78]. Moreover, after the sample was cooled to a temperature below $T_m$ of the low-$T_m$ component, the crystallization of the low-$T_m$ component is often confined to, or initiated in, relatively small spaces. It can crystallize in any of the following modes: as another set of spherulites, in the spaces between first-formed spherulites, between growth arms in the first-formed spherulites, or between crystalline lamellae of the first-formed spherulites. Another possibility is cocrystallization of the two species, if the components are sufficiently similar in chemical structure. The complicated mutual influence between the components leads to the morphology and structure of binary miscible crystalline polymer blends far away from a full understanding. In the following section, the context as to how one component may influence the crystallization process of the other one will be provided with elaborately selected examples.

## 3.3.1
## Crystallization Behavior of the High-$T_m$ Component in Miscible Polymer Blends

Many crystalline/crystalline blends such as polycarbonate (PC)/polycaprolactone (PCL), poly(3-hydroxybutyrate) (PHB)/PEO, poly(vinylidene fluoride) (PVDF)/

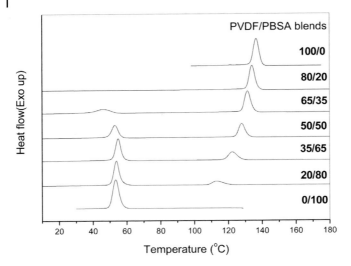

**Figure 3.7** DSC cooling traces of PVDF/PBSA blends at 10 °C/min from the homogeneous melt. PVDF ($M_w$ = 140,000 g/mol) and PBSA (a copolymer containing 20 mol% butylene adipate with $M_w$ = 14,400 g/mol) were purchased from Polysciences, Inc. and Aldrich Chemical Co., respectively. (Reproduced from Ref. [74] with permission, © 2007, American Chemical Society.)

poly(butylene adipate) (PBA), poly(butylene terephthalate) (PBT)/polyarylates (PAr), PVDF/PHB, and PBS/PEO have been reported [50–69]. For the miscible crystalline/crystalline polymer blends, much attention has been paid to the crystallization and morphology of the high-$T_m$ component. It takes place usually at temperature above $T_m$ of the low-$T_m$ component. As an example, Figure 3.7 shows the DSC cooling traces of PVDF/PBSA blends at 10 °C/min from the homogeneous melt [74,75]. Both polymers are crystalline in their neat state, which undergo crystallization over a wide range of temperature. The PVDF/PBSA blends are miscible in the whole composition range, which was determined by the depression of crystallization and melting temperatures of the blends [74,75]. Identification of its miscibility through measuring the glass transition temperature ($T_g$) of the blends is impossible, owing to the close $T_g$ values of PVDF and PBSA (−43.1 °C versus −44.8 °C).

From Figure 3.7, neat PVDF shows a crystallization peak temperature ($T_p$) at 136.6 °C, while neat PBSA has a $T_p$ at 53.4 °C. The difference in $T_p$ between neat PVDF and PBSA is over 80 °C. In this case, they cannot crystallize simultaneously. PVDF and PBSA in the blends display individual exothermal peaks for crystallization. It is obvious that the $T_p$ of PVDF shifts downward to low-temperature range with an increase in the PBSA content. The $T_p$ of PVDF is only around 112.7 °C in the 20/80 blend, which is lower than that of neat PVDF by over 20 °C. Nevertheless, the PVDF always crystallizes during the quenching process before $T_c$ of PBSA is reached. This means that the PBSA is still in the melt as an amorphous diluent during the crystallization of PVDF.

The amorphous diluent of low-$T_m$ component shows influence on the crystallization kinetics of the high-$T_m$ component. The influence of amorphous low-$T_m$ component on the crystallization kinetics of the high-$T_m$ component is composition and $T_c$ dependent. There exists a temperature $T_c*$, below which a steady linear relationship between the spherulite radius ($R$) and time ($t$), that is, $R \propto t$, could be observed. Otherwise, a state with $R \propto t^{1/2}$ relationship between the spherulite radius and time could be established [89–92]. Figure 3.8 presents a representative example with PBS/PBA blends [76], which exhibits full miscibility

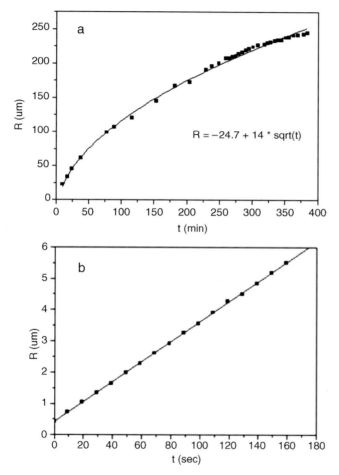

**Figure 3.8** Radius of PBS spherulites versus crystallization time for 80/20 PBS/PBA blend crystallizing at (a) 100 °C and (b) 75 °C. PBA was produced by BASF AG with a weight-average molecular weight of about $3.5 \times 10^4$ g/mol. PBS with an $M_w$ of $2.0 \times 10^5$ g/mol was purchased from Aldrich Chemical. The melting points of PBS and PBA are measured to be 114 °C and 60 °C, respectively. The line in (a) is the least-squares parabolic fit, while the line in (b) is the least-squares linear fit. (Reproduced from Ref. [76] with permission, Copyright © 2008 Elsevier Ltd.)

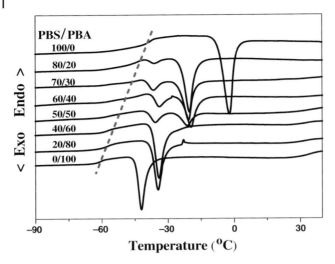

**Figure 3.9** DSC heating curves of PBS/PBA blends with different blend ratios as indicated. The appearance of a single glass transition temperature of all blends indicates the miscibility of the system. The PBS and PBA materials are the same as that used in Figure 3.8. (Copyright © 2008 Elsevier Ltd.)

in the melt and no liquid/liquid phase separation in whole blend composition range due to their very similar chemical structure. The miscibility of PBS/PBA blends is also confirmed by the appearance of single composition dependent $T_g$ as shown in Figure 3.9. Figure 3.8 shows the spherulitic radius of PBS versus crystallization time in a miscible blend with 20% PBA crystallized at 100 °C and 75 °C, respectively. It is seen that the PBS spherulite radius increases with the square root of time when crystallized at 100 °C (Figure 3.8a), while the radius increases linearly with time when crystallized at 75 °C (Figure 3.8b). Similar behavior can be observed for other miscible binary blend systems.

Figure 3.10 shows the spherulitic radius of PVDF versus crystallization time in a miscible blend with 30% PBS crystallized at 162 °C and 100 °C, respectively [43]. A square root relationship between the PVDF spherulite radius and time when crystallized at 162 °C (Figure 3.10a) and a linear relationship between the PBS spherulite radius and time when crystallized at 150 °C are observed (Figure 3.10b). This phenomenon is related to the different diffusion behavior of the noncrystallizable component and can be understood in terms of a diffusion length $\delta = D/G$, where $D$ is the diffusion coefficient of the noncrystallizable species in the blend and $G$ is the radial growth rate of the spherulites. The diffusion length is a quantitative measure of the distance that the noncrystallizable chain can move in the melt ahead of the crystallization front. If the diffusion length $\delta$ is sufficiently small, the noncrystallizable component can only diffuse into the interlamellar region of the crystallizing component. For an intermediate range of $\delta$, the noncrystallizable component should locate between the fibrils known as growth arms. At large $\delta$, the noncrystallizable component will be expelled into the interspherulitic regions [45,89,90,93].

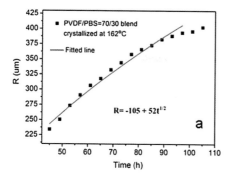
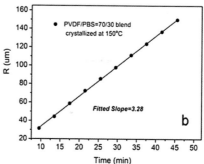

**Figure 3.10** Spherulitic radius of PVDF versus crystallization time in 70/30 blend crystallized at (a) 162 °C and (b) 150 °C. PVDF and PBS were purchased from Sigma-Aldrich Company and have a weight-average molecular weight of about $2.7 \times 10^5$ g/mol and $6.3 \times 10^4$ g/mol, respectively. The melting points were measured to be 186 °C for PVDF and 120 °C for PBS. The line in (a) is the least-squares parabolic fit and the line in (b) is the least-squares linear fit. (Reproduced from Ref. [43] with permission, The Royal Society of Chemistry, 2011.)

Based on the diffusion length $\delta$ of the noncrystallizable component, the above results can be qualitatively explained. To estimate the $\delta$, the crystal growth rate $G$ of PBS in the PBS/PBA blends with different blend ratios at varying crystallization temperatures was measured and shown in Figure 3.11.

It can be found that the crystal growth rate of PBS in different blends at 75 °C is about two orders of magnitude higher than that at 100 °C. At the same time, the diffusivity $D$ of PBA melt in front of growing PBS crystals should decrease with decreasing temperature. Therefore, diffusion length can be increased of

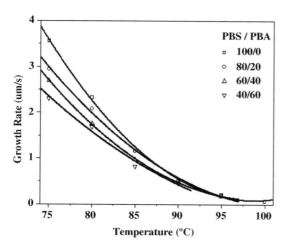

**Figure 3.11** Temperature dependence of spherulitic growth rates of PBS in the PBS/PBA blends. The PBS and PBA materials are the same as that used in Figure 3.8. (Copyright © 2008 Elsevier Ltd.)

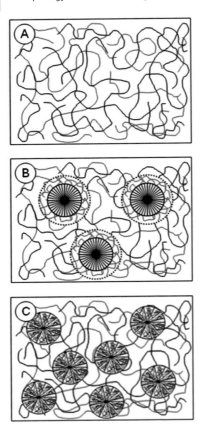

**Figure 3.12** Sketches showing the different diffusion behavior of the PBA melt in the miscible PBS/PBA blend (a) during the crystallization process of PBS at relatively high (b) and low (c) temperatures.

more than a factor of 100 as the crystallization temperature is increased. This leads to the PBA in an 80/20 PBS/PBA blend being excluded from the growing PBS spherulites at 100 °C, as sketched in Figure 3.12b. Taking this into account, the PBA component will build up more and more ahead of the growing PBS spherulite and slow its growth at 100 °C, leading to a $R \propto t^{1/2}$ relationship between the spherulite radius and time [89–91]. At the same time, a PBA surrounding layer will be accumulated at the growth front of PBS, leading to intersherulitic phase separation. On the other hand, the short diffusion length of PBA melt at 75 °C results in all of the PBA retaining within the growing PBS spherulite and crystallizes between PBS lamellae or fibrils (growth arms), as sketched in Figure 3.12c. The inclusion of PBA within the PBS spherulites at 75 °C causes a steady-state $R \propto t$ relationship.

The influence of composition on the crystallization kinetics of the high-$T_m$ component is similar over the whole crystallization temperature range. For this

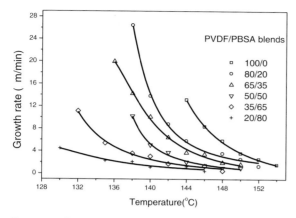

**Figure 3.13** Temperature dependence of spherulitic growth rates of PVDF in the PVDF/PBSA blends. (Reproduced from Ref. [74] with permission, © 2007, American Chemical Society.)

aspect, we recall the PVDF/PBSA system. Figure 3.13 shows the composition dependent spherulitic growth rates of PVDF in the temperature range of 130–155 °C. It can be seen that the spherulitic growth rates of PVDF both in pure state and in the blends decrease with an increase in crystallization temperature, regardless of blend compositions. In the blends, the spherulitic growth rates decrease with an increase in the PBSA content for a given crystallization temperature. The decrease of spherulitic growth rates may be related to the following two factors. One is the depression of $T_m$ of PVDF after blending with PBSA, which reduces the driving force for the crystallization process from the homogeneous melt. For example, $T_m$ decreased from 165 °C for neat PVDF to 146.7 °C for the 20/80 blend. The other is that the PBSA in the melt during the crystallization of PVDF plays a diluent role during the growth process of PVDF spherulites.

The well-known Avrami equation is often used to analyze the isothermal crystallization kinetics; it assumes the development of the relative degree of crystallinity ($X_t$) with $t$ as

$$1 - X_t = \exp(-kt^n), \tag{3.1}$$

where $n$ is the Avrami exponent depending on the nature of nucleation and growth geometry of the crystals, and $k$ is a composite rate constant involving both nucleation and growth rate parameters [94]. The crystallization half-time ($t_{1/2}$), the time required to achieve 50% of the final crystallinity of the samples, is an important parameter for the discussion of crystallization kinetics. Usually, the crystallization rate is easily described by the reciprocal of $t_{1/2}$. The value of $t_{1/2}$ is calculated by the following equation:

$$t_{1/2} = \left(\frac{\ln 2}{k}\right)^{1/n}. \tag{3.2}$$

The Avrami analysis of the PVDF-rich PVDF/PBSA blend reveals the dependence of overall crystallization kinetics of PVDF on the crystallization

**Table 3.1** Crystallization kinetic parameters of PVDF in the PVDF-rich blends.

| PVDF/PBSA blends | $T_c$ (°C) | $n$ | $k$ (min$^{-n}$) | $t_{1/2}$ (min) | $1/t_{1/2}$ (min$^{-1}$) |
|---|---|---|---|---|---|
| 100/0 | 144 | 2.6 | $2.36 \times 10^{-1}$ | 1.51 | $6.62 \times 10^{-1}$ |
| | 146 | 2.7 | $5.64 \times 10^{-2}$ | 2.53 | $3.95 \times 10^{-1}$ |
| | 148 | 2.6 | $1.29 \times 10^{-2}$ | 4.52 | $2.21 \times 10^{-1}$ |
| | 150 | 2.6 | $2.16 \times 10^{-3}$ | 9.20 | $1.09 \times 10^{-1}$ |
| | 152 | 2.6 | $4.57 \times 10^{-4}$ | 16.7 | $5.98 \times 10^{-2}$ |
| 80/20 | 140 | 2.4 | $7.76 \times 10^{-1}$ | $9.53 \times 10^{-1}$ | 1.05 |
| | 142 | 2.5 | $2.19 \times 10^{-1}$ | 1.59 | $6.29 \times 10^{-1}$ |
| | 144 | 2.4 | $5.99 \times 10^{-2}$ | 2.85 | $3.51 \times 10^{-1}$ |
| | 146 | 2.6 | $9.84 \times 10^{-3}$ | 5.07 | $1.97 \times 10^{-1}$ |
| | 148 | 2.5 | $2.80 \times 10^{-3}$ | 9.15 | $1.09 \times 10^{-1}$ |
| 65/35 | 136 | 2.5 | 1.84 | $6.83 \times 10^{-1}$ | 1.46 |
| | 138 | 2.5 | $4.06 \times 10^{-1}$ | 1.24 | $8.06 \times 10^{-1}$ |
| | 140 | 2.4 | $1.50 \times 10^{-1}$ | 1.90 | $5.26 \times 10^{-1}$ |
| | 142 | 2.5 | $3.01 \times 10^{-2}$ | 3.45 | $2.90 \times 10^{-1}$ |
| | 144 | 2.4 | $9.74 \times 10^{-3}$ | 5.83 | $1.72 \times 10^{-1}$ |
| 50/50 | 134 | 2.5 | 1.26 | $7.89 \times 10^{-1}$ | 1.27 |
| | 136 | 2.3 | $4.94 \times 10^{-1}$ | 1.16 | $8.62 \times 10^{-1}$ |
| | 138 | 2.4 | $1.62 \times 10^{-1}$ | 1.86 | $5.38 \times 10^{-1}$ |
| | 140 | 2.5 | $6.12 \times 10^{-2}$ | 2.66 | $3.76 \times 10^{-1}$ |
| | 142 | 2.5 | $2.09 \times 10^{-2}$ | 4.18 | $2.39 \times 10^{-1}$ |

Reproduced from Ref. [74] with permission.

temperature as well as blend composition. Table 3.1 summarizes all the related crystallization kinetics parameters for the isothermal crystallization of PVDF in the PVDF-rich blends at various crystallization temperatures.

The crystallization of PVDF is affected not only by crystallization temperature but also by blend composition in the blends. For a given crystallization temperature, the values of $t_{1/2}$ in the blends are larger than that of neat PVDF and become larger with increasing the PBSA content. On the other hand, the values of $1/t_{1/2}$ in the blends are smaller than those of neat PVDF and become smaller with increasing the PBSA content. The variation trends in the $t_{1/2}$ and $1/t_{1/2}$ values demonstrate that blending with PBSA retards the isothermal crystallization of PVDF, because the former is actually in the melt during the crystallization of the latter and acts as a diluent. The average value of $n$ is around 2.5 for the isothermal crystallization of PVDF, regardless of crystallization temperature and blend composition, indicating that the crystallization mechanism of PVDF corresponds to the spherulitic growth with heterogeneous nucleation [95]. The almost unchanged $n$ in the PVDF-rich blends indicates that the crystallization mechanism of PVDF does not change in the crystallization temperature range, regardless of blend composition [95].

3.3.2

## Crystallization Behavior of Low-$T_m$ Component in Miscible Polymer Blends

After crystallization of the high-$T_m$ component, the low-$T_m$ component will crystallize on further lowering crystallization temperature, corresponding to the transition from the amorphous/semicrystalline to semicrystalline/semicrystalline state of the blend. During this process, the preexisting crystals of high-$T_m$ component will influence the growth behavior of the low-$T_m$ component significantly. As observed in Figure 3.7, the $T_p$ of PBSA in its PVDF/PBSA blends is at around 54 °C and almost unchanged with increasing the PVDF content up to 50 wt%, while decreases with further increase of the PVDF. For example, the $T_p$ of PBSA in a 65/35 blend decreases to 45.6 °C. It should be pointed out that the microscopy observation of the spherulitic growth of the low-$T_m$ component is usually difficult in crystalline/crystalline polymer blends, because tiny crystals of the low-$T_m$ component will usually grow within the spherulites or at the boundaries of the spherulites of the high-$T_m$ component.

Spherulitic growth of the low-$T_m$ component is most often observed for the systems with a relatively high concentration of low-$T_m$ component and crystallized at relatively high temperatures. Figure 3.14 provides an example showing the individual spherulites of PBS formed in a 30/70 PVDF/PBS blend [71]. The crystallization

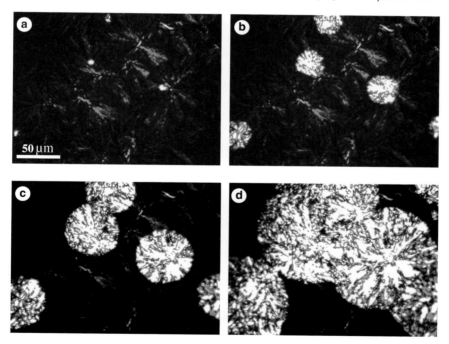

**Figure 3.14** Polarized optical micrographs of 30/70 PVDF/PBS blend crystallized at 130 °C and subsequently quenched to 80 °C for (a) 2 min, (b) 3 min, (c) 4 min, and (d) 5.5 min. The contrast and brightness of the pictures were adjusted to better reveal the PVDF spherulites. (Reproduced from Ref. [71] with permission, © 2011, American Chemical Society.)

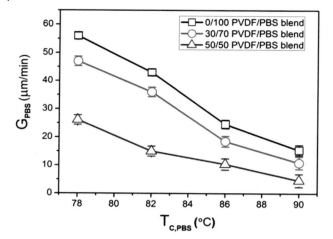

**Figure 3.15** The dependence of the growth rate of PBS in its blends with PVDF on the crystallization temperature and the blend ratio. The crystallization temperature of the PVDF matrix was fixed at 120 °C. (Reproduced from Ref. [71] with permission, © 2011, American Chemical Society.)

temperature of PVDF is 130 °C, while the crystallization temperature of PBS is 80 °C. It can be seen that the nucleus densities of PBS in the 30/70 PVDF/PBS blend remained still very low at $T_c$ ranging from 80 °C despite the presence of the PVDF crystals due to its low supercooling. Furthermore, the concentration of the PBS melt is high enough for a continuous spherulitic growth. Under these circumstances, the spherulitic radius of the low-$T_m$ component in the blends versus crystallization time can be plotted and the crystal growth rate can be obtained.

Figure 3.15 shows the dependence of the growth rate of PBS in its blends with PVDF on the crystallization temperature and the blend ratio. From Figure 3.15, when the PVDF matrix crystallization temperature was fixed at 120 °C, the spherulitic growth rates of PBS and that in the blends decrease with an increase in crystallization temperature. The spherulitic growth rates of PBS in the blends decrease further with increasing PVDF content for a given crystallization temperature, indicating the influence of preformed PVDF crystals on the crystallization process of PBS. It was further found that, in addition to the blend ratio and crystallization temperature of PBS, the PVDF matrix crystallization temperature also influences the growth of PBS in the confined environment.

Figure 3.16 illustrates the influences of the PVDF matrix crystallization temperature, together with the blend ratio, on the growth rates of PBS at a given PBS crystallization temperature, that is, 85 °C for the present case. Two blends of PVDF/PBS with blend ratios of 30/70 and 50/50 were chosen to display the effect of blend ratio on the crystal growth rate of the PBS. It can be seen that in the studied temperature range, the PVDF/PBS 30/70 blend exhibits always a larger growth rate than the 50/50 blend at all $T_{c,PVDF}$. When the PVDF matrix is crystallized above 145 °C, $G_{PBS}$ increases with decreasing $T_{c,PVDF}$ for both blends. When the PVDF matrix is

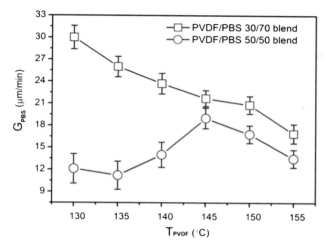

**Figure 3.16** The variation of PBS growth rate at 85 °C as a function of blend ratio and crystallization temperature of the PVDF matrix. (Reproduced from Ref. [71] with permission, © 2011, American Chemical Society.)

crystallized below 145 °C, very different behavior of the two blends is observed. For the 30/70 blend, the growth rate of PBS further increases with decreasing $T_{c,PVDF}$, while the PBS growth rate for the 50/50 blend decreases with further decreasing $T_{c,PVDF}$.

A similar trend is also found in the crystallization half-times for PBS in the blends as a function of $T_{c,PVDF}$ and blend ratio, as shown in Figure 3.17. It is seen that, with $T_{c,PVDF}$ below 146 °C, $t_{1/2}$ of PBS decreases with decreasing $T_{c,PVDF}$ for the 30/70 blend, but increases for the 50/50 blend. This trend fits quite well with the data points presented in Figure 3.16 below 145 °C. It should be pointed out that the $t_{1/2}$ values for the 30/70 blends are all larger than those for the 50/50 blends, indicating a slower crystallization of the 30/70 blends. The values of $G_{PBS}$ for 30/70 blends are, however, larger than those for 50/50 blends. This is caused by the fact that the $t_{1/2}$ value depends not only on crystal growth rate, but also on the induction time for nucleation as well as the nucleation density of the crystallized PBS domain. Taking this into account, the larger $t_{1/2}$ for 30/70 PVDF/PBS blend compared to the 50/50 blend indicates that either the induction time for the crystallization of PBS in the 30/70 blend is longer than that in the 50/50 blend, or the nucleation density of the PBS in the 30/70 blend is lower than that in the 50/50 blend. Morphological observations confirm that the nucleation of PBS in the 50/50 blend is somewhat earlier and denser than that in the 30/70 PVDF/PBS blend. The enhanced nucleation rate and efficiency of PBS, which is normally poor, in the 50/50 blend may reflect an effect of heterogeneous nucleation of PBS on existing PVDF crystals. That is, the larger spatial density of PVDF crystals in the blend provides a larger density of nucleation sites for PBS crystallization.

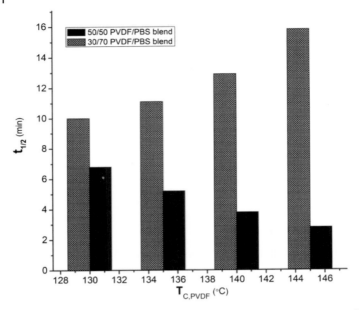

**Figure 3.17** Half-time of PBS crystallized at 85 °C in the blend as a function of PVDF crystallization temperature and the blend ratio. (Reproduced from Ref. [71] with permission, © 2011, American Chemical Society.)

Considering that the nucleation efficiency of one polymer toward the other depends on the interaction between the polymer pairs, the influence of preformed polymer crystal on the crystallization kinetics is blend system dependent. For example, Figure 3.18 summarizes the effects of blend composition and crystallization temperature on the overall crystallization rate of PBSA for the PBSA-rich

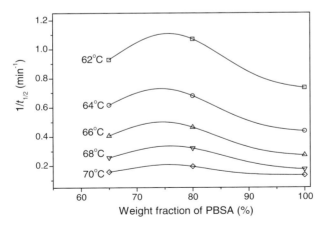

**Figure 3.18** Temperature dependence of the overall crystallization rate of PBSA for the PBSA-rich blends. (Reproduced from Ref. [74] with permission, © 2007, American Chemical Society.)

**Table 3.2** Crystallization kinetic parameters of PBSA for the PBSA-rich blends.

| PVDF/PBSA blends | $T_c$ (°C) | $n$ | $k$ (min$^{-n}$) | $t_{1/2}$ (min) | $1/t_{1/2}$ (min$^{-1}$) |
|---|---|---|---|---|---|
| 0/100 | 62 | 2.7 | $3.03 \times 10^{-1}$ | 1.36 | $7.34 \times 10^{-1}$ |
| | 64 | 2.5 | $8.66 \times 10^{-2}$ | 2.28 | $4.38 \times 10^{-1}$ |
| | 66 | 2.6 | $2.40 \times 10^{-2}$ | 3.68 | $2.72 \times 10^{-1}$ |
| | 68 | 2.7 | $6.06 \times 10^{-3}$ | 5.71 | $1.75 \times 10^{-1}$ |
| | 70 | 2.6 | $3.49 \times 10^{-3}$ | 7.42 | $1.35 \times 10^{-1}$ |
| 20/80 | 62 | 2.6 | $8.35 \times 10^{-1}$ | $9.31 \times 10^{-1}$ | 1.07 |
| | 64 | 2.6 | $2.57 \times 10^{-1}$ | 1.47 | $6.80 \times 10^{-1}$ |
| | 66 | 2.5 | $1.02 \times 10^{-1}$ | 2.15 | $4.65 \times 10^{-1}$ |
| | 68 | 2.6 | $3.52 \times 10^{-2}$ | 3.12 | $3.21 \times 10^{-1}$ |
| | 70 | 2.5 | $1.13 \times 10^{-2}$ | 5.12 | $1.95 \times 10^{-1}$ |
| 35/65 | 62 | 2.8 | $5.62 \times 10^{-1}$ | 1.08 | $9.28 \times 10^{-1}$ |
| | 64 | 2.7 | $1.90 \times 10^{-1}$ | 1.62 | $6.17 \times 10^{-1}$ |
| | 66 | 3.1 | $4.24 \times 10^{-2}$ | 2.46 | $4.07 \times 10^{-1}$ |
| | 68 | 3.0 | $1.21 \times 10^{-2}$ | 3.91 | $2.56 \times 10^{-1}$ |
| | 70 | 2.6 | $5.32 \times 10^{-3}$ | 6.33 | $1.58 \times 10^{-1}$ |

Reproduced from Ref. [74] with permission.

PVDF/PBSA blends after PVDF crystallizing first at 120 °C for 5 min. The data points presented in Figure 3.18 were obtained by DSC measurements through an Avrami analysis. Table 3.2 lists all the related crystallization kinetics parameters of PBSA grown in PVDF matrix. The average values of $n$ are close to 3, indicating that the crystallization of PBSA corresponds to the spherulitic growth with heterogeneous nucleation in the PBSA-rich blends. The overall crystallization rate of PBSA decreases with increasing crystallization temperature, regardless of blend composition, illustrating the dominant role of crystallization temperature on determining the overall crystallization rate. The effect of blend composition on the overall crystallization rate of PBSA is somewhat complicated. It seems that blending a small amount of PVDF with BSA accelerates the crystallization rate of PBSA, while the overall crystallization rate of PBSA decreases with further increase of the PVDF content. The overall crystallization rates of PBSA in 20/80 and 35/65 blends are even faster than the neat PBSA. This implies the enhanced nucleation ability of PBSA due to the presence of the PVDF crystals. The slower overall crystallization rate of PBSA in 35/65 blend compared with that in 20/80 blend is mainly caused by the effect of the preformed PVDF crystals.

### 3.3.3
### Morphology and Structure of Blend Systems with Large Melting Point Difference

From the above discussion, it is clear that, blending one polymer with other one, mutual influence on the crystallization behavior was found during separate

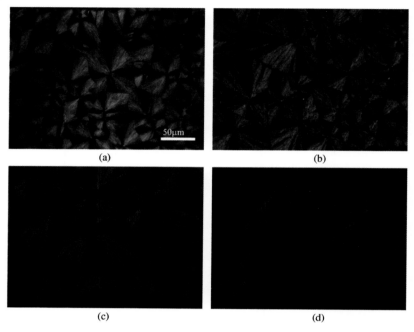

**Figure 3.19** PVDF spherulites at 144 °C in the PVDF-rich blends (same magnification with bar = 50 μm): (a) 100/0 for 5 min, (b) 80/20 for 15 min, (c) 65/35 for 16 min, and (d) 50/50 for 30 min. (Reproduced from Ref. [74] with permission, © 2007, American Chemical Society.)

crystallization. Different crystallization process may affect the morphology of the blend system or even the crystal structure of the components. In this section, the influence of crystallization condition on the morphology and crystal structure of the miscible binary crystalline blend systems will be discussed.

The morphology of miscible binary crystalline blends is composition and crystallization temperature dependent. It has been already mentioned in Section 3.1 that there is a temperature ($T^{*}$) that determines the $R \propto t$ and $R \propto t^{1/2}$ relationships between the spherulite radius and time [89–92]. This is related to the diffusion behavior of the noncrystallizable low-$T_m$ component and therefore determines the phase separation behavior of the blends. It was demonstrated that crystallization of the high-$T_m$ component at temperatures lower than $T^{*}$ results only in intraspherulitic phase separation of the blends. Taking the PVDF/PBSA system as an example again, intraspherulitic phase separation will always be observed when crystallizing the PVDF at temperatures lower than 155 °C. Figure 3.19 illustrates the blend composition effect on the morphologies of PVDF crystallized at 144 °C. Now the PVDF in the blends grew in spherulitic form with radius increasing linearly with time. The PVDF spherulites are found to fill the whole space at any composition and become larger with increasing the PBSA content, indicative of the decrease of nucleation of PVDF and the occurrence of intraspherulitic phase separation. This kind of intraspherulitic phase

**Figure 3.20** Optical micrographs of 80/20 PBS/PBA blend prepared by a two-step crystallization conditions: (a) crystallization of the PBS at 75 °C for 2 h and (b) crystallization of the PBA at 0 °C. (Reproduced from Ref. [76] with permission, Copyright © 2008 Elsevier Ltd.)

separation has been frequently observed in miscible binary crystalline blend systems.

Figure 3.20 presents another example with PBS/PBA blends. Figure 3.20a shows an optical micrograph of the PBS/PBA (80/20) blend after heat treatment at 140 °C for 15 min and then cooled to 75 °C. It is clear that just as in the case of PVDF in the PVDF/PBSA blend, the PBS crystallizes in compact spherulites after the completion of its crystallization at 75 °C for 2 h. The PBS spherulites cover the whole area. The existence of a well-defined spherulite boundary demonstrates that no PBA melt is expelled into the interspherulitic regions. In this case, the PBA is still well mixed within the amorphous part of PBS, leading to the formation of crystalline PBS and miscible amorphous PBS/PBA blends. The blends transfer into crystalline/crystalline polymer blends after the crystallization of low-$T_m$ component. Now the crystallization of PBA can only take place within the preformed PBS spherulites. Figure 3.20b illustrates the morphological change after cooling the PBS/PBA blend from 75 °C to 0 °C. We can recognize an overall intensity increase in birefringence of the PBS spherulites. This confirms that the crystallization of PBA indeed takes place within the PBS spherulitic regions.

To find the exact location of the PBA crystals, atomic force microscopy (AFM) can be used to examine the fine structure of the blend after the completion of crystallization for both components. Figure 3.21a shows a representative AFM phase image of an 80/20 PBS/PBA blend in which the PBS was crystallized isothermally at 75 °C prior to the crystallization of the PBA at 0 °C. Figure 3.21c is a higher resolution image of a portion of Figure 3.21a. There we see mainly edge-on lamellae in the radial direction of the PBS spherulite. In some places, as indicated by the white arrows, the PBA edge-on lamellae can be distinguished from the PBS ones by careful inspection. Selective melting of the PBA lamellae is much helpful for identifying the PBS lamellae more clearly. Figure 3.21b shows the same sample area as Figure 3.21a but scanned at 70 °C. The disappearance of the PBA lamellae at this temperature makes it clear that some of the PBA

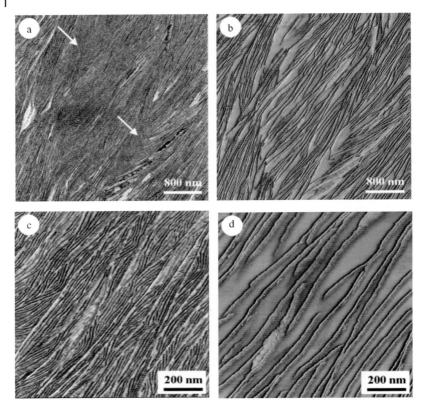

**Figure 3.21** AFM phase images of 80/20 PBS/ PBA blend with the PBS crystallized at 75 °C for 2 h and then cooled at a rate of 30 °C/min to 0 °C. The images were scanned at 25 °C (a and c) and 70 °C (b and d). (Reproduced from Ref. [76] with permission, Copyright © 2008 Elsevier Ltd.)

component crystallizes as thinner lamellae (with respect to the PBS ones) inserted in the PBS interlamellar regions, while the rest accumulates between the lamellar bundles or fibrils of PBS and forms microdomains of stacked edge-on lamellae. This is more clearly exhibited in the magnified AFM phase images (Figure 3.21c and d).

The PBS spherulites are obviously larger in a 50/50 blend than in an 80/20 blend when they are crystallized at 75 °C, and the similar morphologies were also found in the PVDF/PBSA blends. These spherulites also exhibit fairly straight impinging edges (see Figure 3.22). This indicates that no PBA melt accumulates in the boundary areas of the PBS spherulites. The more open structure of the PBS spherulites compared with that formed in pure PBS and its 80% blend implies that the increase of PBA component within the spherulites results in an even sparser arrangement of the PBS crystals or lamellae. A uniform intensity increment of the birefringence of the PBS spherulites after the crystallization of

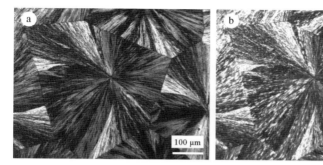

**Figure 3.22** Optical micrographs of 50/50 PBS/PBA blend prepared by a two-step crystallization condition: (a) crystallization of the PBS at 75 °C for 2 h and (b) crystallization of the PBA at 0 °C. (Reproduced from Ref. [76] with permission.)

PBA at 0 °C helps us to conclude that the PBA component is evenly dispersed in the PBS spherulites. AFM observation confirms that the PBA lamellae are partly inserted between the PBS lamellae and partly segregated between the PBS lamellar bundles with a larger domain size than that in the 80/20 PBS/PBA blend.

Even in a PBA-rich blend, for example, 20/80, the PBS spherulites still cover the whole area of the sample. As shown in Figure 3.23a, the open spherulitic structure makes it difficult to identify the impinging edges of the spherulites. Nevertheless, an even dispersion of the PBA melt in the blend can be judged from this optical micrograph. The morphological change of the blend after the subsequent crystallization of PBA (Figure 3.23b) also supports this conclusion.

Figure 3.24a and b present the fine structure of a 20/80 PBS/PBA blend scanned by AFM in phase mode scanned at 25 °C and 70 °C, respectively. It can be seen that, in the 20/80 PBS/PBA blend, the PBS crystalline lamellae construct only a scaffold framework, while the PBA lamellae crystals grow in the PBS scaffold and fill in the whole space. The domains of the aggregated PBA lamellar crystals are now of micrometer size. According to the above microscopy observation, it is

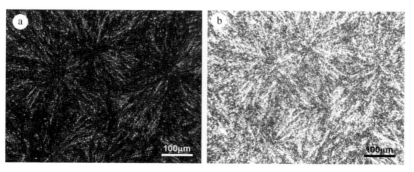

**Figure 3.23** Optical micrographs of 20/80 PBS/PBA blend prepared by a two-step crystallization condition: (a) crystallization of the PBS at 75 °C for 2 h and (b) crystallization of the PBA at 0 °C. (Reproduced from Ref. [76] with permission, Copyright © 2008 Elsevier Ltd.)

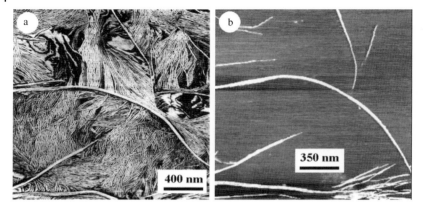

**Figure 3.24** AFM phase images of 20/80 PBS/PBA blend with the PBS crystallized at 75 °C for 2 h and then cooled at a rate of 30 °C/min to 0 °C. The images were scanned at (a) 25 °C and (b) 70 °C. (Reproduced from Ref. [76] with permission, Copyright © 2008 Elsevier Ltd.)

concluded that during the crystallization of PBS at 75 °C, the PBS forms spherulites covering the whole area, regardless its concentration in the blends. Therefore, the PBA melt is dispersed in the intraspherulitic regions of the PBS at all blend rations. At low PBA ratio, for example, 20%, the PBA is mainly dispersed in the interlamellar regions of PBS. With increasing amount of PBA in the blend (e.g., 50%), except for the part of PBA inlaid into the interlamellar regions of the PBS, most of the PBA aggregates to form small homogeneous domains with edge-on lamellar orientation. In a PBA in-rich blend, for example, a 20/80 PBS/PBA blend, the PBS forms only a spherulitic scaffold covering the whole area, while the PBA crystals fill in the scaffold, forming domains with aggregates of edge-on lamellae.

When crystallizing the high-$T_m$ component at temperatures higher than the $T^*$, different phase separation behavior will be observed. For example, the PBS in the PBS/PBA blends crystallized at temperatures over 100 °C. Figure 3.25

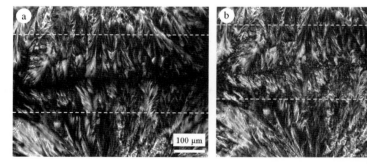

**Figure 3.25** Optical micrographs of 80/20 PBS/ PBA blend under two-step crystallization condition. (a) PBS crystallized isothermally at 100 °C for 2 weeks, and (b) after crystallization of the PBA at 30 °C for 2 h. (Reproduced from Ref. [76] with permission, Copyright © 2008 Elsevier Ltd.)

shows representative optical micrographs of an 80/20 PBS/PBA blend taken during the crystallization process. Keeping the sample at 100 °C for 2 weeks, the crystallization of PBS completes, while the PBA remains molten. It should be mentioned that the spherulites in this 80/20 blend are larger than those for the neat PBS, indicating a reduction of nucleation density caused by the diluting effect of the PBA. In Figure 3.25a, one sees two spherulites, one growing down from above the field of view and the other growing upward from below the field of view. The birefringence in the optical micrograph comes from the PBS crystals solely. When the sample was cooled to 30 °C, PBA crystallized, which resulted in an increment of the birefringence of the sample. With careful comparison of Figure 3.25a with 3.25b, it can be found that the enhancement of the birefringence occurs mainly in the boundary area of the PBS spherulites, that is, between the two dashed lines. This means that during the isothermal crystallization of PBS at 100 °C, the PBA component in the PBS/PBA (80/20) system has mainly been expelled into the interspherulitic regions of the PBS spherulites.

The interspherulitic phase separation can be more clearly recognized with increasing PBA content. Figure 3.26 shows polarized optical micrographs of two areas of a 50/50 PBS/PBA blend crystallized first isothermally at 100 °C for

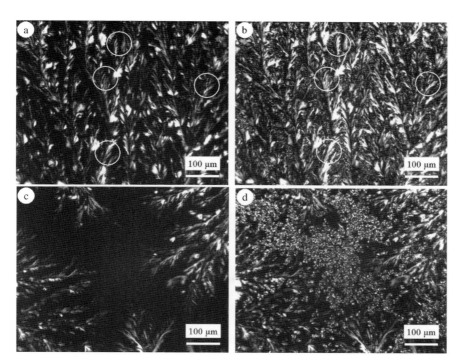

**Figure 3.26** Optical micrographs illustrate the morphologies of 50/50 PBS/PBA blend crystallized first 100 °C for 2 weeks (a and c), and then at 30 °C for 2 h (b and d). (Reproduced from Ref. [76] with permission, Copyright © 2008 Elsevier Ltd.)

2 weeks and then at 30 °C for 2 h. An increase of the content of PBA changes the morphology of the PBS. Nevertheless, the PBS always grows in spherulitic form. The PBS spherulites do not fill the entire space.

The area of Figure 3.26a and b are interior to a PBS spherulite, whereas Figure 3.26c and d are at the intersection of three PBS spherulites. The PBS spherulites have become more open, especially in their margin areas. As shown in Figure 3.26c, a large domain of hundreds of microns remains nonbirefringent between the PBS spherulites. This indicates that there exist no crystals in this area at 100 °C. During the cooling, no birefringence changes have been recognized in both parts a and c of Figure 3.26 at temperatures above the melting point of PBA, indicating the completion of crystallization of PBS at 100 °C. This means that the morphological change during further cooling process is caused by the crystallization of the PBA component. As illustrated in parts b and d of Figure 3.26, the birefringence of the sample changed markedly when the sample was cooled to and held at 30 °C for 2 h, indicating the occurrence of PBA crystallization. The crystallization of the PBA component leads to an increment of the overall birefringence of the sample. One can notice that the initially dark interfibrillar and interspherulitic areas both exhibit birefringence now. This indicates that during isothermal crystallization of PBS at 100 °C, the PBA melt is preferentially expelled to the interfibrillar of the PBS spherulites and PBS interspherulitic regions. With careful inspection, an intensity increment of some microdomains already showing birefringence before crystallization of PBA can be identified; compare the circled parts in Figure 3.26a and b. This implies the existence of some PBA in this 50/50 PBS/PBA blend crystallizing in the interlamellar regions of the PBS spherulite.

AFM observation confirms the above optical microscopy result. It was found that the morphology in the interspherulitic area has a close resemblance to that of neat PBA, reflecting the crystallization of the excluded PBA materials between the PBS spherulites. Figure 3.27 shows the AFM phase images of a 50/50 PBS/PBA blend within the PBS spherulite crystallized at 100 °C for 2 weeks and then cooled at a rate of 30 °C/min to 0 °C. Figure 3.27a and c were scanned at 25 °C, while Figure 3.27b and d were scanned at 70 °C, at which temperature the crystalline PBA should be molten. In the inner part of the PBS spherulite, as shown in Figure 3.27a, we can find individual PBA spherulites (indicated by white arrows), which contribute to the appearance of the birefringence in the initial dark interfibrillar parts. Selective melting of the PBA crystals at 70 °C helps us to locate where the PBA has been (See Figure 3.27b). From Figure 3.27b, it is clear that the PBA crystalline lamellae have been melted from the interlamellar regions of the PBS. This is more clearly seen in the enlarged micrograph shown in parts c and d of Figure 3.27. It is these crystalline lamellae that result in the increase of the birefringence intensity relative to what existed before the crystallization of the PBA. The AFM micrographic features demonstrate that there is indeed some PBA crystallizing in the interlamellar regions of the PBS spherulite, while most of it crystallizes in the interfibrillar and interspherulitic areas.

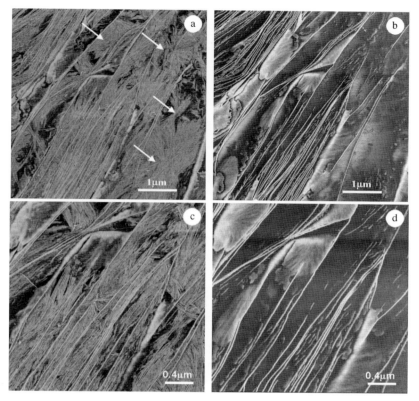

**Figure 3.27** AFM phase images of a 50/50 PBS/PBA blend with the PBS crystallized at 100 °C for 2 weeks and then cooled at a rate of 30 °C/min to 0 °C. The images were scanned at 25 °C (a and c) and 70 °C (b and d), respectively. (Reproduced from Ref. [76] with permission, Copyright © 2008 Elsevier Ltd.)

With further increase in the PBA content to 80%, the PBS forms sparsely dispersed spherulitic structures during isothermal crystallization at high temperature, see the optical micrograph shown in Figure 3.28.

AFM observation (Figure 3.29) shows that after crystallizing the PBA component at 30 °C, a profusion of PBA microcrystallites fill the interspherulitic regions, while there are also some PBA crystals dispersed within the PBS spherulitic regions. Within the PBS spherulites, thinner PBA lamellae grow either between the PBS crystalline lamellae or between the PBS lamellar stacks.

According to the above observations, it is concluded that during the crystallization of PBS in a PBS/PBA blend at 100 °C, when the PBA content is low, for example, 20%, it is mainly expelled from the PBS spherulites and accumulates in the interspherulitic regions of the PBS. With increasing amount of PBA in the blend, some of the PBA has been embedded into the interlamellar and interfibrillar regions of the PBS, while most of the PBA is still expelled to the

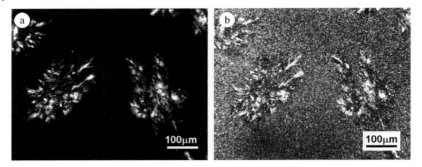

**Figure 3.28** Optical micrographs of 20/80 PBS/PBA blend, which was crystallized at (a) 100 °C for 2 weeks and (b) 30 °C for 2 h. (Reproduced from Ref. [76] with permission, Copyright © 2008 Elsevier Ltd.)

interspherulitic regions of the PBS. For a PBA-rich blend, while most of the PBA crystals fill in the remaining open areas, some of the PBA crystalline lamellae are dispersed in the interlamellar as well as the interfibrillar regions of the PBS.

The above described phase separation behavior of PBS/PBA blends holds true for most of other miscible blend systems. There may exist some variation about the $T*$ or even different phase separation behavior depending on the compounds used in the blends. As an example, Figure 3.30 shows the crystallization process of a PVDF/PBSA 50/50 blend under two-step crystallization condition. The sample was first crystallized at 148 °C (which is lower than the $T^*$ determined for the PVDF/PBS blend system) to investigate the crystallization of PVDF, and then cooled to 70 °C to study the crystalline morphology of PBSA. As shown in Figure 3.30a, PVDF shows a typical birefringence pattern of negative spherulites

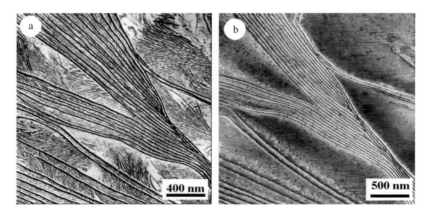

**Figure 3.29** AFM phase images of 20/80 PBS/PBA blend, which was first crystallized at 100 °C for 2 weeks and then cooled at a rate of 30 °C/min to 0 °C. The images were scanned at (a) 25 °C and (b) 70 °C. (Reproduced from Ref. [76] with permission, Copyright © 2008 Elsevier Ltd.)

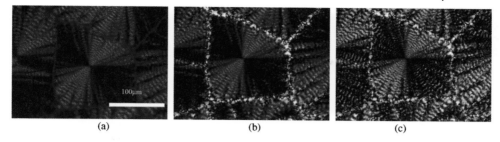

(a)                    (b)                    (c)

**Figure 3.30** Crystallization of PBSA in 50/50 blend under two-step crystallization condition: (a) crystallization of PVDF at 148 °C for 100 min, (b) crystallization of PBSA at 70 °C for 2 min, and (c) crystallization of PBSA at 70 °C for 40 min. (Reproduced from Ref. [75] with permission, © 2007, American Chemical Society.)

with concentric extinction bands after complete crystallization at 148 °C for 100 min. PVDF spherulites are more or less volume filling with nonbirefringent space can be observed at the spherulitic boundaries, suggesting that some of the PBSA is rejected into the interlamellar and interfibrillar regions of the PVDF spherulites during the crystallization process of PVDF and some of PBSA resides in the interspherulitic region. On lowering crystallization temperature to 70 °C, below $T_m$ of PBSA, the crystallization of PBSA happens in the presence of the preexisting PVDF spherulites. Parts b and c of Figure 3.30 show the crystallization of PBSA at 70 °C for 2 and 40 min, respectively. The commencement of the crystallization of PBSA occurred in the interspherulitic regions of the preexisting PVDF spherulites by cooling the sample from 148 to 70 °C. The crystallization of PBSA continued to proceed in these interspherulitic domains, up to the point where more or less filled with PBSA crystallites (Figure 3.30b). PBSA crystallizes even inside the PVDF spherulites by prolonging the crystallization time. The crystallization of PBSA occurred within the PVDF spherulites, which was confirmed by the change of the birefringence of the PVDF spherulites, that is, the apparent increase of the brightness of the PVDF spherulites (Figure 3.30c).

PBSA does not crystallize through spherulitic growth in the constrained space of the preexisting PVDF spherulites, unlike the bulk crystallization of PBSA from the homogeneous melt. Only tiny crystals of PBSA were found within the spherulites or at the boundaries of the spherulites of the PVDF component. Similar confined crystalline morphology was also found for PVDF/PBSA samples with different blend composition and at different crystallization temperatures of both PVDF and PBSA, such as at 142 and 150 °C for PVDF and from 68 to 75 °C for PBSA. For this blend system, the PVDF spherulites crystallized in the PBSA rich samples at relative low temperatures can even not fill the whole space. This will result in quite different crystalline morphology of PBSA compared with that for the PVDF-rich blends. Figure 3.31 displays the crystallization of a 20/80 blend under two-step crystallization. Figure 3.31a shows the PVDF spherulites with sparse fibrils or concentric bands after complete crystallization at 146 °C

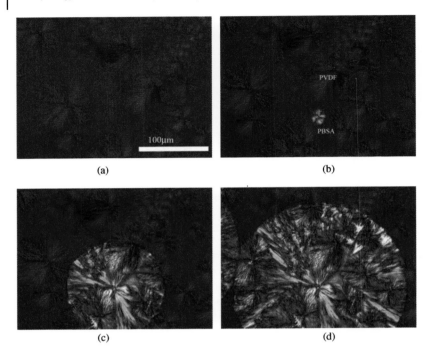

**Figure 3.31** Formation process of interpenetrated spherulites in 20/80 blend under two-step crystallization condition: (a) complete crystallization of PVDF at 146 °C for 400 min, (b) crystallization of 20/80 blend at 75 °C for 4 min, (c) process of interpenetration of PVDF spherulites by PBSA spherulite at 75 °C for 20 min, and (d) process of interpenetration of PVDF spherulites by PBSA spherulite at 75 °C for 32 min. (Reproduced from Ref. [75] with permission, © 2007, American Chemical Society.)

for 400 min. It is clear that PVDF spherulites do not fill the whole space even for a very long time because of the depletion of high-$T_m$ component. On cooling the sample to 75 °C, PBSA crystallized in the left space. Parts b, c, and d of Figure 3.31 represent a time sequence of the crystallization of PBSA at 75 °C after complete crystallization of PVDF at 146 °C for 400 min. As shown in Figure 3.31b, a PBSA spherulite first nucleated and continued to grow in the left space when the blend sample was cooled to 75 °C. The PBSA spherulite with strong birefringence was brighter than those PVDF spherulites with weak birefringence. Both PVDF and PBSA show spherulitic growth for the 20/80 blend under the two-step crystallization condition. With increasing crystallization time, the growth front of the PBSA spherulite reached the PVDF spherulites and continued to grow inside the spherulites instead of the termination of the growth (Figure 3.31c). The brightness of the PVDF spherulites increased in the region where PBSA crystallized and remained unchanged in the area where the growth front did not reach. The growth of the PBSA spherulite stopped until

it impinged on another PBSA spherulite that also penetrated the PVDF spherulites (Figure 3.31d).

The increase in brightness obviously indicated that PBSA must crystallize
inside the PVDF spherulites. Otherwise the brightness of the PVDF spherulites
would have been constant if the PBSA spherulites grew only outside of the
PVDF spherulites and just engulfed them as in the case of $\alpha$ and $\beta$ forms of isotactic polypropylene [96]. The change of birefringence inside the PVDF spherulites also confirms the penetration of them by the PBSA spherulites instead of
forming a layered structure where the spherulites of the two components merely
superposed on each other as two separate layers in a film. In the latter case, the
image should be a result of the superposition of the birefringence patterns of
PVDF and PBSA spherulites. Thus, the retardation would increase where they
have the same birefringence and decrease where they have the opposite
birefringence, resulting in the change of the color of the spherulites. However,
the experimental results demonstrated the increase of the brightness and the
unchanged color of the PVDF spherulites after the PBSA spherulites crystallized
inside them, indicating that the penetration process happened instead of two
separate layers crystallization. The shape of the growing PBSA spherulite
remains circular although it penetrates the PVDF spherulites, indicating that the
growth rate is almost the same for the left melt on the one hand and for
the internal of the PVDF spherulites (Figure 3.31c). The observed morphologies
for the PVDF/PBSA blends are clearly different from those observed for the PBS/
PBA system, which may reflect the differences in miscibility and crystallization
behaviors of the two systems.

## 3.4
## Concluding Remarks

Melt-miscible polymer blends with crystalline components may provide various
possibilities to investigate the crystallization behavior and crystalline morphology
of crystalline polymer blends. Depending on differences in melting points and
crystallization rates of each component in their neat state and in the blends,
each component can crystallize simultaneously or separately. Both components
are able to crystallize simultaneously under some crystallization conditions,
when the difference in their melting points is small. In this case, an interesting
crystalline morphology, that is, interpenetrating spherulites, is often developed,
because spherulites of one component can continue to grow in the spherulites
of the other component after they contact with each other. To form interpenetrating spherulites, the crystallization rates (including nucleation and crystal
growth) of the two components should be comparable so that the growth of
spherulites of each component may be observed simultaneously at the same
crystallization temperature. Both components can only crystallize separately,
when the difference in their melting points is large and the crystallization rate of
one component (usually the high-$T_m$ component) is significantly faster than that

of the other (usually the low-$T_m$ component). Under most crystallization conditions, the high-$T_m$ component crystallizes first, and the low-$T_m$ component melt may be expelled to the interlamellar, interfibrillar, or interspherulitic regions of high-$T_m$ component spherulites during the crystallization process, depending on blend composition and crystallization temperature. When crystallization temperature is further lowered below melting point of low-$T_m$ component, the crystallization of low-$T_m$ component can take place. It crystallizes in the presence of the preexisting crystals of high-$T_m$ component in the blends, which influence obviously both its crystalline morphology and crystallization kinetics, showing often some different features to those of neat low-$T_m$ component. Relative to amorphous/amorphous and amorphous/crystalline polymer blends, only a few melt-miscible polymer blends consisting of two crystalline polymers have been investigated; however, some of them are fully biodegradable polymer blends. The crystallization behavior and crystalline morphology studies are interesting and important from both academic and practical viewpoints, because they will further affect the final physical properties and biodegradation behavior of these biodegradable blends.

## Acknowledgment

The authors thank the National Natural Science Foundation, China (20504004, 51573016, 51521062, 51373020, and 51221002) for the financial support of this research.

## References

1 Tanaka, H. and Nishi, T. (1985) *Phys. Rev. Lett.*, **55**, 1102.

2 Tashiro, K., Stein, R., and Hsu, S. (1992) *Macromolecules*, **25**, 1801.

3 Tashiro, K., Izuchi, M., Kobayashi, M., and Stein, R. (1994) *Macromolecules*, **27**, 1221.

4 Tashiro, K., Izuchi, M., Kobayashi, M., and Stein, R. (1994) *Macromolecules*, **27**, 1228.

5 Tashiro, K., Izuchi, M., Kobayashi, M., and Stein, R. (1994) *Macromolecules*, **27**, 1234.

6 Tashiro, K., Izuchi, M., Kaneuchi, F., Jin, C., Kobayashi, M., and Stein, R. (1994) *Macromolecules*, **27**, 1240.

7 Xing, P., Dong, L., An, Y., Feng, Z., Avella, M., and Martuscelli, E. (1997) *Macromolecules*, **30**, 2726.

8 Zhang, L., Goh, S., and Lee, S. (1998) *Polymer*, **39**, 4841.

9 Kuo, S., Huang, C., and Chang, F. (2001) *J. Polym. Sci. Polym. Phys.*, **39**, 1348.

10 Ha, C. and Cho, W. (2002) *Prog. Polym. Sci.*, **27**, 759.

11 Wang, H., Shimizu, K., Hobbie, E., Wang, Z., Meredith, J., Karim, A., Amis, E., Hsiao, B., Hsieh, E., and Han, C. (2002) *Macromolecules*, **35** 1072.

12 Wang, H., Shimizu, K., Kim, H., Hobbie, E., Wang, Z., and Han, C. (2002) *J. Chem. Phys.*, **116**, 7311.

13 Shimizu, K., Wang, H., Wang, Z., Matsuba, G., Kim, H., and Han, C. (2004) *Polymer*, **45**, 7061.

14 Yoshie, N., Asaka, A., and Inoue, Y. (2004) *Macromolecules*, **37**, 3770.

15 Matkar, R. and Kyu, T. (2006) *J. Phys. Chem. B*, **110**, 16059.

16 Jiang, L., Wolcott, M., and Zhang, J. (2006) *Biomacromolecules*, **7**, 199.

17 Papageorgiou, G and Bikiaris, D. (2006) *J. Polym. Sci. Polym. Phys.*, **44**, 584.

18 Bikiaris, D., Papageorgiou, G., Achilias, D., Pavlidou, E., and Stergiou, A. (2007) *Eur. Polym. J.*, **43**, 2491.

19 Papageorgiou, G. Bikiaris, D., and Panayiotou, C. (2011) *Polymer*, **52**, 4553.

20 Nurkhamidah, S. and Woo, E. (2012) *Macromolecules*, **45**, 3094.

21 Wu, D., Yuan, L., Laredo, E., Zhang, M., and Zhou, W. (2012) *Ind. Eng. Chem. Res.*, **51**, 2290.

22 Qiu, J., Guan, J., Wang, H., Zhu, S., Cao, X., Ye, Q., and Li, Y. (2014) *J. Phys. Chem. B*, **118**, 7167.

23 Pan, P., Shan, G., and Bao, Y. (2014) *Ind. Eng. Chem. Res.*, **53**, 3148.

24 Qiu, Z., Ikehara, T., and Nishi, T. (2003) *Polymer*, **44**, 2503.

25 Qiu, Z., Ikehara, T., and Nishi, T. (2003) *Polymer*, **44**, 3101.

26 Qiu, Z., Ikehara, T., and Nishi, T. (2003) *Polymer*, **44**, 7519.

27 Qiu, Z., Komura, M., Ikehara, T., and Nishi, T. (2003) *Polymer*, **44**, 7749.

28 Qiu, Z., Komura, M., Ikehara, T., and Nishi, T. (2003) *Polymer*, **44**, 8111.

29 Qiu, Z., Fujinami, S., Komura, M., Nakajima, K., Ikehara, T., and Nishi, T. (2004) *Polymer*, **45**, 4515.

30 Qiu, Z. and Yang, W. (2006) *Polymer*, **47**, 6429.

31 Lu, J., Qiu, Z., and Yang, W. (2007) *Polymer*, **48**, 4196.

32 Miao, L., Qiu, Z., Yang, W., and Ikehara, T. (2008) *React. Funct. Polym.*, **68**, 446.

33 Yang, F., Qiu, Z., and Yang, W. (2009) *Polymer*, **50**, 2328.

34 Yang, F. and Qiu, Z. (2011) *Ind. Eng. Chem. Res.*, **50**, 11970.

35 Yang, F. and Qiu, Z. (2011) *Thermochim. Acta*, **523**, 200.

36 Cai, H., Yu, J., and Qiu, Z. (2012) *Polym. Eng. Sci.*, **52**, 233.

37 Liang, Y., Yang, F., and Qiu, Z. (2012) *J. Appl. Polym. Sci.*, **124**, 4409.

38 Yang, F., Li, Z., and Qiu, Z. (2012) *J. Appl. Polym. Sci.*, **123**, 2781.

39 Weng, M. and Qiu, Z. (2013) *Ind. Eng. Chem. Res.*, **52**, 10198.

40 Weng, M. and Qiu, Z. (2014) *Thermochim. Acta*, **575**, 262.

41 Gao, M., Ren, Z., and Yan, S. (2012) *J. Phys. Chem. B*, **116**, 9832.

42 Wang, T., Li, H., and Yan, S. (2012) *Chinese J. Polym. Sci.*, **30**, 269.

43 Wang, T., Li, H., Wang, F., Schultz, J., and Yan, S. (2011) *Polym. Chem.*, **2**, 1688.

44 Stein, R., Khambatta, F., Warner, F., Russell, T., Escala, A., and Balizer, E. (1978) *J. Polym. Sci. Polym. Symp.*, **63**, 313.

45 Keith, H.D. and Padden, F. Jr. (1964) *J. Appl. Phys.*, **35**, 1270.

46 Russell, T., Ito, H., and Wignall, G. (1988) *Macromolecules*, **21**, 1703.

47 Runt, J., Zhang, X., Miley, D., Gallagher, K., and Zhang, A. (1992) *Macromolecules*, **25**, 3902.

48 Defieuw, G., Groeninckx, G., and Reynaers, H. (1989) *Polymer*, **30**, 595.

49 Crevecoeur, G. and Groeninckx, G. (1991) *Macromolecules*, **24**, 1190.

50 Cheung, Y., Stein, R., Wignall, G., and Yang, H. (1993) *Macromolecules*, **26**, 5365.

51 Cheung, Y. and Stein, R. (1994) *Macromolecules*, **27**, 2512.

52 Cheung, Y. and Stein, R. (1994) *Macromolecules*, **27**, 2520.

53 Avella, M. and Martuscelli, E. (1988) *Polymer*, **29**, 1731.

54 Avella, M., Martuscelli, E., and Greco, P. (1991) *Polymer*, **32**, 1647.

55 Avella, M., Martuscelli, E., and Raimo, M. (1993) *Polymer*, **34**, 3234.

56 Liu, A., Liau, W., and Chiu, W. (1998) *Macromolecules*, **31**, 6593.

57 Penning, J. and Manley, R. (1996) *Macromolecules*, **29**, 77.

58 Penning, J. and Manley, R. (1996) *Macromolecules*, **29**, 84.

59 Fujita, K., Kyu, T., and Manley, R. (1996) *Macromolecules*, **29**, 91.

60 Liu, L., Chu, B., Penning, J., and Manley, R. (1997) *Macromolecules*, **30**, 4398.

61 Chiu, H., Chen, H., and Lin, J. (2001) *Polymer*, **42**, 5749.

62 Liu, J. and Jungnickel, B. (2003) *J. Polym. Sci. Polym. Phys.*, **41**, 873.

63 Liu, J. and Jungnickel, B. (2004) *J. Polym. Sci. Polym. Phys.*, **42**, 974.

64 Liu, J. and Jungnickel, B. (2007) *J. Polym. Sci. Polym. Phys.*, **45**, 1917.

65 Qiu, Z., Ikehara, T., and Nishi, T. (2003) *Polymer*, **44**, 2799.

66  He, Y., Zhu, B., Kai, W., and Inoue, Y. (2004) *Macromolecules*, **37**, 3337.

67  He, Y., Zhu, B., Kai, W., and Inoue, Y. (2004) *Macromolecules*, **37**, 8050.

68  Wang, H., Schultz, J., and Yan, S. (2007) *Polymer*, **48**, 3530.

69  Ikehara, T., Kurihara, H., Qiu, Z., and Nishi, T. (2007) *Macromolecules*, **40**, 8726.

70  Wang, T., Wang., H., Li, H., Gan, Z., and Yan, S. (2009) *Phys. Chem. Chem. Phys.*, **11**, 1619.

71  Wang, T., Li, H., Wang, F., Yan, S., and Schultz, J. (2011) *J. Phys. Chem. B*, **115**, 7814.

72  Yang, J., Pan, P., Hua, L., Feng, X., Yue, J., Ge, Y., and Inoue, Y. (2012) *J. Phys. Chem. B*, **116**, 1265.

73  Lee, J., Tazawa, H., Ikehara, T., and Nishi, T. (1998) *Polym. J.*, **30**, 327.

74  Qiu, Z., Yan, C., Lu, J., and Yang, W. (2007) *Macromolecules*, **40**, 5047.

75  Qiu, Z., Yan, C., Lu, J., Yang, W., Ikehara, T., and Nishi, T. (2007) *J. Phys. Chem. B*, **111**, 2783.

76  Wang, H., Gan, Z., Schultz, J., and Yan, S. (2008) *Polymer*, **49**, 2342.

77  Qiu, Z., Fujinami, S., Komura, M., Nakajima, K., Ikehara, T., and Nishi, T. (2004) *Polymer*, **45**, 4355.

78  Chen, H. and Wang, S. (2000) *Polymer*, **41**, 5157.

79  Lee, J., Tazawa, H., Ikehara, T., and Nishi, T. (1998) *Polym. J.*, **30**, 780.

80  Hirano, S., Nishikawa, Y., Terada, Y., Ikehara, T., and Nishi, T. (2002) *Polym. J.*, **34**, 85.

81  Qiu, Z., Ikehara, T., and Nishi, T. (2002) *Macromolecules*, **35**, 8251.

82  Lu, J., Qiu, Z., and Yang, W. (2008) *Macromolecules*, **41**, 141.

83  Ikehara, T., Kurihara, H., and Kataoka, T. (2012) *J. Polym. Sci. Polym. Phys.*, **50**, 563.

84  Ikehara, T., Kimura, H., and Qiu, Z. (2005) *Macromolecules*, **38**, 5104.

85  Zeng, J., Zhu, Q., Li, Y., Qiu, Z., and Wang, Y. (2010) *J. Phys. Chem. B*, **114**, 14827.

86  Yang, Y. and Qiu, Z. (2012) *Ind. Eng. Chem. Res.*, **51**, 9191.

87  Lin, J. and Woo, E. (2006) *Polymer*, **47**, 6826.

88  Schultz, J. (2010) *J. Front. Chem. China*, **5**, 262.

89  Tanaka, H. and Nishi, T. (1989) *Phys. Rev. A*, **39**, 783.

90  Schultz, J. (1991) *Polymer*, **32**, 3268.

91  Schultz, J. (2001) *Polymer crystallization*, Oxford University Press, New York.

92  Okada, T., Saito, H., and Inoue, T. (1990) *Macromolecules*, **23**, 3865.

93  Hudson, S., Davis, D., and Lovinger, A. (1992) *Macromolecules*, **25**, 1759.

94  Avrami, M. (1940) *J. Chem. Phys.*, **8**, 212.

95  Wunderlich, B. (1976) *In Macromolecular Physics*, vol. **2**, Academic Press, New York.

96  Varga, J. (1992) *J. Mater. Sci.*, **27**, 2557.

# 4
# Rubber–Plastic Blends: Structure–Property Relationship

*Sudhin Datta*

*ExxonMobil Chemical Co., Baytown, TX 77520, USA*

## 4.1
## Introduction: Key Challenges

Blends of rubbers and plastics have both an extensive history and a variety of applications, since blends retain most of the stiffness of thermoplastics and impart some of the resilience and the impact toughness of rubbers. Most useful blends have morphology of rubber dispersed in the thermoplastic matrix since the reverse does not lead to same degree of utility. The properties of these blends depend on the relative ratio of the rubber and the plastic and, more importantly, on the morphology of the dispersion. A very large research effort has been made in an attempt to control and determine the morphology and the interface of the rubber and the plastic as well as understand the effect of the rubber dispersion on the properties of the blend. Blends are important commercially since they allow the development of unique properties from the same constituents with small changes in the relative amounts or the dispersion. The conventional practice of blend formation is the melt mixing of the components although, in some cases, a polymerization process that enables the direct production of the binary blend is possible. These direct production processes are efficient since they produce the desired morphology and the required polymers in a single step. Blends have become extremely important for the development of polyolefin rubbers and plastics since a large variety can be mixed in a wide composition ratio to produce novel properties. These blends are easily made and have stable dispersion during processing due to only small thermodynamic differences between the various polyolefins compared with engineering thermoplastics. This chapter describes the physics and the chemistry and more importantly the relationship between the structure of the blend and its properties.

This area has been continually progressed [1] and reviewed [2]. Progress arises from new materials and new technologies for making rubber–plastic blends. The area of polyethylene blends cited above is an important development. The

*Encyclopedia of Polymer Blends: Volume 3: Structure*, First Edition. Edited by Avraam I. Isayev.
© 2016 Wiley-VCH Verlag GmbH & Co. KGaA. Published 2016 by Wiley-VCH Verlag GmbH & Co. KGaA.

development is ultimately driven by increasing material demands of lower cost and higher performance and their relevance to the less developed but growing economies of the world. Reviews such as this are important since they seek to define the past history as the stepping stones of the future developments. A good example of these divergent trends is shown in the description of polylactide plastic–acrylate rubber blends [3].

Blending of plastics and rubbers is a commercially feasible and frequently practiced method to access a combination of properties that are commercially desirable but inaccessible from a single component. These properties can be classified as the ease of fabrication and the stiffness of plastics and the resilience and energy absorption of rubbers. An appropriate example of this novel combined property is impact toughness, which is typically absent in brittle plastics whereas rubbers dissipate energy by easy deformation. In addition, blends in general provide a continuous variation of properties by an adjustment of the ratios of the individual dissimilar components. Thus, the potential applicability of blends is large. This continuous variability allows, in theory, an exact match of the property profile of the blend to the desired use. This is in contrast to the prospect of both the delay, uncertainty and the expense for the synthesis of new polymer. Rubber–plastic blends also encompass a range of postblending processes and a corresponding art of property modification of both the individual components and the composite. This is done by the addition of fillers and plasticizers to either component as well as the vulcanization of the rubber. In summary, rubber–plastic blends are valuable because they offer a simple mechanical process of achieving a continuously variable property for the composite, intermediate between the properties of the rubber and the plastic. Their utilization requires the development of the science of the formation and the stabilization of the blend and the morphology of the blend.

The vast majority of rubber–plastic blends are largely immiscible. For these blends, this appears as a morphological coexistence of at least two domains of dissimilar composition. Domains are aggregations of principally compositionally similar or identical molecules. They are uniform in composition across any macroscopic region. Domains are bounded by an interface. In immiscible rubber–plastic blends, the different domains contain plastics or rubbers, fillers, plasticizers, and other additives in different amounts. The average size and the size distribution of the domains are dictated by the contrary factors of mixing intensity that favors smaller domains and interfacial energy between dissimilar polymers that favors segregation and thus larger domains. The typical high viscosity of the elastomers (compared with the typical thermoplastics) favors the persistence of the current morphology. In short, in these blends it is difficult to achieve a well-dispersed morphology and equally difficult to completely phase separate since the viscosity of the rubber slows down the formation or collapse of the dispersed domains. In practice, this effect can be used to generate well-dispersed, stable morphologies of rubber–plastic blends by chemically cross-linking the rubber after dispersion in the plastic. Chemical cross-linking slows down the agglomeration of the rubber domains to immeasurable rates. This technology is practiced commercially and forms an important segment of this blend family: the thermoplastic vulcanizates (TPVs). An alternative visualization for these immiscible blends is that the thermodynamic steady state is segregation, which is somewhat disrupted by the kinetics of dispersive actions such as mixing. The resultant morphology is thus an arbitrary transit point between these points, dictated by mixing parameters and time.

The science and technology of blending plastic and rubbers has given rise to three very distinct developments that otherwise would be less investigated. First, the process of melt polymer blending and the machinery (e.g., twin-screw extruders) for both industrial and research investigations have benefitted from the commercial importance of these blends. Second, since the properties of the blends are nonlinear function of the amounts of components and depend in addition on the morphology, orientation, and surface area of the dispersed phase, the science of blends that address this phenomenon has gained attention both in academic investigations and in industrial practice. Third, the existence of interfacial areas in blends of immiscible polymers, which are typical for rubber–plastic blends, has given rise to synthetic efforts to deliver surface-active agents, "compatibilizers," which stabilize these labile interfaces and, presumably, improve properties.

Notwithstanding the discussion above, the dispersed phase in most rubber–plastic blends is typically nonspherical: the effect of shear and the persistence of the morphology due to high viscosity and long relaxation times for either of the components tend to favor elongated domains with preferential orientation along the shear axes during mixing and fabrication. They are formed by the unidirectional shear mixing of the components and persist due to high viscosities of the dispersed phase as well as due to only a small thermodynamic loss of free energy for the formation of the isotropic, spherical domains from the elongated nonspherical morphology. These blends have a much higher interfacial surface area between the rubber and the plastic compared with blends of the same composition with spherical domains. A manifestation of this nonisotropic morphology of these blends, particularly in areas where the interfacial surface has contributions, such as impact strength or flexural stiffness, is observed as differences in directions perpendicular to and along the orientation direction. Furthermore, within the specimen, the properties are anisotropic between the directions that are transverse to or parallel to the direction of the elongated dispersed domains.

In addition to changes and variation in morphology, rubber–plastic blends show additional, more complex changes due to different chemical properties of the rubber and the plastic. This difference in the chemical structure manifests as two distinct but interrelated properties. First, the rubber and the plastic differ in the retention of the fillers (talc or glass fiber) and plasticizers because of the inherent differences in the solubility parameters. It is expected that the solubility parameters are different to account for the phase-separated morphology. Both of these effects are driven by thermodynamic effects, but the distribution of fillers is slower to reach equilibrium. Several procedures, referred to as phase blending, seek to ensure filler retention in the desired phase, typically the plastic, by initially adding all of the fillers to that phase before blending with the rubber. These procedures are industrially quite successful, though a rational academic investigation is feasible but challenging. Second, for the specific blends, where the rubber phase is vulcanized, the chemical reactants for the vulcanization and including curatives, cure accelerators, and oxidation retarders – all of which are polar, organic small molecules – distribute between the two constituent rubber–plastic phases according to thermodynamic solubility parameters. These additives are added during blending of the rubber and the plastic and distribute between the phases to minimize free energy. This distribution will determine the vulcanization of the rubber and the subsequent utility of these blends. The diffusion and the consumption of these chemicals during the vulcanization process leads to a gradient in the concentration of the ingredients. The effect of the gradient is most pronounced at the boundaries of the rubber–plastic phases where the diffusion distances between the plastic and the rubber are the least. This complex set of distribution and diffusion dictated by the difference of chemistry of the rubbers and the plastic and the diffusion of additives leads to different vulcanization of the elastomer phases between the bulk and the periphery of the rubber phase.

While immiscible rubber–plastic blends that comprise the vast majority have most of the considerations outlined above, the primary expectations of an economic advantage, wide and continuous variability in properties, and inherently simple processing are largely met in the cases when the two components are miscible and mix to form a single phase. The key is the formation of a single phase with intermixing at a molecular level. Using our earlier definitions, there is only a single domain in these systems, and the thermal and mechanical properties of the blend such as glass transition temperature and tensile properties are a composition weighted function of the components. This miscibility could be because of a close match of the solubility parameters and a corresponding match of the constitutive thermodynamic parameters of the two polymers or due to specific interactions between the component polymers. Both circumstances are rare. In addition, the small number of binary plastic–rubber pairs limits availability of pairs of components with specific favorable attractive interaction. These concepts for formation of a miscible blend with similar or near equivalence of solubility parameters require the components to be similar in properties. Thus, a wide variation in the properties of the blends by changing the relative amounts of the two is not typical.

Miscible blends persist through the process of fabrication and cross-linking with the desirable properties of the miscible blend essentially intact and manifest in the mechanical and thermal properties of the blend. The final fabrications are typically isotropic, even under the most demanding shear history of processing. They will have, in addition, uniform distribution of reinforcing agents and other additives across the entire composition. The properties of the constituents are, to a first approximation, a good measure of the properties of the final blend. Miscible blends are sometimes used though they have been very rarely recognized since an analytical deconstruction of such a system is challenging.

The discussion shows that the benefit of achieving blends of rubbers and plastics that are different in solubility parameters is a sufficiently attractive target.

Various synthetic strategies have been adopted to achieve some degree of miscibility in blends that are ordinarily not miscible. These blends are not thermodynamically miscible but they are compatible in the polymer technology usage
of the concept. The common synthetic strategy is the addition of low molecular
weight but involatile solvent during compounding, which produces a single, ternary phase including both polymers and solvent. The solvent reduces
unfavorable interactions between the rubber and the plastic molecules. Often
the solvent is the monomer used to make the plastic, which is then polymerized in the presence of the rubber. This strategy is useful when the solubility
parameter differences between the rubber and the plastic are not very large.
Another conceptual procedure is for both the rubber and the plastic to interact with the same particulate filler. These procedures use the stabilizing interaction of both the plastic and the rubber on a filler to form a compatible
blend. Another synthetic strategy is to use complementary interacting functionality such as acid and base functionality on the polymeric components to
form a single compatible blend. A deficiency of this procedure is the need for
complementary functionality on each of the components: the generation of
this pair of structural feature may require additional synthetic steps. Another
utilitarian deficiency of the last procedure is the unavoidable rise in viscosity
and consequent loss of processing and fabrication when the rubber and the
plastic interact chemically.

In spite of these difficulties, miscible blends of different plastics and rubbers do
exist. It is illustrative of the concepts we have described earlier that saturated
polyolefin plastics (such as isotactic polypropylene (iPP)) and elastomers (ethylene–
$\alpha$-olefin elastomers) with a minimum of intermolecular enthalpic interactions,
compared with more polar or polarizable polymers, form the large and probably
predominant family of miscible blends. Miscible blends are achieved between polymers where the inserted monomer units are structural isomers, have the same
empirical formula, or are closely related such as ethylene and propylene where the
small difference of solubility parameters required for miscibility can be maintained
over a wide range of difference in composition and crystallinity. Common examples
of miscible blends are certain narrow composition range of ethylene/octene-1
copolymers (such as Engage™ from Dow Chemical Co. or Exact™ from ExxonMobil Chemical Co.) that are elastomers with isotactic polypropylene, which is a plastic.

## 4.2
## Rubber Toughening of Thermoplastics

### 4.2.1
### Mechanism

The predominant morphology of rubber–plastic blends is a continuous plastic
phase with small dispersed rubber domain. There are only a few examples of the
reverse since these blends are of relatively little utility as they function as stiff,

inelastic rubbers. Within the first group, there is large variation in the average size of these domains from tens of nanometers to several micrometers. In addition, these limits are averages of a distribution of sizes and individual rubber particles can be much smaller or bigger. The largest utility of the preferred rubber in plastic matrix morphology of the blend is impact and tensile toughening, due to slowing down or eliminating failure, under stress, of the plastic matrix. The mechanism of this is the absorption or dissipation of the stress by the rubber particles. This delay also indicates a parallel process that perpetuates this impact and tensile toughness to low temperatures that are below the glass transition of the thermoplastic. This toughening at room temperature and the extension of the toughness at reduced efficacy to low temperatures are typically linked and either is an example of this phenomenon. In general, the energy absorption and dissipation processes are vastly more efficient with increasing amounts of rubber in the blend: this improvement is at the expense of the typical thermoplastic properties of both stiffness and lowering of the temperature required for mechanical creep.

The mechanical failure under impact or tensile deformation for thermoplastics occurs because there are only a few mechanisms for the dissipation of energy within a thermoplastic. At temperatures and timescales where the plastic flows, it fails by a ductile flow mechanism, whereas under conditions, such as low temperatures and/or slow deformation, where neither of the above conditions applies, the failure occurs by a brittle mechanism. In general, during brittle failure the molecules do not have time to rearrange, while during ductile failure the flow is the mode of failure. The presence of the rubber in dispersed phase within the plastic changes these exclusive or unique mechanisms because the mechanical response of the rubber to stress is different from the thermoplastic. The difference can be ascribed to difference in the tensile or compressive modulus for the two modes of failure of the rubber–plastic blend between the dispersed rubber and the surrounding thermoplastic. This difference in the mechanical response leads to, at the boundary and the immediate region around the dispersed phase, a field of differential stress concentration. It is important to understand that the rationale for the formation of the differential stress concentration is the difference in the compressive and tensile modulus of the rubber and the thermoplastic. In very simple terms, the rubber is soft and easily deformed while the plastic is neither [3]. This deformation of the plastic and the rubber leads to additional deformation processes in the rubber and the plastic that instead of leading to a mechanical failure of the plastic lead to dissipation of the energy. Failure processes are thus retarded leading to higher energy absorption at elevated temperatures or the energy absorption can persist even at low temperatures where the plastic is normally brittle leading to a dramatic improvement in impact or tensile strength [4–7]. This mechanism shows the salient features of this principle in that (a) rubbers are excellent at impact modification only under conditions that they are different in modulus compared with the thermoplastic that is usually much above their $T_g$ and (b) the dispersion of the rubber has to be fine enough that it can hinder the failure process and initiate the energy absorption at a morphological scale in the blend to prevent propagation of the failure of the thermoplastic.

In commerce, various rubbers have been used including the synthetic butadiene- or isoprene-based unsaturated rubbers and natural rubber has been used to toughen a variety of thermoplastics. A common example is the use of these rubbers for HIPS (high-impact polystyrene) or ABS (acrylonitrile–butadiene–styrene). Polyolefin rubbers such as EPR (ethylene–propylene rubber), butyl, and ethylene/octene-1 copolymers have shown a dramatic growth in this category over the last few decades because of the multitude of blends with different properties that can be accessed through blending with polyolefin thermoplastics such as polypropylene (PP) and polyethylene (PE). However, certain common themes are present for all and these are based on the discussion above. The rubbers are only effective as long as the modulus mismatch between the rubber and the thermoplastic is present – almost invariably, the rubber is the soft phase and the thermoplastic is the hard phase. Semicrystalline rubbers or rubbers where the $T_g$ is close to the use temperatures are thus not the ideal choice since the difference in the modulus is minimized with consequent loss of the energy dissipation. As rule of general use, rubbers within 25 °C of their $T_g$ or having greater than 10% crystallinity are not recommend for this utility; conversely, low-crystallinity rubbers such as polyisobutylene or polybutadiene (PBD) that do not crystallize and have very low $T_g$ are used to form the rubber–plastic blends with the greatest utility and application.

### 4.2.2
### Morphology

One of the key items of the discussion is the amount, dispersion, and size of the rubber domains that should be adequate to promote alternative energy dissipation mechanisms and lead to enhanced toughness and tensile strength for the blend compared with the thermoplastic. In general, this means a large amount of rubber, finely dispersed in small particles. However, in reality the control of the dispersion of the rubber particles is never perfect and there is an inherent discrepancy in the average size of the rubber particles and the average size represents a distribution of sizes where some of the particles are smaller and some larger. The intensity and the duration of the melt mixing can be largely used to determine the average size but other more finer structural details of the rubber particles such as the distribution in the sizes, the breadth of this distribution, and the shape or aspect ratio of the particles are hard to predetermine, control, or retain through the subsequent processing stages. In addition to the mixing process mechanical shear, the most useful descriptors of the rubber particle size are the relative interfacial energy compared with the thermoplastic matrix and the host of viscous relaxation processes. The first determines the relative thermodynamics of the rubber dispersion to increase the interfacial area of the rubber and the plastic and low differences enhance the ability to disperse. The second ensures that particles under shear do break up and, conversely, do not coalesce readily during subsequent processing. Ideally, this condition is fulfilled by rubbers that have a low enough high-shear viscosity to readily break up into droplets in regions of high-shear mixing and correspondingly a high viscosity under low-shear conditions to retain the small size and the elongated, high aspect ratio that are presumed to be most beneficial to this toughening.

Differences in these morphological descriptors affect the impact and tensile toughness of the blends. Thus, changes in the size, size distribution, and aspect ratio of the rubber lead to both changes in the tensile and impact energy absorption before failure and changes in the brittle-to-ductile transition (BDT) of the composite. In general, the distribution of rubber particles that maximize the interfacial area between the rubber and the plastic leads to the most dramatic improvement in the toughness and the low-temperature properties by having a large diminution of the BDT of the blend. Larger surface areas are obtained by an increasing amount of the rubber, smaller particles, or particles with a highly elongated aspect ratio.

There is considerable effort to control or predetermine the morphology of the rubber inclusions. One of the processes is the addition of a compatibilizer that resides and stabilizes the interface of the dissimilar rubber and plastic. This is achieved by the addition of or the creation *in situ* by chemical reaction of a block molecule where one block is soluble in the rubber and the other is miscible with the thermoplastic. These molecules, present in low concentration, lead to energy minimization at the interface and thus more extensive interfacial area through a smaller particle size or more elongated particles. The other process is to constrain the geometry of the rubber particles after the morphology is fully developed by increasing the viscosity, typically by cross-linking. These particles that are partly or fully vulcanized inhibit volume or shape changes leading to a stable morphology and thus invariant toughened properties. Of course, the key point of this technology is to achieve this cross-linking after the morphology is fully developed. Another important process for predetermining the morphology of the rubber particle is to inhibit their coalescence by surrounding them ("decorated" in the literature) by a rigid shell of thermoplastic. These particles give the most reliable control of the morphology since the morphology descriptors are unchanged during the mixing process. A generic description of the last process is core–shell particles.

Another important factor in the morphology of the rubber–plastic blends is the inherent loss of the processability of the blend by the addition of the elastic rubber. The addition of the rubber as a phase-separated blend component allows the dissipation of the mixing energy through transitory elongation and retraction as the rubber particles enter and exit the high-shear process zone. This energy dissipation leads to less energy available for the breakup of the rubber into small particles.

### 4.2.3
### Failure Process

The presence of soft rubber inclusions that are easily deformed in the plastic continuous phase improves tensile and impact toughness by imposing a new distribution of strain fields. These new stress fields are characterized by having novel stress concentrations during deformation by impact or elongation. The effect of these localized stress fields around the rubber particles leads to localized flow and fracture as a method to absorb energy that otherwise would be available for sample fracture or rupture. These localized zones of flow or fracture are commonly referred to as process zones. This localized flow is observed as local whitening in the fracture process zones. Whitening, due to diffraction of light at the polymer–air interface, occurs due to cavities formed by debonding of the rubber and the plastic domains, the fiber formation in the plastic matrix, or extensive crazing of the plastic matrix [4–9]. The rubber inclusions can undergo an internal cavitation through either debonding at the rubber–plastic interface or tearing of the rubber due to triaxial stresses leading to void formation inside the rubber. Thus, elastomeric inclusions with high tensile strength, such as in cross-linked rubbers, typically debond while those with low tensile strength such as low molecular weight uncross-linked rubbers form intraparticle voids. Notwithstanding the exact dissipation mechanism, it is important to understand that the soft rubber inclusions lead to a novel stress field that is locally intense. This results in the dissipation of energy and strain by the development of internal faults, crevices, fibers, and porous areas during the deformation as a necessary step to increase toughness in rubber–plastic blends.

This process for the reinforcement of the plastic phase by localized failure and by the induction of voids in either of the phases actually is presumed to proceed in a very logical manner. The chain of events starts with the process that is at the lowest stress for activation. This may be as simple as the debonding of the plastic and the rubber phase. In blends of iPP with rubbers, Mouzakis *et al.* [10] and Nitta *et al.* [11] showed experimentally that blends with rubbers that are the least miscible with iPP and thus have the weakest interface and would be expected to debond at the lowest activation stress have surprisingly the higher fracture energy on impact. This is counterintuitive in that the weakness in the interface provides the mechanism for the strength of the blend. Mechanistically, the debonded rubber pieces provide the thermoplastic matrix the room to deform and transmit the deformation process into the surrounding regions to start the process of diffusion of the energy. This diffusion is the process for the dissipation of the energy. In consideration of the unit steps of this process, the initial void formation requires a process for the nucleation and the growth of the defect. This process must have an activation step. Rubbery inclusions that are bonded to the plastic can form initial internal voids not at the interface but in the bulk through homogeneous cavitation. For this cavitation to occur during the deformation, the size of inclusions has to be larger than a critical value [12–14]. Larger particles cavitate easier than the smaller particles and the presence of only very small particles is not beneficial. Such morphology requires that the cavitation process produce a stress concentration greater than the critical value of the self-yielding process initiated in the thermoplastic. This type of shear yielding will most frequently lead to brittle failure without energy dissipation by the rubber particles. Thus, contrary to the predominant teaching of this section, the particles of the rubber can be too small to be effective in improving the toughness of the blend. In practice, a distribution of sizes and shapes of the rubber particles is frequently needed as each morphological entity is effective in retarding a specific kind of failure process.

There are two competing effects of the size of the rubber dispersion that determine the fracture and tensile toughness of the blend. When the adhesion of the rubber and the plastic phases is significant but the particles are large and easily deformed and ruptured, the energy relieving intraparticle internal cavitation can occur via void nucleation and propagation. In a contrary case with poor adhesion between the rubber and the plastic, but with small particles, the cavitation stress is higher than the yield stress of the matrix. This promotes brittle failure of the matrix. The determination of the impact energy for the rubber–plastic blend shows an increase of this for small to large particles, followed by a decrease when the particles are too large. This was rationalized as due to effects associated with the "matrix ligament size" by Bartzak *et al.* [6]. This theory indicates that when the interparticle distances are too large than a critical value the average matrix ligament is too large and too stiff to fully relieve all of the cavitation stresses. Thus, blends with widely separated rubber particles are not very efficient in relieving their stresses by matrix yielding. Thus, impact toughness is low because even though the energy dissipation is initiated, the process is not complete or effective before the residual stress levels lead to brittle failure of the plastic.

The key to the difference in the properties of the rubber and the plastic phases for impact and tensile toughening in a blend requires a persistent modulus mismatch between the two phases under the conditions, timescale, and the temperature of the impact failure experiment. Under these conditions, the large matrix stresses are relieved by the cavitation to release the stresses before brittle fracture of the plastic matrix occurs. The postprocess morphology of the impact piece appears to have large cavities and discontinuities in the structure both due to the debonding of the rubber from the plastic matrix and due to the fibrillation of the plastic matrix. Other dissipation mechanisms also result in the formation of porous thermoplastic and rubber domains by stress-induced cavitation. On cavitation, the matrix near the cavities is susceptible to large-scale plastic deformation. The asymmetrical stress field allows lamellae to undergo irreversible processes such as chevroning, slippage, and twisting occurring across macroscopic volumes. All these processes are irreversible and consume the bulk of the energy. Gurson [15] has predicted the yield of a cavitated material. This model has been partly verified for polymers by Kim and Kim [16] and Bagheri and Pearson [17]. The former study used mechanical stirring to incorporate air holes into an epoxy resin before curing. Testing fracture energy showed a toughening effect only when the air was finely dispersed so that the matrix ligament size was less than 100 μm. The latter study used hollow rubber spheres as an additive to epoxy resin and showed that these composites had a decrease of yield strength and improved fracture toughness. Rubber inclusions of controlled size have been used to increase fracture toughness of thermoplastics. Analysis of the stress concentration around spherical inclusions has been obtained by Goodier [3].

In practical systems that are generated by the melt mixing of rubber and plastic, the rubber inclusions span a range of sizes and shapes. In a system composed of multiple particles, as is the case of toughened plastic, the stress concentration fields are different around each particle because of the size and orientation yet they are near and can interact with each other. Finite-element computer models are able to analyze stress fields in various composite systems [18–20]. Huang and Kinloch [20] have done this for polymers containing air or rubber void spherical inclusions. Even more complex geometries of this type have been analyzed. These include structures where the rubber domains include microscopic embedded domains of the plastic giving rise to microscopic representations of a salami structure. Pukanszky and coworkers [14,19] showed that a "salami" structure where the rubbery void is composed of hard subparticles encased in a soft rubber that is in contact with the thermoplastic matrix is equivalent to a single soft particle problem. An extensive amount of finite-element analysis (FEA) results obtained for multiparticle computer models have been generated by Chen and Ma [21,22]. In addition, these results show that a negative hydrostatic stress field was generated around the rubbery particles even under uniaxial extension.

Beyond linear elastic deformation behavior of polymer composites that can initiate deformation but absorb very little energy, improved fracture behavior of a polymer blend requires inelastic processes for energy dissipation. Polymers under triaxial stress conditions undergo volumetric deformation as the most accessible inelastic deformation. This leads to the development of crazes, which is a localized plastic deformation typically resulting in a failure. Rubber inclusions in the plastic undergo cavitation at early stages of the deformation to induce plane stress. This allows the matrix to utilize inelastic deformation mechanisms on a larger scale and relax significant portion of the stress. Such deformation requires little or no volume change under given conditions.

The central message of this discussion is that voids or low-strength inclusions into the plastic matrix during the processing provide pronounced improvements in the fracture behavior. The presence of the defects induces cavitation that is most pronounced at the early stages of the deformation. This allows the toughening mechanism to be fully energy dissipative [23]. Thus, the cavitation is a necessary initial step. The bulk of the energy dissipation occurs by the flow of the matrix material resulting in a uniaxial deformation to yield ligaments when the interparticle distances are less than the critical ligament size, which is a variable parameter based on the plastic and the experimental conditions.

The discontinuous properties of rubber–plastic blends at some critical dimension of interparticle distance produce questions about the comparability of the mechanical properties in the bulk and the ligament for the plastic. Crystallization of polymers in thin layers produces morphologies distinct from those observed in bulk polymers [24]. Second, interfacial effects such as miscibility of the rubber and the plastic are more important in thin ligaments than in the bulk. Such interaction may cause various effects during solidification of the plastic including nucleation or cause an immobilization of chains decreasing the crystallinity in the interfacial layer of the matrix plastic. The morphology of a semicrystalline polymer (high-density polyethylene, HDPE) crystallized among rubbery inclusions (ethylene–octene rubber) has been studied by Bartzak *et al.* [6] using wide-angle X-ray scattering (WAXS) and atomic force microscopy (AFM). The HDPE–rubber interface induces a strong specific crystallization of HDPE in the near-interface layers. Crystallites were oriented preferentially edge-on, with their lamellar normal being parallel or slightly tilted with respect to the interface plane.

During solidification upon cooling, the plastic and the rubber that have different volumetric thermal expansion coefficients lead to the formation of persistent stress fields causing either compressive or dilatational pressure acting on inclusions [25–27]. This is determined by the materials as well as by the thermal history. Dilatational stresses are useful for higher fracture toughness. They promote initialization of the internal cavitation process at considerably lower stress field. Jang *et al.* [26] have shown that blends of PP with SEBS (styrene–ethylene–butylene–styrene) rubbers form high dilatational stresses within the inclusions. SEBS displays high coefficient of thermal expansion above its $T_g$. This compensates the effect of thermally induced shrinkage of PP matrix during crystallization. This continues even at low temperatures and possibly causes thermally induced interfacial debonding of SEBS particles from the matrix. The thermal behavior of PP–SEBS systems was investigated by Mulhaupt and coworkers [28] using a PVT method to measure specific volume of individual components as well as blends. They found that the $T_g$ of SEBS decreased by an average of about 10 °C.

## 4.3
## Models for Rubber Toughening of Plastics

Impact and tensile toughness improvement in rubber–plastic blends indicates that several factors, all of which are important, contribute to an effective process of stress dissipation in the thermoplastic. These factors are varied and, in essence, every combination of the plastic and the rubber requires an evaluation of the different physical and chemical attributes of both the rubber and the plastic. We can summarize that the factors that affect the extent and the magnitude of the toughening are the following:

- Matrix thermoplastic material:
  - modulus;
  - failure mechanism;
  - proximity to thermal transitions;
  - orientation of crystals and molecules.

- Dispersed rubber:
  - amount;
  - size and shape of the particles;
  - distribution of the particles;
  - interparticle distance.
- Compatibilizers:
  - amount;
  - efficacy.

Several models have been proposed for these failure phenomena and the resulting observations. The following are the most widely applicable.

## 4.3.1
### Crazing Model

Plastic–rubber blends under tensile or impact deformation show stress whitening in the deformation zone. This crazing of the matrix is triggered by the stress field overlap of neighboring particles [29]. Consistent with this observation is the deformation behavior of HIPS or supertough nylon where the crazing is thought to be similar. According to this model, the function of the rubber particles is to create these stress fields in the matrix. Even at high rubber concentrations, the rubber particles do not interact though the crazing fields do overlap. The creation of the stress fields requires that the modulus of the dispersed phase has to be lower than that of the matrix. Low rubber modulus is an important parameter [30] and a very low modulus of the rubber is thus an advantage. Wu [31] suggested that crazing was one of the energy absorbing mechanisms and calculated that the amount of energy absorbed in this way was about a quarter of the total impact strength of the rubber–plastic blend. An area where the crazing model is difficult to use is at the low rubber concentration. At low rubber concentrations, the stress field overlap is minimal though stable crack propagation responsible for energy dissipation is experimentally obtained. There have been reports of the failure to observe any crazes in these low-rubber systems [32], which indicates that other mechanisms, not accounted by this local model, may be as important. In the more complex system, both crazing and shear yielding may take place [33]. As the crack propagates, the craze zone is transformed into a shear zone. This mechanism may be the process for the hydrostatic tension to be relieved and allow the shear yielding mechanism to be activated.

The observed stress whitening in the deformation zone of rubber–plastic blends is due to fibrillation of the matrix, crazing in the field of the thermoplastic, and cavitation of the rubber [32]. It is important to stress that in this model, conceptually, the deformation process is initiated in the soft and easily deformed rubber and not in the matrix or at the interface. The onset of cavitation is at low strains [34]. These indicate that the rubber cavitates while the plastic matrix yields. After the cavitation, the stress state is locally relieved and the yield strength of the plastic is lowered. The cavitation of the rubber is governed by the cavitational stress, which must be a function of the cohesive energy density, the entanglement density of the rubber, and inhomogeneity in the material [35,36]. The entanglement density depends on the chemical composition and the molecular weight of the material. Cavitation is helped by rubbers that fail or debond easily: this corresponds to rubbers whose molecular weight is only a few multiples of the entanglement molecular weight and whose viscosity is low [37]. The cavitation is also initiated by heterogeneities in the material. The observed lower limit in average particle size in improving the impact strength might be where a significant proportion of the particles does not contain inhomogeneity and hence does not cavitate. Another explanation for the lower limit in rubber particle size is that it is found where the interface thickness is a significant part of the particle volume [38]. It is expected that a graft polymer formed by the chemical interaction of the rubber and the plastic resides at the interface and is responsible for the formation of the small particles having a modulus not suitable for impact toughening. The data suggest that the cavitation of particles is not influenced by neighboring particles. The main effect of void formation at all test speeds is an acceleration of the shear yielding mechanism [39,40]. The rubber in a blend can be precavitated slightly by tensile straining at a low strain rate. In this way, the cavitation is not optimal as the triaxial stress state is not fully achieved. On precavitated samples, notched Izod studies were made. It was found that the impact behavior of the precavitated samples is comparable to that of the unstrained samples. This suggests that the cavitation process is not the major energy absorbing mechanism and that precavitation even helps to improve the impact behavior [41].

### 4.3.2
### Interparticle Distance Model

The BDT temperature is a function of rubber concentration, particle size, and type of rubber and this model correlates these descriptors into a single set of parameters. This best fit is the interparticle distance or the ligament thickness model. This model indicates that the ligament thickness is important for BDT, which corresponds to the transition from unstable to stable crack propagation. It is less insightful in predicting the value of the impact strength in the ductile region. This is because the level of impact strength in this region is mainly dependent on the rubber concentration and not on the particle size and the type of rubber. Thus, the impact strength values are not directly dependent on the ligament thickness. It was first anticipated that ligament thickness could be explained as a stress field overlap effect similar to overlapping crazing zones in the previous model. More research indicate that the stress field overlap is independent of the particle size and thus the ligament thickness effect cannot be explained by the stress field overlap model. An explanation is given by Bucknall [40] who indicated that the major energy absorption mechanism is the extensive formation of fibrils in the matrix. The fibrillation process initiates as a consequence of cavitation and the diameters of the fibrils are controlled by the interparticle spacings. The field of the zone that ends being fibrillated depends on the fibril diameter and thus the interparticle spacing.

### 4.3.3
### Percolation Models

Two percolation models have been proposed to explain the BDT in rubber-toughened plastics [42–44]. This model described by Margolina and Wu [42,43] is based on the concept of interconnecting ligaments of the plastic that form a connected web able to transfer stresses. The model has been used to predict both the BDT and the value of the impact toughening for both brittle and tough samples. The verification with the impact levels is still questionable as the impact strength is not correlated with the ligament thickness [45]. Another area where the prediction of the model and the experimental data diverge is the effect of broadening of particle sizes and particle distributions, in effect leading to poor mixing of the rubber and the plastic. Theoretically, inhomogeneous distributions always favor the percolation. A poorly distributed blend, however, gives poorer impact properties. A modified version of the model was proposed in which the impact data were modified for the rubber concentration effect on the macro-scopic yield strength [43]. The alternative percolation model of Sjoerdsma is the percolation of overlapping stress fields [44] who assumed the width of the defor-mation zone to be constant.

### 4.4
### Characterization of Rubber–Plastic Blends

### 4.4.1
### Glass Transition

The characteristic transition to the amorphous polymers is the glass transition. The glass transition temperature is a frequency-dependent property, which indi-cates a discontinuous change in internal degrees of freedom of motion for the polymer and its segments. A number of property changes occur at the $T_g$, including the change in activation energy of gas diffusion, creep rate, and mechanical properties such as toughness and strength. Single components exhibit a single $T_g$ as do miscible blends where the final $T_g$ is a composition weighted average, intermediate between the two constituents. The Fox equa-tion [46] has been used to predict the $T_g$ of miscible mixtures, while the Wil-liams, Landel, and Ferry equation [47] has been used to predict the time–temperature superposition for blends of different compositions that have $T_g$ at different temperatures.

The determination of the $T_g$ uses various methods including dynamic mechanical analysis (DMA), dielectric analysis, and calorimetry. Variation in $T_g$, more particularly the breadth of the peak in either thermal or mechanical

spectra, of the blend can lead to fine distinction between miscible, partly miscible, or compatible blends.

In polymer blends near the LCST or the UCST, the broadening of the $T_g$ is
observed. In rare cases, resolution of separate $T_g$ might be observed [48]. Miscible blends with strong interaction between the components lead to positive deviation of the $T_g$ from the expected values for a miscible blend with little
interaction between the components. These general guidelines for polymer
blends are often modified for each particular system of polymer blends depending on chemical interactions, physical form, and the rate of the experimental
observation. In the particular case of rubber–plastic blends, temperature changes
are often accompanied by selective crystallization of the plastic phase: the thermodynamic miscibility of the rubber and the plastic changes under these transformations. Polyolefin blends where the plastic component is iPP is a common
example of this behavior. In the particular case of rubber–plastic blends, selective crystallization of the plastic blends can also lead to migration and enrich-
ment of the rubber phase by the amorphous or less crystalline components of
the plastic during the cooling process with consequent changes in the thermal
characteristics of this rubber phase. This is typically the case for blends with
polyolefins such as polyethylene and polypropylene.

## 4.4.2
### Dynamic Mechanical Characterization

The analysis of the absorption of energy as well as the phase change between the
incident and induced low-amplitude cyclic deformation has been used to determine the elastic and viscoelastic properties of polymers. For rubber–plastic
blends where the two components are substantially different in mechanical
properties, this is a powerful and well-applied method. This oscillatory mechanical analysis has been extended to provide useful information not only on the
components but also on the morphology and the phase behaviors of polymer
blends, though with some reservations. The analysis spans both temperature and
frequency; thus, both thermal and time-dependent polymer processes can be
studied. This method maps the data over both a broad temperature range and a
broad frequency range. This analysis identifies polymer transitions including secondary relaxation processes, crystalline melting transitions, and the glass transition. In skilled analytical hands, this provides information about components
as well as the molecular weight and the phase morphology. Plots of $\tan \delta$, $E''$,
or $G''$, which are characteristics of the frequency-dependent absorption of
energy, as well as the phase lag between the impinged and induced cyclic motion
versus temperature can correlate to the phase separation, miscibility, or other
thermodynamic processes of the blend.

Free vibration and forced vibration methods have been developed for dynamic
mechanical analyses. Free vibration methods include the torsion pendulum, the
vibrating reed, and the torsional braid analyzer (TBA). Momentary angular
deformation of the inertial source creates a damped sinusoidal curve of the

resultant deformation of the sample [49,50]. The TBA uses a glass braid coated with the sample. This allows for measurements of liquid, low molecular weight materials as well as brittle solids [51]. This is useful to determine physical processes such as spinodal decomposition of a rubber–plastic blend with time or temperature or chemical processes such as rubber vulcanization [52].

Forced vibration geometry fixes both ends of a sample and applies a sinusoidal tensile or shear strain, which yields corresponding sinusoidal stress response with a phase angle difference. For forced vibration, the frequency range of 0.01–100 Hz is commonly available with a temperature range of −150 to 400 °C. A discussion of the details and utility of dynamic mechanical analysis is available [50,53].

The immiscible rubber–plastic blends exhibit the glass transition temperatures and the phase transitions of the unblended components. The partially miscible blends exhibit $T_g$ shifted to the mean reflecting the incorporation of minor concentrations of the other polymer constituent as predicted from the phase diagram. Dynamic mechanical methods are more sensitive than calorimetric methods and can detect low concentrations of specific phases in polymer blends. In addition, low-temperature secondary loss transitions, such as the movement of small segments of the chain or side groups, can be easily resolved via dynamic mechanical testing.

### 4.4.3
### Calorimetric Methods

The differential scanning calorimeter is typically employed for these studies. Classically, this consists of a polymer sample with a compensating thermal control reference material that are heated or cooled equivalently: the difference in the temperature change for the sample and the reference material measures internal thermal transitions within the sample. Transitions due to phase change, crystallization, and relaxation in the sample appear as the differences in the temperature between the sample and the reference material and the extent and the temperature of the difference can be used to determine specific heat, glass transition temperature, melting and crystallization points, heat of fusion or crystallization, crystallization kinetics, onset of thermal degradation, and heat of reaction. The data are a function of heating (or cooling) rate, and the sample temperature can be rapidly changed. This method is thus particularly valuable for characterization of rubber–plastic blends with crystalline components.

Differential scanning calorimetry (DSC) provides rapid analysis, but also offers lower sensitivity/resolution for glass transitions or secondary relaxation transitions than dynamic mechanical methods. Temperature-modulated DSC has been used to increase the sensitivity and to allow improved resolution of polymer blend transitions with some miscibility [54]. A modulated temperature (such as an oscillatory heating rate) can resolve transitions not seen with conventional DSC procedures. A calorimetric method for the heat of mixing of polymer blends by mixing low molecular weight compounds has been developed [55] to extrapolate data to high molecular weight polymers since low molecular weight polymers with similar repeat units have similar heats of mixing with little entropic correction for high molecular weight polymers.

### 4.4.4
### Dielectric Characterization

Dielectric characterization measures inherent or induced dipole responses to an applied electric field. Typically, the electrical field is oscillating with a variable frequency and this allows a separation of the processes by the timescale of the oscillating field. In a typical experiment when a frequency due to an alternating current is applied, the dipole polarization is induced and attempts to keep up with the oscillating filed. At low frequency, the dipole movement can follow the electric field changes. With increasing frequency, the dipole movement cannot keep up with the incident electric field changes and the induced dipole orientation lags in phase. This transient delay results in dissipation of power and generation of heat. As the frequency is increased further, there is insufficient time for dipole movement/orientation; thus, power losses occur and the dissipation decreases. This technique is very useful for polar molecules or molecules that are easily polarized: comparatively for nonpolar or symmetrical molecules, the influence of temperature and frequency of electrical oscillation is low. The dielectric properties of polymers exhibit an analogy with dynamic mechanical characterization though specifically in the electrical dipole character of the molecules.

This method is most suited for polar polymers because nonpolar polymers lack the required sensitivity. Dielectric relaxation has been applied to polymer blends to determine the extent of concentration fluctuations. These measurements are conducted at different temperatures. Broadening of the resultant curves relative to the unblended controls demonstrates increasing concentration fluctuations. Secondary relaxation transitions involving polar groups can be easily detected by dielectric measurements and parallel similar low-temperature transitions noted in dynamic mechanical measurements.

An alternative method to observe dielectric properties is thermal stimulated currents (TSC). This method involves polarization of a sample at high temperature (relative to $T_g$) and quenching to a temperature where depolarization is kinetically prevented in the timescale of the experiment. An alternative name for this is poling. The temperature of the specimen is then increased and the depolarization current is measured. Depolarization is most rapid at temperatures corresponding to polymer transitions. The TSC spectrum is useful for secondary relaxations, glass transitions, and liquid crystalline phase transitions. TSC has been applied to PA6–ABS blends to study the intermixing of the components of the respective blends [56].

### 4.4.5
### Morphology/Microscopy

Perhaps the most commonly used tool for the characterization of rubber–plastic blends is microscopy as a measure of the morphology. The rubber–plastic matrix provides a sufficient contrast in optical or mechanical properties within the components that useful information about the morphology can be obtained. The morphology data of phase-separated rubber–plastic blends that are almost uniquely derived from microscopy are required to understand the structure–property relationships in the blend.

Contrast between the phases of the rubber–plastic blend is required and has given rise to a number of complementary techniques to distinguish one phase from the other. For phase-separated rubber–plastic block copolymers, the covalent bond between the blocks residing at the interface yields a long-range ordered uniform morphology. With simple polymer blends, the morphology is less ordered and far less structured.

Morphology characterization addresses the particle size and distribution of rubbery impact modifiers in a polymer matrix, the distribution of components in a ternary blend, the effect of interfacial addition on the particle size, crystalline phase, and morphology, the effect of reactive compatibilizers on morphology, the fracture pattern of polymer blends, the dispersion/agglomeration of particles, and the distribution of fillers in polymer blends.

### 4.4.6
### Optical Microscopy

Optical microscopy is readily available and generally offers the starting point for morphological characterization of polymer blends. Optical microscopy is useful for larger dimensions involving polymer blends since the resolution is limited to 0.2 μm. Thin sections can be prepared to view the morphology. The sections can be viewed in transmission with the contrast between phases provided by differences in opacity, color, refractive index, and topology. These features can be enhanced by staining, preferential adsorption of dye or fluorescent species, and solvent treatment to swell or remove the phases. Osmium tetroxide treatment can be employed to provide phase contrast, as different phases will often exhibit different levels of staining. Variations of optical microscopy include phase contrast microscopy, polarized optical microscopy, differential interference microscopy, fluorescence microscopy, transmitted optical microscopy, reflected optical microscopy, and confocal microscopy.

In transmission light microscopy, a light beam is transmitted through the specimen into the objective lens/eyepiece of the microscope. The thickness of samples employed can be in the range of 1–50 μm. Samples are prepared as either thin cast films or microtomed sections [57]. Reflected light microscopy has a light beam reflected from the surface into the objective lens. This method is employed for opaque samples. Surface etching by organic solvents and oxidizing acids or bases is used to enhance the surface contrast between the phases. Polarizing light microscopy employs crossed polarizers to view the sample. With isotropic specimens, the field is dark, while anisotropic, birefringent areas of a sample will appear bright. Polarizing microscopy is employed to view spherulitic structures [58–60] and deformation morphologies (crazes, shear banding) in polymer blends. The samples slides are used in a hot-stage microscope to allow for controlled crystallization and observation of spherulitic growth for polymer blends comprising crystalline components. This method is particularly useful for studying the morphology of liquid crystalline polymer-containing blends.

Phase contrast and polarized light microscopy was employed to observe the morphology resulting from the spinodal decomposition and crystallization processes in PP/EPR blends [61]. The application of polarized light optical microscopy for the morphology of polymer blends has been described in Ref. [62]. Confocal microscopy, popular in characterization of biological systems, offers the ability to view the three-dimensional structure of materials. A pinhole is placed in the back focal plane that suppresses light from other planes from reaching the detector. This yields a significant improvement in the depth of focus over conventional light microscopy. A volume of the sample is scanned layer by layer and reconstruction of the layers allows for the three-dimensional image. A variation of this method (laser scanning confocal microscopy) was employed to observe the morphology of PP/PE/EPR with EPR labeled with a fluorescent dye [63]. Laser scanning confocal microscopy was used to observe the spinodal decomposition of a styrene–butadiene rubber (SBR)/polybutene (PB) blend cooled from above the UCST to the unstable region showing the development of bicontinuous morphology [64]. The utility of laser scanning confocal microscopy for polymer blend morphology characterization has been reviewed by Ribbe and Hashimoto [64]. Optical microscopy methods have been reviewed in Ref. [65].

### 4.4.7
### Transmission Electron Microscopy

Transmission electron microscopy (TEM) provides one of the major microscopy techniques employed for polymer blends. TEM is analogous to transmission optical microscopy except that an electron beam instead of a light beam is employed. The wavelength of electrons allows for higher resolution (resolution down to 0.2 nm) and thus much smaller dimensional resolution than optical methods. TEM samples must be thinner to allow electron beam penetration (in the range of 100 nm). Energy-dispersed X-ray detectors can be added to the TEM analysis to conduct element mapping for polymers containing resolvable elements.

Low atomic number elements such as in typical hydrocarbon polymers have limited scattering ability and it is hard to distinguish the boundaries of the different phases. Thus, for our cases the key to success employing TEM involves developing phase contrast of the components by selectively absorbing a heavy element in one of the phases. Osmium tetroxide is the most common staining material employed and is particularly useful for polymers with unsaturation, because it selectively and rapidly reacts with the double bond to yield an ester providing excellent contrast for that phase over the other. Samples are typically ultramicrotomed and exposed to the vapor, aqueous, or organic solutions. This staining and observation technique has been used for HIPS, ABS, polyvinyl chloride (PVC) impact modifiers, rubber-toughened epoxy, and ethylene–propylene–diene monomer (EPDM).

One problem with staining is that the chemical reaction can promote phase separation; thus, the procedure may enhance phase differentiation more than is really present for a rubber–plastic blend. This is an artifact in observation of interfaces where limited phase mixing is present, such as in partially miscible blends, blends with a broad transition, and block copolymers. TEM has been particularly useful for observing the detailed and ordered morphologies available with block copolymers.

The use of TEM in following morphology development of spinodal decomposition phase separation as well as nucleation and growth processes in polymer blends has been widely employed and was initially reported by McMaster [66]. The morphologies of ternary blends of PA6/PC/PS and PA6/PC/SEBS triblock polymers were compared with blends where the PS or SEBS component contained maleic anhydride (MA) for improved compatibilization [67]. The samples were stained with $RuO_4$, providing sufficient contrast. TEM of PP/LLDPE (linear low-density polyethylene) blends containing a SEBS block copolymer stained with osmium tetroxide showed the SEBS concentrated in the PP phase with some concentration at the blend interface [68]. Energy-filtering TEM allows elemental composition determination with electron energy loss spectroscopy (EELS). This method can provide information on the polymer composition in the interfacial region of a phase-separated polymer.

### 4.4.8
### Scanning Electron Microscopy

Scanning electron microscopy (SEM) is another popular technique for assessing the structure in a polymer blend. This method focuses an electron beam onto a surface, and the emission of electrons from the specimen is detected and amplified to obtain an image. The accelerating voltage is typically in the range of 1–40 kV, lower than TEM. The resolution achievable ranges down to 2 nm. SEM exhibits a relatively large depth of field; thus, it can show topological features better than other microscopy methods (except perhaps atomic force microscopy). Samples require surface conductivity; thus, a thin layer (~10 nm) of a conductive metal (gold or platinum) is sputtered onto the surface. Staining and etching processes can be employed to provide improved contrast. SEM is particularly useful for observation of fracture surfaces.

In order to accentuate the surface features, chemical (solvent or acid exposure) or plasma etching is often employed. Cryogenic fracture is also employed to provide surfaces with better defined topological features than possible with higher temperature fracture. Particles, such as impact modifiers, will debond during cryogenic fracture providing the desired contrast. SEM is employed for viewing orientation effects in injection-molded specimens. This is particularly important for liquid crystalline polymer reinforcement of nonliquid crystalline polymers, as orientation during molding or extrusion leads to a highly oriented skin structure comprised of liquid crystalline fibrils [69].

Polyolefin blends were stained with $RuO_4$ to provide differentiation in low-voltage SEM characterization [70]. The SEM characterization on cryofractured surfaces of injection-molded specimens of PET/liquid crystalline polymer (LCP) blends showed different LCP morphologies depending on the position in the molded specimen and LCP concentration [71]. PP/EPR and PP/PE/EPR blends were injection molded and fractured in liquid nitrogen in the flow direction and perpendicular to the flow direction of the specimen [72]. Samples were etched with hexane and the SEM results showed phase orientation and fibrillation as a function of the viscosity ratio of the dispersed phase to continuous phase. Transmission electron microscopy and scanning electron microscopy applied to polymer blends have been reviewed [73–76].

### 4.4.9
### Atomic Force Microscopy

A more recent microscopy technique, termed atomic force microscopy, has become a popular method for characterizing specific features of polymer blend morphology. AFM is also referred to as scanning force microscopy (SFM) or scanning probe microscopy (SPM). An extremely sharp probe (of diamond or silicon), near atomic dimensions at the tip, is tracked across the surface with the contact force held constant. The deflection resulting from the specimen surface contour is measured by the deflection of a laser beam focused on the cantilever beam holding the probe. The AFM can be operated in a contact mode, where the probe contacts the surface continuously or in a tapping mode, in which a high-frequency oscillation yields intermittent contact of the probe with the specimen surface. Another variation of the AFM measurement procedure involves lateral force (or frictional force) imaging to measure viscoelastic surface properties, which translate into structural differentiation. The torsion/bending of the cantilever is measured to determine the lateral force on the probe tip. AFM provides the ability to image insulating polymers and provides surface details not observable by other methods (other than SEM). AFM offers the unique advantage of quantifying the surface roughness. More detailed description of AFM details can be found [77,78].

AFM has the ability to identify surface phase separation mechanisms (nucleation and growth versus spinodal decomposition) of polymer blends [79]. Phase separation of mixtures of perdeuterated poly(ethylene–propylene) with hydrogenated poly(ethylene–propylene) was observed with AFM (inside the metastable and unstable regions) [80]. AFM thus offers a microscopy method that allows resolution of the phase separation process, as the process results in surface roughness.

### 4.4.10
### Scanning Tunneling Microscopy

Scanning tunneling microscopy (STM), initially reported in 1982 [81], offers the ability to scan surfaces down to atomic levels. The technique employs a very sharp W or Pt–Ir probe, scanning less than 1 nm above a conducting surface.

The distance is kept constant and maintains constant current as the probe scans across the surface. STM requires conductivity of the surface being scanned, and thus is not as versatile as AFM, which can be employed for insulating polymers.

### 4.4.11
### X-Ray Microscopy

X-ray microscopy employs focused soft X-rays (200–600 eV) onto thin (100–200 nm) samples and can resolve phase structure in biological as well as polymeric systems. X-ray microscopy for polymer blends was initially demonstrated by Ade *et al.* [82,83].

### 4.4.12
### Scattering Methods: Light, X-Ray, and Neutron

Three different scattering techniques have been commonly employed to study small-scale inhomogeneity in polymer blends, namely, light scattering, X-ray scattering, and neutron scattering. Light scattering relies on refractive index variations, X-ray scattering measures electron density variation, and neutron scattering relies on the scattering of free neutrons by matter.

The earliest form of light scattering involves turbidity determination. Amorphous miscible polymers are transparent, whereas phase-separated systems are translucent to opaque, when the phase separation dimensions are larger than the wavelength of visible light and the refractive indices of the components are different. This method is used for phase separation of mixtures that exhibit LCST or UCST. This observation is referred to as cloud point determination, and cloud point measurements can determine the phase diagram of polymer mixtures. Typically, the light transmission of a thin film of the polymer blend is measured as a function of temperature and composition.

A variation of light scattering, termed pulse-induced critical scattering (PICS), has also been employed to measure the spinodal temperature for blend. This method involves a small quantity of a sample rapidly heated or cooled into the metastable region and rapid measurement of the light scattering. The thermal pulse allows measurement before nucleation and growth phase separation can occur. This allows for a more accurate method for the spinodal temperatures to be determined, as measurements can be made much closer to the actual spinodal temperature. This method was initially reported in 1974 [84] and was shown to be applicable for polymer–solvent mixtures and oligomeric mixtures, such as low molecular weight polyisobutene/PS blends [85].

### 4.4.13
### X-Ray Scattering

X-rays are produced by the bombardment of anodes with high-energy electrons. The laboratory equipment employed with X-ray scattering generally involves an

X-ray tube with Cu Kα radiation. The angular dependence for small-angle X-ray scattering (SAXS) is in the range of 10°. The length scale capable of characterization is 1–2 to 20–40 nm. WAXS (angle = 10–50°) yields structure determination of less than several nanometers. SAXS scattering has been useful to obtain data on crystalline polymer blends, including crystalline–amorphous and crystalline–crystalline polymer combinations, in the area of crystal layer thickness and amorphous layer thickness [86].

Chlorinated polyethylene (CPE)/polycaprolactone (PCL) blends were investigated using SAXS [87]. At 42 and 49 wt% chlorine for CPE, miscibility with PCL was observed.

A study of HDPE/LLDPE (ethylene-*co*-butene-l) blends involving DSC, WAXD, SAXS, and light scattering was conducted with the observation that no segregation occurs in the crystalline, lamellar, or spherulitic features of the blend [88]. SAXS studies on HDPE/LDPE (low-density polyethylene) showed an interfibrillar-scale separation within the spherulites containing both HDPE and LDPE [89]. Quenching showed HDPE/LDPE separation at the interlamellar level. Fully deuterated HDPE blends with HDPE and LLDPE studied with time-resolved SAXS indicated similar crystallization rates of deuterated HDPE and a specific LLDPE of intermediate comonomer content could be the governing factor related to the cocrystallization behavior observed [90]. SAXS measurements have also been useful in studies of the morphology and phase behavior of block copolymer/homopolymer blends. SAXS data on PS blends with styrene diblock copolymers showed the presence of microdomains, which disappeared as the order–disorder transition was approached with temperature [91].

WAXS of crystalline polymers exhibits sharp concentric rings, indicating a high degree of order. Amorphous polymers exhibit a diffuse halo, indicating lack of order. WAXS was employed to study the miscibility of elastomer blends [92]. Phase-separated blends exhibited weight-averaged results of molecular interchain spacing.

### 4.4.14
### Neutron Scattering

The unique advantage of neutron scattering is its ability to observe lighter elements (e.g., hydrogen). H versus D isotopes can provide contrast. This has been particularly important for polyolefin blends. Another advantage for neutron scattering is the nondestructive nature of the technique, allowing for evaluation without structural changes. The length scale for small-angle neutron scattering (SANS) covers four orders of magnitude, from several angstroms to the micrometer range. Neutron scattering characterization methods include SANS, neutron reflection, inelastic neutron scattering (INS), quasielastic neutron scattering (QENS), and neutron spin echo (NSE).

The details of neutron scattering can be found in Refs [93,94]. SANS has proven very useful in the characterization of polyolefin blends. The phase behavior of polyolefins of different compositions can often be difficult to analyze using more conventional techniques (dynamic mechanical or calorimetry). In one of the earlier studies of polyolefin blends, PE/PP blends show no scattering as non-deuterated species [95]. With partial or complete deuteration of one or the other component, Lohse [96] reported neutron scattering studies on deuterated iPP with aPP (atactic polypropylene) and EPR copolymers. Blends of iPP/EPR were immiscible with as low as 8 wt% ethylene in EPR. Crist and coworkers [97,98] studied blends of linear PE with hydrogenated polybutadiene with SANS to determine the interaction parameter as a function of ethyl branches.

## 4.4.15
### Neutron Reflectivity

Neutron reflectivity methods have advantages over SANS for studying the interfacial region between phase-separated polymers and have shown utility in measuring polymer–polymer interdiffusion. Neutron reflectivity offers the advantage of probing surfaces and has a length scale in the range of 0.5–500 nm. The application to polymer interfaces has been reviewed by Richards and Penfold [99] and Bucknall [100].

## 4.4.16
### Neutron Spin Echo Spectroscopy

Neutron spin echo spectroscopy offers a method to cover a large range of dimensions and timescales of polymer chain motion from relaxation processes to polymer melt reptation. This method has been reviewed for relevance to polymer blends [101].

## 4.4.17
### Nuclear Magnetic Resonance

Nuclear magnetic resonance (NMR) is based on the magnetic moment of nuclei possessing a spin quantum number. This exists when either the mass number or the atomic number is odd; thus, hydrogen, deuterium, carbon-13, nitrogen-14, nitrogen-15, fluorine-19, silicon-29, and phosphorus-31 can be considered [102]. NMR has been a powerful tool for the characterization of structure in polymers, including structural confirmation, tacticity, monomer repeat units, and molecular motion involving secondary relaxation processes. Solid-state NMR offers the ability to observe molecular homogeneity from the molecular levels to over 20 nm. Chemical shifts in the $^{13}$C NMR spectra can provide evidence of specific interaction between dissimilar components of the blend. A method termed cross-polarization magic angle spinning (CPMAS-$^{13}$C NMR) has shown promise in identification of the environment of carbon nuclei in blends along with an assessment of the degree of homogeneity. $^{129}$Xe NMR has been employed to determine structural order in amorphous blends. Xenon as a probe molecule will show separate resonances for a phase-separated blend, but singular values for miscible blends. Reviews on the use of solid-state NMR to study the miscibility and characteristics of polymer blends are given in Ref. [103].

Two-dimensional NMR methods including the nuclear Overhauser enhancement spectroscopy (NOESY) and the heteronuclear correlation (HETCOR) methods study intermolecular interactions between polymers. 2D-NOESY has been utilized primarily for biological macromolecules but has shown utility for polymer solutions in appropriate deuterated solvents.

### 4.4.18
### Spectroscopic Methods

Electromagnetic radiation and the interaction with specific groups of a rubber–plastic blend as a function of wavelength is an important method to determine specific interactions between interacting species. Infrared spectroscopy is the most important spectroscopic method employed in polymer blends. UV–visible spectroscopy has been employed but to a lesser extent.

### 4.4.19
### Infrared Spectroscopy

Infrared spectroscopy that detects transitions between vibrational or rotational states is a very valuable tool for detecting specific structural groups in organic and polymeric materials. Hydrogen bonding interactions can also be detected by observing the shift in frequency of the absorption peak for the hydrogen bonded unit. Specific groups capable of hydrogen bonding include the hydroxyl group, the carbonyl group C=O (in acid or ester units), and amide groups, which exhibit both N—H and C=O stretching. These absorption spectra can shift in the presence of hydrogen bonding environments (either solvents or miscible polymers) or even exhibit unique peaks. The review of the use of infrared spectroscopy in ascertaining interactions in polymer blends is given by Coleman *et al.* [104].

### 4.4.20
### UV–Visible Spectroscopy

In the UV–visible frequency range, valence electrons can be excited and observed. UV–visible spectroscopy is most useful for blends that exhibit electron transfer interactions, such as observed with charge transfer complexation involving electron acceptor–electron donor polymer blends. This method was applied to electron acceptor polyesters blended with electron donor polymers containing tertiary amine groups, where a shift in the absorption maxima in the visible region was observed [105].

### 4.4.21
### Raman Spectroscopy

Raman spectroscopy and Fourier transform Raman spectroscopy (FTRS) can detect vibrational motion in polymers but are less employed in polymer blend characterization than Fourier transform infrared spectroscopy (FTIR);

nevertheless, they offer utility in characterization of crystalline polymer morphology and surface modification.

## 4.4.22
### Fluorescence Spectroscopy: Nonradiative Energy Transfer and Excimer Fluorescence

A method termed nonradiative energy transfer (NRET) has been used to qualitatively determine the level of mixing in polymer blends. The polymer components are labeled with chromophores (donor and acceptor), with label concentration kept low so that the miscibility is not altered. Donor–acceptor pairs are chosen such that overlap occurs between the emission spectra of the donor and the adsorption spectra of the acceptor. The energy transfer can occur up to several nanometer separation. The excitation energy absorbed by the donor molecule can be transferred to the acceptor molecule via dipole–dipole interactions if the donor–acceptor pairs are in close proximity. The fluorescence emission technique allows naphthalene (donor)-labeled PVC and anthracene-labeled chlorinated PVC (CPVC) blends to show blend immiscibility [106].

## 4.4.23
### X-Ray Photoelectron Spectroscopy and Secondary Ion Mass Spectroscopy

X-ray photoelectron spectroscopy (XPS), known as electron spectroscopy for chemical analysis (ESCA), is commonly employed to determine the surface composition of polymeric materials. Upon X-ray irradiation of the surface, the inner shell electrons can be emitted. The measurement of the kinetic energy of the emitted electrons can yield the binding energy and thus the identification of the source of the emitted electrons.

## 4.4.24
### Vapor Sorption and Solvent Probe Techniques

Inverse gas chromatography (IGC) and vapor sorption methods utilize solvent probes to ascertain the interactions existing in polymer blends [107–109]. The procedure for IGC involves coating the stationary inert support material with the polymers and the blends and measuring the retention time according to GC protocols. The retention time is a function of the activity of the probe and can be used to measure the interactions in the polymer blend. Probes that compete with predominant polymer–polymer interaction in the blend such as similarity in solubility parameter or hydrogen bonding are useful for this diagnostic. One of the problems noted in several studies is that the results obtained by IGC are probe dependent. Further details on the experimental procedures and application to polymer blends can be found in Ref. [103]. Different solvent probes were employed by Olabisi [107] to study the nature of PVC/PCL interactions by IGC in one of the early investigations using this approach.

4.4.25
**Characterization of Interfacial Properties**

The key interfacial property for phase-separated polymers is the interfacial tension, $X_{12}$. While the measurement of interfacial tension for low molecular weight liquids is simple, the high viscosity of polymer melts precludes rapid analysis. Nevertheless, several methods can be employed using model polymers of oligomeric molecular weights. The low viscosity enhances the sensitivity of this technique. The techniques are the pendant drop, the sessile drop, and the spinning drop methods.

The pendant drop method involves a drop of the more dense polymeric component immersed in a matrix of the lower density polymer component (both above the $T_g/T_m$). The equilibrium drop shape is a balance between gravitational and surface forces. The interfacial tension can be determined from the drop shape. The pendant drop technique has been reviewed by Wu [110] and experimental data were compared with theory for polydimethylsiloxane (PDMS)/PB oligomers by Anastasiadis et al. [111]. The sessile drop method also involves a more dense polymer drop immersed in a less dense polymer matrix that is then allowed to settle on a flat surface. The spinning drop method determines the cylindrical profile of a polymer drop dispersed in a denser polymer matrix under constant rotation, thus yielding a balance between centrifugal and interfacial forces. This method has been described by Elmendorp and de Vos [112] for molten polymers and by Princen et al. for low molecular weight liquids [113].

Another method to determine the interfacial tension of immiscible polymer blends is termed the breaking thread method. Thin fibers (about 10 µm) are embedded in a matrix polymer. Heating to the viscous liquid state results in sinusoidal fiber distortion, to yield regularly dispersed ellipsoidal to spherical domains. This method was applied to LDPE/PS blends with an interfacial tension of 8.26 mN m$^{-1}$ [114]. The value rapidly decreased to 3.6 mN m$^{-1}$ above 1 wt% SEBS. The breaking thread method was applied to LDPE/PA6 blends with several different compatibilizers added to determine the change in interfacial tension (EAA, EAA: Zn ionomer, SEBS, and SEBS-g-MA) [115]. EAA and SEBS-g-MA addition yielded the largest decrease in interfacial tension and thus led to decreased domain size of the dispersed phase. The interfacial tension for LDPE/PA6 was 12.5 mN m$^{-1}$ and 2 wt% EAA gave a value of 3.0 mN m$^{-1}$, while 2 wt% SEBS-g-MA gave a value of 3.8 mN m$^{-1}$. An example with similarity to this method (but conducted during melt fiber orientation) involved blends of polypropylene in a thermoplastic polyvinyl alcohol (PVOH) matrix. The conditions employed yielded ellipsoidal PP particles with similar lengths and diameters, even though the unextracted blend was highly concentrated in the dispersed phase.

Neutron reflectivity and ellipsometry are two methods commonly employed to determine the interfacial thickness of phase-separated polymer blends. Ellipsometry involves spin coating a layer (several hundreds of nanometers) on a thicker polymer film supported on a rigid substrate (e.g., silicon wafer). From refractive index values of both polymers and the thickness of the spin-coated layer, data analysis of incident light reflection and refraction can determine the refractive index and the thickness of the interfacial layer. This procedure and application to the determination of interfacial thickness have been noted by Yukioka *et al.* [116,117].

## 4.5
## Experimental Rubber–Plastic Blends

### 4.5.1
### Early Work

Cellulose nitrate, which was produced by Schonbein [118] in the 1847, was the first thermoplastic that could be made processable and impact resistant by incorporation of a plasticizer, camphor, by the Hyatt brothers [119]. A similar approach was successful for PVC that was rendered easily processable by addition of phthalate ester plasticizers by Clarke [120], Wilkie [121], and others. This work, which is a backdrop for the subject of this chapter, indicated that variation of the properties of the composite could be achieved by changing the proportions of the plasticizer and the plastic. An alternative approach was promoted by Reid [122], Voss and Dickhauser [123], and others who recognized that copolymerization would reduce the intractability of PVC. They produced commercial copolymers of vinyl chloride with vinyl acetate, vinyl alkyl ethers, and vinylidene chloride that, in varying degrees, were processable and impact resistant.

PVC was made impact resistant and soft by the incorporation of acrylonitrile–butadiene copolymer elastomers (NBR) [124]. However, this improvement in toughness was noted only when the difference in solubility parameters of the PVC and the NBR was small [125]. Implicit in this invention was the discovery that a successful impact toughening of a rigid plastic needed dispersion of the elastomer in the plastic with a large interfacial surface. A requirement for that finely dispersed morphology to be formed on melt mixing and be stable was a reasonable match of the solubility parameter and the viscosities of the two components. These requirements of kinetic accessibility during mixing and thermodynamic stability of the dispersed morphology of the two phase rubber–plastic blends are fundamental underpinnings of the technology of forming rubber–plastic blends that have commercial utility.

### 4.5.2
### Blends of Polyvinyl Chloride

The blends of PVC and NBR that were made by the *in situ* emulsion polymerization of NBR in the presence of a latex of PVC [126] were the first successful ones and had both very small particles of NBR with dimensions of about $0.1\,\mu m$ and close match of the refractive indices of the two components, leading to an almost translucent composite that was both impact resistant and easily processed.

Improved dispersion between the phases has been obtained by reducing the molecular weight of NBR and by the addition of compatibilizing agents [127]. Reviews of rubber-toughened PVC are available [128,129]. However, the NBR elastomers have been replaced, to a large extent, by other elastomers that have solubility parameters similar to NBR. These include ABS copolymers, methacrylate–butadiene–styrene (MBS) terpolymers, chlorinated polyethylene, polyacrylates [130], polyurethanes, and ethylene–vinyl acetate (EVA) [131–133] polymers. Cohan and Pittenger [134,135] discussed the advantages of addition of PVC to NBR and cited several applications for the blend, including wire and cable jacketing, printing roll covers, gaskets, valve disks, and football covers. At higher concentrations (>50 wt%), various applications for this permanently plasticized PVC are available, including shoe uppers, shoe welts, book bindings, vinyl adhesives, and tubing. Of many applications of PVC/NBR blends, one of the most successful was flame-resistant conveyer belting, discussed by Hay *et al.* [136].

Blends of NBR and PVC were reported by Nielsen [137] and others [138–140] to have a single broad glass transition over a temperature range between the $T_g$ values of the components, indicating a high degree of miscibility. The temperature dependence of the dynamic storage modulus $E'$ and the dynamic loss modulus $E''$ for the NBR/PVC blends as a function of composition was reported by Matsuo [141]. In this system, the effects of changing the concentration of acrylonitrile (AN) and hence the degree of phase compatibility were studied. In the blends of PVC and NBR (containing 20 wt% AN), the relaxation phenomenon of the blends was distinctly a composite of both components, indicating limited miscibility. Blends of PVC and NBR (containing 40 wt% AN) exhibit only one sharp glass transition temperature, which monotonically lowers as the amount of NBR is increased. These blends are more miscible than the ones where the NBR contains 20 wt% AN. Landi [142,143] observed that two-phase behavior could be obtained in the PVC/NBR blends if the composition distribution in the NBR composition were large. The material with the higher AN content was miscible with PVC, whereas the material with the lower AN content was not. In blends with PVC, only the first of these transitions shifted with varying PVC content. Zakrzewski [144], using calorimetric data, observed miscibility with PVC in all compositions at levels of 23–45% AN content in NBR. More recently, DSC, IR dichromism [145], and pulse NMR [146] methods have confirmed the miscibility of PVC and NBR at 31% AN content in the NBR. This compatibility of PVC and NBR of a selected composition has been extended to ternary blends of NBR, PVC, and poly(vinylidene chloride-co-vinyl chloride) by Wang and Chen [147] who showed some compatibility in these blends.

Electron microscopy studies by Matsuo [148,149] focused on the morphology of PVC/NBR blends in which the AN content of the NBR was varied. In the blend containing PBD (AN = 0), dispersed rubber particles are several microns in diameter, which indicate incompatibility. As the AN level of the NBR was increased to 20% (NBR20), the rubber particles were finer indicating enhanced compatibility. Increase of the AN to 40% (NBR40) renders a homogeneous system with the rubber in domains less than 100 Å. Inoue *et al.* [150] showed that

NBR/PVC blends cast from tetrahydrofuran solutions when studied by X-ray diffraction, radio thermoluminescence, DSC, and light scattering demonstrated a fairly regular two-phase structure.

The properties of blends of PVC and NBR vary as a function of the relative amounts of the two components. At PVC contents of up to about 50%, the properties of the NBR predominate. These blends are processed as rubbers and can be vulcanized. The main advantage of NBR/PVC blends over NBR is that they have excellent resistance to ozone. At PVC contents of more than 50%, the properties of the thermoplastic are predominant. These blends are processed and used as thermoplastics. The NBR serves as a polymeric plasticizer and impact modifier for the PVC [151,152].

A further advance in the technology, over mere melt blending PVC and NBR of PVC/NBR blends, was reported by Sperling *et al.* [153] who prepared interpenetrating networks (IPNs) by copolymerizing AN and BD in the presence of preformed PVC latex. They concluded that the polymer had a composition gradient with portions being entirely PVC and entirely NBR with some segments intermediate compositions. The properties of PVC/NBR blends are sensitive to the variations of the blend components and a number of studies have addressed this for PVC/NBR blends [154–157]. Factors studied are (i) AN content of the NBR, (ii) viscosity (~molar mass) of the NBR, (iii) molar mass of the PVC, (iv) type of PVC (emulsion, suspension), (v) PVC/NBR ratio, (vi) type of NBR (linear, crosslinked, carboxylated, acrylate comonomer), (vii) use of acrylate and vinyl acetate comonomers in the PVC, and (viii) type of filler (carbon blacks, mineral fillers).

In blends containing predominant amount of NBR, mechanical properties improved substantially as the composition of the NBR became progressively richer in AN due to improved compatibility with PVC [155,156]. This change in compatibility was accompanied by corresponding changes in the volume of mixing, as well as gas permeability. Admixture of the rubber and the plastic led to changes in the rebound resilience and low-temperature brittleness in the expected pattern. The same study highlighted the effects of changing the molar mass of the NBR [158]. Higher molecular weight of the NBR resulted in better physical properties and abrasion resistance. These studies also reported that molecular mass of the PVC did not affect properties in any significant manner except that the compression set, cold flexibility, and rebound became lower, as the molecular mass of the PVC increased. A significant increase in the ozone resistance was observed in the intermediate range of molar mass of 1000–2000 for the PVC. The same studies extended the understanding to the replacement of AN in NBR with methyl, ethyl, and *n*-butyl acrylates. This improved the low-temperature brittleness but reduced the resistance to ozone degradation and resulted in poorer physical properties.

In blends containing a predominant amount of PVC, the properties of the thermoplastic dominate. These blends are processed and used like thermoplastics; the NBR serves as a polymeric plasticizer or impact modifier. The important characteristics of these blends are the AN content of the NBR and the degree of branching of the NBR and the particle size of the NBR in the blend. Schwarz *et al.* [159]

reported a study of NBR/PVC design as thermoplastic elastomers wherein NBR has different molecular masses and polarities, PVC has different molecular masses, and the blend has different filler contents. Deanin and Sheth [160] studied the properties of 0.75 PVC/0.25 NBR blends with variation of the AN content and molecular mass of NBR. The low-AN, high molecular mass sample gave higher tensile strength, in contrast to the NBR predominant case. Schwarz *et al.* [159] showed that an increased AN level improved such properties of NBR/PVC as elongation, tear and abrasion resistance, and tensile strength and melt index. Matsuo [161] determined that the Charpy impact strength of PVC/NBR blends increases as the AN content is increased with increased compatibility. The phase boundaries become indistinct, and some mixing takes place. At very high AN (>40 wt%) content, mixing becomes too extensive, and impact resistance again declines. The compositions that are semicompatible exhibit the greatest toughening. As the molecular mass of the NBR increased, the tensile strength and modulus increased and the fluidity of the blend decreased. The influence of the increased molecular mass of PVC is evident in the improved low-temperature flexibility and compression set. Increasing the NBR proportion causes the tensile strength and modulus to decrease due to increased rubber phase along with increased ultimate elongation and low-temperature flexibility increase. Cross-linking the NBR increases the tensile strength and the ultimate elongation [162,163].

Two separate product areas have evolved from these blends: the addition of PVC to NBR, followed by vulcanization, and the utility of NBR as a permanent plasticizer for PVC. Addition of plasticized PVC to NBR vulcanizates resulted in improved ozone and UV resistance, flex cracking, heat aging, and chemical resistance that was mitigated by lower tensile strength, lower abrasion resistance, and higher compression set. NBR liners and chloroprene (CR) covers have been used for fuel hose construction. PVC predominant blends are used in footwear and cable insulation.

PVC/NBR thermoplastic elastomers (TPEs) made by dynamically cross-linking the NBR selectively in the PVC matrix are an extension of the blend technology and compete in slightly more advanced applications. Stockdale [164] studied TPEs produced by using varying levels of NBR in a PVC and compared PVC/NBR TPEs with conventionally cured polychloroprene and NBR compounds. Kobayashi and Nakamura [165] introduced an ionically cross-linked NBR/PVC thermoplastic elastomer that showed much less deterioration in storage modulus ($G'$) with temperature compared with the conventional TPE. Blends of highly saturated nitrile (HSN) rubber and PVC are another extension of this technology.

HSN was first commercialized in 1984 by Nippon Zeon Co., Ltd [166]. HSN/PVC blends are less affected by peroxides than the corresponding NBR/PVC blends.

## 4.5.3
### Blends of Polystyrene and Styrene Copolymers

Polystyrene and styrene–acrylonitrile (SAN) copolymers are transparent and brittle thermoplastic materials with $T_g$ above room temperature. The addition of rubber increases impact strength particularly at low temperatures. This leads to the development of rubber–plastic blends for PS and SAN, which are designated as HIPS and ABS. Ostromislensky [167] made rubber-toughened polystyrene by polymerizing a solution of natural rubber in styrene monomer. Seymour [168] produced rubber-toughened HIPS in 1952. Boyer produced HIPS by blending emulsions of SBR and polystyrene [169]. Hayward and Elly made HIPS by the polymerization of a solution of SBR and styrene [170]. A review of these early efforts in rubber modification of polystyrene by Amos exists [171]. Daly [172] produced the first ABS blends by melt blending the components. ABS was produced by *in situ* solution polymerization, in which the SAN was grafted as a shell to the rubber emulsion. Current ABS is a graft copolymer that Childers and Fisk [173] and Calvert [174] produced by emulsion copolymerization. Some ABS is also produced by a suspension polymerization process [175]. Other rubbers have been used in blends with SAN to impart specific improvements in properties of the blend over ABS. Polybutadiene and EPDM have been used to impart low-temperature impact or weathering resistance [176]. The most common alternative formulation of ABS is the use of acrylic elastomers [177] and chlorinated polyethylene [178] as the dispersed rubber phase that fill niche applications. Variants of the structure of ABS to impart transparency by the copolymerization of methyl methacrylate [179] or to enhance the temperature resistance of SAN by replacing the styrene with *o*-methylstyrene [180] have been developed. Applications for HIPS are mainly packaging, appliances, housewares, and toys, while ABS appears in more engineered applications such as automobile manufacture, appliances, electronics, and pipes and fittings.

HIPS and ABS are two-phase materials composed of the predominant continuous plastic phase with a dispersed rubber phase. In the rubber component of HIP and ABS, the preferred material is polybutadiene (BR) with a glass transition temperature ($-85\,°C$), much lower than that of SBR ($-50\,°C$). This provides a much better low-temperature impact strength. Buchdahl and Nielsen [181] showed this by dynamic mechanical testing of PS–SBR mixtures. Two phases were observed from TEM images obtained from microtomed films stained by osmium tetroxide, according to Kato [182]. Most importantly, the imaging proved the dispersed rubber phase to be of a two-phase nature, especially for HIPS and ABS made by polymerization of styrene in the presence of dissolved rubber. The rubber particles contain relatively large inclusions of plastic material. Because of this appearance, these are referred to as the "salami" structure.

Polymerizing styrene (or styrene–acrylonitrile) in the presence of dissolved rubber leads to the formation of the plastic in a homogeneous rubber solution that separates into two incompatible phases with the monomer partitioned between the two. As the amount of PS/SAN increases with polymerization while the amount of rubber remains constant, phase inversion occurs when the plastic phase is the predominant to form the final dispersed rubber morphology. This morphology is essentially invariant after about 30 wt% of the plastic phase has been polymerized though the final polymer product contains very little residual monomers. Phase inversion has been observed with a phase contrast microscope [183] and monitored by viscosity measurements [184]. Phase inversion does not lead to the clean separation of rubber and the plastic domains, but rather the rubber phase contains subinclusion of the plastic. This subinclusion is the origin of the heterogeneous structure of the final rubber particles.

Viscosity changes of the polymerizing solution pass through a maximum near the point of the phase inversion. The reaction pathway can be monitored by viscosity–conversion measurements. Structural changes within the continuous rubber phase and phase inversion both manifest themselves in the sigmoidal shape of the viscosity–conversion curve. In the case of added block copolymers, structural changes in the rubber phase are superimposed on the shape of the viscosity–conversion plot.

The properties of HIPS and ABS depend to a large extent on the molecular weight of the matrix plastic phase. This is controlled by the polymerization conditions and the addition of thiol chain transfer agents. An increasing molecular weight of the plastic correlates to a higher melt viscosity and slower relaxation during injection molding; thus, these materials tend to be highly oriented but have anisotropic properties. In most cases, the rubber is BR, primarily because of its low glass transition temperature of $-85\,°C$.

Amorphous medium-*cis*-BR has slight advantages over semicrystalline high-*cis*-BR that has a melting point of about $-40\,°C$, and this degrades the low-temperature impact properties of the final HIPS/ABS. The solution viscosity of the rubber used determines the particle size of the rubber phase and for BR rubber this is determined by the degree of branching and the molecular weight. Most BR rubbers used commercially have molecular weights between 180 000 and 260 000 (viscosity average). Other rubbers used occasionally are SBR and EPDM, the latter yielding products with improved weatherability [185].

The volume ratio of the rubber and the plastic phases is influenced by the rubber content, by the number and size of occlusions of the plastic inside the rubber, and to a lesser extent by the degree of grafting. The work of Cigna [186] indicated that the volume ratio, as well as the rubber content, is a key determinant for the properties of rubber-toughened HIPS and ABS. A more detailed investigation was done by Bucknall *et al.* [187–189]. Grafting of the plastic to the rubber at an early stage of polymerization converts the rubber solution in styrene into an emulsion of PS solution in styrene dispersed in the rubber solution. These PS droplets will coalesce only partially, and the phase inversion takes place between a polystyrene solution and the unchanged emulsion. The dispersed rubber phase still contains inclusions of polystyrene solution that eventually polymerize to form the subinclusions. At a constant amount of rubber, the volume ratio of rubber to plastic is strongly dependent on the amount of these subinclusions. The rubber particle size is determined during the polymerization between the phase inversion and the solidification of the composite. The size is governed by the shear stress from the agitator, the viscosity ratio of the two phases, and the interfacial tension between them. Particle size decreases with increasing agitator speed [183,190,191]. The influence of the viscosity ratio with respect to drop breakup in a system of two immiscible fluids was studied by Rumscheidt and Mason [192] and Karam and Bellinger [193]. Rumscheidt and Mason demonstrated the mechanism of drop breakup at different viscosity ratios, and Karam and Bellinger showed that the critical deformation for a drop to split off is at a minimum when the viscosity ratio approaches unity. The third parameter for adjusting particle size is interfacial tension. It can be varied by suitable surfactants in this oil-in-oil emulsion, as has been shown by Molau [194].

Graft copolymers and block copolymers of styrene and butadiene act as such surfactants; thus, an increase in grafting or the addition of block copolymers in the early stages of polymerization reduces the particle size. Block copolymers induce a change in the internal structure of the particles. Polymerization of styrene in the presence of preformed styrene–butadiene di-, tri-, or multiblock copolymers produces different particle structures [195–197]. These structural changes map the structure of the styrene–butadiene block polymer, which exhibit domain arrays depending on their composition. Additional grafting onto the polybutadiene part of the block copolymers leads to a higher polystyrene content in the rubber micelles during polymerization. If this polystyrene content exceeds a certain critical limit, the structure of the micelles changes according to the corresponding structural change in the parent block copolymer. In the presence of styrene–butadiene–styrene (SBS) triblock copolymers, the polymerization details are identical; however, because of the presence of persistent solution network in the rubber phase, the shear stress due to the agitator is insufficient to rupture the rubber micelles leading to very large rubber domains.

Most of the HIPS polymers contain partly cross-linked BR domains. Rubber particles withstand a triaxial stress field around the particles due to differential thermal contraction of the rubber and the matrix when the melt cools to ambient temperatures. Mechanical integrity of the particles, in avoiding voiding inside the particles, is promoted by cross-linking. The cross-linking is thermal and is mostly created during the devolatilization step of the manufacturing processes.

ABS is the largest rubber–plastic blend of SAN, styrene copolymer. It can be made by bulk polymerization, and both discontinuous mass suspension and continuous bulk processes are in use. The most common ABS process is emulsion polymerization, since it provides products with much smaller particles than other processes: it is thus dissimilar to the preferred bulk mode of polymerization for HIPS. The emulsion process is run discontinuously in several steps [197]. Initially, polybutadiene latex containing small particles is produced. Because these particles are not of the optimum size, a subsequent agglomeration step is required. A grafting step follows particle size redistribution. In this step, a seeded polymerization is started to cover the rubber particles with a matrix polymer shell. The graft rubber particle dispersion is mixed with SAN dispersion and recovered.

The properties of these materials depend on the same molecule and morphology as HIPS with important differences. Matrix molecular weight, rubber type, phase volume ratio, rubber particle size, particle structure, cross-link density, and additive content determine the engineering performance. ABS is tougher than HIPS and deforms by both crazing and shear yielding [198,199], which is promoted by smaller particles than are usually present in HIPS. The compositions of both matrix and graft shell have to be closely matched within 5% to prevent incompatibility between the graft and the matrix and incompatibility in SAN [200]. Matrix composition and molecular weight are determined by feed composition and process conditions. Styrene and acrylonitrile have an azeotrope in emulsion polymerization conditions with 28.5% (by weight) acrylonitrile [201]. Since the matrix and the graft shell are conducted independently, the composition of the polymer is confined to a relatively narrow range of 28.5 ± 3% to avoid shifts in composition and thus properties as well as incompatibility of the shell and the matrix. This is critical in the grafting shell formation reaction that is conducted where the two monomers have different solubilities in the rubber particles [202].

The rubber used in ABS is polybutadiene. The microstructure of the polybutadiene as well as the cross-link density and the level of branching depends on the reaction temperature. While polybutadiene is susceptible to aerobic oxidation, acrylate rubbers and EPDM are used to replace polybutadiene in order to obtain better weatherability. Acrylate rubbers are made in emulsion polymerization similarly to polybutadiene and are used to make acrylonitrile–styrene–acrylic ester (ASA) copolymer; EPDM is used only in a bulk precipitation process to provide acrylonitrile–(ethylene–propylene–diene terpolymer)–styrene (AES) copolymer.

The rubber–plastic ratio, which is determined by the blending of the rubber graft and the plastic emulsions prior to recovery, has the greatest influence on impact strength. Impact strength increases with increasing rubber content. In a series with a constant particle size and a constant degree of grafting, impact strength increases with the rubber–plastic volume ratio [203,204]. Unlike HIPS, the rubber particles contain little or no matrix material occlusions. Thus, the total amount of polybutadiene in ABS polymers is three or four times as high as in corresponding HIPS [205]. Interfacial bonding by grafts and particle size are most important in determining toughness.

In emulsion polymerization, the rubber particle size is determined by emulsifier concentration and the redistributed bimodal particle size distribution of unagglomerated and agglomerated particles. In ABS polymers, the optimum particle size is much smaller than that in HIPS. However, a single size distribution of very small particles makes the ABS resins stiff, glossy, and less tough, while big particles impair stiffness and gloss, but improve toughness. The optimum mean particle size for a combination of toughness, stiffness, and surface gloss is about 0.3–0.5 µm [206]. Gloss is largely dependent on particle size [207].

The particle structure of the rubber of emulsion-polymerized ABS is a core–shell structure. Particles of polybutadiene are anchored in the surrounding matrix by a graft shell to provide coupling of the two phases. The rubber particles should be completely covered to have better coupling and more uniform distribution in the matrix. The proportion of grafted SAN in the interface depends on the total amount of AN added and on the distribution of the grafted polymer between internal and external grafts. The proportion of graft copolymer included rises with increasing particle size and decreases with cross-link density [206]. Emulsion-grafted small particles exhibit a closed shell of graft copolymer on their surface. Bigger particles with low cross-link density have hardly any detectable graft shell, and the mixture with matrix polymer shows that the particles contain many large inclusions. If the cross-link density is enhanced, the graft shell becomes more clearly marked, but a majority of the matrix polymer is still located inside the particle.

If the particles are not completely covered by the graft shell, they will tend to agglomerate during the mixing process. If the primary particles are small (0.1 µm in diameter), an increase in grafting will increase impact strength. With larger particles, less than 50% grafting is sufficient to obtain very tough materials, and products with bimodal particle size distribution are even more impact resistant [208]. Most ABS types have more solid rubber particles with few and small inclusions, depending on particle size and cross-link density.

The importance of HIPS and ABS has led to the development of many styrenic materials for specialized applications such as transparency (with high impact strength) and high heat distortion temperature. The dispersed-phase structure of HIPS and ABS causes haze by light scattering at the dispersed phases that are larger than the wavelength of the incident light and have dissimilar refractive indices from the matrix [209].

One strategy for this is to use poly(methyl methacrylate) (PMMA) that is partly compatible with SAN and has a refractive index less than SAN as a blend component in the plastic phase to approach the lower refractive index of BR leading to translucent ABS-like plastic–rubber blends. However, PMMA and SAN are compatible only if the SAN contains 12–22% AN, which is less than the ideal amount. The thermal resistance of the styrene blends depends on the glass transition temperature of the plastic matrix.

α-Methylstyrene has been used as comonomer in SAN since it has a glass transition temperature of 115 °C. An alternative approach is to replace the AN with maleic anhydride, which raises the $T_g$ by about 3 °C per weight percent of MA in the styrene copolymer [210]. The most versatile method of preparing styrenic materials that are temperature resistant is blending with other polymers. Polystyrene is miscible with poly(2,6-dimethylphenylene ether) (PPE) [211]; HIPS–PPE blends have glass transition temperatures of 100–200 °C. These blends are isomorphic to the HIPS but are much more tolerant of high temperatures. The second blend of commercial interest consists of ABS and polycarbonate (PC). SAN, the matrix material of the ABS, and PC are only semicompatible and thus the morphology is dependent on the mixing though both SAN continuous and PC continuous blends have been constructed.

## 4.5.4
### Blends of Polyamides

Polyamides (PAs) are ductile at room temperature and tough above their glass transition temperature $(T_g)$. Above $T_g$, excessive shear yielding can take place owing to the lowering of the yield strength and modulus. Polyamides are impact toughened without proportionate loss of modulus by incorporation of rubbers such as block polymers or EPR elastomers. The toughness can be increased manyfold [212] while the tensile strength and modulus decrease proportionally to the rubber concentration [213–215]. The impact strength measured by notched Izod is 50–100 kJ m$^{-2}$. This is due to formation of a fine dispersion of chemically reactive rubber in the PA matrix. These plastic–rubber blends show two transitions, due to the rubber, one near $T_g$ of the rubber and the other at the BDT between $T_g$ of the rubber and $T_g$ of the PA. For these impact tested samples, the notch region stress whitening is apparent. The energy absorption is mainly due to deformation during the crack initiation. Higher impact strength is associated with a larger area of deformation in the sample. Experimentally, several factors determine the toughness of the blends – the rubber content, the rubber particle size, the rubber type, and the size distribution of the rubber particles. A host of other properties depend on the rubber–plastic ratio; thus, the modulus of the blends and the yield strength decrease linearly with the rubber volume fraction. Sue and Yee found that the impact toughness of the blends was inversely dependent on the testing speed [216]. In addition to the impact strength, the PA–rubber blends also have excellent flexing fatigue characteristics [217,218].

The PA–rubber blends are mostly made by a reactive blending process where the physical blending leading to dispersion is accompanied by the chemical reaction of the rubber and the PA leading to dispersion stabilizing graft polymer of the PA and the rubber. PA can be impact toughened with a rubber if the particle size of the rubber dispersion is about 0.1–2.0 μm [219–221]. The need for the chemical reaction arises because of the difference of interfacial tension between PA and olefinic rubbers, this is too high for the persistence of a fine dispersion [222]. The interfacial tension can be lowered by increasing the polarity of the rubber, by adding or creating an interfacial agent to generate chemical bonding at the interphase. Fine dispersions can be obtained from thermoplastic elastomers (segmented block copolymers) with polyester- and polyamide-containing segments [223–225]. A chemical bonding leading to graft polymer can be obtained by functionalization of the PA with unsaturated groups that can react with the rubber or by modifying the rubber with groups, typically maleic anhydride, that are reactive toward PA.

Functionalization of the rubber ensures the persistence of the finely dispersed morphology, which is achieved by dispersive mixing. The mixing leads to the breaking up of the particles (dispersion) and the uniform distribution of the particles in the PA matrix. The distribution of the particles is determined by melt mixing [226]. The mixing is typically conducted in internal mixers, twin-screw extruders, and single-screw extruders.

The dispersion is governed by the viscous forces and the interfacial forces [227–229]. Wu [230] found, for the PA66/EPR system, the effect of interfacial tension, viscosity ratio, and shear rate on the dispersion process in a corotating twin-screw extruder. The dispersion improved linearly with the shear rate, matrix viscosity, and the drop interfacial tension. If the rubber is functionalized with MA, the interfacial tension at 280 °C in the melt can be lowered from 10 to

0.25 mJ m$^{-1}$. This decrease in interfacial tension by a factor of 40 also decreases the particle size of the rubber by the same factor [231]. PA melt blends in a twin-screw extruder are efficient because both dispersion and distribution occur. Oostenbrink and Gaymans [220,232] studied the particle size of the system PA6–EPM-*g*-MA at different positions in the twin-screw extruder and concluded that the dispersion is extremely rapid. Functionalization of the rubber is typically with maleic anhydride [231,233], maleic esters [234], fumaric acid, and other unsaturated acids [235] in the melt state [232]. It is expected that the functional rubber segregates to the interface and reacts with the PA. Lower concentrations of the functionality, such as by diluting the functional rubber with nonfunctional one, lead to an increased particle size [236], indicating that interfacial tension increases with decreasing reactive functionality. The reactions take place between the free terminal amine of PA and the diacid or the anhydride of the rubber and higher amine levels lead to higher impact toughness [237].

Assay of the graft PA layer on the rubber indicates that the amount of PA grafted is a function of the functionality and the particle size [233] and can be as high as 40% of the rubber [237]. Lawson *et al.* [237] concluded from the analysis of the molecular weight of the nongrafted PA that the grafting was dominated by the reaction with the terminal amine end groups of the PA.

The melt viscosity of blends increases compared with the weighted average of the components when grafting reactions occur. Rheology provides an insight into the progress of the chemical processes. The deviation depends both on the extent of the graft and on the particle size of the dispersed phase [238]. Oshinski *et al.* [221,231] measured the mixer torque during blending of PA with styrene–hydrogenated butadiene–styrene triblock copolymers (SEBS) and observed a strong increase in torque with rubber concentration with an MA-modified rubber (SEBS-MA) and hardly any effect with an unmodified rubber.

The modification of PA has been attempted with various kinds of rubbers used in a large variety of procedures. One method of rubber modification is using segmented block copolymers with soft blocks such as SEBS. The types of rubber segments that are used are polyethers, aliphatic polyesters, polybutadiene, and ethylene copolymer. These block polymers are typically phase separated and thus are tough. The elongation at break increases and the blends are tough [225,239]. If the longer blocks are used [240], more complex microphase separation is evident though the high toughness is retained. Rubber can be introduced and incorporated into the PA with concomitant intimate reaction of the functional groups by dissolving the rubber in the PA monomer or monomer mixture or in a common solvent. In particular, caprolactam, the starting monomer for PA6, is an excellent solvent. Martuscelli and coworkers studied the solution blending of PA6 with EPM and EPM-*g*-MA [241,242]. The rubbers were dissolved in xylene before being added to the caprolactam. In the absence of functionalization of the EPM, the dispersion of the rubber was poor with domains of 2.5–20 μm. In the presence of minor amounts of EPM-*g*-MA within the EPM, the dispersion got progressively finer and at 20% of the EPM-*g*-MA the dispersion was about 1–5 μm.

Another way obtaining stable, high-impact rubber–plastic blends is by using core–shell particle impact modifiers. Several core–shell systems have been developed consisting of a rubbery core, for example, acrylate rubber–PMMA and polybutadiene–SAN. Blends where the core–shell modifier is just dispersed for the agglomerates provide good impact behavior [243]. The dispersion could be improved by changing the shell to be more compatible with the PA [244,245].

Several structural factors affect the efficacy of the impact toughening by the rubber phase. If the rubber phase contains inclusions [231,246], the dispersed-phase effective volume fraction increases and the modulus and yield strength of the blend decrease with proportionate increase in the impact strength. Increase of the rubber concentration increases impact strength [243,247]. This increase is monotonic to about 30 vol.% rubber both at −40 °C and above the BDT temperature. At the same time, BDT of the blend decreases although the $T_g$ values of the PA and EPDM phases remain constant. Blends with high rubber concentrations (more than 30 vol.%) show a marked drop in impact strength as the rubber becomes the continuous phase [248]. Decreasing particle size decreases the BDT to lower temperatures. For the rubber to be effective, the particle size has to be within a given range (0.1–2 µm) since very small particles are not effective because it is possible that for small particles all of the volume is the graft copolymer without any free rubber. Wu [31,230] found the interfacial thickness to be on the order of 50 nm. Oostenbrink et al. [236] observed no shift in the glass transition temperatures of either the PA or the rubber for these dispersions. Oshinski et al. [231] found that the tan $\delta$ peak of the rubber was smaller and the tan $\delta$ peak of the PA larger, both by about 15%, for these fine dispersions. These effects suggest but are not enough to conclude that in fine blends the interfacial layer concentration is so large that it has changed the properties of the system drastically. BDT decreases with decreasing particle size though there is lower limit at about 0.1 µm. The surprising result is that the impact strength in the range between the upper and lower limits is unaffected by the actual size of the dispersed particles. However, at the lower end of the size scale the impact strength in the tough region falls off with temperature. Another hypothesis is that fine particles are more difficult to cavitate and thus relieve the stress [249,250]. Gent and Tompkins [251] showed that the cavitation starts from a defect and that a decreasing defect size increases the cavitational stress. The combination of the observations indicates that with increasing temperature the very finely dispersed rubber in the blend cavitates with greater difficulty. The particle size distribution in PA–rubber blends is usually small and in the range of 1.4–2.1 µm. The particle size distribution can be artificially broadened in two rubbers – one functionalized and the other nonfunctionalized are used in conjunction with each other.

The Izod impact strength increases approximately linearly with rubber concentration both in the brittle and in the ductile region. It is independent of the type of rubber [41] and the particle size as long as it is between the limits. BDT, however, depends on the type of rubber, rubber concentration, and particle size. For this transition from ductile to brittle failure, the rubber concentration and particle size effect can be reduced into a single structural parameter. Wu [222] expressed this as the interparticle distance model that is effective in correlating properties, in particular, for the tough blends with high rubber concentrations. In the interparticle distance model, the ligament thickness is the determining factor. Tough, rubber-modified PA blends have similar relationship of the particle size and the BDT. Thus, the interfacial tension influences the dispersion process but not the impact behavior at constant particle size. Wu [252] suggested that the interfacial strength for adhesion in blends has only to be $1000 \, J \, m^{-2}$, which is the tearing strength of the rubber. This level of interfacial strength is easily obtained by van der Waals bonding. Under these conditions, the rubber fails by cavitation before the particle delaminates. In fracture surfaces of blends, delamination of the blend at the rubber–PA interface was only observed with unmodified EP rubbers [252]. The rubber induces toughening by lowering the BDT and improving the impact strength. Different rubbers have a profound effect on the former, but little effect on the latter [41]. These data indicate that the deformation that triggers the stable crack propagation is a different process from the deformation that determines the toughness level. The effect of the type of rubber on BDT could not be correlated with the tensile strength or elongation at break of the elastomers though the tensile modulus does affect the BDT since the lower the modulus of the rubber [253] the lower the BDT at constant rubber concentration and particle size. Another explanation was provided by Wu [254]. He suggested that different rubbers may generate different amounts of internal stresses and may thus cavitate differently. There is a correlation of the blend with the best impact behavior and the ability of the rubber to cavitate [34,255].

### 4.5.5
### Blends of Isotactic Polypropylene

Isotactic polypropylene is a stiff, high melting polyolefin thermoplastic that is widely used in automotive parts, electronics, medical and food containers, housing, housewares, and woven and nonwoven garments. A limitation on these uses is the poor impact toughness of iPP, particularly at subambient temperatures. This necessary improvement of mechanical properties of iPP has been attempted by the blending of rubbery polymers and has been successful with blends of ethylene–propylene random copolymers (EPR), styrene–ethylene–butylene–styrene triblock copolymers (SEBS), butyl rubbers (IIR), and ethylene–propylene–diene terpolymers (EPDM). In particular, binary blends of iPP and EPR have been extensively investigated. Miscibility and/or compatibility in polyolefin blends dominate their material characteristics and basic properties. The lack of favorable and attractive polarity, however, often leads to the phase-separated state in most polyolefins. However, because of weak interactions these blends are only weakly phase separated and morphology changes can be easily achieved through mechanical intensive mixing or shear. iPP is immiscible with PE and EPR. One of the research areas is an attempt to control the miscibility and/or compatibility of iPP with rubbery polyolefins. Promising leads appear from polyolefin catalyst technology that allow for the controlled introduction of comonomers as well as stereoregularity of propylene or α-olefin sequence in the copolymers. Furthermore, various types of random copolymers covering nearly entire composition range using metallocene-based catalyst systems are being prepared. Their morphology and thus their properties differ from those of the commercially available Ziegler–Natta iPP. The new types of iPP polymers are expected to lead to an improvement in mechanical toughness.

### 4.5.5.1 With Ethylene–Propylene Copolymer Rubber

It is expected that a predominant amount of propylene in EPR would lead to miscibility with iPP. Lohse used SANS and found that EPR with 92 wt% of propylene is immiscible with iPP in the molten state [96]. Yamaguchi *et al.* also showed phase-separated morphology in the blends with EPR37 and EPR67 [256] where the numerical suffix is the composition of the rubber expressed as ethylene content. The immiscible state in the amorphous region is also consistent with the dynamic mechanical spectra showing double $T_g$ peaks. Contrary data appear from TEM data of blends of iPP/EPR89 and the iPP/EPR84 blends exhibit homogeneous morphology, suggesting that both EPR89 and EPR84 are miscible with iPP, whereas the iPP/EPR77 and iPP/EPR52 blends are not. According to WAXD studies of the iPP and iPP/EPR blends, the diffraction patterns of all the blend samples have five sharp diffraction lines ascribed to the α-form of iPP [257] indicating that the blending of EPR copolymers does not affect the crystalline region of iPP irrespective of composition of EPR copolymers. An explanation for this duality of results is that while the EPRs are essentially immiscible with iPP, the high miscibility of above-mentioned EPR92, EPR89, and EPR84 is due to their isotactic propylene sequence, which can be included in the iPP crystal lattice. Thus, the miscible state in iPP/EPR copolymers is different from the miscible iPP/EHR or iPP/EBR copolymers depending on the tacticity of the propylene residues. The polarized optical micrographs (POMs) of spherulite morphologies obtained by isothermal crystallization at 403 K for miscible iPP/EPR84 and immiscible iPP/EPR blends indicate that only for the miscible EPR copolymers are they incorporated in the intercrystalline regions of iPP spherulites [258]. The miscible blends of EPR and iPP where both species have isotactic propylene sequences exhibit a single β-relaxation peak associated with the glass transition $T_g$ between $T_g$ values of the iPP and EPR copolymers. On the contrary, immiscible blends showing the heterogeneous morphology exhibit two separate peaks in the β-relaxation temperature range from 200 to 300 K. Furthermore, these dynamic mechanical behaviors also demonstrated that the EPR92, EPR89, and EPR84 are incorporated in the amorphous region of iPP in the solid state, whereas the EPR77, EPR70, and EPR53 copolymers are immiscible with the iPP.

Blends of isotactic polypropylene and EPR or EPDM rubber have been principally used for impact toughening the thermoplastic. The inertness of the polyolefin rubber, similar solubility parameter to the plastic, and the ease of mixing contribute to this development. Most commercial EPR rubbers containing more than 50 mol% of ethylene, at molecular weights of several fold the entanglement molecular weight, are elastomers. The solubility parameter of EPR rubbers is intermediate to those of PE and iPP and depends on ethylene content. This intermediacy in properties manifests in two ways. An alternative use of EPR rubbers is as a compatibilizer in blends of PE and iPP [259,260]. The difference in solubility parameters is large enough that iPP–EPR blends are generally immiscible [96,261,262]. The miscibility increases with decrease of ethylene content. iPP was miscible with EPR with ethylene content lower than around 17 mol %, based on both DMA and TEM. Chen *et al.* [263] have reported both miscibility and the existence of an LCST in blends of isotactic polypropylene and an ethylene–propylene–diene terpolymer (EPDM). More recently, Kamdar *et al.* [264] and Nitta *et al.* [257] reported that the miscibility between iPP and EPR, at a molecular weight higher than $100\,000\,\mathrm{g\,mol^{-1}}$, is dependent on ethylene content in EPR.

The blend of iPP and EPDM has been prepared by different processes, such as melt mixing, dynamic radiation curing, and ultrasonic curing, among others. These iPP and EPDM blends have been widely studied for structure–property relationship, morphology, mechanical properties, rheology, and thermal properties. A typical example of the effect of EPDM addition on the mechanical properties of iPP can be seen in the work of Da Silva and Coutinho [265] who described the effect of EPDM amount and also processing conditions on the mechanical properties of PP and EPDM blends. As EPDM content increases, the impact strength of the blends increases but the tensile strength and Young's modulus decrease and the elongation at break increases. The toughness of the PP and EPDM blend was investigated by Huang *et al.* over a wide range of EPDM contents and temperatures [266]. The Izod impact strength of the blend increases with increasing EPDM content and temperature. The notched failure of these blends is sensitive to the radius of the notch tip since the toughness tended to decrease with increasing radius while the BDT temperature increased with increasing radius. The thermal and morphological behaviors of these EPDM blends were studied by Da Silva and Coutinho [267] using DSC and POM, respectively. Crystallization kinetics of PP and EPDM blends were also investigated. Addition of EPDM resulted in increase of spherulite size while the heat of fusion and crystallinity of the blend systems decreased.

Though the solubility parameter of these two components is similar, there are improvements in both the morphology of the blend and the resulting mechanical properties to be obtained by the compatibilization of these systems. Two independent studies by Lopez-Manchado *et al.* [268] and by Datta and Lohse [269] looked at the effect of grafted PP on the compatibility and properties of PP/EPDM thermoplastic elastomer blends. Both approaches depend on the generation and the reaction of complementary functionality on the iPP and the EPDM to create the compatibilizing graft copolymer. In the first, the functionalization of iPP was performed by melt blending through grafting itaconic acid derivatives. In the second, the functionalization was performed by the use of maleic anhydride-grafted iPP that was reacted with an amine functionality on the EPDM.

The rheological properties during mixing as well as after the formation of the blend showed that the blends made with grafted PP have better processing. Good interaction between two phases is evident from DMA and tensile properties. Su *et al.* [270] studied the mechanical properties and structure–property relationship of sulfonated EPDM ionomer (zinc salt) and PP and determined that $Zn^{2+}$-neutralized EPDM ionomer and PP blends have better dispersion and correspondingly better mechanical properties than those of the PP–EPDM blend.

Similarly, Ha *et al.* [271] blended the zinc ionomer with maleic anhydride-grafted PP (PP-*g*-MAH). Using light scattering, DMA, and FTIR spectroscopy, they found that the compatibility of the PP/EPDM blend was improved over the nonfunctional parents. The $T_g$ data support a strong interaction of the grafted polymers, leading to a near-miscible system.

Ternary blends of iPP and EPDM, containing small amounts of PE, have substantially improved properties over the binary parents. For instance, Sanchez *et al.* [272] prepared ternary blends of PP, high-density polyethylene, and EPDM at several compositions. They showed that addition of an elastomer to the polyolefin blends changes the shape of the viscosity dependence with composition and supported the general conclusions in similar systems by Ha and Kim [273,274].

### 4.5.5.2  With Ethylene–Isotactic Propylene Copolymers

Recently, a new technology that manipulates the propylene sequence in EP copolymers has been proposed and commercialized using stereospecific polymerization catalysts, which leaves most of the propylene residues in the isotactic stereochemistry. These novel EP random copolymers, with high isotacticity in propylene sequence, show essentially different properties from those of conventional EPRs [275] where the propylene residues have no predominant tacticity. In particular, they are semicrystalline, thermoplastic elastomers. The elastomeric polypropylene materials have the physical properties of elastomers along with the processing characteristics of thermoplastics. These materials are characterized by a low degree of crystallinity [96,258,276], where the crystalline regions dispersed in the amorphous matrix essentially provide physical cross-links to the amorphous elastomeric segments of the chain. The size and distribution of these crystalline regions in the amorphous matrix thus have important influences on the mechanical properties.

### 4.5.5.3  With Higher α-Olefin Rubber

The morphology and the mechanical properties of iPP/PE blends depend on the density and thus the crystallinity of the PE. Dumoulin *et al.* [277] demonstrated that blends of iPP and LLDPE show superior mechanical properties, and thus better compatibility to LDPE or HDPE [278]. These mechanical properties correlate to the interfacial tension between iPP and PE where the interfacial tension in iPP/LDPE and iPP/HDPE blends is higher than that in iPP/LLDPE [279,280] indicating better adhesion between iPP and LLDPE. This interfacial tension between iPP and LLDPE decreases with the increase in the α-olefin content in

LLDPE [256]. In this paper, Yamaguchi *et al.* [256] who studied the miscibility of isotactic propylene with ethylene copolymers of hexene-1 and butene-1 (EBR) copolymers found that iPP was miscible with hexene-1 copolymers with more than 50 mol% of hexene as shown by both TEM and DMA. In addition, DMA indicated predominant miscibility between iPP and ethylene–butene-1 copolymers with 56 and 62 mol% of butene-1. These results suggest that the miscibility with iPP depends on chemical composition of polyolefins. Thomann *et al.* [281] showed that ethylene–butene-1 copolymers are miscible with iPP when the butene-1 content was higher than 78 mol%. In addition, Yamaguchi *et al.* later investigated the mechanical properties of the blends of polypropylene with ethyl-ene–hexene-1 copolymers [282,283] and showed an unexpected improvement in tensile properties when the rubber and the plastic reached near compatibility. Two distinct β-transitions have been noted in the iPP–plastomer blends that are thermodynamically immiscible though they are mechanically quite compatible. Blends of iPP containing some ethylene and butene-1 as comonomers with the same plastomer containing octene-1 as comonomer showed an increasing degree of compatibility [284,285].

Since about 1990, high molecular weight, narrow compositional distribution, and near most probable molecular weight distribution of ethylene copolymers of higher α-olefins such as butene-1, hexene-1, and octene-1 with substantially high level of olefin such that the copolymers are almost amorphous rubbers have become available. These copolymers are being designed to be blended with iPP as impact tougheners in competition to EP [286,287]. The miscibility between these copolymers, which are rubbers and iPP, is strongly dependent on the comonomer type and content in the copolymer. Examples of such blend compo-nents are ultralow-density ethylene copolymer that is equivalent to a high comonomer content. Examples of these copolymers that are both amorphous (or barely crystalline for ease of mechanical handling as pellets) and high molec-ular weight are ethylene–octene-1 copolymers. These in blends with iPP form soft to hard blends. Soft blends typically contain a large amount of the plastomer while the reverse is true for the hard blends. These blends offer excellent ther-mal and weathering aging since they are saturated [288].

### 4.5.5.4 With Ethylene–Butene-1 Copolymer

Considering that poly(butene-1) (PB-1) and iPP are at least partly miscible, ran-dom copolymer (EBR) with large amount of butene-1 is expected to be miscible with iPP in the amorphous region. Yamaguchi *et al.* delineated this composition by showing that EBRs with 62 mol% of butene-1 dissolve in the amorphous iPP region, whereas the EBRs with 36 and 45 mol% of butene-1 are immiscible with iPP [256]. Weimann *et al.* also found that the EBRs with 73 wt% (57 mol%) and 90 wt% (81 mol%) of butene-1, which are obtained by hydrogenation of polybuta-dienes with various amounts of 1,2-addition, are miscible with iPP in the molten state as determined by SANS [289]. According to the electron microscopic observation of the morphology in iPP/EBR blends by Mader *et al.* [290], there is no phase separation in the blends with the EBRs with 82 wt% (69 mol%) and

90 wt% (81 mol%) of butene-1. The experimental results are consistent with the theoretical prediction proposed by Lohse and Graessley [291] that EBRs require at least 58 mol% of butene-1 to be miscible with iPP. TEM micrographs for the blends composed of iPP and EBRs with a different butene-1 content in the weight ratio of 75 : 25 show that the blend with EBR45 (EBR with 45 mol% of butene-1) has finer dispersion particles than iPP/EBR36 (that with 36 mol% of butene-1). The micrographs also demonstrate that the iPP lamellae are inserted into EB-rich amorphous region, resulting from the thick interfacial region in the molten state. Dynamic mechanical spectra of the same blends show that the blends with phase-separated morphology have double peaks in the tensile loss modulus $E''$ curve distinctly in the temperature range from 200 to 300 K, while the iPP/EBR56 shows only single peak in the temperature range. Notably, all blends show a broad relaxation in the temperature range from 320 to 380 K, which is associated with the molecular mobility in the crystalline phase of iPP. The storage modulus drops at the melting point of iPP, which is independent of blending with EBR, suggesting that EBRs do not affect the crystalline region of iPP. These results also suggest that the amorphous regions of iPP and EBR45 are partially dissolved in each other in the blends. Moreover, there is only single intermediate $T_g$ peak between pure components for the blend of iPP with EBR56 (56 mol% of butene-1) suggesting that EBR56 molecules are completely incorporated in the amorphous region of iPP.

### 4.5.5.5 With Ethylene–Hexene-1 Copolymer

Ethylene–hexene-1 copolymers (EHR) with more than 50 mol% of hexene-1 are miscible with iPP [258,276], based on DMA experiments using EHR with 33 and 51 mol% hexene contents. The values of dynamic shear moduli of the iPP/EHR (51 mol%) are intermediate between those of the pure components. Miscibility in the molten state affects the lateral growth rate of secondary crystallization of iPP component. In both cases, the spherulite radii of all samples increase linearly with time over the entire crystallization process suggesting that the composition of the melt is constant in front of spherulites. However, blending EHR (51 mol%) depresses the spherulite growth indicating that EHR51 molecules are dissolved into iPP matrix and act as a diluent. In contrast, immiscible blends of the other EHRs show phase separation in the molten state with little effect on the crystallization rate. Although the crystallization conditions affect the morphology, the miscibility in the molten state has strong influence on the morphology in the solid state.

The iPP/EHR51 shows a single β-relaxation peak ascribed to the glass transition in the temperature region between $T_g$ of the pure components indicating that most of EHR51 chains are incorporated in the iPP. On the contrary, the iPP/EHR33 shows two separated peaks in the temperature region and each peak is located at the $T_g$ of each pure component (279 K for iPP and 214 K for EHR33). TEM micrographs of the blends show homogeneous morphology in the iPP/EHR51 at any blend composition, whereas the iPP/EHR33 shows apparent phase separation, which corresponds to two separate $T_g$ peaks in the dynamic mechanical spectra. WAXD measurements indicate that the location of the $2\theta$

peaks is unchanged by blending EHR, indicating exclusion of EHR from the crystalline region of iPP.

### 4.5.5.6 With Ethylene–Octene-1 Copolymer

Carriere and Silvis determined the interfacial tension between iPP and ethylene–octene-1 copolymers (EOR) with various amounts of octene-1 [292]. They found that the interfacial tension decreases monotonically with increasing amounts of octene-1 up to 14 mol%. Yamaguchi and Nitta demonstrated that the EOR with 52 mol% of octene-1 is miscible with iPP, whereas the blend with EOR32 having 32 mol% of octene-1 shows phase-separated morphology [283].

### 4.5.6
### Tensile Properties

Tensile testing in conjunction with optical spectroscopy affords an insight into the molecular aspects of the tensile deformation for these blends. Rheo-optical techniques [293] afford information on the strain dependence not only of stress but also of optical quantities associated directly with the structure. In practice, a tensile tester was set in FTIR spectrometer to allow an infrared beam through a film specimen mounted on the tensile tester.

Blends of iPP and EPR have been widely investigated. While the incompatibility of the blends where the EPR does not have the same tacticity of the iPP limits their use and the ability to modify morphology, the blends of EPR where the propylene is *meso*-stereoregular and thus can, in theory, epitaxially cocrystallize with the polypropylene are very more diverse. For these polymers, EPR92, EPR89, and EPR84 are miscible with iPP, while EP75, EP70, and EP53 are not. The iPP sample shows a defined yield peak and formed a neck in the postyield region [294]. For the miscible blends, the ductility is pronounced and the stress whitening occurs in a higher strain region compared with iPP. The ductility is due to the fact that EP copolymers are incorporated into the amorphous region, and they act as a plasticizer as shown in miscible iPP/EBR and iPP/EHR blends [295]. On the contrary, for the immiscible blends the stress whitening occurs due to separation at the interface between matrix and dispersed domains [296]. The yield energy of iPP/EP blends, estimated from the area under the stress–strain plot from the origin to the stress drop, shows that the yield energy of the miscible blends is greater than that of iPP and immiscible blends. Thus, the addition of EP92, EP89, and EP84 toughens the spherulite structure composed of iPP lamellar crystals.

In the miscible polymers, the length and number of crystallizable propylene sequences increases with the decrease in the ethylene unit content [297]. Therefore, in the miscible blends the isotactic propylene sequence in the EP copolymer chains having relatively high P-unit contents such as EP92, EP89, and EP84 can participate in the crystallization process of iPP during solidification, resulting in that the EP copolymer chains are incorporated partly in the crystal lattice and partly in the amorphous region. Consequently, these molecules act as the additional tie molecules linking between adjacent lamellae and lead to the enhancement of the yield toughness of the spherulitic structure. The EP chain portions that are not available to participate in the crystallization process are trapped into the interlamellar region, leading to the reduction of $T_g$ of iPP as well as the increment of amorphous layer thickness.

The stress–strain curves of iPP and iPP/EHR blends at room temperature, which is above $T_g$ of all samples, are characterized as a well-defined yielding and neck formation with the stress whitening. The immiscible blends such as iPP/EHR33 immediately show the intense stress whitening, whereas the miscible iPP/EHR51 blends show no stress whitening and more ductile behavior. The temperature is above $T_g$ of EHR but is below $T_g$ of PP in the immiscible blends. The pure iPP and immiscible iPP/EHR33 are brittle at 253 K, while the miscible iPP/EHR51 blends are ductile. The temperature dependence of tensile behavior shows that the ductile-to-brittle transition in the immiscible blends is independent of temperature, since the fracture is associated with the phase separation at the interface between iPP matrix and EHR. In contrast, the EHR51 that is miscible lowers the $T_g$ by the incorporation of the EHR molecules into the amorphous region of iPP. The similar results were obtained in the iPP/EBR blends [296]. The light scattering pictures taken for the sample indicate lack of the plastic deformation, cracks, and voids and are consistent with the absence of stress whitening. The yield stress of the miscible blends is found to increase linearly with the tie molecule concentration. The linearity suggests that tie molecules act as stress transmitters for the external force required for the lamellar fragmentation that takes place in the yielding process.

The overall stress level in iPP is considerably greater than those of the blends with EHR. The iPP failure is just beyond the yield point, while the blend samples show a higher extensibility. Furthermore, the iPP/EHR53 shows less anisotropy of the yield stress. The strain at the yield point for the iPP/EHR53 is larger than those for the iPP and iPP/EHR30. Furthermore, it is also found during the tensile testing that many cracks appear on the surface of the TD samples before the yield.

### 4.5.7
### Structure in Injection-Molded Specimens

The flow-induced molecular orientation of the crystalline phase and/or the distorted shape of the dispersed rubber phase have to be considered in the injection molding. This distortion affects the mechanical performance of iPP/rubber blends [282]. Injection-molded specimens have a distinct skin core structure and the thickness of skin layer is almost the same among the samples. The TEM micrographs in the core region of the blend samples of the injection-molded iPP/EHR53 show homogeneous morphology, distinct from the skin. The TEM micrographs of the skin layer show homogeneous morphology in the miscible systems, whereas the thin layers of ERR phase orient to the flow direction in the immiscible systems. The birefringence values in the skin layer of the miscible blends are higher than those for immiscible blends. Furthermore, the orientation of the crystallites of iPP, determined from the infrared dichroism, indicates that the skin layer in the miscible blends exhibits a higher orientation of iPP compared with the iPP in the bulk. The injection-molded sheets exhibit the mechanical anisotropy in the flow direction (MD) and its perpendicular direction (TD). Fujiyama *et al.* [298] suggested that the skin layer of an injection-molded iPP displays the shish-kebab structure proposed by Keller and Machin [299]. Kalay and Bevis have also demonstrated the shish-kebab morphology by WAXD and TEM observations [300].

## 4.5.8
### Impact Performance

A study on the effectiveness of these low-density plastomers as an impact modifier for polypropylene in blends conducted using solid-state $^{13}$C NMR spectroscopy, SEM, and DSC showed that while these blends were similar to EPR, they were much less viscous at equivalent mechanical properties, indicating a better melt compatibility for the plastomers in iPP than EPR [301]. The same school has measured the rheology of these blends, in addition to the coarsening of the dispersion in a shear field [302]. A non-Newtonian flow behavior was observed in all samples in the shear rate range from 27 to 2700 s$^{-1}$, whereas at shear rates in the range from 0.01 to 0.04 s$^{-1}$, the flow was Newtonian. The SEM showed a co-continuous morphology at about 50 wt% of the plastomer. The dynamic mechanical analysis showed that plastomer/PP, with 5 wt% of plastomer, presented an anomalous increase in the mechanical properties indicating some level of compatibility. This was verified with DMTA analysis that showed single peak in the presence of small amounts of the plastomer indicating single phase. Thermal analysis showed that there was no change in the crystallization of the matrix when different elastomer contents were added [303]. Similar dynamic measurements and quantification of the hysteresis of the blends have been done [304]. Blends of plastomers of two different molecular weights, but having similar compositions, have also been used to generate blends with iPP. These iPP blends have both good impact strength and good weld line strength [305]. Ternary blends of iPP and hydrogenated styrene block polymer with plastomer as the dominant rubber phase have been described to have properties superior to binary blends of iPP and plastomer alone [306].

In blends of the plastomer and iPP, crystallization of the iPP forms spherulites that have been characterized as a rounded aggregate of radiating lamellar crystals with a fibrous appearance, which originates from a nucleus such as particle of contaminant or a catalyst residue. The crystal structure of PP greatly depends on crystallization conditions such as supercooling and crystallization time as well as chain stereoregularity, molecular weight, and its distribution. The microstructure of polypropylene has a profound effect on its morphology and physical properties. Blends of semicrystalline polymer iPP and plastomer are complex than pure iPP due to possible involvement of the crystalline–amorphous, amorphous–crystal, and crystal–crystal interactions in addition to the conventional amorphous–amorphous chain interaction of the binary amorphous blends.

Blends of iPP and POE demonstrate, in addition to the spherulitic morphology, a crosshatched structure at lower crystallization temperatures [307,308]. Optical micrographs demonstrate the effect of supercooling (or crystallization temperature) on the crystalline structure in the 10/90 iPP/plastomer blend. At a lower crystallization temperature such as 120 °C, a lamellar branching pattern was observed. Ternary blends of iPP/EPDM and plastomers were investigated by time-resolved light scattering for phase separation followed by nonisothermal crystallization to the conditions for a molten blend after injection molding. The study [309] showed that decreased chain mobility, due to phase separation, leads to increased crystallization half-life for the iPP in the blends.

### 4.5.9
### Poly(butene-1) as Semicrystalline Rubber

Blends of iPP and PB-1 have been investigated for miscibility. In the dynamic mechanical analysis by Piloz *et al.* [310], PB-1 is miscible with iPP in the amorphous region. iPP/PB-1 blends showed only one tan $\delta$ peak in a DMA as evidence of miscibility [278,311,312]. Boiteux *et al.* [313] confirmed the miscibility between iPP and PB from dielectric relaxation measurements. The observation that blending PB depresses the crystallization rate of iPP since PB-1 acts as a diluent for iPP supports this view [277,314]. However, the miscibility has an LCST since in iPP/PB-1 blends made by melt mixing, Cham *et al.* reported that liquid–liquid demixing occurred at a range of temperatures above 180 °C [278]. Consistent with this observation, Bartczak *et al.* [315] reported that PB-1 dispersed in the iPP spherulites at 125 °C. Cham *et al.* [278] have found some evidence of partial miscibility. Observation of the crystallization of blends of PB-1 and iPP indicates partial mixing of the blends since each phase contains the other. The data suggest that iPP and PB-1 are partially miscible and phase separation occurs at up to 250 °C. Melt-mixed blends should have a two-phase structure with a PP-rich phase and a PB-rich phase.

### 4.5.10
### Styrene Block Polymer Rubber

Hydrogenated styrene block polymers have an elastomer block consisting functionally of ethylene–butene (for the insertion of butadiene in 1,4- and 1,2-fashion) or strictly alternating ethylene–propylene (from the insertion of isoprene) [316]. Usually SEBS contains less than 35 wt% PS. This structure forms a microphase-separated thermoplastic elastomer where the PS blocks at both ends are associated in rigid regions that form a regular lattice while the rest of the space is filled by the elastomeric, hydrogenated polyolefin. The elastomer properties of styrene block polymers reflect this structure. In addition, the block structure is ideal for the reduction of interfacial tension and an improvement in interfacial adhesion in blends of polyolefins with other more polar plastics.

The morphology of blends of SEBS with iPP and the correlation of the morphology with mechanical properties of the blend arises from the compatibility between SEBS and PP. This is ascribed to random copolymer character of the EB block that can be considered as a random copolymer with ethylene and butene-1 segments. A quantitative understanding of the miscibility of polyolefin model copolymers can be achieved by a solubility parameter approach [317–319]. Experimentally, PP is miscible with copolymers equivalent to the EB block, only when the content of the 1,2-enchainment is high. But blends of PP with SEBS must always remain heterogeneous because the PS block is highly incompatible with PP. The morphology of the blend consists of finely divided SEBS within the iPP matrix. There is evidence of intermixing of the phases in both the iPP- and the SEBS-rich phase as well as crystallization of the iPP emanating from the interface of the SEBS [320]. Strong adhesion between the laminates of the SEBS and the iPP supports this interdiffusion mechanism that was observed directly by TEM.

Addition of SEBS to iPP decreases the yield strength, the modulus, and the elongational flow plateau. The yield at maximum stress becomes broader and elongation at break becomes higher with increasing content of the thermoplastic elastomer. Adding SEBS leads to an improvement of the impact strength of iPP. An addition of 10 wt% SEBS to iPP makes the specimens unbreakable in Charpy impact testing [321].

## 4.6
## Thermoplastic Vulcanizates

TPVs are a particular class of rubber–plastic blends that have beneficial and desirable engineering properties. In these blends, the plastic is the continuous phase while the rubber that is dispersed phase is chemically cross-linked. They have become commodities of commerce since their introduction in 1981. These materials have the processing character of a thermoplastic and the utility and performance of a conventional thermoset rubber. TPVs are incompatible rubber–plastic blends that have mechanical properties significantly better than either the unvulcanized blend or the individual components. TPVs are made by dynamic vulcanization, wherein the elastomer is preferentially vulcanized under dynamic shear to generate fine, cured rubber particles in a thermoplastic polymer matrix. The process was first discovered by Gessler and Haslett [322] in their attempt to improve the impact toughness of PP. The technology was widely improved, refined, adapted, and extended to a number of rubber–plastic pairs by Coran et al. [323–328] by vulcanizing the rubber phase under dynamic shear. This discovery was improved by Abdou-Sabet and Fath [329] and Coran et al. [330] through the use of phenolic resin curatives to achieve improvement in elastomeric properties and flow characteristics. Dynamic vulcanization can be applied to combinations of elastomer and plastic polymer blends; however, only a limited number of these are useful. Even within this limited range of useful components, the number of blends of different composition allows a wide range of chemical, mechanical, and rheological properties to be accessible. These improvements include a high upper service temperature, resistance to hydrocarbon fluids, and resistance to compression setting [329].

TPVs possess significantly improved properties over simple blends in the areas of

- stable phase morphology and consistent processing,
- higher ultimate tensile strength,
- improved upper service temperature performance,
- improved fatigue resistance, and
- greater resistance to attack and swell by fluids.

The improvements in properties are generally obtained when (a) the surface energies of the elastomers and plastic are similar, (b) the entanglement molecular length of the elastomer is low, and (c) the thermoplastic polymer is at least semicrystalline. For rubber–plastic pairs with large difference between the solubility parameters, a compatibilizer can be used to improve the morphology of the blend and the properties of TPVs. The presence of the compatibilizer allows the formation of very small rubber particles in the plastic matrix.

Dynamic vulcanization involves the melt mixing of an elastomer with a thermoplastic and the subsequent vulcanization under shear where continuous mixing of the rubber and the plastic phase occurs. The temperature must be above the melting point of the plastic and sufficiently high to activate the vulcanization. The vulcanization system generates the same cross-links or three-dimensional polymer structure as in static vulcanization though in small domains that are nominally about 1 μm in diameter. The size of these rubber particles determines the properties of the composite with larger average particle size being less effective.

Important blend pairs for TPVs are those where the plastic and the rubber are thermodynamically compatible. This principle is best illustrated with blends of ethylene–propylene–diene monomer rubbers and isotactic polypropylene. Lohse [331] described these blends as accessible during mixing but unstable. Under molten and static conditions, the rubber phase coalesces and agglomerates. Thermodynamically incompatible polymers do not melt mix, and the blends are visually inhomogeneous. In such cases, a compatibilizer that makes the surface energy difference to be less dramatic is needed. An example of the large differences in surface energy or solubility parameters occurs between acrylonitrile–butadiene copolymer rubber (NBR) and polypropylene.

Coran *et al.* [323] demonstrated the beneficial effects of complete vulcanization over partial dynamic vulcanization [332]. Cross-linking the rubber phase to complete gelation leads to improvement of mechanical properties of the blend, both at room temperature and, more importantly, at elevated temperature. This is also seen for resistance to solvents. A measure of the degree of cross-linking is the rubber that is normally determined by solvent extraction. An approximate measure of the target cross-link density is higher than $7 \times 10^{-5} \, \mathrm{mol \, cm^{-3}}$ and/or the elastomer is at least 97% nonextractable.

Vulcanization of rubber can access a wide choice of vulcanization chemistries and procedures [333]. Sulfur systems are most well studied and used because of its dominance in tire manufacturing. Coran *et al.* developed sulfur vulcanization for a majority of their experiments. The most obvious disadvantage of this curative is the sulfurous odor. The use of peroxide, on the other hand, with iPP as the plastic phase, leads to chain scission side reactions. In the case of polyethylene, the peroxy radicals lead to cross-linking of the polyethylene, yielding very viscous products that are difficult to process. The degradation of PP leads to a loss of properties. Abdou-Sabet and Fath [329] demonstrated that this disadvantage can be overcome by using phenolic curatives to cross-link the EPDM phase. This improves the cross-link density at the EPDM without affecting the polypropylene or the polyethylene phase. Finally, a conventional silicone rubber curative, multifunctional organosilicon hydride [334], was used in the partial vulcanization of elastomers containing carbon–carbon double bonds in a saturated plastic matrix.

The preferred morphology for TPV is an elastomeric dispersed phase and a thermoplastic continuous phase. Such morphology should allow for flow in a shear field and thus fabrication by molding or extrusion. The final morphology of the TPV is affected by the morphology of the blend at the outset of dynamic vulcanization. For a given polymer pair, melt blending is most efficient when the viscosities of the phases, at the shear rate of mixing, are matched [335]. Other parameters that affect the morphology and the properties of the TPV are the shear rates of the mixing process, the polymer ratio, the surface energies of the polymer pairs, the cross-link density, the type of cross-link, the molecular weight distribution of the rubber, and the presence of compounding additives (e.g., fillers, plasticizers, etc.).

The most widely used combinations of polymer blends are combinations of olefinic elastomers and olefinic thermoplastics. This type of combination is best exemplified by blends of EPDM and PPs that have similar solubility parameters. Romanini *et al.* [336] studied the effect of molecular weight (i.e., the viscosity ratio) on the phase morphology of simple blends prior to dynamic vulcanization. In the investigation [337], an 80/20 EPDM/iPP composition was vulcanized, where PP is the minor component that is initially in the dispersed phase in an EPDM matrix. In the initial stages of dynamic curing, two co-continuous phases are generated, and as the degree of cross-linking of the elastomer advances during mixing, it breaks up into polymer droplets. During dynamic vulcanization of such a blend, the EPDM and the PP undergo a phase inversion where the PP is the continuous phase to maintain plasticity of the blend. Much effort continues to be devoted to the morphology and cross-linking of this blend. A paper by Ellul *et al.* [338] describes the progress that was made in this area by using combinations of techniques, including scanning transmission electron microscopy (STEM) of sections and network visualization by TEM imaging of TPV. In another paper by O'Connor and Fath [339], a study of the effect of the compatibilization of the inorganic filler on the properties of the blend has concluded that morphological changes predominate in these blends on change of the level of compatibilizers. A similar level of effort has also been expended in understanding the physics and the physical processes within these intricate blends. These physical processes include the crystallization [339] and the orientation of the polypropylene in molded articles, the distribution of the diluents and fillers, and the extent of aggregation of the rubber particles.

TPVs are preferred over simple blends because of their stable morphology. Uncross-linked elastomers are viscous liquids that will flow and coalescence, which leads to changes in the morphology and the performance of the product. The dynamic vulcanization of these rubber particles eliminates this potential for the change in the morphology. The choice of cure systems, however, can lead to the same phenomenon. Cure systems that exchange, such as sulfidic, diisocyanate, and transesterification systems, are less stable kinetically than a C—C or a C—Si bond and lead to slow agglomeration of the rubber particles.

Fillers and plasticizers can be added to TPVs to obtain the desired change in properties. The addition of plasticizers, on the other hand, allows the preparation of softer compositions while leading to significant improvement in processability and elastic recovery. In the melt, the oil partitions between the phases. This interphase transfer of the oil lowers the viscosity of the PP, which allows for enhanced flow. On crystallization of the polypropylene phase, most of the oil is expelled and is believed to be predominantly in the rubber phase. The use of paraffinic oil has a moderate effect on the $T_g$ of both the EPDM and the PP.

There are four principal classifications of thermoplastic vulcanizates which are discussed below.

### 4.6.1
### Nonpolar Rubber with Nonpolar Thermoplastic

#### 4.6.1.1  EPDM Elastomer with iPP Thermoplastic

TPR, from Uniroyal, was the first (1972) commercial thermoplastic olefin elastomer based on dynamic vulcanization but has been supplanted since 1981 by Santoprene® thermoplastic rubber by ExxonMobil Chemical Co. [339]. EPDM–PP TPVs can be prepared in a large composition range of the relative amounts of iPP and EPDM. As the ratio of EPDM to PP is increased, the hardness decreases and the properties become more rubbery and more flexible.

#### 4.6.1.2  Natural Rubber Elastomer with PE Thermoplastic

Natural rubber (NR) blends with polyolefin plastics that can be dynamically vulcanized to form TPVs that have the rubber dispersed-phase morphology. Natural rubber and polyethylene can be dynamically vulcanized to form a TPV by using a peroxide or an efficient sulfur vulcanization system for NR. The properties of TPVs can be changed significantly through the selection of the crosslinking agent. Several cross-linking agents, such as sulfur, sulfur donors, peroxides, dimaleimides, diurethane, and phenolic resin, may be used. However, natural rubber is a highly unsaturated rubber and suffers from poor oxidative stability leading to degradation during repeated exposure to high temperatures during mixing, fabrication, and recycling.

### 4.6.1.3 Natural Rubber Elastomer with Polypropylene Thermoplastic

NR–iPP TPVs are made by intensive melt mixing of a mixture of NR and iPP after the NR is masticated to reduce the viscosity in the presence of a curative. These compositions were developed to use NR as a more economical rubber that would have superior mechanical properties, for example, green strength, due to strain-induced crystallization. These have not been verified. It is also believed, without experimental verification, that the cross-linking reaction of NR in the iPP matrix leads to the formation of block polymers by grafting of iPP to the NR. Coran and Patel [340] describe NR–PP compositions cured with sulfur, urethane, and phenolic resin curatives. Tinker *et al.* [341] describe thermoplastic NR–iPP blends that are peroxide cured.

### 4.6.1.4 Butyl Rubber Elastomer with Polypropylene Thermoplastic

Thermoplastic butyl rubber thermoset compounds are used when impermeability to gases, namely, oxygen and water vapor, is desired. TPVs with butyl rubber [342] are capable of duplicating the inherent low gas permeability of thermoset butyl rubber. These TPVs contain butyl, halogenated butyl rubber, or any of the newer halogenated *p*-methylstyrene–isobutylene copolymers [343]. TPVs using copolymers of isobutylene as the component of the elastomer phase are capable of outstanding long-term heat aging [344].

### 4.6.2
### Polar Rubber with Nonpolar Plastic

### 4.6.2.1 NBR Elastomer with iPP Thermoplastic

TPVs based on NBR as the elastomer and iPP as the thermoplastic should have the ideal combination of elastomeric properties and resistance to hydrocarbon solvents. However, these TPVs are difficult to make because of the large difference in the solubility parameters of iPP and the NBR, which leads to poor dispersion and large rubber particles. This requires a compatibilizer to improve the interfacial adhesion between the two phases by decreasing differences in surface energy. This enables the formation of a fine dispersion prior to vulcanization. A secondary effect of using a compatibilizer is a stronger interfacial interaction between the two phases, leading to improved tensile strength of the TPV.

Two methods of compatibilizing NBR with iPP are available: (1) the use of a block copolymer and (2) the formation of polymer–polymer grafts between the NBR and PP. Maleic anhydride-modified PP, formed by free radical grafting, can be used to react with amine-terminated NBR, formed by the condensation of diamine with carboxyl-terminated NBR. PP and NBR TPVs have much lower volume swell on exposure to hydrocarbon fluids. This is an expected outcome for greater polarity of NBR compared with polyolefin rubber.

### 4.6.2.2 Acrylate Rubber with iPP Thermoplastic

The acrylate rubber used in this kind of TPV specifically consists of ethylene, methyl acrylate, and a monomer with a pendant carboxylic acid. This acrylate rubber can be cross-linked with multivalent amines and is used for its low $T_g$ and thermal resistance.

In a representative synthesis, a 50/50 blend of acrylate rubber and polypropylene was dynamically vulcanized [345] with 4,4-methylenedianiline (MDA) as a vulcanizing agent. A graft copolymer compatibilizer was created *in situ* by replacing 5 wt % of the PP with maleated PP that would also react with the diamine.

### 4.6.3
### Nonpolar Rubber with Polar Thermoplastic

#### 4.6.3.1   EPDM Rubber with PA6 Thermoplastic

Blending of polyamides with polyolefin elastomers has been practiced to improve the impact strength of polyamides. However, the incompatibility of these systems requires compatibilization, the most common being based on the reactivity of maleic anhydride and epoxide functionality with the terminal amine of polyamides. Olivier [346,347] has extended this reactivity to the preparation of TPVs wherein the functionalized EPDM reacts prior to vulcanization.

#### 4.6.3.2   EPDM Rubber with PBT Thermoplastic

The need for TPVs that adhere to engineering thermoplastics has led to the use of PBT as the continuous thermoplastic phase of the TPVs. Moffett and Dekkers [348] and, later, Campbell *et al.* [349] demonstrated the formation of TPVs by using EPDM grafted with 3% acrylate monomers (e.g., butyl acrylate or glycidyl acrylate). The elastomer phase was cured with peroxide to form TPVs with excellent tensile and compression set properties. Formation of compatibilizers, by grafting the EPDM, improves the dispersion of the EPDM and the mechanical properties of the TPVs.

#### 4.6.3.3   EPDM Rubber with iPP + PA6 Thermoplastic

Higher use temperatures for TPVs of EPDM rubber and iPP thermoplastic and of butyl rubber and iPP thermoplastic can be obtained using a compatibilized blend with PA6 [350]. In these TPVs, maleic anhydride-grafted PP is the compatibilizer and the compatible blend of iPP and PA6 is the thermoplastic phase.

### 4.6.4
### Polar Rubber with Polar Thermoplastic

#### 4.6.4.1   Acrylate Elastomer with Polyester Thermoplastic

The resistance to hydrocarbon solvents has led to the development of TPVs that contain polar elastomer and thermoplastic components. Patel [351] and, later, Venkataswamy [352] have shown TPVs of such compositions to exhibit low hydrocarbon absorption and minimal loss of properties. The thermoplastic phase could be PET, PBT, or PC. The flexibility in the choice of the acrylic rubber phase leads to TPVs with very low $T_g$ values. Compatibilization of the blend is achieved by having either a carboxylic acid or an epoxy functionality on the elastomer.

**4.7**
**Blends Made during Polymerization**

**4.7.1**
**Gum Elastomers**

A large number of unvulcanized or gum rubbers are available for use in blends with the engineering polymers. Because of the high temperatures and high shear encountered in blending with plastics to form the required fine dispersion of rubber in the plastic matrix, thermal and shear stability of the elastomer is of importance and often the determining factor in the applicability. Unsaturated elastomers that are prone to chemical reactions during processing or later in the presence of oxygen, ozone, or light are less desirable than the saturated elastomers (polyolefins) that are not. In these cases, blends containing butadiene-based elastomers are subject to thermal or oxidative degradation and consequent loss of strength and discoloration. The preferred rubber-in-plastic morphology, however, slows down this degradation of the rubber and reduces the extent of the problem. Polycarbonate–ABS and PPO–HIPS blends are examples where polybutadiene-based rubbers are used. Engineering polymers that require low-temperature toughness benefit from the use of butadiene-based elastomers, because their glass transition temperatures are significantly lower than those of most saturated elastomers such as those based on $n$-butyl acrylate or ethylene–propylene.

In all cases where the postpolymerization production of the blends is required, it is often desirable for the ease of handling to have the gum elastomer as a pellet instead of as a bale during the mixing process to make the elastomer as a minor blend with a plastic that can be let down into a larger amount of the plastic at a later stage to form the final blend. The gum elastomer is thus a concentrate or a master batch. There are a variety of known methods for making these master batch gum elastomers that span the range of making slightly hard or crystalline rubbers by manipulation of the composition to the external admixture of a small amount of thermoplastic.

**4.7.1.1  Diene Rubbers**
Diene rubbers are quite common [353]. In addition to polybutadiene, isoprene and chloroprene are used, along with comonomers such as styrene and acrylonitrile.

Free radical emulsion polymerization was first developed during World War II. Anionic initiation, which predated and postdated this effort, was a precursor to the development of the Ziegler–Natta and butyl lithium processes in the 1950s that are now used. These processes form diene rubbers with extremes of microstructures ranging from 1,2- to the *cis*-1,4 addition. Hydrocarbon solutions of these rubbers have been used for HIPS and certain ABS polymers. In this process, 5–15% of rubber is dissolved in styrene or styrene–acrylonitrile. Rubber particles that are formed in this process have a complex structure and contain numerous PS or SAN occlusions. The presence of occlusions in the particles produces a large increase of rubber phase volume compared with the actual amount of rubber used. Thus, mass-made HIPS and ABS make very efficient use of the rubber. Diene rubbers were used as impact modifiers for polyamides by Murdock *et al.* [354] who used butadiene–methyl methacrylate–methacrylic acid rubber for toughening a mixed polyamide matrix.

Low molecular weight butadiene–acrylonitrile copolymers with carboxyl, hydroxyl, or amine end groups are frequently used for rubber toughening epoxy resins [355]. These are telechelic, solvent-free polymers with a number-average molecular weight in the range of 3600–4200 Da. Free radical polymerization is typically used for these polymers.

Diene rubbers can be easily handled by increasing the amount of the styrene, acrylonitrile, or the acrylate portion of the polymer at the expense of the diene. Although these materials made by either admixture or polymerization slowly deviate away from the ideal rubber properties, their benefits in handling are substantial enough that the practice is widely used.

### 4.7.1.2 Ethylene-Based Elastomers

Ethylene–propylene and ethylene–higher α-olefin elastomers with butene, hexene, and octene as the comonomer are random copolymers that have many desirable features that make them attractive as toughening agents for engineering polymers, for example, excellent thermal and UV stability and $T_g$ of about −50 °C. Chemical modification of EP copolymers, for example, by grafting maleic anhydride, to enhance interaction or to induce chemical reactivity between the elastomer and the plastic, further enhances the utility of these elastomers. EPR is synthesized by Ziegler–Natta processes [356] that are intolerant of chemical functionality. In connection with their use as modifiers for plastics, several aspects of this synthesis are important. Amorphous elastomers that have a low $T_g$ require a uniform composition distribution and an average ethylene content of 40–55 wt%. EPDM is a terpolymer that contains a small amount (typically 2 mol%) of a nonconjugated diene monomer, such as 5-ethylidene-2-norbornene. This incorporates into the backbone without undue interference of the polymerization as conjugated dienes do. This structure provides a double bond pendant from the main chain. Thus, chemical reaction at this site does not lead to chain scission and reduction of molecular weight. This pendant double bond is also a point for introducing chemical functionality to react with plastics. Typically, EPR with chemical functionality is used in conjunction with plastics. A typical use for MA-grafted EPR is for toughening of polyamides with controlled and stable rubber phase adhesion, particle size, and uniformity of dispersion [357]. EPM rubbers may also be functionalized by reacting with glycidyl methacrylate. Such polymers are used to toughen polyamides and polyesters [358,359].

These gum elastomers are made pellet stable and easily processed for further blending by making them slightly crystalline at higher than expected ethylene contents. This is one of the favorable benefits of using ethylene–higher α-olefin impact modifiers for propylene blends compared with amorphous ethylene–propylene copolymers.

### 4.7.1.3 Ethylene Copolymers

Ethylene can be copolymerized with a variety of comonomers by using a high-pressure free radical process. Many polar or functional monomers, for example, vinyl carboxylic acids, esters, and anhydrides, carbon monoxide, and sulfonates [360,361], are suitable as minor comonomers to provide sites for chemical reactions or adhesive bonding when blended with plastics. Typical copolymerization of ethylene and acrylic or methacrylic acid requires the use of about 3–6 mol% acid. Ethylene copolymers and terpolymers containing vinyl acetate are also widely produced including terpolymers with monomers such as glycidyl methacrylate and half esters of maleic acid [362].

### 4.7.1.4 Ionomers

Ionomers useful for toughening plastics are based on ethylene and acrylic or methacrylic acid copolymers. They contain less than about 10 mol% of acid monomer, and are partially neutralized with zinc or sodium cations. In these polymers, the ionic groups interact to form ion-rich domains that act as physical cross-links and have a strong influence on the mechanical properties of the ionomer. The structure and properties of the ionomers are influenced by the type of the polymer, ionic content (amount of carboxyl functionality and degree of neutralization), and the type of cation [363].

### 4.7.2
### Emulsion Rubbers

Emulsion rubbers are those that are made as cross-linked rubber particles or microgels – they are dispersible in plastics, but insoluble in solvents. They were initially used for impact toughening polystyrene [364] both as microgel of butadiene rubber and as styrene–butadiene rubber. ABS was made by the incorporation of broad size distribution butadiene rubber microgel into styrene–acrylonitrile copolymer [365] where a substantial amount of SAN was grafted to the latex particles.

### 4.7.3
### Core–Shell Graft Polymers

Core–shell modifiers for toughening plastics are specifically designed for this purpose and have minimal ungrafted shell material. The elastomer core is a cross-linked copolymer based on either butadiene or *n*-butyl acrylate. This provides an easily dispersible but stable rubber–plastic morphology. In addition, this process delivers a predetermined and known particle size and particle size distribution of the modifier. Butadiene-based core elastomers have a lower $T_g$ in the range from −60 to −85 °C, while those based on *n*-butyl acrylate have a $T_g$ of 30 °C higher. However, the acrylate-based core–shell modifiers have better thermal and UV stability. The latex particle size for impact modification is of crucial importance. Toughening of styrenic copolymers requires that some particles be relatively large, for example, >0.2 µm, while rubber modification of engineering polymers usually is effective with smaller particles in the range of <0.2 µm. Also larger particles above 1 µm are possible through various agglomeration techniques. They include chemical approaches, pressure agglomeration, and addition of solvents, electrolytes, and water-soluble polymers. The formation of the grafted shell can be accomplished by using a number of different synthetic techniques. A number of commercially important methyl methacrylate-grafted core–shell polymers are manufactured by such processes [366]. Core–shell impact modifiers for engineering thermoplastics often have a methyl methacrylate copolymer outer shell [367]. This shell composition is quite useful since PMMA is thermodynamically miscible with a large number of polymers. Thus, these modifier particles disperse readily during melt blending and are a nearest approach to a universal modifier.

### 4.7.4
### Block Polymers

#### 4.7.4.1 Butadiene–Styrene Block Copolymers
Styrene-based block copolymers made by living anionic polymerization are used for toughening plastics. Controlling of the initiator concentration and the amount and the sequence of monomer addition leads to a large variety of structures. Block sequence and size may be changed to obtain polymers of controlled phase structure, chemical nature, and physical properties. The soft block is typically based on polybutadiene or polyisoprene. These blocks may be subsequently hydrogenated to eliminate substantially double bonds and to render the polymer more stable. Structures resembling ethylene–propylene or ethylene–butene can be made in this way by controlling the microstructure of the diene block during its formation. Styrene is the most common monomer to produce the hard block.

### 4.8
### Conclusions

Rubber–plastic blends provide a way of attaining useful properties intermediate between the parents by combining the stiffness of plastics with the resilience of rubbers. The variety in ultimate properties by simple changes in the ratio of the components leads to wide commercial and industrial practice. Blends drive a large proportion of commercial development, industrial technology, and academic science. This manifests as development of mixing tools, characterization processes, and academic science of polymer blends. Not all morphologies are useful; thus, plastic in rubber leads to an undesirable combination of stiff rubbers and thus only rubber in plastic has been widely investigated. The miscibility of rubbers such as the butadiene rubbers and the polyolefin rubbers such as ethylene–α-olefin copolymers was found to be controlled by the structural composition and the primary structure of the copolymers.

Models of the rubber in plastic morphology to correlate the structure with mechanical properties such as impact and tensile strength as well as the BDT are quite primitive. No single theoretical concept works for all systems, nor do the particular systems account for anisotropic changes in properties due to extension or the elongation of the rubber due to shear. Most profoundly, the theoretical direction of having weak plastic–rubber interface to maximize impact is diametrically opposed to the synthetic and experimentally directed efforts at making interfacial agents such as graft polymers that maximize plastic–rubber adhesion as a way to enable the highest impact properties. Improvements in this area should be forthcoming with improved characterization tools that are rapidly coming on line both in sensitivity, ability to observe fine structure, and in the ability to process data. In a sign of progress in this area there is growth in the number of publications describing processes and techniques to study and quantify rubber–plastic interfaces.

Perhaps the greatest area of development of blends of rubber and plastic is the area of polyolefins. New developments in polyolefin architecture and catalysis, coupled with precise reaction engineering and design, have opened an entirely new area for development. Perhaps the greatest incentive is the lack of polarity and reactivity in polyolefins. In the absence of these interactions, the solubility parameter of all polyolefins is similar and all possible rubber–plastic blends not only are feasible but also have some reasonable set of properties. In addition, new developments such as block polymers, cyclic olefin polymers, and EPR with isotactic polyolefin crystallinity make possible subtle variation in morphology that leads to changes in properties. One of the effects of this development is that the blends involving the polar thermoplastic are becoming less numerous as the polyolefin blends progress slowly to match them in properties. The other is the relative segregation of the polar and nonpolar systems: there are only a handful of examples of polar rubbers with nonpolar plastic and vice versa.

It is expected that the next area of improvement in the blends is the ability to fabricate these materials with known and understood effects on the morphology and thus the properties in the direction and away from the direction of flow. These dynamic studies on the material properties are not yet very mature. These advances will be possible only if greater attention is paid to understanding the interface and the role of the physical process in enabling and extending the interfacial effect of rubber–plastic blends.

## References

1 Utracki, C.A. (2014) Polyethylenes and their blends, in *Polymer Blends Handbook*, Springer, Dordrecht, p. 1559.

2 Parameswaranpillai, J., Thomas, S., and Grohens, Y. (2014) Polymer blends: state of the art, new challenges, and opportunities, in *Characterization of Polymer Blends: Miscibility, Morphology and Interfaces*, Wiley-VCH Verlag GmbH, Weinheim, Germany.

3 Goodier, J.N. (1933) *J. Appl. Mech.*, **55**, 39.

4 Paul, D.R. and Bucknall, C.B. (1999) *Polymer Blends*, John Wiley & Sons, Inc., New York.

5 Paul, D.R. and Bucknall, C.B. (2009) *Polymer*, **50**, 5539.

6 Bartzak, Z., Argon, A.S., Cohen, R.E., and Weinberg, M. (1999) *Polymer*, **40**, 2331.

7 Bartzak, Z., Argon, A.S., Cohen, R.E., and Weinberg, M. (1999) *Polymer*, **40**, 2347.

8 Michle, G.H., Adhikari, R., and Henning, S. (2004) *J. Mater. Sci.*, **39**, 3281.

9 van der Wal, A., Mulder, J., Thijs, H., and Gaymans, R. (1998) *Polymer*, **39**, 5467.

10 Mouzakis, D.E., Mader, D., Mulhaupt, R., and Karger-Kocsis, J. (2000) *J. Mater. Sci.*, **35**, 1219.

11 Nitta, K., Kawada, T., Yamahiro, M., Mor, H., and Terano, M. (2000) *Polymer*, **41**, 6765.

12 Kim, G.-M. and Michle, G.H. (1998) *Polymer*, **39**, 5699.

13 Vollenberg, P.H.T. and Heikens, D. (1990) *J. Mater. Sci.*, **25**, 3089.

14 Renner, K., Yang, M.S., Moczo, J., Cho, H.J., and Pukanszky, B. (2005) *Polymer*, **41**, 25320.

15 Gurson, A.L. (1977) *J. Eng. Mater. Technol. Trans. ASME*, **99**, 2.

16 Kim, N.H. and Kim, H.S. (2005) *J. Appl. Polym. Sci.*, **98**, 1290.

17 Bagheri, R. and Pearson, R.A. (1995) *Polymer*, **36**, 4883.

18 Voros, G. and Pukanzsky, B. (2001) *Composites A*, **32**, 343.

19 Pukanszky, B. and Voros, G. (1996) *Polym. Compos.*, **17**, 384.

20 Huang, Y. and Kinloch, A.J. (1992) *J. Mater. Sci.*, **27**, 2253.

21 Chen, X.-H. and Ma, Y.-W. (1998) *J. Mater. Sci.*, **33**, 3529.

22 Chen, X.-H. and Ma, Y.-W. (1998) *Polym. Eng. Sci.*, **38**, 1763.

23 van der Wal, A. and Gaymans, R.J. (1999) *Polymer*, **40**, 6045.

24 Reiter, G. and Sommer, J.-U. (2003) *Polymer Crystallization*, Springer, Berlin.

25 Wu, S., Bosnyak, C.P., and Sehanobish, K. (1997) *J. Appl. Polym. Sci.*, **65**, 2209.

26 Jang, B.J., Uhlmann, D.R., and Vander Sande, J.B. (1985) *J. Appl. Polym. Sci.*, **30**, 2485.

27 Mader, D., Bruch, M., Maie, R.-D., Stricke, F., and Mulhaupt, R. (1999) *Macromolecules*, **32**, 1252.

28 Mader, D., Bruch, M., Dieter-Maier, R., Stricker, F., and Mulhaupt, R. (1999) *Macromolecules*, **32** (4), 1252–1259.

29 Hobbs, S.Y., Bopp, R.C., and Watkins, V.H. (1983) *Polym. Eng. Sci.*, **23**, 380.

30 Gaymans, R.J., Borggreve, R.J.M., and Oostenbrink, A.J. (1990) *Makromol. Chem., Macromol. Symp.*, **38**, 125.

31 Wu, S. (1983) *J. Polym. Sci.*, **21**, 699.

32 Ramsteiner, F. (1983) *Kunststoffe*, **73**, 148.

33 Sue, H.J. and Yee, A.F. (1988) Churchill Conference Papers, Cambridge.

34 Borggreve, R.J.M., Gaymans, R.J., and Eichenwald, H.M. (1989) *Polymer*, **30**, 78.

35 Wu, S. (1989) *J. Polym. Sci. B*, **27**, 723.

36 Kramer, E.J. (1984) *Polym. Eng. Sci.*, **24**, 219.

37 Brown, N. and Ward, I.M. (1983) *J. Mater. Sci.*, **18**, 1405.

38 Michler, G.H. (1990) *Makromol. Chem., Macromol. Symp.*, **38**, 195.

39 Bucknall, C.B. (1988) *Makromol. Chem., Macromol. Symp.*, **16**, 209.

40 Bucknall, C.B. (1990) *Makromol. Chem., Macromol. Symp.*, **38**, 1.

41 Dijkstra, K., Oostenbrink, A.I., and Gaymans, R.J. (1991) PRI Conference on Deformation Yield and Fracture of Polymers, Cambridge.

42 Margolina, A. and Wu, S. (1988) *Polymer*, **29**, 2170.

43 Wu, S. and Margolina, A. (1990) *Polymer*, **31**, 972.

44 Sjoerdsma, S.D. (1989) *Polym. Commun.*, **30**, 106.

45 Gaymans, R.J. and Dijkstra, K. (1990) *Polymer*, **31**, 971.

46 Fox, T.G. (1956) *Bull. Am. Phys. Soc.*, **1**, 123.

47 Williams, M.L., Landel, R.F., and Ferry, J.D. (1955) *J. Am. Chem. Soc.*, **77**, 3701.

48 Lodge, T.P., Wood, E.R., and Haley, J.C. (2006) *J. Polym. Sci. B: Polym. Phys.*, **44**, 756.

49 Nielson, L.E. and Landel, R.F. (1994) *Mechanical Properties of Polymers and Composites*, Marcel Dekker, New York.

50 Menard, K.P. (1999) *Dynamic Mechanical Analysis: A Practical Introduction*, CRC Press, Boca Raton, FL.

51 Gillham, J.K. and Roller, M.B. (1971) *Polym. Eng. Sci.*, **11**, 231.

52 Gillham, J.K. (1976) *Polym. Eng. Sci.*, **16**, 353.

53 Nakayama, K. (2003) *Polymer Characterization Techniques and Their Application to Polymer Blends* (ed. G.P. Simon), Oxford University Press, New York, p. 68.

54 Shanks, R.A. and Amarasinghe, G. (2003) *Polymer Characterization Techniques and Their Application to Polymer Blends* (ed. G.P. Simon), Oxford University Press, New York, p. 23.

55 Cruz, C., Barlow, J.W., and Paul, D.R. (1979) *Macromolecules*, **12**, 726.

56 Kolesov, I.S. and Radusch, H.-J. (1999) *J. Macromol. Sci. Phys.*, **B38** (5&6), 1055.

57 Parker, D.S. and Yee, A.F. (1989) *J. Mater. Sci. Lett.*, **8**, 921.

58 Parker, D.S. and Yee, A.F. (1989) *J. Mater. Sci. Lett.*, **8**, 921.

59 Bulakh, N. and Jog, J.P. (1999) *J. Macromol. Sci. Phys.*, **B38** (3), 277.

60 Smit, L. and Radonjic, G. (2000) *Polym. Eng. Sci.*, **40**, 2144.

61 Inaba, N., Yamada, T., Suzuki, S., and Hashimoto, T. (1988) *Macromolecules*, **21**, 407.

62 Bretas, R.E.S. (2003) *Polymer Characterization Techniques and Their Application to Blends* (ed. G.P. Simon), Oxford University Press, New York, p. 121.

63 Moffitt, M., Rharbi, Y., Tong, J.-D., Farhina, J.P.S., Li, H., Winnik, M.A., and Zahalka, H. (2003) *J. Polym. Sci. B: Polym. Phys.*, **41**, 637.

64 Ribbe, A.E. and Hashimoto, T. (1997) *Macromolecules*, **30**, 3999.

65 Dibbern-Brunelli, D., Atvars, T.D.Z., Joekes, I., and Barbosa, V.C. (1998) *J. Appl. Polym. Sci.*, **69**, 645.

66 McMaster, J.P. (1975) *Adv. Chem. Ser.*, **142**, 43.

67 Horiuchi, S.M.N., Yase, K., and Kitano, T. (1997) *Macromolecules*, **30**, 3664.

68 Cser, E., Rasoul, E., and Kosior, E. (1999) *Polym. Eng. Sci.*, **39**, 1100.

69 Da Silva, J. and Bretas, R.E.S. (2000) *Polym. Eng. Sci.*, **40**, 1414.

70 Brown, G.M. and Butler, J.H. (1997) *Polymer*, **38**, 3937.

71 Turcott, E., Nguyen, K.T., and Garcia-Rejon, A. (2001) *Polym. Eng. Sci.*, **41**, 603.

72 Kim, B.K. and Do, I.H. (1996) *J. Appl. Polym. Sci.*, **60**, 2207.

73 Bassett, D.C. and Vaughan, A.S. (2003) *Polymer Characterization Techniques and Their Application to Blends* (ed. G.P. Simon), Oxford University Press, New York, p. 436.

74 Bassett, D.C. (1989) *Comprehensive Polymer Science: Polymer Characterization*, vol. 1 (eds C. Booth and C. Price), Pergamon Press, Oxford, p. 841.

75 Wood, B. (1999) *Polymer Blends and Alloys* (eds G.O. Shonaike and G.P. Simon), Marcel Dekker, New York, p. 475.

76 Hobbs, S.Y. and Watkins, V.H. (2000) *Polymer Blends, vol. 1: Formulation* (eds D.R. Paul and C.B. Bucknall), Wiley–Interscience, New York, p. 239.

77 Hobbs, J.K., Winkel, A.K., McMaster, T.J., Humphris, A.D., Baker, A.A., Blakely, S., Aissaoui, M., and Miles, M.J. (2001) *Macromol. Symp.*, **167**, 1.

78 Vanlandingham, M.R., Villarrubia, J.S., Guthrie, W.E., and Meyers, G.F. (2001) *Macromol. Symp.*, **167**, 15.

79 Wen, G., Li, X., Liao, Y., and An, L. (2003) *Polymer*, **44**, 4035.

80 Heier, J., Kramer, E.J., Revesz, P., Battistig, G., and Bates, E.S. (1999) *Macromolecules*, **32**, 3758.

81 Binning, G., Rohrer, H., and Gerber, C. (1982) *Phys. Rev. Lett.*, **49**, 57.

82 Ade, H., Zhang, X., Cameron, S., Costello, C., Kirz, J., and Williams, S. (1992) *Science*, **258**, 972.

83 Ade, H., Smith, A.P., Cameron, S., Cieslinski, R., Mitchell, G., Hsiao, B., and Rightor, E. (1995) *Polymer*, **36**, 1843.

84 Derham, K.E., Goldsbrough, J., and Gordon, M. (1974) *Pure Appl. Chem.*, **38**, 97.

85 Koningsveld, R. and Kleintjens, L.A. (1977) *Br. Polym. J.*, **9**, 212.

86 Chen, H.-L., Li, L.-J., and Lin, T.-L. (1998) *Macromolecules*, **31**, 2255.

87 Defieuw, G., Groeninckx, G., and Reynaers, H. (1989) *Polymer*, **30**, 595.

88 Hu, S.-R., Kyu, T., and Stein, R.S. (1987) *J. Polym. Sci. B: Polym. Phys.*, **25**, 71.

89 Song, H.H., Wu, D.Q., Chu, B., Satkowski, M., Ree, M., Stein, R.S., and

Phillips, J.C. (1990) *Macromolecules*, **23**, 2380.

90 Tashiro, K., Imanishi, K., Izumi, Y., Kobayashi, M., Kobayashi, K., Satoh, M., and Stein, R.S. (1995) *Macromolecules*, **28**, 8477.

91 Zin, W.-E. and Roe, R.-J. (1984) *Macromolecules*, **17**, 183.

92 Halasa, A.F., Wathen, G.D., Hsu, W.L., Matrava, B., and Massie, J.M. (1991) *J. Appl. Polym. Sci.*, **43**, 183.

93 Bucknall, D.G. and Arrighi, V. (2000) *Polymer Blends, vol. 1: Formulation* (eds D.R. Paul and C.B. Bucknall), John Wiley & Sons, Inc., New York, p. 349.

94 Schwahn, D. (2003) *Polymer Characterization Techniques and Their Application to Blends* (ed. G.P. Simon), Oxford University Press, New York, p. 346.

95 Wignall, G.D., Child, H.R., and Samuels, R.J. (1982) *Polymer*, **23**, 957.

96 Lohse, D.J. (1986) *Polym. Eng. Sci.*, **26**, 1500.

97 Nicholson, J.C., Finerman, T.M., and Crist, B. (1990) *Polymer*, **31**, 2287.

98 Rhee, J. and Crist, B. (1991) *Macromolecules*, **24**, 5663.

99 Richards, R.W. and Penfold, J. (1994) *Trends Polym. Sci.*, **2** (1), 5.

100 Bucknall, D.G. (2004) *Prog. Mater. Sci.*, **49**, 713.

101 Richter, D., Monkenbusch, M., Arbe, A., and Colmenero, J. (2005) *Adv. Polym. Sci.*, **174**, 1.

102 Bovey, R.A. and Jelinski, L.W. (1987) *Encyclopedia of Polymer Science and Engineering*, 2nd edn, vol. **10** (eds H.E. Mark, N.M. Bikales, C.G. Overberger, and G. Menges), John Wiley & Sons, Inc., New York.

103 Nishi, T., Wang, T.T., and Kwei, T.K. (1975) *Macromolecules*, **8**, 227.

104 Coleman, M.M., Graf, J.F., and Painter, P.C. (1991) *Specific Interactions and the Miscibility of Polymer Blends*, Technomic Publishing Co., Lancaster.

105 Sulzberg, T. and Cotter, R. (1970) *J. Polym. Sci. A-1*, **8**, 2747.

106 Zhao, Y., Levesque, J., Roberge, P.C., and Prud'homme, R.E. (1989) *J. Polym. Sci. B: Polym. Phys.*, **27**, 1955.

107 Olabisi, O. (1975) *Macromolecules*, **8**, 316.

108 Deshpande, D.D., Patterson, D., Schreiber, H.P., and Su, C.S. (1974) *Macromolecules*, **7**, 530.

109 Al-Saigh, Z.Y. (1997) *Trends Polym. Sci.*, **5** (3), 97.

110 Wu, S. (1982) *Polymer Interface and Adhesion*, Marcel Dekker, New York.

111 Anastasiadis, S.H., Chen, J.K., Koberstein, J.T., Sohn, J.E., and Emerson, J.A. (1986) *Polym. Eng. Sci.*, **26**, 1410.

112 Elmendorp, J.J. and de Vos, G. (1986) *Polym. Eng. Sci.*, **26**, 415.

113 Princen, H.M., Zia, I.Y.Z., and Mason, S.G. (1967) *J. Colloid Interface Sci.*, **23**, 99.

114 Nam, G.J., Kim, K.Y., and Lee, J.W. (2005) *J. Appl. Polym. Sci.*, **96**, 905.

115 Minkova, L., Yordanov, H., Filippi, S., and Grizzuti, N. (2003) *Polymer*, **26**, 7925.

116 Yukioka, S., Nagato, K., and Inoue, T. (1992) *Polymer*, **33**, 1171.

117 Yukioka, S. and Inoue, T. (1994) *Polymer*, **35**, 1182.

118 Schonbein, C.F. (1847) *Philos. Mag.*, **31**, 7.

119 Hyatt, V.W. and Hyatt, I.V. (1870) US Patent 105,338.

120 Clarke, H.T. (1920) US Patent 398,939.

121 Wilkie, H.F. (1922) US Patent 1,449,159.

122 Reid, E.W. (1928) US Patent 1,935,517.

123 Voss, A. and Dickhauser, E. (1928) US Patent 2,012,177.

124 Kulich, D.M., Kelley, P.D., and Pace, J.E. (1985) *Encyclopedia of Polymer Science and Technology*, John Wiley & Sons, Inc., New York, pp. V1, 338.

125 Seymour, R.B. and Stahl, G.A. (1985) *Macromolecular Solutions*, Pergamon Press, Elmsford, NY.

126 Semon, W.A. and Stahl, G.A. (1982) *History of Polymer Science and Technology* (ed. R.B. Seymour), Dekker, New York.

127 Giles, H.F. (1986) *Mod. Plast.*, **63**, 10.

128 Bucknall, C.B. (1977) *Toughened Plastics*, Applied Science Publishers, London.

129 Carlson, A.W., Jones, T.A., and Martin, J.L. (1957) *Mod. Plast.*, **44**, 155.

130 Dunkelberger, D.L. (1987) *Polymeric Composites: Their Origin and Development* (ed. R.B. Seymour), VNU Science Press, Utrecht, The Netherlands.

131 Nakamura, Y. (1928) *J. Polym. Sci., Polym. Chem. Ed.*, **16** (1981), 2055.

132 Riddle, E.H. (1954) *Monomeric Acrylic Esters*, Van Nostrand Reinhold, New York.

133 Jennings, G.B. (1953) US Patent 2,646,417.

134 Cohan, J.E. and Pittenger, C.F. (1947) *Mod. Plast.*, **25**, 81.

135 Cohan, J.E. and Pittenger, C. (1947) *Rubber Age*, **61**, 536.

136 Hay, D., Disterer, H.A., and Storey, E.B. (1955) *Rubber Age*, **94**, 77.

137 Nielsen, L. (1953) *J. Am. Chem. Soc.*, **75**, 1435.

138 Breuers, W., Hild, W., and Wolff, H. (1954) *Plaste Kautsch.*, **1**, 170.

139 Takayanag, M. and Manabe, S. (1965) *Rep. Prog. Polym. Phys. Jpn.*, **8**, 285.

140 Zakrzewski, G.A. (1973) *Polymer*, **14**, 348.

141 Matsuo, M. (1969) *Polym. Eng. Sci.*, **9**, 197.

142 Landi, V. (1972) *Rubber Chem. Technol.*, **45**, 222.

143 Landi, V. (1972) *Rubber Chem. Technol.*, **45**, 1684.

144 Zakrzewski, G.A. (1973) *Polymer*, **14**, 348.

145 Wang, C. and Cooper, S. (1983) *J. Polym. Sci., Polym. Phys. Ed.*, **21**, 11.

146 Fukumori, K., Sato, N., and Kurauchi, N. (1988) 2nd Elastomer Meeting, The Society of Rubber Industry, Tokyo, December 1988.

147 Wang, Y.-Y. and Chen, S.-A. (1981) *Polym. Eng. Sci.*, **21**, 47.

148 Matsuo, M. (1969) *Polym. Eng. Sci.*, **9**, 197.

149 Matsuo, M. (1968) *Jpn. Plast.*, **2**, 6.

150 Inoue, T., Kobayash, T., Hashimoto, T., Tanigami, T., and Miyasaka, K. (1984) *Polym. Commun.*, **25**, 148.

151 Kronman, A.G. and Kargain, V.A. (1966) *Polym. Sci. USSR*, **8**, 1878.

152 Mann, J. and Williamson, R. (1968) *Physics of Glassy Polymers*, John Wiley & Sons, Inc., New York.

153 Sperling, L.H., Thomas, D.A., Lorenz, J.E., and Nagel, E.J. (1975) *J. Appl. Polym. Sci.*, **19**, 2225.

154 Abrams, W.J. (1962) *Rubber Age*, **91**, 255.

155 Horvath, J., Wilson, W., Lundstrom, H., and Purdun, J. (1968) *Appl. Polym. Symp.*, 7, 95.

156 Jorgensen, A.H. and Frazer, D.G. (1968) *Appl. Polym. Symp.*, 7, 83.

157 Schwarz, H.F. and Edwards, W.S. (1974) *Appl. Polym. Symp.*, **25**, 643.

158 Jorgensen, A.H. and Frazer, D.G. (1968) *Rubber World*, **157**, 57.

159 Schwarz, H.F., Bley, J.W.F., and Hansmann, J. (1987) *Kunstoffe Ger. Plast.*, 77, 761.

160 Deanin, R.D. and Sheth, K.B. (1980) *Org. Coat. Plast. Chem.*, **43**, 23.

161 Matsuo, M. (1968) *Jpn. Plast.*, **2**, 6.

162 Giudici, P. and Milner, P.W. (1978) Conference on Technology of Plastics and Rubbers Interface, Brussels.

163 Woods, M.E., Morsek, R.J., and Whittington, W.H. (1973) *Rubber World*, **167**, 42.

164 Stockdale, M.K. (1988) Rubber & Plastics News, vol. 12, May 30.

165 Kobayashi, T. and Nakamura, M. (1986) ACS Rubber Division Meeting, Atlanta, GA, October 1986.

166 Watanabe, N., Hashimoto, K., and Oyama, M. (1984) Annual Meeting of the Swedish Institute of Rubber Technology, Gothenburg.

167 Ostromislensky, I.I. (1927) US Patent 1,613,673.

168 Seymour, R.B. (1952) US Patent 2,574,438.

169 Boyer, R. (1982) Chapter 19, in *History of Polymer Science and Technology* (ed. R.B. Seymour), Dekker, New York.

170 Hayward, R.M. and Elly, J. (1954) US Patent 2,668,806.

171 Amos, J.L. (1974) *Polym. Eng. Sci.*, **14**, 1.

172 Daly, L.E. (1948) US Patent 2,439,202.

173 Childers, C.W. and Fisk, C.F. (1958) U.S. Patent 2,820,773.

174 Calvert, W.C. (1959) US Patent 2,908,991.

175 Lee, L.H. (1966) US Patent 3,238,275.

176 Grawboski, T.S. and Irvin, H.H. (1962) US Patent 3,953,800; US Patent 3,130,177 (1964).

177 Ogawa, M. and Takezoe, S. (1973) *Jpn. Plast. Age*, **11**, 39.

178 Pavelich, W.A. (1986) Chapter 13, in *High Performance Polymers: Their Origin and Development* (eds R.B. Seymour and G.S. Kirchenbaum), Elsevier, New York.

179 Frazier, W. (1966) *Chem. Ind.*, **8**, 1399.

180 Basdekis, C.H. (1964) *ABS Plastics*, Van Nostrand Reinhold, New York.

181 Buchdahl, R. and Nielsen, L.E. (1950) *J. Appl. Phys.*, **21**, 482.

182 Kato, K. (1967) *Polym. Eng. Sci.*, **7**, 38.

183 Echte, A. (1977) *Angew. Makromol. Chem.*, **58/59**, 175.

184 Molau, G.E., Wittbrodt, W.M., and Meyer, V.E. (1969) *J. Appl. Polym. Sci.*, **13**, 45.

185 Freund, G., Lederer, M., and Strobel, W. (1977) *Angew. Makromol. Chem.*, **58/59**, 199.

186 Cigna, G. (1970) *J. Appl. Polym. Sci.*, **14**, 1781.

187 Bucknall, C.B., Cote, F.F.P., and Partridge, I.K. (1986) *J. Mater. Sci.*, **21**, 301.

188 Bucknall, C.B., Davies, P., and Partridge, I.K. (1986) *J. Mater. Sci.*, **21**, 307.

189 Bucknall, C.B., Davies, P., and Partridge, I.K. (1987) *J. Mater. Sci.*, **22**, 1341.

190 Freeguard, G.F. (1974) *Br. Polym. J.*, **6**, 205.

191 Ide, F. and Sasaki, I. (1970) *Kobunshi Kagaku*, **27**, 617.

192 Rumscheidt, F.D. and Mason, S.G. (1961) *J. Colloid Sci.*, **16**, 238.

193 Karam, H.J. and Bellinger, J.C. (1968) *Ind. Eng. Chem. Fundam.*, **7**, 576.

194 Molau, G.E. (1965) *J. Polym. Sci. A-1*, **12**, 67.

195 Echte, A., Gausepohl, H., and Lutje, H. (1980) *Angew. Makromol. Chem.*, **90**, 95.

196 Riess, G., Schlienger, M., and Marti, S.J. (1980) *Macromol. Sci. Phys. B*, **17**, 355.

197 Echte, A. (1982) *Chemische Technologie*, 4th edn, vol. **6** (eds K. Harnisch, R. Steiner, and K. Winnacker), Hanser Publishers, Munich, p. 373.

198 Bucknall, C.B. and Drinkwater, I.C. (1973) *J. Mater. Sci.*, **8**, 1800.

199 Bucknall, C.B. (1978) *Adv. Polym. Sci.*, **27**, 121.

200 Molau, G.E. (1965) *J. Polym. Sci., Polym. Lett. Ed.*, **3**, 107.

201 Lin, C.-C., Ku, H.-C., and Chiu, W.-Y. (1981) *J. Appl. Polym. Sci.*, **26**, 1327.

202 Locatelli, J.L. and Riess, G. (1974) *Makromol. Chem.*, **175**, 3523.

203 Bucknall, C.B. (1977) *Toughened Plastics*, Applied Science Publishers, London, p. 298.

204 Morbitzer, L., Kranz, D., Humme, G., and Ott, K.H. (1976) *J. Appl. Polym. Sci.*, **20**, 2691.

205 Bucknall, C.B. (1977) *Toughened Plastics*, Applied Science Publishers, London, p. 104.

206 Haaf, F., Breuer, H., Echte, A., Schmitt, B.J., and Stabenow, J.J. (1981) *J. Sci. Ind. Res.*, **40**, 659.

207 Lednicky, F. and Pelzbauer, Z. (1986) *Angew. Makromol. Chem.*, **141**, 151.

208 Morbitzer, L., Kranz, D., Humme, G., and Ott, K.H. (1976) *J. Appl. Polym. Sci.*, **20**, 2691.

209 Conaghan, B.F. and Rosen, S.L. (1972) *Polym. Eng. Sci.*, **12**, 134.

210 Dean, B.D. (1985) *J. Elastomers Plast.*, **17**, 55.

211 Jelenic, J., Kirste, R.G., Oberthur, R.C., Schmitt-Strecker, S., and Schmitt, B.J. (1984) *Makromol. Chem.*, **185**, 129.

212 Kohan, M.L. (1973) *Nylon Plastics*, SPE Monographs, John Wiley & Sons, Inc., New York.

213 von Flexman, E.A. (1979) *Kunststoffe*, **69**, 172.

214 Epstein, B.N. and Adams, G.C.A. (1985) PRI Conference on Toughening of Plastics, London.

215 Epstein, B.N. (1979) US Patent 4,174,385.

216 Sue, H.J. and Yee, A.F. (1989) *J. Mater. Sci.*, **24**, 1447.

217 Hahn, M.T., Hertzberg, R.W., and Manson, J.A. (1983) *J. Mater. Sci.*, **18**, 3552.

218 Hahn, M.T., Hertzberg, R.W., and Manson, J.A. (1986) *J. Mater. Sci.*, **21**, 39.

219 Wu, S. (1985) *Polymer*, **26**, 1855.

220 Oostenbrink, A., Molenaar, L., and Gaymans, R.J. (1990) PRI Conference on Polymer Blends, Cambridge.

221 Oshinski, A., Keskkula, H., and Paul, R. (1992) *Polymer*, **33**, 268.

222 Wu, S. (1987) *Polym. Eng. Sci.*, **27**, 335.

223 Wu, S. (1987) *Polym. Eng. Sci.*, **27**, 135.

224 Xanthos, M. (1988) *Polym. Eng. Sci.*, **28**, 1392.

225 Borggreve, R.J.M., Gaymans, R.J., and Schuijer, I. (1989) *Polymer*, **30**, 71.

226 Gale, G.M. (1982) *Plast. Rubber Proc. Appl.*, **2**, 347.

227 Taylor, G.I. (1932) *Proc. R. Soc. Lond. Ser. A*, **41**, 138.

228 Taylor, G.I. (1934) *Proc. R. Soc. Lond. Ser. A*, **146**, 50.

229 Karam, H.I. and Bellinger, J.C. (1986) *Ind. Eng. Chem. Fundam.*, **7**, 576.

230 Wu, S. (1997) *Polym. Eng. Sci.*, **27**, 335.

231 Oshinski, A., Keskkula, H., and Paul, D.R. (1992) *Polymer*, **33**, 268.

232 Oostenbrink, A.L. and Gaymans, R.J. (1992) *Polymer*, **3086**, 33.

233 Borggreve, R.J.M. and Gaymans, R.J. (1989) *Polymer*, **30**, 63.

234 Greco, R., Lanzetta, G., Maglio, G., Malinconico, M., Martuscelli, E.P.R., Ragosta, G., and Scarinzi, G. (1986) *Polymer*, **27**, 299.

235 Sunderland, P. and Kausch, H.H. (1988) *Makromol. Chem., Macromol. Symp.*, **16**, 365.

236 Oostenbrink, A.L., Molenaar, L.J., and Gaymans, R.J. (1990) PRI Conference on Polymer Blends, Cambridge.

237 Lawson, D.F., Hergenrother, W.L., and Matlock, M.G. (1988) *ACS Polym. Prepr.*, **29** (2), 193.

238 Serpe, G., Jarrin, L., and Dawans, F. (1990) *Polym. Eng. Sci.*, **30**, 553.

239 Borggreve, R.J.M. and Gaymans, R.I. (1988) *Polymer*, **29**, 1441.

240 Cimmino, S., D'Orazio, L., Greco, R., Maglio, G., Malinconico, M., Mancarella, C., Martuscelli, E., Musto, P., Palumbo, R., and Ragosta, G. (1985) *Polym. Eng. Sci.*, **25**, 193.

241 D'Orazio, L., Mancarella, C., and Martuscelli, E. (1988) *J. Mater. Sci.*, **23**, 161.

242 Greco, R., Malinconico, M., Martuscelli, E., Ragosta, R., and Scarinzi, G. (1988) *Polymer*, **29**, 1418.

243 Sederel, L.C., Mooney, J., and Weese, R.H. (1987) Polymer Blends and Alloys, Strasbourg.

244 Udipi, K. (1988) *J. Appl. Polym. Sci.*, **36**, 117.

245 Lefelar, J.A. and Udipi, K. (1989) *Polym. Commun.*, **30**, 38.

246 Ban, L.L., Doyle, M.J., Disko, M.M., and Smith, G.R. (1988) *Polym. Commun.*, **29**, 163.

247 Neuray, D. and Ott, K.H. (1981) *Angew. Makromol. Chem.*, **98**, 213.

248 Hobbs, S.Y., Dekkers, M.E., and Watkins, V.H. (1989) *J. Mater. Sci.*, **24**, 2025.

249 Bowman, C. (1990) PRI Conference on Polymer Blends, Cambridge.

250 Gent, A.N. (1990) *Rubber Chem. Technol.*, **63**, 949.

251 Gent, A.N. and Tompkins, D.A. (1969) *J. Polym. Sci. A-2*, **7**, 1483.

252 Wu, S. (1983) *J. Polym. Sci.*, **21**, 699.

253 Gaymans, R.J., Borggreve, R.J.M., and Oostenbrink, A.J. (1990) *Makromol. Chem., Macromol. Symp.*, **38**, 125.

254 Wu, S. (1990) *Polym. Eng. Sci.*, **30**, 753.

255 Hobbs, S.Y. and Dekkers, M.E.J. (1989) *J. Mater. Sci.*, **24**, 1316.

256 Yamaguchi, M., Miyata, H., and Nitta, K.H. (1996) *J. Appl. Polym. Sci.*, **62**, 87.

257 Nitta, K.H., Shi, Y.W., Hashiguchi, H., Tanimoto, S., and Terano, M. (2005) *Polymer*, **46**, 965.

258 Yamaguchi, M., Miyata, H., and Nitta, K. (1997) *J. Polym. Sci., Polym. Phys.*, **35**, 953.

259 Nolley, E., Barlow, J.W., and Paul, D.R. (1980) *Polym. Eng. Sci.*, **20**, 364.

260 Bartlett, D.W., Barlow, J.W., and Paul, D.R. (1982) *J. Appl. Polym. Sci.*, **27**, 2351.

261 Dorazio, L., Mancarella, C., Martuscelli, E., and Sticotti, G. (1991) *J. Mater. Sci.*, **26**, 4033.

262 Martuscelli, E., Silvestre, C., and Abate, G. (1982) *Polymer*, **23**, 229.

263 Chen, C.Y., Yunus, W.M.Z.W., Chiu, H.W., and Kyu, T. (1997) *Polymer*, **38**, 4433.

264 Kamdar, A.R., Hu, Y.S., Anserns, P., Chum, S.P., Hiltner, A., and Baer, E. (2006) *Macromolecules*, **39**, 1496.

265 Da Silva, A.L.N. and Coutinho, F.M.B. (1996) *Polym. Test.*, **15**, 45.

266 Huang, L., Pei, Q., Yuan, Q., Li, H., and Jiang, W. (2003) *Polymer*, **44**, 3125.

267 Da Silva, A.L.N. and Coutinho, F.M.B. (2001) *J. Appl. Polym. Sci.*, **81**, 3530.

268 Lopez-Manchado, M.A., Kenny, J.M., and Pedram, M.Y. (2001) *Macromol. Chem. Phys.*, **202**, 1909.

269 Datta, S. and Lohse, D.J. (1993) *Macromolecules*, **26**, 2064.

270 Su, Z., Jiang, P., Li, Q., Wei, P., Wang, G., and Zhang, Y. (2004) *J. Appl. Polym. Sci.*, **94**, 1504.

271 Ha, C.S., Cho, Y.W., Cho, W.J., Ki, Y., and Inoue, T. (2000) *Polym. Eng. Sci.*, **40**, 1816.

272 Sanchez, F.H., Olayo, R., and Manzur, A. (1999) *Polym. Bull.*, **42**, 481.

273 Ha, C.S. and Kim, S.C. (1988) *J. Appl. Polym. Sci.*, **1988**, 2211.

274 Ha, C.S. (1990) *J. Kor. Inst. Rubber Ind.*, **25**, 203.

275 Shin, Y.W., Uozumi, T., Terano, M., and Nitta, K. (2001) *Polymer*, **42**, 9611.

276 Yamaguchi, M., Nitta, K., Miyata, H., and Masuda, T. (1997) *J. Appl. Polym. Sci.*, **63**, 467.

277 Dumoulin, M.M., Carreau, P.J., and Utracki, L.A. (1987) *Polym. Eng. Sci.*, **27**, 1627.

278 Cham, P.M., Lee, T.H., and Marand, H. (1994) *Macromolecules*, **27**, 4263.

279 Yamaguchi, M. (1998) *J. Appl. Polym. Sci.*, **70**, 457.

280 Yamaguchi, M. (2000) *Surf. Sci. Jpn.*, **21**, 226.

281 Thomann, Y., Suhm, J., Thomann, R., Bar, G., Maier, R.D., and Mulhaupt, R. (1998) *Macromolecules*, **31**, 5441.

282 Yamaguchi, M., Suzuki, K., and Miyata, H. (1999) *J. Polym. Sci. B: Polym. Phys.*, **37**, 701.

283 Yamaguchi, M. and Nitta, K. (1999) *Polym. Eng. Sci.*, **39**, 833.

284 Rana, D., Lee, C.H., Cho, K., Lee, B.H., and Choe, S. (1998) *J. Appl. Polym. Sci.*, **69** (12), 2441.

285 Rana, D., Lee, C.H., Cho, K., Lee, B.H., and Choe, S. (1998) *J. Appl. Polym. Sci.*, **69** (12), 2441.

286 Stricker, F., Maier, R.D., Thomann, Y., and Mulhaupt, R. (1998) *Kunststoffe Plast. Eur.*, **88**, 527.

287 McNally, T., McShane, P., Nally, G.M., Murphy, W.R., Cook, M., and Miller, A. (2002) *Polymer*, **43**, 3785.

288 Walton, K.L. and Clayfield, T. (2000) *Annu. Tech. Conf. Soc. Plast. Eng.*, **58** (3), 2623.

289 Weimann, P.A., Jones, T.D., Hillmyer, M.A., Bates, F.S., Londono, J.D., Melnichenko, Y., Wignall, G.D., and Almdal, K. (1997) *Macromolecules*, **30**, 3650.

290 Mader, D., Thomann, Y., Suhm, J., and Mulhaupt, R. (1999) *J. Appl. Polym. Sci.*, **74**, 838.

291 Lohse, D.J. and Graessley, W.W. (1999) Chapter 8: Thermodynamics of polyolefin blends, in *Polymer Blends* (eds D.R. Paul and C.B. Bucknall), John Wiley & Sons, Inc., New York.

292 Carriere, P.J. and Silvis, H.C. (1997) *J. Appl. Polym. Sci.*, **66**, 1175.

293 Tanaka, A., Fukuda, M., Nagai, H., and Onogi, S. (1989) *J. Polym. Sci., Polym. Phys.*, **27**, 2283.

294 Darras, O. and Seguela, R. (1993) *J. Polym. Sci., Polym. Phys.*, **31**, 759.

295 Popli, R. and Mandelkern, L. (1987) *J. Polym. Sci., Polym. Phys.*, **25**, 441.

296 Nitta, K., Okamoto, K., and Yamaguchi, M. (1998) *Polymer*, **39**, 53.

297 Shin, Y.W., Uozumi, T., Terano, M., and Nitta, K. (2001) *Polymer*, **42**, 9611.

298 Fujiyama, M., Wakino, T., and Kawasaki, Y. (1988) *J. Appl. Polym. Sci.*, **35**, 29.

299 Keller, A. and Machin, M.I. (1976) *J. Macromol. Sci.*, **16**, 176.

300 Kalay, G. and Bevis, M.I. (1997) *J. Polym. Sci., Polym. Phys.*, **35**, 265.

301 Da Silva, A.L.N., Tavares, M.I.B., Politano, D.P., Coutinho, F.M.B., and Rocha, M.C.G. (1997) *J. Appl. Polym. Sci.*, **66** (10), 2005.

302 Da Silva, A.L.N., Rocha, M.C.G., Coutinho, F.M.B., and Bretas, R. (2000) *Polym. Test.*, **19** (4), 363.

303 Da Silva, A.L.N., Rocha, M.C.G., Coutinho, F.M.B., and Bretas, R. (2000) *J. Appl. Polym. Sci.*, **75** (5), 692.

304 Raue, F. and Ehrenstein, G.W. (1999) *J. Elastomers Plast.*, **31** (3), 194.

305 Srinivasan, S., Szczepaniak, E., Her, J., Laughner, M.K., and Karjala, T.P. (1998) WO9820069.

306 Ratzsch, M., Reichelt, N., and Hesse, A. (2000) European Patent EP972802A1.

307 Tang, L., Qu, B., and Shen, X.J. (2004) *J. Appl. Polym. Sci.*, **92**, 3371.

308 Jain, A.K., Gupta, N.K., and Nagraz, A.K. (2000) *J. Appl. Polym. Sci.*, **77**, 1488.

309 Lee, J.H., Lee, J.K., Lee, K.H., and Lee, C.H. (2000) *S. Kor. Polym. J.*, **32**, 321.

310 Piloz, A., Decroix, J.Y., and May, J.F. (1976) *Angew. Makromol. Chem.*, **54**, 77.

311 Siegmann, A. (1982) *J. Appl. Polym. Sci.*, **27**, 1053.

312 Hsu, C.C. and Geil, P.H. (1987) *Polym. Eng. Sci.*, **27**, 1542.

313 Boiteux, G., Dalloz, J.C., Douillard, A., Guillet, J., and Seytre, G. (1980) *Eur. Polym. J.*, **16**, 489.

314 Gohil, R.M. and Petermann, J. (1980) *Macromol. Sci. Phys.*, **B18**, 217.

315 Bartczak, Z., Galeski, A., and Pracella, M. (1994) *J. Appl. Polym. Sci.*, **54**, 1513.

316 Holden, G. and Legge, N.R. (1987) *Thermoplastic Elastomers* (eds N.R. Legge, G. Holden, and H.E. Schroeder), Hanser Publishers, Munich.

317 Walsh, D.J., Graessley, W.W., Datta, S., Lohse, D.J., and Fetters, L.J. (1992) *Macromolecules*, **25**, 5236.

318 Krishnamoorti, R., Graessley, W.W., Balsara, N.P., and Lohse, D. (1994) *Macromolecules*, **27**, 3073.

319 Graessley, W.W., Krishnamoorti, R., Reichhart, G.C., Balsara, N.P., Fetters, L.J., and Lohse, D.J. (1995) *Macromolecules*, **28**, 1260.

320 Setz, S., Stricker, F., Kressler, J., Duschek, T., and Mulhaupt, R. (1996) *J. Appl. Polym. Sci.*, **59**, 1117.

321 Gupta, A.K. and Purwar, S.N. (1986) *J. Appl. Polym. Sci.*, **31**, 535.

322 Gessler, A.M. and Haslett, W.H. (1962) US Patent 3,037,954.

323 Coran, A.Y., Das, B., and Patel, R.P. (1978) US Patent 4,130,535.

324 Coran, A.Y. and Patel, R.P. (1981) *Rubber Chem. Technol.*, **54**, 91.

325 Coran, A.Y. and Patel, R.P. (1981) *Rubber Chem. Technol.*, **54**, 982.

326 Coran, A.Y. and Patel, R.P. (1982) *Rubber Chem. Technol.*, **55**, 116.

327 Coran, A.Y. and Patel, R.P. (1982) US Patent 4,355,139.

328 Coran, A.Y., Patel, R.P., and Williams, D. (1982) *Rubber Chem. Technol.*, **54**, 1063.

329 Abdou-Sabet, S. and Fath, M.A. (1982) US Patent 4,311,628.

330 Coran, A.Y., Patel, R.P., and Williams-Headd, D. (1985) *Rubber Chem. Technol.*, **58**, 1014.

331 Lohse, D. (1985) *Annu. Tech. Conf. Soc. Plast. Eng.*, **43**, 301.

332 Fischer, W.K. (1973) US Patent 3,758,643.

333 Hofmann, W. (1967) *Vulcanization and Vulcanizing Agents*, Maclaren and Sons Ltd, London.

334 Umpleby, J.D. (1989) US Patent 4,803,244.

335 Avgeropoulos, G.W., Weissert, F.C., Bohm, G.G.A., and Biddison, P.H. (1976) *Rubber Chem. Technol.*, **49**, 93.

336 Romanini, D., Garagnani, E., and Marchetti, E. (1986) International Symposium on New Polymeric Materials,

European Physical Society (Macromolecular Section), Naples, Italy, June 9–13, 1986.

337 Abdou-Sabet, S., Puydak, R.C., and Rader, C.P. (1996) *Rubber Chem. Technol.*, **69**, 476.

338 Ellul, M.D., Patel, J., and Tinker, A.J. (1995) *Rubber Chem. Technol.*, **68**, 573.

339 O'Connor, G.E. and Fath, M.A. (1982) *Rubber World*, **185**, 26.

340 Coran, A.Y. and Patel, R. (1981) US Patent 4,271,049.

341 Tinker, A.J., Icenogle, R.D., and Whittle, I. (1989) *Rubber World*, **199** (6), 25.

342 Wang, H.C., Powers, K.W., Puydak, R.C., and Dharamarajan, N.R. (1991) US Patent 5,013,793.

343 Powers, K.W. and Wang, H.C. (1992) US Patent 5,162,445.

344 Patel, R.P. and Abdou-Sabet, S. (1997) US Patent 5,621,045.

345 Patel, R.P. (1987) US Patent 4,654,402.

346 Olivier, E.J. (1991) US Patent 5,003,003.

347 Olivier, E.J. (1996) US Patent 5,525,668.

348 Moffett, A.J. and Dekkers, M.E.J. (1992) *Polym. Sci. Eng.*, **32**, 1.

349 Campbell, J.R., Khouri, F.F., Hobbs, S.Y., Shea, T.J., and Moffett, A.J. (1993) *ACS Polym. Prepr*, **34** (2), 846.

350 Venkataswamy, K. (1996) US Patent 5,574,105.

351 Patel, R. (1994) US Patent 5,300,573.

352 Venkataswamy, K. (1996) US Patent 5,523,530.

353 Tate, D.P. and Bethea, T.W. (1985) *Encyclopedia of Polymer Science and Engineering*, 2nd edn, vol. 2, John Wiley & Sons, Inc., New York, p. 537.

354 Murdock, J.D., Nelan, N., and Segall, G.H. (1966) US Patent 3,274,289.

355 Siebert, A.R. (1987) *Makromol. Chem., Macromol. Symp.*, **7**, 115.

356 Ver Strate, G. (1986) *Encyclopedia of Polymer Science and Engineering*, 2nd edn, vol. 6, John Wiley & Sons, Inc., New York, p. 522.

357 Roura, M.I. (1984) US Patent 4,478,978.

358 Olivier, E.I. (1990) US Patent 4,948,842.

359 Epstein, B.N. (1979) US Patent 4,172,859.

360 Lavengood, R.E., Patel, R., and Padwa, A.R. (1987) European Patent Application 0220155.

361 Seven, M.K. and Bellet, R.J. (1969) US Patent 3,456,059.

**362** Roura, M.J. (1982) US Patent 4,346,194.

**363** Eisenberg, A. and Kim, J.-P. (1988) *Introduction to Ionomers*, John Wiley & Sons, Inc., New York.

**364** Keskkula, H. (1989) *Adv. Chem. Ser.*, **222**, 289.

**365** Calvert, W.C. (1966) US Patent 3,238,275.

**366** Goldman, T.D. (1984) US Patent 4,443,585.

**367** Witman, M.W. (1983) US Patent 4,378,449.

# 5
# Morphology of Rubber/Rubber Blends

*Avraam I. Isayev and Tian Liang*

*The University of Akron, Department of Polymer Engineering, Akron, OH 444325-0301, USA*

## 5.1
## Introduction

Rubbers have the glass transition temperature well below the room temperature, with their soft feel and modulus being lower compared to plastics. The presence of double bonds in their macromolecular chains enables rubbers to be cured, which creates a three-dimensional (3D) chemical network allowing them to maintain their permanent shape, if not under stress. While a rubber is subjected to stress, deformation occurs. However, upon removal of stress the rubber can almost fully recover to its undeformed state due to its high elasticity. The high elasticity and softness of rubbers provide a possibility for rubbers to be applied in many fields, including tire, footwear, belt, flooring, gloves, and so on. For many applications, various properties from different rubbers are required; hence, different rubbers can be combined to achieve the performance. Generally, two approaches are used to realize this goal: copolymerization and blending. The former approach can provide rubbers with required properties, and dispersion of components is not a problem, since different monomer units are polymerized into macromolecular chains. Although this approach is beneficial in many cases, it is not cost-effective for mass production, since a whole new synthetic production line needs to be developed. In contrast, blending of different rubbers can provide blends with desired properties in an easier way, if dispersion is acceptable. However, a good dispersion is not always easy to achieve. Macromolecules of different rubbers in blends could result in a substantial heterogeneous morphology of the blend with domain size being large enough to result in poor performance properties. In addition, curing state in different rubbers could also be different, which could cause one component of the blend being under-cured [1–3]. Different distribution of fillers in different rubbers is another problem, which also needs to be solved [4].

*Encyclopedia of Polymer Blends: Volume 3: Structure*, First Edition. Edited by Avraam I. Isayev.
© 2016 Wiley-VCH Verlag GmbH & Co. KGaA. Published 2016 by Wiley-VCH Verlag GmbH & Co. KGaA.

This chapter aims to provide the current state of knowledge on the structure and morphology of rubber blends. Characterization techniques suitable for the study of rubber blends will be introduced first. Then the effect of material parameters and processing conditions on the structure and morphology of rubber blends will be discussed, as well as issues related to the filler distribution and curing in blends. Finally, blends containing different rubbers will be discussed.

## 5.2
## Characterization Techniques for Rubber Blends

To study the morphology of rubber blends, various characterization techniques can be employed. Since this chapter is dedicated to the structure and morphology of rubber blends, this section will be focused to briefly introduce how these techniques can be used to characterize the features of rubber blends, such as their morphology and filler distribution in each rubber phase.

### 5.2.1
### Optical Microscopy

Phase contrast optical microscopy (PCOM) is one of the characterization techniques used by researchers to study the morphology of rubber blends [5]. It provides a direct illustration about how the morphology of rubber blend looks like. Before examination under the microscope, a very thin sample is prepared, usually around or under the glass transition temperature [4]. The difference in refractive index between different rubber phases of the blend provides necessary contrast. PCOM is a suitable instrument to study heterogeneous rubber blends with the size of separated phases at microscale. However, the maximum magnification of the optical microscope is not very high due to the limitation from light wavelength; thus, features in nanoscale could not be observed clearly.

### 5.2.2
### Transmission Electron Microscopy

Transmission electron microscopy (TEM) can provide morphology images with much higher resolution compared to PCOM due to the much smaller wavelength of electrons. To obtain necessary contrast when using TEM to examine the rubber blends, typically two methods can be employed. The first method is to stain the sample with $OsO_4$ and $RuO_4$, which tend to interact with rubber that is less saturated [5]. Therefore, in the blend containing two rubbers, with one being more saturated than the other, staining can provide contrast between different rubbers. Another way is to swell the sample in a monomer or a mixture of monomers, followed by polymerization of this monomer. The different swelling ratio of rubbers in the blends can result in different thickness after sectioning, which can provide contrast during TEM characterization [6]. The sample preparation process for TEM is not easy since a very small thickness of the sample is required to make sure that electrons can pass through the sample. Therefore, special care needs to be taken to obtain qualified samples for TEM examination.

### 5.2.3
### Scanning Electron Microscopy

Scanning electron microscopy (SEM) is a more recent technique developed compared to PCOM and TEM. SEM can give a very direct illustration on the morphology of the studied surface. Specimen for SEM characterization is usually prepared by cryo-fracturing in liquid nitrogen, or at least well under the glass transition temperature of the sample. The sample is then coated with a conductive layer and examined. Separated phases in rubber blends could be observed under SEM [7]. However, in some cases, the contrast between phases is not strong enough to distinguish between phases. In this case, etching of one phase using suitable solvent can provide better contrast between two phases [8].

### 5.2.4
### Atomic Force Microscopy

The mechanism and procedures of atomic force microscopy (AFM) and its application in rubber research are well documented in the literature [9–11]. For the study of rubber blends, phase images obtained from tapping mode are very important. The difference from the phase image indicates the difference in modulus of each rubber phase. Therefore, for rubber blends with each component having certain difference between their modulus, AFM can be used to measure the morphology of the blends. In addition, since the modulus of fillers is significantly higher compared to that of rubbers, the AFM can also be used to detect fillers and illustrate their distribution and dispersion in rubber blends. Characterization of a sample using AFM requires a relatively flat surface of the sample. The sample is usually prepared by using microtome to get a very flat surface or by fracture of the sample to get a fracture surface at the glass transition temperature of the sample. Obviously, the preparation of a specimen for this technique is easier compared to that needed for TEM characterization, where a specimen with a very small thickness is needed.

### 5.2.5
### Dynamic Testing

Dynamic mechanical analysis (DMA) can also be used to extract information, such as homogeneity of blends and distribution of fillers. One of its applications is to test material under oscillatory strain to measure its storage modulus, loss modulus, and loss tangent as a function of temperature. During a dynamic testing, the temperature at which storage modulus drops drastically or loss tangent shows a peak indicates the glass transition temperature, $T_g$. The change of $T_g$ can be used to determine the compatibility of a blend. For two rubbers that are completely immiscible, their $T_g$ do not change after blending. However, the difference between their glass transition temperatures decreases with the increase of their compatibility.

Dynamic testing can also be used to determine the distribution of filler in each rubber phase. The dynamic properties of EPDM/BR blends and NR/SBR blends were studied and a linear correlation between the concentration of carbon black (CB) in each phase and the maximum value of the loss modulus was found [12]. In particular, this maximum was increased with concentration of CB. This maximum value of the loss modulus in the glass transition region of components was used to evaluate the migration of CB between different phases. DMA was also used to study the preferential location of fillers in polymer blends [13]. In particular, the preferential location of CB in BR/EPDM blends was studied, indicating that the loss tangent peak corresponding to BR was more suppressed, leading to the conclusion that CB was preferentially located in the BR phase.

### 5.2.6
### Thermal Analysis

The rubber miscibility in blends can be detected from the glass transition temperature of each rubber in the blend. Differential scanning calorimetry (DSC) is usually used to measure the glass transition temperature and determine the miscibility of the blend. Thermal analysis can also be used to determine the distribution of filler in rubber blends. The crystallization temperature of BR can be affected after mixing with CB [14]. Specifically, the crystallization temperature increased after filling with CB N330. This was attributed to the alignment of BR molecular chains induced by shearing during the mixing with CB, which facilitated the crystallization of BR. The similar observation was made in NR/BR/CB compounds where a higher structure CB N550 was used [15]. An annealing experiment conducted on CB loaded BR reduced the crystallization temperature of BR and supported the latter observation. An approach to determine the amount of CB in the BR phase by comparison of BR crystallization temperature in filled blends with that of unfilled BR and filled BR was proposed. The same conclusion from research on the crystallization of NR/BR/CB compounds was made [16]. The increased crystallization temperature was attributed to the nucleation effect of CB. This effect was determined to be affected by the distribution of CB in different rubber phases.

TGA was also used to study the NR/SBR/CB compounds after extraction to determine the wetting and location of CB in NR/SBR blends [17]. The difference in the decomposition temperature of the NR and SBR made it possible to determine the amount of each rubber bound to CB particles. These data could be used to calculate the distribution of CB in each rubber phase.

Thermal analysis techniques, unlike microscopy, are indirect methods for the characterization of blend morphology and distribution of filler, and they are relatively easy to perform. This allows one to avoid the time-consuming preparation process for specimen that is necessary for any microscopy.

## 5.3
## Effect of Material Parameters and Processing on Structure and Morphology of Rubber Blends

The morphology of rubber blends can generally be classified into two types. For the first type, one component in rubber blend forms the continuous phase (matrix) and the other component is the dispersed phase in the matrix. In the second type, both rubber phases are continuous throughout the blend. This type of morphology is usually called co-continuous morphology. The first type morphology is commonly observed in blends where one component is present at high volume fraction. In this case, the component with higher fraction is the matrix while the component with lower fraction is dispersed in the matrix. This is illustrated in Figure 5.1, in which the phase microscope images of SBR/NBR blends at blending ratios of 75/25 and 25/75 are shown [18]. However, it should be noted that although the composition plays an important role in the determination of which rubber is the matrix and which is not, there are some

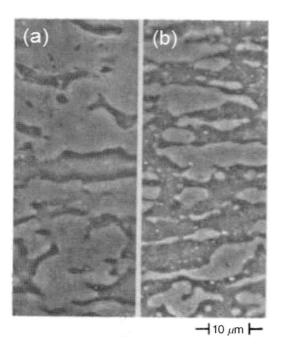

$\dashv$10 $\mu$m $\vdash$

**Figure 5.1** Phase microscope images of 75/25 (a) and 25/75 (b) SBR/NBR blends [18], with permission © 1968 Rubber Division, ACS.

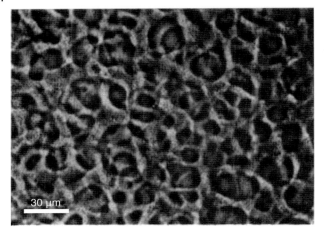

**Figure 5.2** Phase microscope image of butyl/EPDM 50/50 blend [19], with permission © 1970 Rubber Division, ACS.

exceptions, where the minor component can become the matrix [18]. The second type can be observed when the composition of both components is close to each other, as shown in Figure 5.2 [19]. In this figure the phase morphology of butyl/EPDM blend at a ratio of 50/50 is displayed and obtained using a phase contrast microscope. But similarity in composition does not always guarantee this kind of morphology [18]. In some cases, a third kind of morphology is observed where one rubber phase droplet is dispersed in a larger droplet of another rubber phase, as shown for SBR/EPDM 50/50 blend in Figure 5.3 [19].

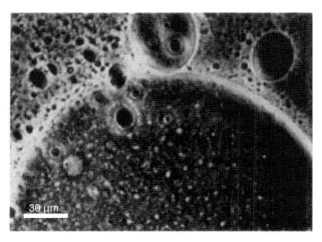

**Figure 5.3** Phase contrast optical microscopy image of SBR/EPDM 50/50 blend [19], with permission © 1970 Rubber Division, ACS.

The morphology of rubber blends has a profound influence on the properties of rubber products. Hence numerous efforts have been devoted to changing the morphology in order to develop materials with preferred properties. An important approach is to reduce the size of the dispersed phase.

The size of the dispersed phase is affected by many factors, including the solubility parameter difference between rubber components in blends, their critical surface tension, viscosities ratio, volume fraction, mixing temperature, and mixing time. The solubility parameter determines whether the two rubber components are miscible or not. According to the well-known Flory–Huggins theory, the free energy difference between components before and after mixing is composed of entropic and enthalpic parts [20,21]. The insufficient increase of polymer chain configuration after mixing of different rubbers resulted in the difficulty to obtain homogeneity at the molecular level, in most cases. Therefore, the miscibility of rubber blends could only be achieved when the solubility parameters of the two rubber components are close to each other. Corish [22] reported that the solubility parameter of rubbers changed with their molecular structure. An increase of vinyl group in BR can reduce its solubility parameter, while an increase of bound styrene in SBR can result in a higher solubility parameter. An increase of acrylonitrile content in NBR can also increase the solubility parameter.

Gardiner [19] studied various solution-mixed rubber blends and summarized the relation between miscibility and solubility parameters of these blends. Table 5.1 provides the solubility parameter difference between various rubbers [19]. The immiscibility between blends increased with the increase of difference between their solubility parameters. This is understandable since larger differences in solubility parameters would result in a higher interaction parameter, leading to a larger positive difference of enthalpy of mixing, which offsets the negative difference from entropy of mixing and hinders dispersion. Also, Table 5.2 lists the critical surface tension difference and miscibility [19]. It

**Table 5.1** Solubility parameter difference between various rubbers[a] unit: $(cal\ cm^{-3})^{1/2}$ [19].

|  | Butyl | Chlorobutyl | EPDM | SBR | NR | cis-1, 4-Polybutadiene |
|---|---|---|---|---|---|---|
| Chlorobutyl | >0[b] | — | — | — | — | — |
| EPDM | 0.11[c] | 0.11[c] | — | — | — | — |
| SBR | 0.49 | 0.49 | 0.38 | — | — | — |
| NR | 0.26 | 0.26 | 0.15 | 0.23 | — | — |
| cis-1,4-Polybutadiene | 0.60 | 0.60 | 0.49 | 0.11[c] | 0.34[c] | — |
| Neoprene | 1.01 | 1.01 | 0.90 | 0.52 | 0.75 | 0.41 |

a) Benzene values are from Sheehan and Bisio [23]; it is estimated that chlorobutyl is the same as butyl.
b) Indicates miscibility.
c) Indicates borderline miscibility.

**Table 5.2** Difference in critical surface tension of various rubbers[a] (unit: dynes cm$^{-1}$) [19].

|  | Butyl | Chlorobutyl | EPDM | SBR | NR | *cis*-1,4-Polybutadiene |
|---|---|---|---|---|---|---|
| Chlorobutyl | >0[b] | — |  |  |  |  |
| EPDM | 1[c] | 1[c] | — |  |  |  |
| SBR | 6 | 6 | 5 | — |  |  |
| NR | 4 | 4 | 3 | 2[c] | — |  |
| *cis*-1,4-Polybutadiene | 5 | 5 | 4 | 1[c] | 1[c] | — |
| Neoprene | 11 | 11 | 10 | 5 | 7 | 6 |

a) Values are from L.H. Lee [24]; it is estimated that chlorobutyl is the same as butyl.
b) Indicates miscibility.
c) Indicates borderline miscibility.

was found that the immiscibility of rubber blends increased with the increase of critical surface tension.

Schuster *et al.* [25] prepared blends of NR/SBR with different solubility parameter differences and studied their morphology. As indicated from TEM images of NR/SBR blends of different blending ratios and solubility parameter differences, depicted in Figure 5.4, the domain size of the dispersed phase decreased with the decrease of the solubility parameter difference [25].

## NR / SBR

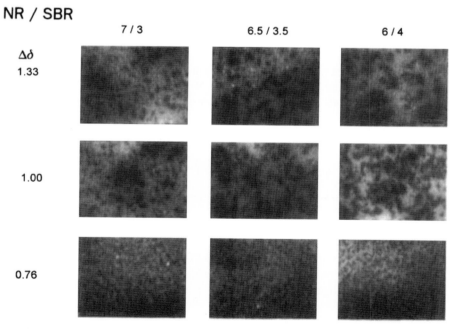

**Figure 5.4** TEM images of NR/SBR blends of various blending ratios and solubility parameter differences [25], with permission © 1996 Rubber Division, ACS.

Besides solubility parameters of rubbers, the viscosity ratio also plays a critical role in the determination of domain size. The best mixing effect, that is, smallest droplet dimension, is usually achieved when the viscosities of both components are close to each other [26]. Otherwise the component with higher viscosity would become less dispersed. Walters and Keyte [4] prepared blends of NR and BR of various molecular weights and found that the blends composed of BR with a molecular weight being closest to that of NR had the smallest domain size. For BR having the molecular weight higher or lower, blends with NR had a larger domain size. Therefore, they concluded that the viscosity ratio of components, which could be implied from the molecular weight ratio, is a very important factor affecting the domain size. They also tried an approach to achieve better mixing for rubber blends by using processing oil to reduce the viscosity of the more viscous rubber. A successful reduction of domain size with the aid of processing oil was achieved. In addition, due to the significant increase of viscosity by introducing filler, premixing of one rubber component with filler could result in a larger domain size, since the viscosity ratio in this case becomes very high [27]. This increase of domain size is even more obvious when the rubber with higher viscosity is premixed with filler [27].

Volume fraction of the dispersed phase can also affect the size of dispersed rubber domains. Tokita [26] observed an increase of domain size with the increase of volume fraction of the dispersed phase. He attributed this observation to a higher chance of coalescence between dispersed domains.

The effect of mixing temperature on rubber domain size is not straightforward. On one hand, an increase of the mixing temperature reduces the viscosity and shear stresses in the system resulting in a worse mixing. But on the other hand, the interfacial tension, which hinders the breakdown of the dispersed phase, is also reduced at higher temperature [28]. Therefore, it is not certain what result on domain size will be brought by the change of temperature, because there is the competition between change of viscosity and change of interfacial tension. In some cases, an increase of mixing temperature may cause a reduction of domain size as a result of reduced interfacial tension, as reported by Tokita [26].

Another important factor affecting the morphology and structure of rubber blends is the mixing time. Hamed [29] studied the morphology evolution of BR/EPDM blends during mixing. BR and EPDM were mixed in a Brabender mixer at a low rotation speed of 5 rpm, the morphology of blends after mixing for 2, 5, 15, and 30 min was studied. It was found that strands of rubber phases can be observed by naked eyes even after mixing for 2 min. After further mixing for 5, 15, and 30 min, the dispersed phase BR was formed and stretched, followed by the break-up into droplets. Hess *et al.* [6] showed that an increase of mixing time can reduce the domain size in NR/BR blends. This is shown in Figure 5.5, in which TEM images of NR/BR blends at blending ratios of 75/25 and 50/50 mixed for different time are displayed [6]. Furthermore, Fujimoto and Yoshimiya [30] reported that the two dielectric loss peaks corresponding to BR and SBR before mixing tended to merge into one with the increase of mixing time. In particular, mixing for 3 to 5 min seemed to be enough for a good mixing of

**75 NR/25 BR**

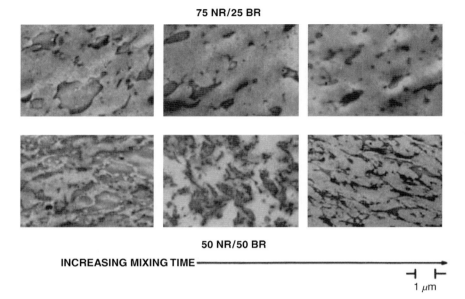

**50 NR/50 BR**

INCREASING MIXING TIME ⟶

1 μm

**Figure 5.5** TEM images of NR/BR blends of blending ratios of 75/25 (top) and 50/50 (bottom) [6], with permission © 1967 Rubber Division, ACS.

this rubber blend. This was supported by Tokita [26], who studied the blends of NR/EPDM. The author also found that for EPDM with Mooney viscosity lower than 47, the equilibrium state of mixing was reached very fast, within 5 min, while for EPDM with a higher Mooney viscosity of 150, even 20 min was not sufficient to reach the equilibrium state.

## 5.4
### Distribution of Fillers and Cure Balance in Rubber Blends

This section describes distribution of fillers and state of cure of rubber components in blends. It also discusses migration of curatives from one component into the other one.

### 5.4.1
### Distribution of Fillers in Rubber Blends

Most rubbers are compounded with fillers before they can be utilized for any practical applications. Therefore, the dispersion of fillers and its effect on the morphology of filled blends is worthwhile to discuss here.

The morphology of filled rubber blends has several important aspects. The first is the selectivity of fillers toward different rubber phases. The second is the dispersion of fillers in each rubber phase. The third is the influence of compounding sequence on the final distribution of fillers.

The selectivity of filler distribution in rubber blends is related to factors such as the viscosity of each rubber component and the interaction between the filler and rubber. Walters and Keyte [4] observed that CB tended to locate in the phase having lower viscosity. This was confirmed by Hess *et al.* [6], who found that the preference of CB location in one phase increased with the decrease of viscosity of that phase.

Klüppel *et al.* [12] found that CB in BR/EPDM blends is preferred to be located in the BR phase. Also, CB in NR/SBR blends is preferred to be in the SBR phase. They suggested that this could be related to good interaction between double bonds in BR with CB and a better interaction between styrene and CB. Maiti *et al.* [31] studied the distribution of CB and silica in the blend of NR and epoxidized natural rubber (ENR). They used dynamic testing to characterize the distribution of silica in the vulcanizates of filled blends. They found that silica tended to migrate toward ENR when each rubber was first mixed with filler and then blended together. They explained the migration of silica toward ENR as a result of its lower viscosity or some interaction between ENR and silica. They also found that at higher epoxidization level, the viscosity of ENR increased and the amount of silica migrating to the ENR phase reduced. This finding supported the hypothesis that filler favored lower viscosity phase. When silica filler replaced by CB, they found that CB migrated toward the NR phase due to better NR–CB interaction.

Le *et al.* [32] considered the wetting process during the preparation of NR/SBR/EPDM blends filled with silica. They observed that at first silica dispersed more in the NR phase due to the fact that NR molecules wetted silica fastest. The fastest wetting process of NR was attributed to its linear molecular chain structure, while the slower wetting rate of SBR and EPDM could be due to the relatively rigid SBR macromolecules and branching of EPDM macromolecules. After that, the preferential location of silica in SBR was observed due to better interaction between SBR and silica. Only a little amount of silica was detected in EPDM due to its poor interaction with silica.

The sequence of adding rubbers and fillers during the mixing process also has a profound influence on the distribution of fillers in the blends. Lee and Singleton [33] studied the effect of compounding sequence on filler distribution. They prepared SBR/BR/CB 80/20/20 compounds in three ways: (1) mixed SBR and CB first and then mixed the masterbatch with BR; (2) mixed BR and CB first and then mixed the masterbatch with SBR; (3) added SBR, BR, and CB to mixer at the same time. They found that there was no transfer of CB through interface for compounds prepared by the first and second methods. For compounds prepared by the third method, at early stage of mixing CB was located at the interface. At later time, it was migrated mainly to the SBR phase.

For practical purposes, it is interesting to know the relation between the preferential location of filler and the properties of compounds. Hess and Chirico [27] prepared NR/SBR/CB and NR/BR/CB compounds. They observed that in

NR/SBR/CB compounds, CB tended to locate in the SBR phase. In NR/BR/CB compounds, CB tended to locate in the BR phase. However, this preferential location of CB brought different effect on vulcanizates prepared from these compounds. For NR/SBR/CB compounds, when most CB was located in SBR, the tear strength of vulcanizates was improved, compared to NR/SBR/CB vulcanizates prepared from masterbatch, where CB was located almost evenly in both rubber phases. For the case of NR/BR/CB compounds, the situation was different. When most CB was located in BR, the tear strength was worse compared to the NR/BR/CB compounds prepared by the masterbatch, where CB was located evenly in both rubber phases. The authors explained this observation in the following way. The viscosity of BR was highest and the viscosity of SBR was lowest, and the viscosity of NR was between those of SBR and BR. Also SBR was the continuous phase in NR/SBR/CB compounds and NR was the continuous phase in NR/BR/CB compounds. Therefore, the preferential location of CB in the continuous phase would result in a higher tear strength since the growth of tearing path would be diverted.

### 5.4.2
### Migration of Curatives in Rubber Blends

Vulcanization is essential for manufacturing of rubber products. The vulcanization process in rubber blends is different than that in one-component rubber compound. The solubility of curatives in different rubber components could be different [34,35]. In addition, after mixing and before the completion of vulcanization, curatives could migrate from one rubber component to another [30,36].

Since the vulcanization process can create crosslinks between different rubber phases [37,38], acting similarly as compatibilizer [39], the influence of these crosslinks on the structure and properties of rubber blends is also an important issue. Clarke *et al.* [40] argued that for the case of NR/BR blends, with a good co-vulcanization system for the rubber blends, it is even not necessary to put extra efforts to compatibilize the rubber blends. In fact, co-vulcanization can provide chemical crosslinks at the interface between different rubbers. Yoshimura and Fujimoto [41] studied the blend of an amorphous SBR and a crystallizable BR, and characterized their homogeneity using dynamic testing. They found that the crystallization of BR in unvulcanized BR/SBR blend was not affected. There were two temperature transitions in the storage modulus, indicating that the unvulcanized BR/SBR blend was a heterogeneous system. After vulcanization, BR still showed a peak of the storage modulus, which indicated the occurrence of crystallization. However, the crystallization in BR/SBR blends was absent after vulcanization, which indicated a good interaction between BR and SBR molecules. They found that the two separated temperature at the transition of storage modulus started to merge into one with the increase in the vulcanization time. Accordingly, they concluded that vulcanization of BR/SBR blends can create crosslinks between BR and SBR, which led to a good interaction between these two rubber molecules.

Use of an accelerator having almost the same solubility toward both rubber phases of the blend may help to minimize the diffusion of curatives and achieve a good vulcanization for both phases [42]. Choosing curatives other than sulfur also helps to improve the vulcanization state in rubber blends [43,44].

Hamed [29] prepared BR/EPDM blends of a low rotation speed (5 rpm) at different mixing times. The prepared blends were cured using radiation of different doses to obtain vulcanizates with different degree of crosslinking. The size of the dispersed phase decreased with the increase of mixing time, but the effect of this change of morphology on mechanical properties was different when the irradiation doses (or degree of crosslinking) were different. It was found that at lower dose, morphology had less effect on mechanical properties, while at higher dose the effect of morphology on mechanical properties was significant. At lower irradiation dose this was attributed to the weaker BR rubber phase having low crosslink density. Therefore, in this case the strength could not be improved just by the enhancement of mixing and improvement of morphology. The obvious effect of mixing and morphology on mechanical properties at higher irradiation dose was explained as a result of an increased crosslink density and strengthening of the BR phase. In this case the propagation of cracks was deviated at the interface. This could be related to the morphology of BR/EPDM blends, such as the size of BR phase.

## 5.5
## Morphology and Properties of Different Rubber Blends

The structure and morphology of blends containing different rubbers will be discussed in this section. For rubbers exhibiting the ability to crystallize, their crystallization behavior such as the crystallization rate and crystallinity could be affected after mixing with other rubbers. Also, the distribution of fillers in different rubbers could be different for different blends, as was mentioned in the previous section. Moreover, the unsaturated rubbers, such as EPDM, are often undercured in blends with other rubbers. Considering the diversity of rubbers and their unique properties, blends containing different rubbers deserve a separate discussion. This is done in the present section.

### 5.5.1
### Blends Containing NR

Natural rubber (NR) is the first rubber utilized by humans and had dominated the world rubber market until the commercialization of SBR. While many different types of synthetic rubbers have come to mass production and widely used in products from tires to shoes, NR is still one of the most important rubbers. Its ability of crystallization under strain gives NR excellent mechanical properties, such as tensile strength and elongation. But NR possesses certain drawbacks including low aging resistance, low ozone resistance, and low oil resistance.

**75 NR/25 BR**     **50 NR/50 BR**     **25 NR/75 BR**

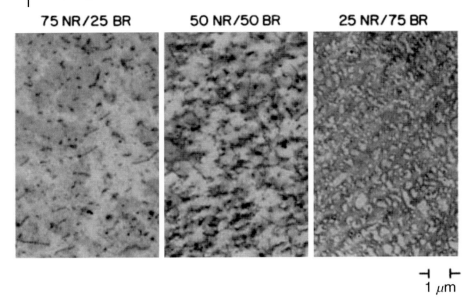

1 µm

**Figure 5.6** TEM images of NR/BR blends with different blending ratios [6], with permission © 1967 Rubber Division, ACS.

Meanwhile, rubbers with good aging, ozone, or solvent resistance often need to be reinforced to meet certain requirements in mechanical properties that are affected by service environment. Therefore, NR is blended with other rubbers to improve these properties and widen the scope where NR can be utilized. NR is immiscible with most of the synthetic rubbers, such as BR, SBR, EPDM, and NBR [45]. Hence, phase separation is expected in the blends of NR with these rubbers and the morphology of the phase separated blends would be an interesting point to discuss.

Studies of blends using electron microscope can help to get a direct understanding of the morphology on properties. In fact, this was shown in the work by Hess *et al.* [6]. They studied the morphology of blends using electron microscopy, as shown in Figure 5.6 [6]. It shows the TEM micrographs of NR/BR blends at blending ratios of 75/25, 50/50, and 25/75. The morphology of the blends changed with the composition, with the major component forming the continuous phase and the minor component forming the dispersed phase, while at NR/BR composition of 50/50 co-continuous phase morphology was obtained.

While the TEM observation of microsectioned vulcanizates of blends can provide a two-dimensional (2D) image on how the phase morphology of NR/BR blends looks like, SEM observation of cryo-fracture surface can provide a 3D image to deepen our knowledge on the phase structure of NR/BR blends. The study by Yan *et al.* [46] provided the SEM image of fractured surface of vulcanized NR/BR 80/20 blends. Their image clearly revealed the fact that BR particles were surrounded by NR and dispersed in the NR matrix.

Besides phase separation, another important factor affecting NR after blending with other rubbers is the change in the crystallization behavior. Mixing NR with other rubbers may affect the crystallization behavior of NR. Gent and Kawahara [47] prepared NR/SBR blends through solution and mechanical mixing. They observed that in the former case the crystallization rate was lower than that of pure NR. However, the final degree of crystallization of NR in the blend was the same as that of pure NR. They related this observation to less nucleation in the NR phase of reduced size. However, in the case of mechanical mixing, the crystallization rate of NR at the beginning was almost not affected, but the degree of crystallization was lower than that in pure NR. The authors proposed that this finding could be a result of a residual strain introduced during mechanical mixing. A later research by Kawazura *et al.* [48] discussed the effect of NR droplet size in NR/SBR blend on crystallization of NR. They observed the decrease of the crystallization rate with the decrease of NR droplet size, which was also attributed to a slower nucleation in smaller size of NR droplets. The heterogeneous nucleation of pure NR diminished after blending with SBR. In the extreme case, where the droplet size of NR was the smallest, Avrami exponent indicated a value corresponding to a homogeneous nucleation.

The crystallization behavior of NR under strain is different. The strain introduces orientation and decreases the entropy. This provides possibilities for NR to crystallize at a temperature higher than the melting point of unstrained NR. The crystallization is the reason of superior mechanical performance of NR over many synthetic rubbers. This strain-induced crystallization behavior could be affected when NR is blended with other rubbers. Qu *et al.* [49] studied the effect of *trans*-1,4-polyisoprene (TPI) on the strain-induced crystallization behavior of NR. They found that the strain-induced crystallization of NR/TPI blends and crystal size of NR was reduced with the increase of TPI. This could be attributed to the reduced molecular mobility caused by TPI and smaller distance between nuclei caused by the preferential location of curatives in the NR phase. The TPI did not show strain-induced crystallization until its content reached more than 70%. This effect was due to the co-vulcanization with NR at lower contents. Manzur and Rubio [50] studied the effect of the TPI molecular weight on the strain-induced crystallization of NR in NR/TPI blends. They suggested that TPI acted as nuclei site and showed the existence of an optimized value of TPI molecular weight maximizing the strain-induced crystallization in NR.

Most rubber products, especially tires, are only useful when CB, silica, and other fillers are incorporated in the rubber. However, due to difference in the filler–rubber interaction and viscosity of rubbers, the distribution of filler is usually not uniform. Hence, how to make different fillers homogeneously distributed in blends containing NR is a very important factor affecting the performance of products. Hess *et al.* [6] found that CB in NR/BR blends distributed more in BR than in NR, when the composition of both rubbers was the same. They used various approaches to balance the distribution of CB in NR/BR blends, and compared the results. In one approach, CB was added into NR/BR blends. This resulted in the highest amount of CB distributed in BR phase. In another approach, they prepared masterbatches of NR/CB and BR/CB. Then they mixed these masterbatches and found that reduced amount of CB was located in the BR phase. In the third approach, NR/CB masterbatch was prepared and mixed with BR. Interestingly, a further reduction of the amount of CB was found in BR. However, preferential location of CB in different phases is not universal for all CB-filled NR/BR compounds [16]. In particular, Massie *et al.* [15] prepared and studied NR/BR/CB compounds. In these compounds, CB was N330 and BR was a high cis-BR. The Mooney viscosity of NR was almost twice of that of BR. Even in this case, selectivity of CB toward either component was not affected when CB was added to the premixed blend. However, when the viscosity of NR was reduced, more CB tended to locate in the NR phase when mixed with the pre-mixed NR/BR blend. In addition, the preparation of the masterbatch of NR/CB or BR/CB followed by addition of the other rubber to prepare NR/BR/CB compounds did not cause a significant migration of CB between phases.

The difference in polarity between NR and the other elastomer component in blends containing NR is also an important factor. When the difference is large, the preferential location of fillers can be affected. Bandyopadhyay *et al.* [51] studied the preferential location of two kinds of nanoclays in NR/ENR and NR/BR blends. Nanoclays were Cloisite Na$^+$ (NA) and Cloisite 30B (30B). In the case of NR/ENR blends having a large difference in polarity of components, only 7% NA was located in NR. In the case of less polar nanofiller, nanoclay 30B, 42% of nanoclay was located in NR. The authors concluded that the polarity was the dominant factor defining the location of nanoclay. In the case of NR/BR blends, due to their close polarity, it was found that viscosity was a more determining factor and nanoclay preferred less viscous NR phase.

Zhang *et al.* [52] compounded CR with organo-modified montmorillonite (OMMT), followed by mixing with NR. They studied the effect of OMMT on NR/CR blends. No observable migration of OMMT to the NR phase was found and the interface between NR and CR was not observed. This could be related to the good affinity of OMMT to CR due to smaller difference in their polarity. Antireversion and mechanical properties of NR/CR blends were improved with the addition of OMMT. This was attributed to the ability of OMMT to hinder hydrochloride from damaging NR macromolecules.

Varghese *et al.* [53] studied the structure and properties of NR/polyurethane/layered silicate compound. They mixed prevulcanized NR latex, polyurethane latex, and aqueous layered silicate solution, and studied the structure and morphology of compounds using XRD and TEM. They found that there was poor compatibility between NR and PU. Although the polarity of PU facilitated the exfoliation of the layered silicate, the interaction between the NR and layered silicate was poor. Certain amount of the layered silicate was located at the boundary between NR and PU forming a "skeleton" structure. This is depicted in Figure 5.7 [53] showing the morphology of NR/PU blends containing 10 phr of layered silica. This "skeleton" structure is believed to enhance the mechanical performance of the vulcanizate.

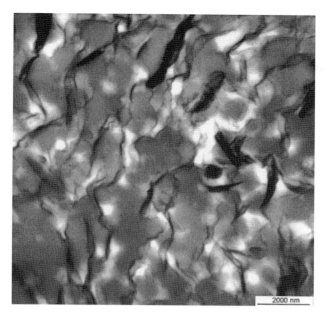

2000 nm

**Figure 5.7** TEM images of casted NR/PU/layered silica 50/50/10 [53], with permission © 2004 Wiley Periodicals, Inc.

To achieve better performance properties of the blends containing NR, many approaches to enhance the compatibility of NR with other elastomers have been developed. In particular, research efforts were directed to the development of molecules that can function as compatibilizers. To some extent better compatibility can be achieved by the utilization of common polymers. In particular, Saad and El-Sabbagh [54] used SBR and PVC to compatibilize NR/EPDM, NR/NBR, and NR/CR blends. SEM images showed an improved morphology after the incorporation of SBR or PVC. The mechanical properties were also improved. El-Sabbagh [55] also studied the compatibilization of NR/EPDM blends using various approaches, including gamma ray irradiation of blends, gamma ray irradiation treated EPDM grafted with maleic anhydride, BR, chlorinated rubber, chlorosulfonated polyethylene, and PVC. SEM analysis showed that in all cases the domain size of compatibilized blends reduced and properties improved compared to NR/EPDM blends.

Bualek *et al.* [56] used an isoprene-butadiene diblock copolymer to compatibilize NR/BR blends. Although a single phase was not observed, the homogeneity of the blends was increased, as indicated by an increased modulus measured by DMA.

To compatibilize the blends of NR and acrylic rubber, poly(isoprene-butyl acrylate) block copolymer was used [57]. It was reported that the compatibilization occurred in the presence of block copolymer. In particular, acrylic rubber was the dispersed phase in the blend, which was the case when the weight

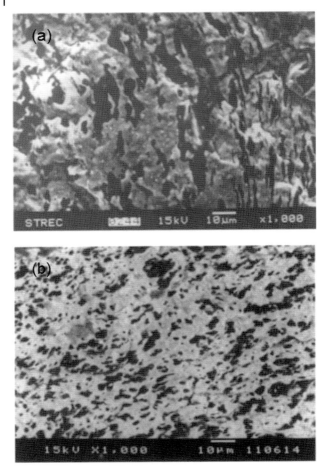

**Figure 5.8** SEM images of NR/acrylic rubber 50/50 blends without (a) and with compatibilizer (b) [57], with permission © 2003 Wiley Periodicals, Inc.

composition of NR was 50% and 80%. The dispersion was also improved leading to an increase of the tensile strength. Figure 5.8 shows the SEM images of NR/acrylic rubber 50/50 blends without (a) and with (b) compatibilizer [57]. These micrographs clearly show improved dispersion after the incorporation of the compatibilizer. At NR content of 20 wt%, a co-continuous morphology was formed. In this case, compatibilization at the interface only occurred using the block copolymer of a lower molecular weight. This difference of compatibilization effect in blends of different compositions was attributed to the diffusivity of block copolymer in different rubbers. The block copolymer of the lower molecular weight can more easily diffuse through acrylic rubber to the interface. The latter also explained why when acrylic rubber was the major component, only block copolymer with the lower molecular weight acted as a compatibilizer.

Besides polymers, fillers can also improve the compatibility and dispersion of rubber blends due to an increase of the shear stresses during mixing or a reduction of the interfacial tension. On the other hand, the dispersion of filler in the matrix rubber could also be enhanced by incorporation of the other rubber in the compound, through increasing the interaction between the matrix rubber and filler. CR can improve NR-silica interaction in silica-filled NR/SBR blends [58]. It was also found that the bound rubber content was increased with the incorporation of CR. In addition, dispersion of silica was improved as observed by the optical microscope. Also, the crack resistance, as well as wear properties, was improved. The compatibility in NR/CR blends was also improved by addition of silica [59]. This study claimed that silica increased the viscosity of blend during mixing, causing a smaller size of the dispersed NR phase. As a result, the oil and aging resistance of the blends were improved in the presence of silica.

Phadke *et al.* [60] studied the blends of NR and cryoground rubber. They found that the poor adhesion between the NR and cryoground rubber can be improved by the addition of CB. Yan *et al.* [46] studied co-compatibilization of NR/BR blends using modified CNTs and liquid isoprene. They found that after the incorporation of modified CNTs and liquid isoprene, the BR domain size was reduced. The chemical modification of CNTs with bis-(triethoxysilylpropyl)-tetrasulfane (a widely used coupling agent for rubbers and fillers with hydroxyl group, such as silica) makes CNTs to act as a bridge between NR and BR. Incorporation of liquid isoprene lowered the viscosity of blends, which further facilitated the interaction of CNTs with rubbers. Arroyo *et al.* [61] prepared compounds of organoclay with NR/ENR blends. They found a better interaction between organoclay and ENR. This was attributed to the polarity of ENR. Morphology study also showed that the most exfoliated organoclay stayed at the NR–ENR interface. In addition, the dispersion of ENR in NR was improved, possibly due to the compatibilization effect of organoclay.

## 5.5.2
### Blends Containing BR

BR used in the rubber industry typically contains units from 1,2-addition, *trans*-1,4-addition and *cis*-1,4-addition. Depending on application, the amount of each type of units in BR varies. BR used in the tire industry typically contains a significant amount of cis content providing high abrasion resistance [22] and low hysteresis [62].

Just like NR, BR with a high cis content can also crystallize. The crystallization of BR in blends is influenced by the other rubber component in the blends. Morris [63] studied the crystallization behavior of BR/SBR and BR/IR blends. It was found that for solution mixed cis-BR/SBR blends, the crystallization rate decreased with the increase of the SBR content in the blend. However, the crystallinity did not change significantly. For blends mixed on mill, the crystallization rate did not change significantly at short milling time. This was attributed to the fact that the particle size of BR was large enough to have significant number of impurities needed for nucleation. However, a further increase of milling time reduced the crystallization rate. Eventually, the crystallization rate reached a plateau because no further change of size of BR domains occurred. However, there was no interaction between BR and IR in BR/IR blends as can be judged from crystallization. Portal *et al.* [16] observed the reduced crystallization temperature of BR in the BR/NR blends when NR was the major phase. This finding was explained as follows: when NR in blends was the major phase, most CB was located in NR. This increased the viscosity of the NR phase, which led to a reduction in the BR domain size during mixing and reduced its crystallization temperature.

In the tire industry, BR is usually blended with other elastomers and mixed with fillers. The distribution of fillers in blends containing BR can be critical to the properties of tires. The distribution of fillers can be affected by the properties of elastomer components such as polarity and viscosity. Sircar and Lamond [64] studied the CB-filled blends of BR with NR, SBR, and CIIR. They found that the transfer of CB between the phase of BR, phase of NR, and phase of SBR did not happen. However, a migration of CB from CIIR to the interface between CIIR and BR in the BR/CIIR blends was observed. In fact, Figure 5.9 shows electron micrographs of BR/CIIR/CB 50/50/50 compounds prepared using different methods as indicated in the figure, which support the authors' observation [64].

Ibarra-Gomez *et al.* [13] studied CB-filled BR/EPDM blends. They found the preferential location of CB in the BR component. It was related to the interfacial energy between CB and each rubber. A lower interfacial energy between CB and

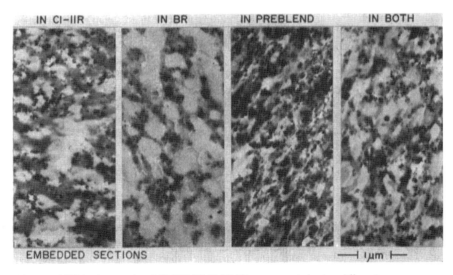

**Figure 5.9** TEM micrographs of BR/CIIR/CB 50/50/50 compounded using different approaches [64], with permission © 1973 Rubber Division, ACS.

BR resulted in more CB located in BR. This was confirmed by bound rubber measurement showing increased bound rubber with increased content of BR.

Sirisinha and Prayoonchatphan [65] studied the distribution of CB and silica in BR/NBR blends. The preferential distribution of CB in the BR phase was observed. This was attributed to the lower viscosity and lower polarity of BR. At the blending ratio of BR/NBR 50/50, silica also preferred to locate in the BR phase.

In contrast to some observations described above, Marsh *et al.* [66] argued that the migration of the filler between different rubbers was impossible. They prepared BR/IR/CB compounds using three different methods. In the first method, CB was mixed with BR and then IR was added to the premixed compound. In the second method, CB was mixed with IR and then mixed with BR. In the third method, CB was mixed with the BR/IR blend. For compounds prepared using any of the three methods, TEM images of CB were indicated by a darker phase. They attributed the presence of the darker phase due to a denser CB population and not separated rubber phases. However, it does not mean that the transfer of CB particles from one rubber phase to the other could not occur.

### 5.5.3
### Blends Containing SBR

SBR is an important synthetic rubber and its mixing with other rubbers is an attractive way to affect the performance of the blend. SBR/BR blends are widely used in tires. Figure 5.10 shows a SEM image taken on a cryo-fractured surface of SBR/BR 50/50 vulcanizate. This image clearly indicates that there is no phase separation between SBR and BR. This confirms that the miscibility between SBR

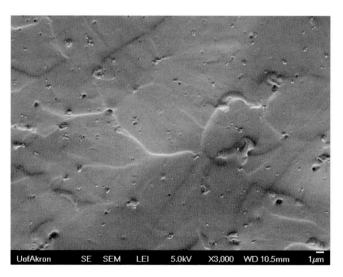

**Figure 5.10** SEM image of cryo-fractured surface of SBR/BR 50/50 vulcanizate. Particles shown in the image are zinc oxide added during compounding.

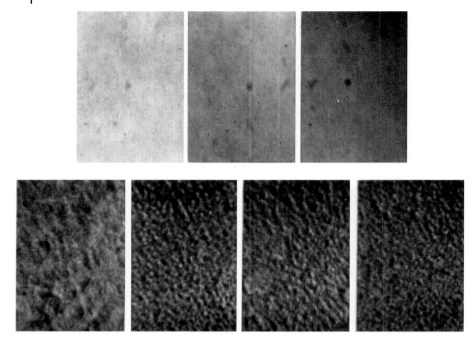

**Figure 5.11** Phase contrast optical microscopy images of SBR/BR 60/40 blends, of different amount of styrene content (from top left to bottom right): 8.6%, 25.5%, 31.4%, 41.4%, 57.3%, 88.0%, 100% [67], with permission © 1968 Rubber Division, ACS.

and BR is quite good. Since SBR is a copolymer of styrene and butadiene, a variety of molecular structure can be introduced in SBR by adjusting styrene and vinyl contents. Therefore, the effect of content of monomer units on the morphology and structure is worthwhile to discuss.

Fujimoto and Yoshinura [67] studied the effect of the styrene content on the microstructure of SBR/BR blends. A noncrystallizable BR was chosen to eliminate the effect of crystallization. They studied the morphology of the blends using PCOM. Various images of these blends are shown in Figure 5.11 [67]. In particular, this figure shows PCOM images obtained from SBR/BR 60/40 blends at the styrene content of 8.6%, 25.5%, 31.4%, 41.4%, 57.3%, 88.0%, and 100%. Clearly, the heterogeneity of the blends increased with the increase of the styrene content. Specifically, at the styrene content of above 40%, an obvious heterogeneity was observed. The authors also measured physical properties of the blends and found that at low styrene content the tensile strength, tear strength, and elongation at break were only slightly affected by the addition of BR. At the same time with the increase of the styrene content these properties increase. However, with addition of BR, the properties decrease. It is interesting to note that no drastic effect on properties was observed when the phase morphology was changed from homogeneous to heterogeneous. However, in the case of SBR/NR blends the miscibility between SBR and NR is not good. Accordingly, phase separation

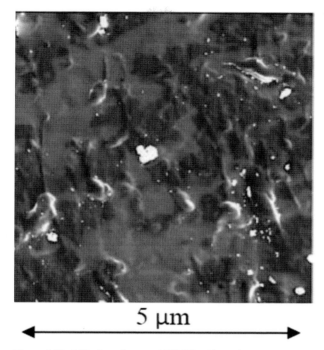

5 μm

**Figure 5.12** AFM phase image of SBR/NR 50/50 vulcanizate [69].

was observed in these blends [68,69]. Figure 5.12 is an AFM phase image taken from the SBR/NR 50/50 vulcanizate [69]. The blend surface for AFM was prepared using a microtome. The phase with a darker color is SBR and the phase with a lighter color is NR. The white particles in the image are zinc oxide particles. This figure clearly displays the phase separation between SBR/NR and confirms the immiscibility of SBR with NR. To enhance the compatibility in the SBR/NR blend, Zhao *et al.* [70] prepared a block copolymer of star styrene-isoprene-butadiene (SIB) rubber consisting blocks of IR and SBR to compatibilize. They compared the structure and properties of this block copolymer with a random copolymer of star SIB rubber and NR/star SBR blends. A better dispersion was observed along with improved wear resistance for the block copolymer.

Lee and Singleton [33] observed that there was no transfer of CB filler from one component to another in SBR/BR blends if CB was only premixed with SBR or BR. However, a preference of CB to SBR was observed when CB was mixed with a preblended SBR/BR blend for a long time. Cotton and Murphy [71] found that CB preferred SBR in SBR/NR blends, possibly due to the higher unsaturation of SBR. This preference was even more obvious with the increase of the mixing time.

Blending of SBR and NBR is beneficial to obtain rubbers exhibiting good mechanical properties, due to the presence of SBR, and the oil resistance, due to the presence of NBR. Due to a huge difference between the polarity of SBR and

NBR, they are not compatible [72]. Therefore, preparation of this blends is not easy. Accordingly, efforts have been made to improve compatibility in SBR/NBR blends using different compatibilizers, including montmorillonite clay [73] and copolymers of NBR grafted with cellulose acetate and NBR grafted with methyl-methacrylate [74]. In both cases compatibilization effect was observed as indicated by SEM images and an improvement of tensile properties.

## 5.5.4
## Blends Containing EPDM

The highly saturated backbone in EPDM provides excellent heat resistance properties and makes EPDM a good choice for blending with other rubbers to reduce their thermal degradation. Deuri *et al.* [75,76] studied the thermal degradation of IIR/EPDM blends. They found that after the incorporation of EPDM, the heat resistance, ozone resistance, and service life of IIR increased, especially under oxygen atmosphere. The best blend was found to be at a concentration of 15 phr of EPDM, as evidenced by good heat resistance, as well as low gas permeability. Botros [77] studied the effect of aging on physical properties of IIR/EPDM blends. The heat resistance of IIR was improved with the addition of EPDM.

However, due to the highly saturated backbone, EPDM phase in rubber blends may often be undercured, causing negative effect on mechanical properties [78]. To solve this problem, Ghosh *et al.* [79] used bis-(diisopropyl) thiophosphoryl disulfide to precure EPDM, followed by mixing with NR and subsequent vulcanization of the blends. This approach compensated the difference in the curing rate of these two rubbers to some extent, leading to blends with improved physical properties. Coran [80] used maleic anhydride to modify EPDM in order to let EPDM to be cured by zinc oxide. In this case, the problem of undercure of EPDM in the EPDM/NR blend was solved. Experimental results showed improved mechanical properties. Sae-Oui *et al.* [81] studied the effect of accelerator on the structure and properties of the NR/EPDM blend. They found that TBBS, MBTS, and MBT can give the NR/EPDM blend good cure compatibility and mechanical properties. However, the addition of TMTD resulted in poor mechanical properties, due to the fact that curing of the EPDM phase was much less compared to that of the NR phase.

Nanoclay was found to have a compatibilization effect on the blends of EPDM/CR [82]. Nanoclay was compounded with EPDM and then the nanoclay-filled EPDM was mixed with CR. The compounds with a different ratio of EPDM/CR were prepared. They found that a significant amount of nanoclay was intercalated. Also, exfoliation took place, except in the case when CR was the minor component. This was related to the smaller polarity difference between CR and nanoclay, as well as the lower viscosity of CR. They also observed the migration of nanoclay toward the CR phase. Moreover, nanoclay was also located at the interface in EPDM/CR blends. The presence of the nanoclay at the interface helped to build up the filler–filler network, which was stronger in the blends containing more CR. Thus, more nanoclay exfoliated and located at the interface of the blends. This was also confirmed by carrying out strain sweep experiments.

## 5.5.5
### Blends Containing Butyl Rubber

Butyl rubber (IIR) is a copolymer of mostly isobutylene and some amount of isoprene. It is well-known for its excellent gas barrier properties due to the presence of two methyl groups in its repeating unit [83]. Mixing IIR with polar rubbers such as NBR [77] can help to obtain blends with good thermal stability and oil resistivity.

Puskas and Kaszas [84] used thermoplastic elastomers to compatibilize IIR/BIIR blends. The properties of these compatibilized blends, such as green strength, die swell, air permeability, tensile properties, and fatigue life, were improved. They stated that interpenetrating chemical and physical networks were created after vulcanization.

Pandey *et al.* [85] used chlorinated polyethylene to compatibilize IIR/CR blends to improve their physical properties. They characterized the blend with and without compatibilizer using various techniques including SEM, ultrasonic velocity measurement, DSC, TGA, and DMA. SEM of the solution mixed blend without compatibilizer showed the presence of dispersed domains, which was absent in a solution-mixed blend with the compatibilizer. Ultrasonic velocity measurement showed that the velocity did not change linearly with the composition of IIR, but showed an S-shape curve. This could be related to the phase inversion and indicated the immiscibility of the blend. However, after incorporation of the compatibilizer, the velocity changed linearly with the change of IIR composition, confirming the improved dispersion.

Suma *et al.* [86] studied the effect of precuring of IIR before blending with NR. They found that this method can balance the curing between IIR and NR and can improve physical properties of the blend. Figure 5.13

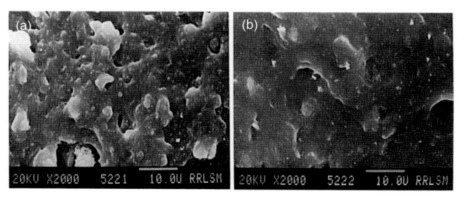

**Figure 5.13** SEM image of fractured surface of vulcanized NR/IIR 50/50 blends without (a) and with (b) IIR precured [86].

shows the SEM images obtained from NR/IIR 50/50 blends without (a) and with (b) precuring of IIR [86]. These SEM micrographs show a more smooth fracture surface for the sample in which IIR was precured before blending.

### 5.5.6
### Blends Containing NBR

NBR is widely used because of its good oil resistivity. However, its mechanical properties are poor. This fact hinders its use in many applications. Blending NBR with other rubbers to improve mechanical properties could help to solve this problem. However, the high polarity of NBR is a big obstacle for application of this approach. Many rubbers with excellent mechanical properties typically have low polarity. The interfacial tension between NBR and other rubbers exhibiting good mechanical properties is typically high. This causes large domain size in the blends.

A good solution to overcome this difficulty is the incorporation of compatibilizers. Various compatibilizers have been developed for different rubber blends containing NBR, including NBR/NR [87], NBR/SBR [88], and NBR/EPDM [89]. Basu and coworkers [90–92] extensively studied the compatibilization effect of thiophosphoryl disulfide for blends of NBR and various rubbers including NR, SBR, and EPDM. Thiophosphoryl disulfide in these blends and blends of NBR with various kinds of other rubbers functioned as a coupling agent. With the aid of this coupling agent, improved morphology in the blends was observed as indicated by SEM observation. Also, better physical properties of these blends were achieved.

The compatibilization effect of CB also works for rubber blends containing NBR [93]. In this research, the authors prepared NBR/NR/CB compounds and found that the domain size of the dispersed phase decreased by the incorporation of CB when CB stayed at the interface. However, it was also found that CB had a tendency to locate in the NBR phase due to its lower viscosity. The latter should be avoided such that more CB can participate in compatibilization at the interface.

Tiwari *et al.* [94] compared NBR, EPDM, and NBR/EPDM blends filled with unmodified silica, silane-modified silica, and plasma-treated silica. The filled blends showed worse tensile properties compared to the filled NBR or filled EPDM, regardless of the choice of the silica filler. This indicated the poor compatibility between NBR and EPDM due to huge difference in their polarity. They stated that the plasma treatment of silica may induce in situ polymerization and grow polyacetylene at the silica surface to reduce the interfacial tension between silica and rubbers. They observed a reduced Payne effect for blends filled with the plasma-treated silica compared to the untreated one. In addition, higher bound rubber content and better mechanical properties were achieved for the plasma-treated silica rubbers than untreated and silane-modified silica rubbers.

5.5.7
**Blends Containing CR**

CR is widely applied in the adhesion industry due to its polarity and crystallinity [95]. Moreover, CR exhibits good mechanical properties due to its ability to crystallize. Therefore, significant efforts have been made to mix CR with other rubbers to develop new products to fully utilize its properties.

The big difference in the polarity of BR and CR results in obvious phase separation in CR/BR blends. The phase separation was observed using phase contrast optical microscope by Zheng *et al.* [96]. This is seen from Figure 5.14 showing the PCOM images of CR/BR blends at blending ratios of 75/25 (a) and 50/50 (b) [96]. As seen, the co-continuous phase in the CR/BR blend at a ratio of 75/25 is observed. However, CR became the dispersed phase in the CR/BR blend at a ratio of 50/50.

Some polymers can be used as a compatibilizer for elastomer blends containing CR, and the choice of the compatibilizer depends on the other elastomer component. Younan *et al.* [97] blended CR with EPDM. They found that the CR/EPDM blend at a ratio of 50/50 had the best mechanical and electrical properties. They also studied the variation of dielectric properties of the blends with respect to the composition. They found that the permittivity of blends did not change linearly with the composition, indicating the incompatibility of the blends. However, after the addition of PVC, the permittivity changed linearly with the composition, showing that PVC is a good compatibilizer for CR/EPDM blends. Ramesan *et al.* [98] modified the double bonds present in SBR by attaching chlorine to obtain chlorinated SBR (CSBR) as a compatibilizer for the preparation of CR/SBR blends. The compatibilizer provided means for the blends to achieve a certain level of molecular homogeneity and improved mechanical properties. They studied the effect of the chlorine content and the amount of the modified SBR and found the optimum recipe. They also prepared NR/CSBR

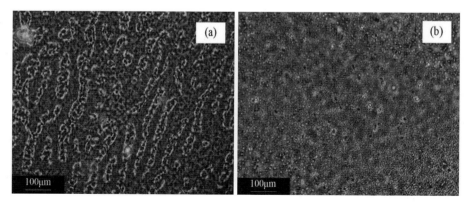

**Figure 5.14** PCOM images of CR/BR blends at a blending ratio of 75/25 (a) and 50/50 (b) [86], with permission © 2014 Rubber Division, ACS.

blends and NR/CR blends and compared their properties [99]. Interestingly, NR/CSBR blends showed better mechanical properties, and ozone and oil resistance than NR/CR blends. The improved mechanical properties were attributed to the compatibility of NR and SBR segments in the presence of CSBR. In addition, the better ozone and oil resistance was attributed to the reduced amount of double bonds and the presence of two chlorine atoms in the SBR segment.

Sae-Oui *et al.* [100] studied NR/CR blends of different blending ratios filled with silica. Their study showed that the good interaction between the CR and silica facilitated the dispersion of silica. In addition, it was found that when CR was the continuous phase in the blends, better mechanical properties, aging, and oil resistivity were achieved. In earlier study, Marsh *et al.* [18] prepared the NR/CB compound. Then this compound was mixed with CR. It was found that CB migrated to the interface of the CR/NR blend, showing a preference of CB to CR.

### 5.5.8
### Blends Containing Silicone Rubber

Silicone rubber exhibits high temperature stability. It also has good resistance to chemicals. However, it is not easy to achieve a good microhomogeneity in blends containing silicone rubber. Ghosh and De [101] studied the blends of silicone rubber and fluororubber. Their SEM study showed the presence of the microheterogeneity in the blends. In the blends, due to its low viscosity, silicone rubber was the continuous phase and fluororubber was the dispersed phase. However, in spite of the microheterogeneity, the tensile testing results indicated that silicone rubber and fluororubber were technologically compatible and suitable to use in applications.

To compatibilize the blends containing silicone rubber, Kole *et al.* [8] used various chemicals for the in situ compatibilization of silicone rubber/EPDM blends. In their study, SEM was used to investigate the morphology of the blends. Figure 5.15 shows the SEM images of etched silicone/EPDM 50/50 blends before vulcanization (a) and compatibilized with EVA and after vulcanization (b) [8]. As

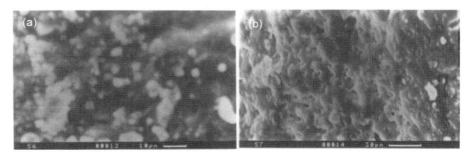

**Figure 5.15** SEM image of etched silicone/EPDM 50/50 blend before vulcanization (a) and compatibilized with 2.5 phr of EVA and after vulcanization (b) [8], with permission © 1994 Rubber Division, ACS.

seen from this figure, the domain size in the compatibilized blend was significantly reduced.

### 5.5.9
### Blends Containing Hydrogenated Nitrile Butadiene Rubber (HNBR)

HNBR has excellent heat stability and oil resistivity. Therefore, the incorporation of HNBR in other rubbers can help to improve their heat stability under extreme conditions of service including under high temperature or immersed in oil.

Elizabeth *et al.* [102] blended NR with HNBR to combine the good mechanical properties and low temperature flexibility of NR and the excellent high temperature stability and oil resistance of HNBR. They introduced chlorine atom into NR molecules to increase its polarity. The results showed that better compatibility was achieved for the blends of chlorinated NR and HNBR, compared to NR/HNBR. Hirano *et al.* [103] blended HNBR with fluorocarbon elastomer at various blending ratios and studied the structure and properties of the blends. Figure 5.16 shows TEM images of blends of fluorocarbon elastomer with HNBR at blending ratios of 70/30, 50/50, and 30/70 [103]. It is seen that at a blending

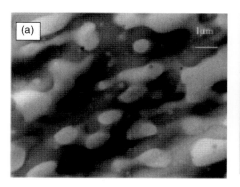
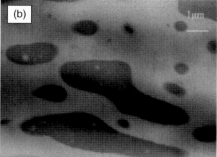
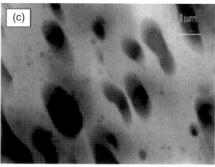

**Figure 5.16** TEM images of fluorocarbon elastomer/HNBR blends at blending ratios of 70/30 (a), 50/50 (b), and 30/70 (c) [103], with permission © 2004 Wiley Periodicals, Inc.

ratio of 70/30 fluorocarbon elastomer in fluorocarbon elastomer/HNBR blend formed the matrix and HNBR was the dispersed phase. At blending ratios of 50/50 and 30/70, the HNBR was the matrix and the fluorocarbon elastomer was the dispersed phase. The authors also observed no cavities or crazes in the stretched blends. It was suggested that such a behavior resulted from the strong interaction between fluorocarbon elastomer and HNBR. To enhance the compatibilization between the fluorocarbon elastomer and HNBR, Yeo *et al.* [104] used a silane compatibilizer (KBM503) and observed improvement in physical properties in comparison with blends without a compatibilizer.

## 5.6
## Conclusions

The structure and morphology of rubber blends have been reviewed. Various techniques for the characterization of rubber blends have been introduced. PCOM can be utilized for blends having the microscale morphology. TEM can provide information at the nanoscale level. Special care needs to be taken to prepare the specimen for these morphological studies in order to obtain good contrast. Sample preparation for SEM is relatively easier, which makes it a good choice to characterize blends with a poor miscibility between the rubber components. AFM can be used to distinguish between different rubbers in a blend, even in cases of good miscibility. However, in this case it is an important requirement to have a difference in the modulus of components. Also, sample preparation procedure for the AFM study is not difficult, compared to that for the TEM study. However, a special care needs to be taken to adjust parameters for operating AFM to get good quality images. Dynamic rheological testing and thermal analysis can also be used to examine the compatibility of blends, as well as the distribution of fillers in each rubber phase. With these powerful techniques, morphology of rubber blends can be obtained to explain how it is related to the properties or processing conditions.

The solubility parameter and critical surface tension of the rubbers are important parameters in the determination of the blend morphology. In addition, the blend morphology is strongly affected by the composition of blend, viscosity ratio, mixing temperature, and mixing time. The distribution of fillers depends on the viscosity ratio of rubbers and the interaction between rubbers and fillers. To control the distribution of fillers, the compounding sequence can be adjusted to prepare the masterbatch. When the saturation of rubber components (amount of double bonds) in a blend is very different, distribution of curatives as well as curing rate balance should be considered in preparing vulcanizates. When dealing with specific rubber blends, the characteristics of each rubber component in the blends, such as crystallization behavior, curing state, and preference of filler distribution should be considered, since all these factors influence the blend morphology.

## References

1 Baranwal, K.C. and Son, P.N. (1974) Co-curing of EPDM and diene rubbers by grafting accelerators onto EPDM. *Rubber Chem. Technol.*, **47** (1), 88–99.

2 Hopper, R.J. (1976) Improved cocure of EPDM-polydiene blends by conversion of EPDM into macromolecular cure retarder. *Rubber Chem. Technol.*, **49** (2), 341–352.

3 Baldwin, F.P. and Strate, G.V. (1972) Polyolefin elastomers based on ethylene and propylene. *Rubber Chem. Technol.*, **45** (3), 709–881.

4 Walters, M.H. and Keyte, D.N. (1965) Heterogeneous structure in blends of rubber polymers. *Rubber Chem. Technol.*, **38** (1), 62–75.

5 Kruse, J. (1973) Rubber microscopy. *Rubber Chem. Technol.*, **46** (3), 653–785.

6 Hess, W.M., Scott, C.E., and Callan, J.E. (1967) Carbon black distribution in elastomer blends. *Rubber Chem. Technol.*, **40** (2), 371–384.

7 Setua, D.K. and White, J.L. (1991) Flow visualization of the influence of compatibilizing agents on the mixing of elastomer blends and the effect on phase morphology. *Polym. Eng. Sci.*, **31** (24), 1742–1754.

8 Kole, S., Santra, R., and Bhowmick, A.K. (1994) Studies of in-situ compatibilized blend of silicone and EPDM rubbers. *Rubber Chem. Technol.*, **67** (1), 119–128.

9 Johnson, L.L. (2008) Atomic force microscopy (AFM) for rubber. *Rubber Chem. Technol.*, **81** (3), 359–383.

10 Yerina, N. and Magonov, S. (2003) Atomic force microscopy in analysis of rubber materials. *Rubber Chem. Technol.*, **76** (4), 846–859.

11 Wang, C.C., Donnet, J.B., Wang, T.K., Pontier-Johnson, M., and Welsh, F. (2005) AFM study of rubber compounds. *Rubber Chem. Technol.*, **78** (1), 17–27.

12 Kluppel, M., Schuster, R.H., and Schaper, J. (1999) Carbon black distribution in rubber blends: A dynamic-mechanical analysis. *Rubber Chem. Technol.*, **72** (1), 91–108.

13 Ibarra-Gomez, R., Marquez, A., Valle, L., and Rodriguez-Fernandez, O.S. (2003) Influence of the blend viscosity and interface energies on the preferential location of CB and conductivity of BR/EPDM blends. *Rubber Chem. Technol.*, **76** (4), 969–978.

14 Fouche, P.M. and McGill, W.J. (1989) A study of the transfer of carbon-black between the phases of polyisoprene polybutadiene blends by differential scanning calorimetry and electron-microscopy. *Plast. Rubber Proc. Appl.*, **12** (4), 227–234.

15 Massie, J.M., Hirst, R.C., and Halasa, A.F. (1993) Carbon-black distribution in NR/polybutadiene blends. *Rubber Chem. Technol.*, **66** (2), 276–294.

16 Portal, J., Carrot, C., Majeste, J.C., Cocard, S., Pelissier, V., and Anselme-Bertrand, I. (2009) Quantification of the distribution of carbon black in natural rubber/polybutadiene blends by differential scanning calorimetry. *Polym. Eng. Sci.*, **49** (8), 1544–1552.

17 Le, H.H., Ilisch, S., Kasaliwal, G.R., and Radusch, H.J. (2008) Filler phase distribution in rubber blends characterized by thermogravimetric analysis of the rubber-filler gel. *Rubber Chem. Technol.*, **81** (5), 767–781.

18 Marsh, P.A., Voet, A., Price, L.D., and Mullens, T.J. (1968) Fundamentals of electron microscopy of heterogeneous elastomer blends. II. *Rubber Chem. Technol.*, **41** (2), 344–355.

19 Gardiner, J.B. (1970) Studies in the morphology and vulcanization of gum rubber blends. *Rubber Chem. Technol.*, **43** (2), 370–399.

20 Flory, P.J. (1942) Thermodynamics of high polymer solutions. *J. Chem. Phys.*, **10** (1), 51–61.

21 Huggins, M.L. (1942) Theory of solutions of high polymers. *J. Am. Chem. Soc.*, **64** (7), 1712–1719.

22 Corish, P.J. (1967) Fundamental studies of rubber blends. *Rubber Chem. Technol.*, **40** (2), 324–340.

23 Sheehan, C.J. and Bisio, A.L. (1966) Polymer/solvent interaction parameters. *Rubber Chem. Technol.*, **39** (1), 149–192.

24 Lee, L.H. (1966) ACS Polymer Preprints, September 1966, N. Y. Meeting, 7(2): p. 916.

25 Schuster, R.H., Issel, H.M., and Peterseim, V. (1996) Selective interactions in elastomers, a base for compatibility and polymer-filler interactions. *Rubber Chem. Technol.*, **69** (5), 769–780.

26 Tokita, N. (1977) Analysis of morphology formation in elastomer blends. *Rubber Chem. Technol.*, **50** (2), 292–300.

27 Hess, W.M. and Chirico, V.E. (1977) Elastomer blend properties—influence of carbon-black type and location. *Rubber Chem. Technol.*, **50** (2), 301–326.

28 Anastasiadis, S.H., Gancarz, I., and Koberstein, J.T. (1988) Interfacial-tension of immiscible polymer blends—temperature and molecular-weight dependence. *Macromolecules*, **21** (10), 2980–2987.

29 Hamed, G.R. (1982) Morphology-property relationships in EPDM-polybutadiene blends. *Rubber Chem. Technol.*, **55** (1), 151–160.

30 Fujimoto, K. and Yoshimiya, N. (1968) Blends of cis-1,4-polybutadiene with natural or styrene butadiene rubber. *Rubber Chem. Technol.*, **41** (3), 669–677.

31 Maiti, S., De, S.K., and Bhowmick, A.K. (1992) Quantitative estimation of filler distribution in immiscible rubber blends by mechanical damping studies. *Rubber Chem. Technol.*, **65** (2), 293–302.

32 Le, H.H., Ilisch, S., Heidenreich, D., Osswald, K., and Radusch, H.J. (2011) Phase selective localization of filler in ternary rubber blends. *Rubber Chem. Technol.*, **84** (1), 41–54.

33 Lee, B.L. and Singleton, C. (1979) Experimental study of the relationship of processing to the morphology in blends of SBR and cis-polybutadiene with carbon-black. *J. Appl. Polym. Sci.*, **24** (10), 2169–2183.

34 Gardiner, J.B. (1968) Curative diffusion between dissimilar elastomers and its influence on adhesion. *Rubber Chem. Technol.*, **41** (5), 1312–1328.

35 Guillaumond, F.X. (1976) The influence of the solubility of accelerators on the vulcanization of elastomer blends. *Rubber Chem. Technol.*, **49** (1), 105–111.

36 Gardiner, J.B. (1969) Measurement of curative diffusion between rubbers by microinterferometry. *Rubber Chem. Technol.*, **42** (4), 1058–1078.

37 Inoue, T., Shomura, F., Ougizawa, T., and Miyasaka, K. (1985) Covulcanization of polymer blends. *Rubber Chem. Technol.*, **58** (5), 873–884.

38 Lee, B.L. (1984) Multiphase polymer processing—controlled-ingredient-distribution mixing and its effect on the covulcanization of elastomer blend compounds. *J. Rheol.*, **28** (4), 307–324.

39 Bauer, R.F. and Dudley, E.A. (1977) Compatibilization of rubber blends through phase interaction. *Rubber Chem. Technol.*, **50** (1), 35–42.

40 Clarke, J., Clarke, B., and Freakley, P.K. (2001) Relationships between mixing method, microstructure and strength of NR:BR blends. *Rubber Chem. Technol.*, **74** (1), 1–15.

41 Yoshimura, N. and Fujimoto, K. (1969) Structure of vulcanized and unvulcanized SBR/BR blends. *Rubber Chem. Technol.*, **42** (4), 1009–1013.

42 Woods, M.E. and Davidson, J.A. (1976) Fundamental considerations for the covulcanization of elastomer blends: II. Lead oxide-activated cures of NBR–EPDM blends. *Rubber Chem. Technol.*, **49** (1), 112–117.

43 Dutta, N.K. and Tripathy, D.K. (1993) Miscibility studies in blends of bromobutyl rubber and natural rubber. *J. Elastomers Plast.*, **25** (2), 158–179.

44 Shi, X.Y., Zhu, Z., Jia, L.Y., Li, Q., and Bi, W.N. (2013) HNBR/EPDM blends: Covulcanization and compatibility. *J. Appl. Polym. Sci.*, **129** (5), 3054–3060.

45 Sirisinha, C., Baulek-Limcharoen, S., and Thunyarittikorn, J. (2001) Relationships among blending conditions, size of dispersed phase, and oil resistance in natural rubber and nitrile rubber blends. *J. Appl. Polym. Sci.*, **82** (5), 1232–1237.

46 Yan, N., Xia, H.S., Zhan, Y.H., Fei, G.X., and Chen, C. (2012) Co-compatibilising effect of carbon nanotubes and liquid isoprene rubber on carbon black filled

natural rubber/polybutadiene rubber blend. *Plast. Rubber Compos.*, **41** (9), 365–372.

47 Gent, A.N. and Kawahara, S. (1998) Crystallization of cis- and trans-1,4-polyisoprene dispersed in SBR. *Rubber Chem. Technol.*, **71** (5), 837–845.

48 Kawazura, T., Kawahara, S., and Isono, Y. (2003) Morphology and crystallization behavior of lightly crosslinked natural rubber in blend. *Rubber Chem. Technol.*, **76** (5), 1164–1176.

49 Qu, L.L., Huang, G.S., Nie, Y.J., Wu, J.R., Weng, G.S., and Zhang, P. (2011) Strain-induced crystallization behavior of natural rubber and trans-1,4-polyisoprene cross linked blends. *J. Appl. Polym. Sci.*, **120** (3), 1346–1354.

50 Manzur, A. and Rubio, L. (1997) Strain-induced crystallization in cis- and trans-polyisoprene blends: Effect of molecular weight of trans-PI. *J. Macromol. Sci. Phys.*, **B36** (1), 103–115.

51 Bandyopadhyay, A., Thakur, V., Pradhan, S., and Bhowmick, A.K. (2010) Nanoclay distribution and its influence on the mechanical properties of rubber blends. *J. Appl. Polym. Sci.*, **115** (2), 1237–1246.

52 Zhang, P., Huang, G.S., and Liu, Z.Y. (2009) An effect of OMMT on the anti-reversion in NR/CR blend system. *J. Appl. Polym. Sci.*, **111** (2), 673–679.

53 Varghese, S., Gatos, K.G., Apostolov, A.A., and Karger-Kocsis, J. (2004) Morphology and mechanical properties of layered silicate reinforced natural and polyurethane rubber blends produced by latex compounding. *J. Appl. Polym. Sci.*, **92** (1), 543–551.

54 Saad, A.L.G. and El-Sabbagh, S. (2001) Compatibility studies on some polymer blend systems by electrical and mechanical techniques. *J. Appl. Polym. Sci.*, **79** (1), 60–71.

55 El-Sabbagh, S.H. (2003) Compatibility study of natural rubber and ethylene-propylene-diene rubber blends. *J. Appl. Polym. Sci.*, **90** (1), 1–11.

56 Bualek, S., Ikeda, Y., Kohjiya, S., Phaovibul, O., Phinyocheep, P., Suchiva, K., Utani, K., and Yamashita, S. (1993) Natural-rubber and butadiene rubber blend using diblock copolymer of isoprene butadiene as compatibilizer. *J. Appl. Polym. Sci.*, **49** (5), 807–814.

57 Wootthikanokkhan, J. and Tongrubbai, B. (2003) Compatibilization efficacy of poly (isoprene-butyl acrylate) block copolymers in natural/acrylic rubber blends. *J. Appl. Polym. Sci.*, **88** (4), 921–927.

58 Choi, S.S. (2002) Improvement of properties of silica-filled natural rubber compounds using polychloroprene. *J. Appl. Polym. Sci.*, **83** (12), 2609–2616.

59 Sae-Oui, P., Sirisinha, C., Wantana, T., and Hatthapanit, K. (2007) Influence of silica loading on the mechanical properties and resistance to oil and thermal aging of CR/NR blends. *J. Appl. Polym. Sci.*, **104** (5), 3478–3483.

60 Phadke, A.A., Chakraborty, S.K., and De, S.K. (1984) Cryoground rubber-natural rubber blends. *Rubber Chem. Technol.*, **57** (1), 19–33.

61 Arroyo, M., Lopez-Manchado, M.A., Valentin, J.L., and Carretero, J. (2007) Morphology/behaviour relationship of nanocomposites based on natural rubber/epoxidized natural rubber blends. *Compos. Sci. Technol.*, **67** (7–8), 1330–1339.

62 Isayev, A.I. (ed.) (2011) *Encyclopedia of Polymer Blends: Processing*, vol. **2**, John Wiley & Sons Inc., Hoboken, NJ.

63 Morris, M.C. (1966) Rates of crystallization of cis-1 4-polybutadiene in elastomer blends. *Rubber Age*, **98** (7), 66.

64 Sircar, A.K. and Lamond, T.G. (1972) Carbon-black transfer in blends of cis-poly (butadiene) with other elastomers. *Rubber Chem. Technol.*, **46** (1), 178–191.

65 Sirisinha, C. and Prayoonchatphan, N. (2001) Study of carbon black distribution in BR/NBR blends based on damping properties: Influences of carbon black particle size, filler, and rubber polarity. *J. Appl. Polym. Sci.*, **81** (13), 3198–3203.

66 Marsh, P.A., Voet, A., and Price, L.D. (1967) Electron microscopy of heterogeneous elastomer blends. *Rubber Chem. Technol.*, **40** (2), 359–370.

67 Fujimoto, K. and Yoshimura, N. (1968) Study of polymer blends. II. Blending SBR varying in styrene content with

polybutadiene. *Rubber Chem. Technol.*, **41** (5), 1109–1121.

68 George, S.C., Ninan, K.N., Groeninickx, G., and Thomas, S. (2000) Styrene-butadiene rubber/natural rubber blends: Morphology, transport behavior, and dynamic mechanical and mechanical properties. *J. Appl. Polym. Sci.*, **78** (6), 1280–1303.

69 Choi, J. (2013) *Ultrasonically Aided Extrusion of Rubber Nanocomposites and Rubber Blends*, University of Akron, Akron, OH.

70 Zhao, S.H., Zou, H., and Zhang, X.Y. (2004) Structural morphology and properties of star styrene-isoprene-butadiene rubber and natural rubber/star styrene-butadiene rubber blends. *J. Appl. Polym. Sci.*, **93** (1), 336–341.

71 Cotten, G.R. and Murphy, L.J. (1988) Mixing of carbon-black with rubber. VI. Analysis of NR/SBR blends. *Rubber Chem. Technol.*, **61** (4), 609–618.

72 Mansour, A.A., Elsabagh, S., and Yehia, A.A. (1994) Dielectric investigation of SBR–NBR and CR–NBR blends. *J. Elastomers Plast.*, **26** (4), 367–378.

73 Essawy, H. and El-Nashar, D. (2004) The use of montmorillonite as a reinforcing and compatibilizing filler for NBR/SBR rubber blend. *Polym. Test.*, **23** (7), 803–807.

74 Khalf, A.I., Nashar, D.E.E., and Maziad, N.A. (2010) Effect of grafting cellulose acetate and methylmethacrylate as compatibilizer onto NBR/SBR blends. *Mater. Design*, **31** (5), 2592–2598.

75 Deuri, A.S., Adhikary, A., and Mukhopadhyay, R. (1992) Degradation of IIR/EPDM blends: Part I. *Polym. Degrad. Stab.*, **38** (2), 173–177.

76 Deuri, A.S., Adhikary, A., and Mukhopadhyay, R. (1993) Degradation of IIR/EPDM blends: Part II. *Polym. Degrad. Stab.*, **41** (1), 53–57.

77 Botros, S.H. (1998) Thermal stability of butyl/EPDM rubber blend vulcanizates. *Polym. Degrad. Stab.*, **62** (3), 471–477.

78 Mastromatteo, R.P., Mitchell, J.M., and Brett, T.J. (1971) New accelerators for blends of EPDM. *Rubber Chem. Technol.*, **44** (4), 1065–1079.

79 Ghosh, A.K., Debnath, S.C., Naskar, N., and Basu, D.K. (2001) NR-EPDM covulcanization: A novel approach. *J. Appl. Polym. Sci.*, **81** (4), 800–808.

80 Coran, A.Y. (1988) Blends of dissimilar rubbers—cure-rate incompatibility. *Rubber Chem. Technol.*, **61** (2), 281–292.

81 Sae-Oui, P., Sirisinha, C., Thepsuwan, U., and Thapthong, P. (2007) Influence of accelerator type on properties of NR/EPDM blends. *Polym. Test.*, **26** (8), 1062–1067.

82 Das, A., Mahaling, R.N., Stockelhuber, K.W., and Heinrich, G. (2011) Reinforcement and migration of nanoclay in polychloroprene/ethylene-propylene-diene-monomer rubber blends. *Compos. Sci. Technol.*, **71** (3), 276–281.

83 van Amerongen, G.J. (1964) Diffusion in elastomers. *Rubber Chem. Technol.*, **37** (5), 1065–1152.

84 Puskas, J.E. and Kaszas, G. (2001) Blends of butyl and bromobutyl rubbers and polystyrene-polyisobutylene-polystyrene (PS-PIB-PS) block copolymers with improved processability and physical properties. *Rubber Chem. Technol.*, **74** (4), 583–600.

85 Pandey, K.N., Debnath, K.K., Rajagopalan, P.T., Setua, D.K., and Mathur, G.N. (1997) Thermal analysis on influence of compatibilizing agents—Effect of vulcanization of incompatible elastomer blend. *J. Therm. Anal.*, **49** (1), 281–292.

86 Suma, N., Joseph, R., and George, K.E. (1993) Improved mechanical-properties of NR/EPDM and NR/butyl blends by precuring EPDM and butyl. *J. Appl. Polym. Sci.*, **49** (3), 549–557.

87 Mounir, A., Darwish, N.A., and Shehata, A. (2004) Effect of maleic anhydride and liquid natural rubber as compatibilizers on the mechanical properties and impact resistance of the NR–NBR blend. *Polym. Adv. Technol.*, **15** (4), 209–213.

88 Botros, S.H., Moustafa, A.F., and Ibrahim, S.A. (2006) Improvement of the homogeneity of SBR/NBR blends using polyglycidylmethacrylate-g-butadiene rubber. *J. Appl. Polym. Sci.*, **99** (4), 1559–1567.

89 Botros, S.H. and Tawfic, M.L. (2006) Synthesis and characteristics of MAH-g-EPDM compatibilized EPDM/NBR

rubber blends. *J. Elastomers Plast.*, **38** (4), 349–365.

90 Naskar, N., Biswas, T., and Basu, D.K. (1994) Polymer blend—a novel method for the preparation of a natural-rubber carboxylated nitrile rubber blend. *J. Appl. Polym. Sci.*, **52** (8), 1007–1014.

91 Biswas, T., Naskar, N., and Basu, D.K. (1995) Thiophosphoryl disulfides—novel coupling agents for styrene-butadiene rubber carboxylated nitrile rubber blends. *J. Appl. Polym. Sci.*, **58** (6), 981–993.

92 Naskar, N., Debnath, S.C., and Basu, D.K. (2002) Effect of bis (diisopropyl) thiophosphoryl disulfide on the co-vulcanization of carboxylic-acrylonitrile-butadiene rubber and ethylene-propylene-diene rubber blends. *Rubber Chem. Technol.*, **75** (2), 309–322.

93 Clarke, J., Clarke, B., Freakley, P.K., and Sutherland, I. (2001) Compatibilising effect of carbon black on morphology of NR–NBR blends. *Plast. Rubber Compos.*, **30** (1), 39–44.

94 Tiwari, M., Noordermeer, J.W.M., Dierkes, W.K., and van Ooij, W.J. (2008) Effect of plasma polymerization on the performance of silica in NBR, EPDM and NBR/EPDM blends. *Rubber Chem. Technol.*, **81** (2), 276–296.

95 Kardan, M. (2001) Adhesive and cohesive strength in polyisoprene/polychloroprene blends. *Rubber Chem. Technol.*, **74** (4), 614–621.

96 Zheng, J., Tan, J., Gao, H., Wang, C., and Dong, Z. (2014) Preparation of low temperature resistant and high electrical insulation chloroprene rubber-butadiene rubber blends. *Rubber Chem. Technol.*, **87** (2), 360–369.

97 Younan, A.F., Abd-El-Messieh, S.L., and Gasser, A.A. (1998) Electrical and mechanical properties of ethylene propylene diene monomer-chloroprene rubber blend loaded with white and black fillers. *J. Appl. Polym. Sci.*, **70** (10), 2061–2068.

98 Ramesan, M.T., Mathew, G., Kuriakose, B., and Alex, R. (2001) Role of dichlorocarbene modified styrene butadiene rubber in compatibilisation of styrene butadiene rubber and chloroprene rubber blends. *Eur. Polym. J.*, **37** (4), 719–728.

99 Ramesan, M.T., Alex, R., and Khanh, N.V. (2005) Studies on the cure and mechanical properties of blends of natural rubber with dichlorocarbene modified styrene-butadiene rubber and chloroprene rubber. *React. Funct. Polym.*, **62** (1), 41–50.

100 Sae-oui, P., Sirisinha, C., and Hatthapanit, K. (2007) Effect of blend ratio on aging, oil and ozone resistance of silica-filled chloroprene rubber/natural rubber (CR/NR) blends. *Express Polym. Lett.*, **1** (1), 8–14.

101 Ghosh, A. and De, S.K. (2004) Dependence of physical properties and processing behavior of blends of silicone rubber and fluororubber on blend morphology. *Rubber Chem. Technol.*, **77** (5), 856–872.

102 Elizabeth, K.I., Alex, R., and Varghese, S. (2008) Evaluation of blends of natural rubber and hydrogenated nitrile rubber containing chemically modified natural rubber. *Plast. Rubber Compos.*, **37** (8), 359–366.

103 Hirano, K., Suzuki, K., Nakano, K., and Tosaka, M. (2005) Phase separation structure in the polymer blend of fluorocarbon elastomer and hydrogenated nitrile rubber. *J. Appl. Polym. Sci.*, **95** (1), 149–156.

104 Yeo, Y.G., Park, H.H., and Lee, C.S. (2013) A study on the characteristics of a rubber blend of fluorocarbon rubber and hydrogenated nitrile rubber. *J. Ind. Eng. Chem.*, **19** (5), 1540–1548.

# 6

# Phase Morphology and Properties of Ternary Polymer Blends

*V.N. Kuleznev and Yu. P. Miroshnikov*

*Lomonosov State University of Fine Chemical Technology, Prospect Vernadskogo 86, Moscow 119571, Russia*

## 6.1
## Introduction

The creation of polymer blends is a quick and cost-effective way of making new materials. The main direction here is the production of the binary mixtures of polymers. The theoretical fundamentals of polymer blends have generated an opinion that the lack of miscibility of polymers is the cause for unsatisfactory complex properties of the resulting blend. Yet the practice of mixing has led investigators to look at immiscible polymers with the introduction of coupling agents – compatibilizers. These additives increase the interaction between mixed polymers at the interphase boundary leading to good dispersion of one polymer in another polymer matrix, as well as to increase the stability of the blends, its mechanical homogeneity and, consequently, to improve the complex of properties. Hence, polymer compositions containing small addition of compatibilizers were generated.

At the same time an interesting problem of increasing the number of basic components in a mixture of polymers is considered. The transition from one polymer to a binary polymer mixture provides a possibility of creating a new complex of properties. Futhermore, the transition from binary to ternary and muliphase mixtures extends the capabilities to develop a new combination of properties. In this section, the binary mixtures containing third component, compatibilizer, are not considered. Attention is given to mixtures of only three "equal rights" polymer components.

Today, the use of ternary blends is not wide spread. The book by L .Utracki *Commercial Polymer Blends* [1] contains a list of 850 commercially available polymer blends with only 11 of them being ternary polymer blends. The latter is due to the lack of a theory that is capable at least roughly to predict the properties of binary blends. It is clear that these difficulties are further increased in case

*Encyclopedia of Polymer Blends: Volume 3: Structure*, First Edition. Edited by Avraam I. Isayev.
© 2016 Wiley-VCH Verlag GmbH & Co. KGaA. Published 2016 by Wiley-VCH Verlag GmbH & Co. KGaA.

of the ternary mixture due to an increase of the number of possible combinations of the three components.

## 6.2
### Miscibility of Polymers in Ternary Polymer Blends

During the mixing of two polymers, the change of combinatorial entropy is very small. With the increase of number components, the entropy of mixing increases (Figure 6.1) [2]. Especially, there is noticeable growth when the number of components increases from 2 to 4. This suggests that the polymer miscibility in ternary mixtures may more frequently occur than in the binary mixtures.

Considering the role of the third component in the mixing of two liquids I. Prigogine showed that in the case of regular solutions "introduction of a third component, which is equally soluble in the first two components, lowers the critical solution temperature, that is, increases their mutual solubility" or "addition of a third component which is much less soluble in one of the first two components, than in another, always increases the critical solution temperature, that is, decreases mutual solubility" [3]. The extent to which the Prigogine rules can be applied to polymer systems can be seen in the examples below. For this study, an equilibrium mixture is required where one can experimentally reliably determine the position of binodal and the spinodal in the phase diagram. To do this, it is convenient to take three solutions of polymers in a common solvent. At low

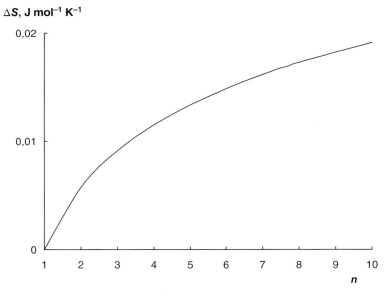

**Figure 6.1** Ideal entropy of mixing as a function of the number of components in the blend at a degree of polymerization of $N = 1000$.

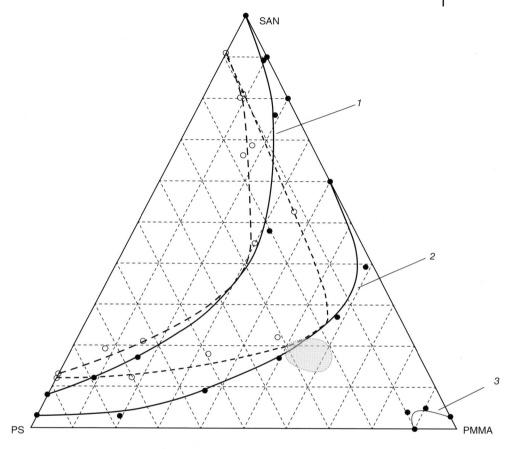

**Figure 6.2** The phase diagram of PS–PMMA–SAN chloroform solutions. The total concentration of the solution of polymer mixtures: (1)7, (2) 8.5 and (3) 10 vol.%. Solid and dashed curves are calculated values of binodals and spinodals, respectively. Points are the experimental results. In the center, the metastable region calculated by the convex hulls is shown [2].

concentration of polymers in the solution, the system is a single-phase mixture. With increasing concentrations, phase separation occurs. By varying the concentration of solutions, one may obtain admixtures with different mutual solubility of the polymer components.

Figure 6.2 shows the phase diagram of the ternary mixture polystyrene (PS)–polymethylmetacrylate (PMMA)–co-(styrene-acrylonitrile) (SAN) (49 mol.% AN), obtained by mixing of the three solutions in chloroform at total concentration of blends being 7 vol.% (curve 1), 8.5 vol.% (curve 2), and 10 vol.% (curve 3). It can be seen that when PMMA is readily soluble in one polymer (PS) and poorly in another (SAN), in the case of ternary mixtures 1 and 2, its introduction into the mixture deteriorates the polymer miscibility, as shown by an increase in area of two-phase systems. In the mixture number 3, the solubility of PMMA in

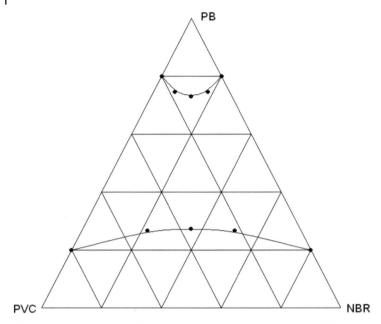

**Figure 6.3** Diagram of turbidity points in chloroform solution for PB–PVC–NBR (butadiene–acrylonitrile copolymer 60:40) system [2].

PS and SAN is small, being practically the same; the introduction of PMMA leads to an increase of the miscibility of PS and SAN in the ternary mixture.

There are other examples of the applicability of the rule of Prigogine to ternary mixtures of polymers in a common solvent. Figure 6.3 shows that the solubility of the polybutadiene (PB), that is equally soluble in both PVC and NBR, increases markedly in their mixture. This is also an example of the equilibrium system (regular solution) similar to those of the chloroform solution of the polymer components.

Application of the theory of Prigogine to polymer systems in the absence of solvent has significant limitations, as polymers are highly associated liquids and their mixtures are not to be regarded as regular solutions.

The approximate equilibrium in ternary polymer blends was obtained in the absence of solvent [4,5]. In particular, a triple mixture of poly(methylmetacry-late)-poly(epichlorohydrin) (PECH)– poly (ethyleneoxide) (PEO) was prepared by mixing the polymer solution and also by blending of the mixture in the melt state [4]. The polymers were completely miscible. Complete miscibility also exists over the entire ternary composition diagram. Heat of mixing for binary pairs of low molecular weight liquids with molecular structures analogous to the three above-mentioned polymers confirms that exothermic mixing is the basis for the miscibility of polymer pairs PEO–PMMA and PEO–PECH. However, similar results suggest that the interaction for the PECH–PMMA pair is rather weak. The results of determination of heat of mixing oligomers are not completely reliable, since in this case end groups play an important role. Probably, some differences in the interaction in these polymer pairs cannot prevent the formation of completely miscible ternary blend.

At the same time, in the ternary blends of styrene (maleic anhydride) copolymers, styrene(acrylonitrile) copolymers and poly(methylmethacrylate), or poly (ethylmethacrylate), or methylmethacrylate/acrylonitril) copolymer, a significant reduction of LCST takes place, which indicates that at certain composition these ternary blends are not miscible. Calculated values of the Flory–Huggins interaction parameter showed that their values are quite different for different pairs. It means that although all binaries are miscible and have exothermic interaction, immiscibility in ternary blends may result from an asymmetry of binary interactions, so-called «$\Delta\chi$» effect [6].

It is possible to theoretically calculate the phase diagrams (binodals and spinodals) using the method of convex hulls [7]. This method is based on the fact that the Gibbs free energy function of the entire system at all compositions is convex toward the negative values in the case of thermodynamic equilibrium of multicomponent systems. The convex hull is constructed together with the many functions describing the thermodynamic potentials for the set of all possible phases of the system. For a given composition, this hull corresponds to the minimum of the thermodynamic potential. It moves the regions of unstable and metastable states of the system to the heterogeneous equilibrium state. If the values of the corresponding convex hull coincide with the values of the thermodynamic potential (i.e., convex hull touches the thermodynamic potential), then this composition corresponds to the homogeneous system. Otherwise the system may be heterogeneous.

In study [2], the convex hull was constructed based on a discrete set of points corresponding to the values of the thermodynamic potential, calculated for the whole composition range of the components with an intervals of $\Delta\varphi = 0.001$. In this way, the theory of convex hulls was used to calculate the position of binodals in ternary polymer blends (Figure 6.2). Solid lines in Figure 6.2 are the coexistence curves calculated by the method of convex hulls. It is seen that the calculated values agree well with the experimental data.

Solution of the direct problem of determining the areas of the thermodynamic stability and instability in the phase diagram is mathematically easier task than the calculation of the areas of homogeneous and heterogeneous state of mixture. The condition of thermodynamic stability with respect to diffusion is a positive determinant of the stability matrix composed of derivatives of the thermodynamic potential. For the three-component system this condition is

$$\begin{vmatrix} \dfrac{\partial^2 \Delta G}{\partial \varphi_1^2} & \dfrac{\partial^2 \Delta G}{\partial \varphi_1 \varphi_2} \\ \dfrac{\partial^2 \Delta G}{\partial \varphi_2 \varphi_1} & \dfrac{\partial^2 \Delta G}{\partial \varphi_2^2} \end{vmatrix} > 0. \tag{6.1}$$

The areas where this determinant is negative are unstable in solution with diffusion separation leading to separate phases. The line formed by the points where the determinant is zero, is the spinodal. The equation of the spinodal in the case of a three-component system takes the form

$$\frac{\partial^2 \Delta G}{\partial \varphi_1^2} \cdot \frac{\partial^2 \Delta G}{\partial \varphi_2^2} - \frac{\partial^2 \Delta G}{\partial \varphi_1 \partial \varphi_2} \cdot \frac{\partial^2 \Delta G}{\partial \varphi_2 \partial \varphi_1} = 0. \tag{6.2}$$

According to Eq. (6.2), it is also possible to estimate the parameters $\chi_{12}$, $\chi_{13}$, and $\chi_{23}$ based on the concentration of solution components that are on the border of the thermodynamic stability, that is, on the spinodal. The calculated position of spinodals also shown in Figure 6.2 as dashed lines. The calculated values are in good agreement with the experimental data. This calculation also shows that in addition to the metastable region between the binodal and spinodal in the diagram, a spot *metastable region in the center of the diagram* appears. In case of the molecular weight of components not being completely identical, it is shifted to compounds containing increased content of low molecular weight component in the polymer mixture (PMMA). The general conclusion is that in ternary systems at certain composition the probability of the presence of such a region is higher than in the binary mixture due to the higher entropy of mixing (Figure 6.1).

The above approach to the evaluation of the mutual solubility of polymers on the basis of the rules of Prigogine is based on the concept of regular solutions. According to Hildebrand and Scott [8] such systems are solutions in which the "thermal stirring is able to overcome all the molecular orientation, the combination or association". Note that the structure and properties of solution of a mixture of polymers in a common solvent are closer to those of regular solution than the mixture in the melt. Only in the solution, thermal motion can destroy and regroup associates of each polymer or mixed associates, if any formed. Obviously, the mixtures of polymers are highly associated liquids, and their noncombinatorial entropy of mixing is significantly higher than the combinatorial entropy. The first one may even be negative.

According to Flory's calculation, an increase in the short-range order in mixed polyethylene and polyisobuthylene at $150\,°C$ reduces the value of $T\Delta S$ to $-22.61\,J/100\,g$, and also reduces the value of $\Delta H$ to $10.47\,J/100\,g$. At the same time the thermodynamic potential is increased to $\Delta G_m = 33.08\,J/100\,g$ [9].

Conditions of association of macromolecules and their resistance to thermal stirring in a ternary mixture can differ greatly from the conditions in binary one. The latter can produce immiscibility in ternary blends, even when one, two, or all three binary blends are miscible. In this case, the estimation of the molecular structure and the intensity polymer–polymer interaction should be carried out with regard to the experimental data for a mixture of polymer components or binary mixtures of polymers (composition sides on ternary diagram). The experimentally determined characteristics of properties of each polymer obviously reflect the characteristics of its supramolecular structure.

The calculation of the phase diagram for a ternary mixture of polymers in the melt, based on experimentally evaluated Flory–Huggins parameters for binary blends, can be carried out as follows [10]. It is based on the investigation carried out in melt of the blend of poly(propylene glycol) (PPG)–copolymer poly (methylmethacrylate co n-butylmethacrylate) (MMAnBMA) and poly(hexame-thyleneadipate) (PHMA). The melts were kept in vacuum by heating to form an equilibrium mixture. The spectral method determines the content of each poly-mer in each phase separated mixtures. It is possible to determine the separation region and the region of single-phase mixtures by measurement of light transmission.

In the formation of a three-phase mixture, the change in the thermodynamic potential $\Delta G_m$ described by the equation:

$$
\begin{aligned}
\frac{\Delta G_m}{RT} = &\left( \frac{\varphi_1 \ln \varphi_1}{n_1} + \frac{\varphi_2 \ln \varphi_2}{n_2} + \frac{\varphi_3 \ln \varphi_3}{n_3} \right) + g_{1,2}\varphi_1\varphi_2 + g_{1,3}\varphi_1\varphi_3 \\
&+ g_{2,3}\varphi_2\varphi_3,
\end{aligned}
\tag{6.3}
$$

where $n$ is the degree of polymerization, $\varphi_i$ is the volume fraction of the $i$th com-ponent, $g_{ij}$ are the binary interaction parameters. If the composition of each phase is quantified, the chemical potential of component 1 in the phase I and phase II is equal to

$$
\begin{aligned}
\frac{\Delta \mu_1^I}{RT} &= \ln \varphi_1^I - \varphi_1^I - \frac{n_1}{n_2}\varphi_2^I + n_1 g_{1,2}(\varphi_2^I)^2 + 1, \\
\frac{\Delta \mu_1^{II}}{RT} &= \ln \varphi_1^{II} - \varphi_1^{II} - \frac{n_1}{n_2}\varphi_2^{II} + n_1 g_{1,2}(\varphi_2^{II})^2 + 1.
\end{aligned}
\tag{6.4}
$$

Similarly, the potentials $\mu_2$ and $\mu_3$ recorded. The transformation of these equa-tions allows us to calculate the parameter $g_{1,2}$:

$$
g_{1,2} = \frac{\ln\left(\dfrac{\varphi_1^I}{\varphi_1^{II}}\right) + (\varphi_1^{II} - \varphi_1^I) + \dfrac{n_1}{n_2}(\varphi_2^{II} - \varphi_2^I)}{n_1\left[(\varphi_2^{II})^2 - (\varphi_2^I)^2\right]}.
\tag{6.5}
$$

Similarly, the parameters $g_{1,3}$ and $g_{2,3}$ are calculated. Thus, the interaction parameters for all three mixtures are calculated. Then the ternary mixture phase diagram is calculated.

Figure 6.4 shows the phase diagram of the studied blend. Although this dia-gram does not show the experimental data points, however, it should be noted that the experimental results are in good agreement with the calculated values in both stable single-phase or two-phase regions. Three single-phase regions, three two-phase regions, and one three-phase region (point line triangle) are seen in this diagram. The two-phase regions are adjacent to three binary axes. The trian-gular region within the two phases is a three-phase region. In general, the dia-gram is characterized by the fact that although the two binary mixtures are partially miscible, significant parts of the ternary diagram are immiscible blends.

The experimental determination of the interaction parameters is a difficult task, as can be seen from the preceding example. Sometimes this task is virtually impossible. Therefore, it is the most important for calculation of phase diagrams according to this method in which it is sufficient to know only the parameters of the three individual components. The main parameter that characterizes the

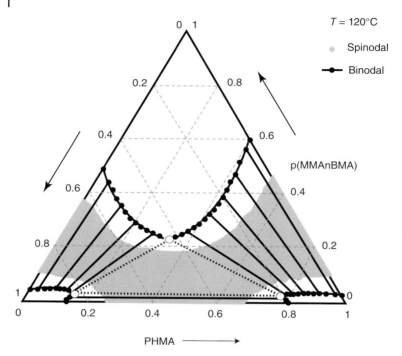

**Figure 6.4** Calculated phase diagram for PPG–PHMA–p(MMA-n-BMA) ternary blends [10], with permission © 2002, Kluwer Academic Publishers.

packing of the macromolecules in the polymer, that is, the nature of the association of macromolecules, is its compressibility. Flory theory indicates that when
the difference in the compressibility coefficients of polymers is more than 4%,
they are immiscible.

For a binary mixture of compressible polymer free energy of mixing per unit
volume can be written as [11,12]

$$\Delta G_{\mathrm{m}} = kT\left[\frac{\varphi_1\tilde{\rho}_1}{N_1\nu_1}\ln\varphi_1 + \frac{\varphi_2\tilde{\rho}_2}{N_2\nu_2}\ln\varphi_2\right] + \varphi_1\varphi_2\tilde{\rho}_1\tilde{\rho}_2(\delta_{1,0}-\delta_{2,0})^2$$
$$+ \varphi_1\varphi_2(\tilde{\rho}_1-\tilde{\rho}_2)(\delta_1^2-\delta_2^2), \tag{6.6}$$

where $\tilde{\rho}_i$ is the reduced density, which is the ratio of the density to its value at
close packing, that is, for hard core density, $\nu_i$ is the hard core molar volume, $\sigma_{i,0}$
and $\sigma_i$ are solubility parameters at $0\,^{\circ}$K and at temperature $T$, respectively, $N_i$ is
the number of repeat units per chain, and $\varphi_i$ is the volume fraction of component $i$. The calculation of hard-core molar volumes can be easily found in
Askadskii monograph [13].

The first term of this equation accounts for the combinatorial entropy of mixing. The second term can be related directly to the Flory–Huggins interaction
parameter. The third term arises solely from the compressibility of the
components.

Equation (6.6) gives a basis for calculating the free energy of mixing in a ternary mixture of compressible polymers [14]. The result may be written as expression for $\Delta G_m$:

$$\Delta G_m = kT\left(\frac{\varphi_1\tilde{\rho}_1}{N_1\nu_1}\ln\varphi_1 + \frac{\varphi_2\tilde{\rho}_2}{N_2\nu_2}\ln\varphi_2 + \frac{\varphi_3\tilde{\rho}_3}{N_3\nu_3}\ln\varphi_3\right) + \varphi_1\varphi_2(\tilde{\rho}_1\delta_{1,0} - \tilde{\rho}_2\delta_{2,0})^2$$
$$+ \varphi_1\varphi_3(\tilde{\rho}_1\delta_{1,0} - \tilde{\rho}_3\delta_{3,0})^2 + \varphi_2\varphi_3(\tilde{\rho}_2\delta_{2,0} - \tilde{\rho}_3\delta_{3,0})^2.$$

$$(6.7)$$

Using this expression for the free energy of mixing and following Scott [12], the boundary points of the spinodal curve for ternary systems may be found by solution of the following equation (see also Eqs. (6.1) and (6.2)):

$$\begin{vmatrix} \dfrac{\partial^2\Delta G_m}{\partial\varphi_1^2} \cdot \dfrac{\partial^2\Delta G_m}{\partial\varphi_1\partial\varphi_2} \\[2ex] \dfrac{\partial^2\Delta G_m}{\partial\varphi_2\partial\varphi_1} \cdot \dfrac{\partial^2\Delta G_m}{\partial\varphi_2^2} \end{vmatrix} = \dfrac{\partial^2\Delta G_m}{\partial\varphi_1^2} \cdot \dfrac{\partial^2\Delta G_m}{\partial\varphi_2^2} - \dfrac{\partial^2\Delta G_m}{\partial\varphi_1\partial\varphi_2} \cdot \dfrac{\partial^2\Delta G_m}{\partial\varphi_2\partial\varphi_1} = 0, \qquad (6.8)$$

where 1 and 2 – components – may be also 1 and 3 or 2 and 3.

In Figure 6.5, spinodal curve for the ternary blend of tetramethylpolycarbonate (TMPC), polycarbonate (PC), and poly(styrene-acrylonitryle) (SAN) is presented.

The pure component properties needed for the calculation of spinodal curve were obtained from experimental PVT measurements [15,16] and by group contribution method [13,17].

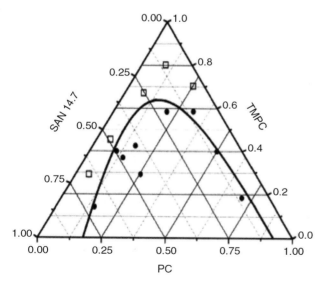

**Figure 6.5** Spinodal diagram at T = 413 K of TMCP–PC–SAN (14.7% AN). Open squares are miscible compositions, filled circles are compositions where two phases exist. Experimental data were obtained Kim *et al.* [18] for a system with equivalent molecular weights [14], with permission, © 2003, American Chemical Society.

As can be seen, agreement with the previously reported experimental data is quite good despite the simplifying assumptions of the model.

There are two methods of phase diagram calculation for ternary polymer blends.

Some examples indicate that even at complete miscibility in three binary mixtures comprising ternary polymer blend, this blend may exhibit region of incompatibility, that is, the twophase region. So the preparation (through the solution where the maximum equilibrium can be achieved) of the ternary mixture PMMA–poly(vinylcinnamate) (PVCN)–polyvinyl(phenol) (PVPh) in the triangle of compositions closed domain of incompatibility is evident. All three binary mixtures are completely miscible. Importantly, the interaction is substantially higher in mixture PMMA–PVPh than the PMMA–PVCN according to the spectral data. Authors emphasize that the reason of closed behavior of immiscibility is likely because of the "$\Delta\chi$ effect" [19].

The difference in the intensity of intermolecular interactions in miscible binary blends as can be seen from the preceding example followed by the appearance of an area of immiscibility in the ternary mixture. In a sense, this is consistent with the Prigogin rules [3]. The relationship between the miscibility of binary and ternary mixtures can also demonstrated by the example of the mixture PS– PC–TMPC [20]. It was found that TMPC, which is miscible with each of other components of the blend, does not solubilize the two immiscible polymers PS and PC. The $\chi_{i,j}$ values used for PS–PC, PS–TMPC, and PC–TMPC are 0.026, −0.029, and −0.034, respectively. The difference between the solubility parameters of two pairs PS–TMPC and PC–TMPC is small; however DSC results indicate a greater interaction in TMPC–PC than in TMPC–PS. Again, there is an element of asymmetry in the intensity of the interaction. In this case, the main field of the ternary diagram is occupied by heterophase blends. Obviously, there is deterioration miscibility in the pair PS–PC when TMPC is added. In this example, it should be emphasized that in the two pairs PS–TMPC and PC–TMPC, values of interaction parameters are very close (−0.029 and −0.034); however, there is more interaction in TMPC–PC than in TMPC–PS. Here the difference in molecular weights plays the important role. The molecular weight of PS, PC, and TMPC is, respectively, 200 000, 31 000, and 67 000.

There are examples of the formation of single-phase ternary mixtures. In this regard, the phase composition of the mixture poly(3-hydroxybutyrate), poly(ethylene oxide), and polyepiclorhidrin was investigated. Binary blends are miscible with interaction parameters of the same magnitude in different binaries: −0.068 (PHB–PECH), −0.096 (PHB–PEO), and −0.092 (PECH–PEO). Ternary blends of all compositions are miscible. Obviously, there is no "$\Delta\chi$" effect [21].

There is an interesting example of determination of the number of layers after delamination of polymer mixture solution as a measure of polymer miscibility. The series of polyacrylates, polymethylacrylates, and also polystyrene and polyvinylacetate were served as the base for preparation of different polymer mixtures in solution in chloroform. Two three-component and three four-component polymer mixtures were prepared. After delamination (1–2 days), it appears that in three-component mixtures only two layers are formed, while in four-component only three layers are formed. It seems to be very simple experimental method to indicate increased miscibility of polymers in three- and four-component mixtures [22].

Another example is a ternary blend of poly(ε-caprolactone), poly(benzilmethacrylate), and poly(vinyl methyl ether) that exhibits one-phase (miscible) at all compositions. All binaries are also miscible. Balanced interactions exist with no offsetting "$\Delta\chi$" among the three binary pairs. At the same time, in the ternary mixtures low critical solution temperature (LCST) equals 75 °C, which is much lower than in the binary mixtures, indicating that the intermolecular interaction is reduced in ternary blends compared with binary blends [23].

The Flory–Huggins interaction parameter depends primarily on the temperature. Its dependence on the composition of the mixture is less pronounced and seldom taken into account. Also, the dependence of molecular weight is generally not considered. Solubility parameter is rarely compared with its critical value, although the latter is strongly dependent on the molecular weight.

The calculations of phase diagrams in ternary mixtures are based, as shown above, on the experimental or calculated values of binary Flory–Huggins interaction parameters. But first of all, the dependence of binary parameters is defined (sometimes approximately) only on the temperature and second, parameter $\chi_{i,j,k}$ should not be considered as an additive value of corresponding binary parameters. Therefore, a simple summation of the contributions of binary interaction parameters in the free energy of mixing in the ternary polymer blends is a simplification. From the examples given above one cannot confidently say that polymer miscibility increases with the transition from binary to ternary mixtures, or, conversely, decreases, as there are different effects. The development of the theory of mixing in the field of ternary polymer blends obviously requires a lot of effort in the future.

## 6.3
## Formation of Phase Morphology

For polymer blends or composite, the term morphology describes the structures and shapes observed, often by microscopy or scattering techniques, of the different phase domains present within the mixture. Phase morphology is formed during blending of immiscible components. After mixing in a melt state, the developed phase morphology can easily be fixed by rapid cooling of the batch.

Polymer blends have appeared mainly as an alternative to the synthesis of new polymers [24]. The situation resembles in some aspects the way in which metal alloys have been developed. Starting with the use of simple binary alloys in the ancient times, people switched to the design of multicomponent combinations. In spite of the limited number of the known metals, the number of alloys is enormous, many of them comprising several components and having sophisticated structures and outstanding properties.

Similarly, there is an obvious tendency toward the increase of the number of components in polymer blends, which enhances their versatility and allows for a more flexible control over final properties of compositions [25]. A simple general principle exists according to which each component added introduces to a blend some portion of its properties or qualities. Unfortunately, this rule cannot be employed in a mechanistic fashion without knowledge of particular phase morphology developed in a blend. The latter is the more complex the larger is the number of polymer components.

In recent years, increasing attention of researchers and technologists has been drawn to the study and use of multiphase blends, which is explained mostly by the need and possibility of obtaining materials with new combinations of properties using the available industrial polymers [26]. In addition, multiphase systems sometimes exhibit synergistic effects not inherent in single-phase polymer blends.

The properties of any heterogeneous system are not only determined by the properties of individual components but also depend to a considerable extent on the intensity of interaction at the interface and the phase morphology as well, the latter factor characterizing the degree of dispersion, the shape, and the mutual arrangement of phases in the bulk of the blend [27]. Experience gained in the work with binary blends led to the development of practical techniques providing a variety of blends with preset phase structures. In particular, by controlling the type of flow, the ratio of blend components, and their viscoelastic and surface properties, it is possible to develop polymer compositions in which the dispersed phase domains could have the shape of discrete spheres, ellipsoids, fibers, layers of different morphology, or continuous networks [28].

Adding the third immiscible component to a binary composition increases the number of possible phase structures [28]. For example, should any of the three components be capable of forming a continuous phase, the two others may either form independent dispersions or encapsulate one another. If two of the three polymers form interpenetrating co-continuous phases, the third component may form domains localized predominantly in one of these phases, in both of them, or at the interface between the two continuous phases. Finally, each of the three polymers may form their own continuous structure developing the blend morphology comprising three interpenetrating continuous networks.

Due to the rapid growth of industrial use of heterogeneous mixtures of polymers there is an urgent need to develop a method for predicting the phase morphology of the blends.

### 6.3.1
### Prediction of Phase Morphologies of Polymer Blends

Since phase morphology produces great impact on final properties of polymer blends, the development of reliable methods of prediction of this parameter is of primary importance. There are a few approaches developed to predict phase morphologies in binary and multiphase polymer blends. All of them are based on surface properties of polymers and interfaces. This section presents a detail description and critical analysis of practical examples of phase morphologies and methods of their prediction.

### 6.3.1.1 Binary Blends

The phase structure of binary and multicomponent polymer systems have a lot in common. Therefore, it is advisable to consider features of the phase morphology and its prediction in binary mixtures the first. The phase structure of immiscible liquid binary systems (classical emulsion) is not very diverse. At rest, there are generally spherical droplets of the dispersed phase in a continuous matrix phase. Thermodynamic demand for minimization of the free surface energy initiates the droplets of the dispersed phase to coalesce so that over time the emulsion separates into two layers with a minimum interface area. Numerous papers on binary blends of immiscible polymers indicate that their phase morphology is characterized by a larger diversity. This is mainly due to the possibility of fixing the phase morphology at any stage of its formation, by, for example, fast quenching the melt. Figure 6.6 shows the well-known examples of possible morphological types in binary blends.

In 1972, VanOene [29] proposed a semiquantitative theory for prediction of the modes of dispersions in polymer blends, which is based on the interfacial and viscoelastic properties of the components. The essence of the theory is the assumption that during flow the differences in elasticity of the components may contribute to the value of the interfacial tension, which may be both positive and negative.

Three modes of dispersions namely droplet-fiber, layered (laminate), or composite droplets are discussed (Figure 6.6). Parameters responsible for the formation of a certain type of morphology are the values of the dynamic ($\gamma'_{12}$) and static ($\gamma_{12}$) interfacial tensions, the radius of the dispersed phase (b), and the first normal stress difference for the dispersed phase $N_{1d}$ and the matrix $N_{1m}$, characterizing the viscoelastic properties of blended polymers:

$$\gamma'_{12} = \gamma_{12} + b_1 \frac{N_{1d} - N_{1m}}{6}. \tag{6.9}$$

In the above formula, subscripts designate corresponding polymer phases. The prediction of the modes of dispersion with Eq. (6.9) is as follows. The formation

    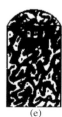

(a)    (b)    (c)    (d)    (e)

**Figure 6.6** Typical samples of phase morphologies in polymer blends prepared by melt extrusion: (a) droplets and (b) droplet-fiber; (c) ribbons or layers (laminate); (d) encapsulated or composite; and (e) co-continuous phases.

of drops or fibers (Figure 6.6a,b) of phase 1 in the phase 2 takes place provided $(N_{1d} - N_{1m}) > 0$ and the dynamic interfacial tension is positive, that is, when $\gamma'_{12} > 0$. When the less elastic phase 2 is dispersed in the more elastic phase 1, $(N_{1d} - N_{1m}) < 0$, $\gamma'_{12} < 0$ and the phase 2 stratifies. Nevertheless, the formation of the droplet-fiber morphology of the less elastic phase will be possible if

$$b_2^{cr} \leq 6 \frac{\gamma_{12}}{(N_{1d} - N_{1m})}. \tag{6.10}$$

For polymer melts, a value of this radius is of the order of $b_2 = 0.1 - 1.0\,\mu m$.

Thus, two cases can be realized when the less elastic phase is dispersed in a more elastic medium. In the initial stages of mixing when the particles of the dispersed phase are still large, that is, $b_2 \gg 1\,\mu m$, $\gamma'_{12} < 0$ and phase 2 forms layers or bands (stratified morphology).

When the mixing process is effective enough and the particle size falls below the critical ($b_2 < 1\,\mu m$), the layers split into discrete droplets and a stable droplet-fiber dispersion is developed. However, the drops of the less elastic phase 2 can capture small fragments of the matrix to form composite (encapsulated) particles (Figure 6.6d).

An important position in the hierarchy of the phase morphologies belongs to the structure of co-continuous phases shown in Figure 6.6e [24–26,28,30–32]. Conditions favoring the formation of this morphology are sufficiently high content (above the percolation threshold) and/or a low viscosity of the dispersed phase.

As a unified classification for the morphology of polymer blends is not yet developed, it is necessary to agree on the terms. Three morphologies, specified by VanOene [29] (Figure 6.6a–d) and the co-continuous phases in Figure 6.6e will be called here as the modes of dispersion. All the rest phase structures including the ones in multicomponent blends will be named "morphological types" or simply "morphology". It should be stressed that virtually any type of morphology can be represented by one of the modes of dispersion. For example, the composite particles of the dispersed phase in the ternary blends can be prepared in the shape of droplet, fibers, layers, or co-continuous structures shown schematically in Figure 6.6.

### 6.3.2
### Ternary Polymer Blends

VanOene's theory demonstrates the paramount importance of the interfacial tension in the formation of the morphology of binary mixtures of polymers. An even bigger role this parameter plays in complex multiphase heterogeneous systems.

In contrast to the binary systems with one interface and the interface type contacts 1–2, three-phase system is characterized by three interfaces, the three corresponding types of contacts 1–2, 1–3, and 2–3, and three interfacial tension $\gamma_{12}$, $\gamma_{13}$, and $\gamma_{23}$.

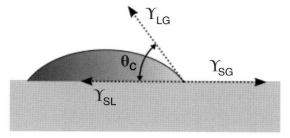

**Figure 6.7** Young's equation equates the surface tension forces at equilibrium.

Prediction of the phase situation in a multiphase mixture is based on the analysis of the mutual wetting (or non-wetting) of phases, which in turn is determined by the interfacial tensions at the interfaces.

The wettability of a liquid governs by how it spreads over a solid substrate or an immiscible liquid phase. It is characterized by a contact angle, which is the angle subtended by the liquid/vapor (gas) interface and the solid or immiscible liquid surface.

The relationship between the equilibrium contact angle θ that the drop makes with the surface (Figure 6.7) and the three surface tensions γ was first established by Thomas Young in 1805 through a force balance in the plane of the surface at the three phase contact line [33–36]:

$$\gamma_{SG} = \gamma_{SL} + \gamma_{LG} \cos \theta, \tag{6.11}$$

where γs indicate various interface tensions and subscripts S, L, and G refer to substrate, liquid, and gas (or vapor, or atmosphere), respectively. Spreading behavior of a liquid drop over the solid surface and a contact angle depends on a balance between adhesive and cohesive forces. Adhesive forces between a liquid and solid cause a liquid drop to spread across the substrate. Cohesive forces within the liquid cause the drop to ball up and avoid contact with the surface. If the three tensions are known, the wetting state of the fluid follows directly [37].

At θ>90° and $\gamma_{SG}<\gamma_{SL}$, liquid does not wet a solid. This case is classified as non-wetting behavior (Figure 6.8a). If $\gamma_{SG}<\gamma_{SL}+\gamma_{LG}$ and 0<θ<90°, a droplet with a finite (equilibrium) contact angle minimizes the free energy of the system leading to partial wetting (Figure 6.8b). If $\gamma_{SG}=\gamma_{SL}+\gamma_{LG}$, the contact angle θ ~ 0. The system will consequently be in equilibrium when a macroscopic uniform

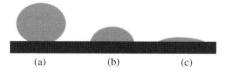

(a)          (b)          (c)

**Figure 6.8** The three different wetting behavior. See text.

liquid layer covers the whole solid surface corresponding to complete wetting (Figure 6.8c).

Further steps toward the prediction of the spreading behavior were made by Cooper and Nuttal in 1915 [33], and by Harkins and Feldman in 1922 [38] by introducing the spreading coefficient. Harkins and Feldman defined this parameter as

$$\lambda = W_A - W_C. \tag{6.12}$$

Thus a liquid will not spread if its work of cohesion $W_C$ is greater than the work of adhesion $W_A$ for the interface of the liquid and another liquid or solid upon which spreading is to occur. The $W_A$ and $W_C$ values are related to interfacial tensions by

$$W_A = \gamma_{SG} + \gamma_{LG} - \gamma_{SL} \tag{6.13}$$

$$W_C = 2\gamma_{LG}. \tag{6.14}$$

Therefore, the spreading coefficient can be defined as

$$\lambda = \gamma_{SG} - \gamma_{LG} - \gamma_{SL}. \tag{6.15}$$

If $\lambda \geq 0$, complete spreading occurs in which the drop, due to a decrease in the free energy, spontaneously forms a liquid film above the solid substrate, as shown in Figure 6.8c. If $\lambda < 0$, $\theta > 0°$ from Eq. (6.11) and the drop spreads to an equilibrium shape, constrained by the contact line where the three phases meet; if $\theta < 90°$, the surface is said to be *hydrophilic* (Figure 6.8b) whereas the surface is said to be *hydrophobic* if $\theta > 90°$ (Figure 6.8a). Strictly speaking, there is no local minimum in the interfacial energy for $\lambda > 0$ and hence the equilibrium contact angle $\cos\theta$ cannot be defined by Eq. (6.11). The minimum is defined at $\lambda = 0$, which from Eqs. (6.11) and (6.15) gives $\cos\theta = 1$ or $\theta = 0°$, corresponding to the complete wetting case.

Torza and Mason [39] were apparently the first who adapted Eq. (6.15) for the ternary mixtures of immiscible liquids. They considered the equilibrium state established after two immiscible liquid drops of phases 1 and 3 were brought into intimate contact when they were suspended in a third immiscible liquid of the matrix phase 2. This state corresponds to a minimum surface free energy $G = \Sigma\gamma_{ij} A_{ij}$, where $A_{ij}$ is the area of the $ij$ interface. As a result, the expressions for the three spreading coefficients $\lambda_{ij}$ enabled to predict the morphological type of composition and wetting one phase by another were evaluated. As an example, consider a ternary liquid heterogeneous system consisting of matrix 2 and two dispersed phases 1 and 3. The values of the three spreading coefficients of Eq. (6.16) define mutual wetting in the system as well as the morphological type of a mixture (Figure 6.6):

$$\begin{aligned} \lambda_{31} &= \gamma_{12} - \gamma_{32} - \gamma_{13}, \\ \lambda_{13} &= \gamma_{32} - \gamma_{12} - \gamma_{31}, \\ \lambda_{21} &= \gamma_{13} - \gamma_{12} - \gamma_{23}, \end{aligned} \tag{6.16}$$

where $\lambda_{31}$, $\lambda_{13}$, and $\lambda_{21}$ are the spreading coefficients signifying spreading of component 3 over component 1, component 1 over 3, and component 2 over 1.

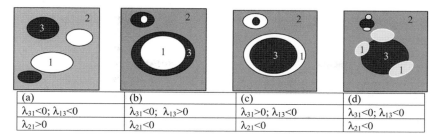

| (a) | (b) | (c) | (d) |
|---|---|---|---|
| $\lambda_{31}<0$; $\lambda_{13}<0$ | $\lambda_{31}<0$; $\lambda_{13}>0$ | $\lambda_{31}>0$; $\lambda_{13}<0$ | $\lambda_{31}<0$; $\lambda_{13}<0$ |
| $\lambda_{21}>0$ | $\lambda_{21}<0$ | $\lambda_{21}<0$ | $\lambda_{21}<0$ |

**Figure 6.9** Possible equilibrium morphological types in ternary polymer blends.

Parameters $\gamma_{ij}$ are the interfacial tensions of polymer pairs 1–2, 1–3, and 2–3. The index 2 refers to the matrix while indices 1 and 3 denote the two dispersed phases. When the spreading coefficients $\lambda_{31}$ and $\lambda_{13}$ are both negative while $\lambda_{21}>0$, the dispersed phases 1 and 3 form separate dispersions in the matrix 2 (Figure 6.9a). If one of the spreading coefficient $\lambda_{31}$ or $\lambda_{13}$ is positive, the core/ shell composite domains are formed (Figure 6.9b,c). For example, at $\lambda_{13}>0$, the dispersed phase 1 (shell) encapsulates the core forming phase 3 (Figure 6.9c). When all the three spreading coefficients are negative, a partial wetting phenomenon is observed (Figure 6.9d).

In classic ternary emulsions depicted schematically in Figure 6.9, wetting or non-wetting between the inner phases 1 and 3 is considered. Therefore, phase arrangement in Figure 6.9a is classified as nonengulfing [39] or complete non-wetting; Figure 6.9b,c refers to a complete engulfing (or complete wetting), and Figure 6.9c denotes the case of a partial wetting (engulfing).

For unknown reasons, the work of Torza and Mason [39], which represents an up-to-date version of the spreading coefficients theory, was forgotten and unclaimed for almost 30 years. Meanwhile, in 1988, Hobbs *et al.* [40] published their work in which they adopted the Harkins equation for the case of three liq-uid phases and obtained the following equation for the spreading coefficient:

$$\lambda_{31} = \gamma_{12} - \gamma_{32} - \gamma_{13}. \tag{6.17}$$

This paper greatly stimulated the study of phase morphology of multicomponent polymer blends. Dozens of papers on the subject that prove the adequacy of the proposed prediction model have been published.

Hobbs *et al.* did not consider the third spreading coefficient $\lambda_{21}$ signaling the partial wetting phenomena. Virgilio *et al.* [41] focused attention on these impor-tant misconceptions that still remain in the polymer blend literature concerning the use of the spreading coefficients. One often encountered misconception is the prediction of the morphology in ternary polymer blends based on the calcu-lation of only one $\lambda_{31}$ or two ($\lambda_{31}$ and $\lambda_{13}$) spreading coefficients. This can lead to erroneous conclusions, especially when these coefficients are negative. Specifi-cally, at both negative values $\lambda_{31}$ and $\lambda_{13}$, the negative $\lambda_{21}$ value predicts partial wetting (partial encapsulation Figure 6.9d, while a positive $\lambda_{21}$ value informs about the development of the morphology of two separated dispersed phases

(Figure 6.9a). Therefore, in ternary blends, three spreading coefficients are necessary to correctly predict the resulting morphology.

The approaches outlined above enable to predict the type of phase morphology from the standpoint of mutual arrangements of internal phases in ternary or multiphase polymer blends. At the same time, as mentioned above, each particular type of phase structure can admit a large variety of morphologies (like shown in Figure 6.6) depending, for a given combination of the surfaces and viscoelastic properties of components, on the type of flow realized upon blending and/or processing, composition, the initial arrangement, and mixing sequence of the components.

When ternary polymer blend has two main co-continuous phases B and C and a third dispersed phase A, a different approach based on the sign of the spreading coefficients is used [41,42]. The wetting is considered complete if at least one of the three coefficients is greater than zero. Therefore, the four morphological types shown in Figure 6.10 pertain to two wetting behaviors: (i) complete wetting (Figures 6.10a–c) or (ii) partial wetting at the interface (Figure 6.10d). In the latter case, a three-phase line of contact exists between all three polymers and the

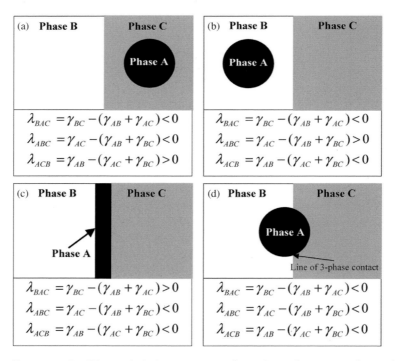

**Figure 6.10** Possible morphologies in a ternary system composed of two major phases B and C (in white and gray) and one minor phase A (black), as predicted by the spreading coefficients. From part (a) to part (c), morphologies displaying complete wetting, in which phases C, B and A, respectively, wet the AB, AC and BC interfaces. The morphology in part (d) displays partial wetting, in which none of the phases locates between the other two, resulting in a line of contact along which the three phases meet [41].

minor phase shares its surface area with both components. One can see that the topologies shown in Figure 6.10a–c imply coalescence, while morphology in Figure 6.10d refers to a partial wetting. The spreading coefficient λA/B/C indicates the thermodynamic tendency of a phase B to spread between the other phases A and C. In order to predict the equilibrium morphology, a set of three spreading coefficients is required. If λA/B/C is positive and the other two negative, then phase B forms a continuous phase between A and C and complete encapsulation is observed (Figure 6.10b). In the case that all three spreading coefficients are negative, a partial wetting is observed (Figure 6.10d).

There have been several attempts to modify the spreading coefficient prediction theory.

Guo *et al.* [43] assume that the equilibrium phase structure of a multiphase system is determined not by interfacial tensions alone, but rather by the interfacial free energy which represents a combination of interfacial tensions and interfacial areas. They modified the spreading coefficient concepts to include both interfacial tensions and interfacial areas, and have used the resulting expressions to predict the phase structures of different ternary polymer blends of polyethylene (PE), polypropylene (PP), polystyrene (PS) and PMMA.

The Gibbs free energy of mixing for any multiphase polymer system having more than two phases:

$$G = \sum_i n_i \mu_i + \sum_{i \neq j} A_i \gamma_{ij}, \tag{6.18}$$

where $\mu_i$ is the chemical potential of component $i$, $n_i$ is the number of moles, $A_i$ is the interfacial area, and $\gamma_{ij}$ is the interfacial tension between components $i$ and $j$.

For a ternary mixture of incompatible polymers, the formation of 9 types of phase morphologies is depicted in Figure 6.9a–c (three for each matrics polymer A, B, and C). The version of partial wetting shown in Figure 6.9d was not considered by the authors. In the case when the matrix phase is A, the G value is calculated for each of the three types of morphology $B + C$, B/C, and C/B, indicated respectively in Figure 6.9a–c. Because $\Sigma\, n_i \mu_i$ terms for all types of morphology are the same, the first term in Eq. (6.18) can be neglected. It is also assumed that the number of B and C particles is the same.

As a result, the interfacial free energies for the all three types of morphology can be calculated by using the following equations:

$$\left(\sum A_i \gamma_{ij}\right)_{B+C} = (4\pi)^{1/3}\left[n_B^{1/3}x^{2/3}\gamma_{AB} + n_C^{1/3}\gamma_{AC}\right](3V_C)^{2/3}, \tag{6.19}$$

$$\left(\sum A_i \gamma_{ij}\right)_{B/C} = (4\pi)^{1/3}\left[n_B^{1/3}(1+x)^{2/3}\gamma_{AB} + n_C^{1/3}\gamma_{BC}\right](3V_C)^{2/3}, \tag{6.20}$$

$$\left(\sum A_i \gamma_{ij}\right)_{C/B} = (4\pi)^{1/3}\left[n_B^{1/3}x^{2/3}\gamma_{BC} + n_C^{1/3}(1+x)^{2/3}\gamma_{AC}\right](3V_C)^{2/3}. \tag{6.21}$$

Here $x = V_B/V_C$; $V_i$ is the volume fraction of each phase.

In accordance with the proposed method of prediction, the type of morphology is formed which has the lower free energy. Calculations show that when high density polyethylene (HDPE) is the matrix phase and PS and PMMA are the minor components, the blend will have a phase structure with PMMA encapsulated by PS (PS/PMMA). When PS is the matrix phase and HDPE and PMMA are the minor components, separate dispersions of HDPE and PMMA has the lowest interfacial free energy. Finally, when PMMA is the matrix phase and HDPE and PS are the minor components, the blend is predicted to have the phase structure of HDPE encapsulated by PS. The authors (Guo *et al.* [43]) found that the predictions made were in good agreement with experimental results. Calculations based on the model suggest that interfacial tensions play the major role in establishing the phase structure, whereas a less significant (but still important) role is played by the surface areas of the dispersed phases.

Similar prediction procedures for the ternary PP/ethylene propylene diene monomer (rubber)(EPDM)/PE and PP/EPDM/PS (minor phase 1/matrix/minor phase 2) were used by Hemmati *et al.* [44,45]. It was shown that in accordance with the model by Guo *et al.* [43] and the spreading coefficients values the first blend exhibited the core (PE)–shell (EPDM) morphology while the second one formed separate dispersions of the EPDM and PS phases in the PP matrix.

Valera *et al.* [46] studied PMMA/PP/PS ternary blends and compared the observed results with spreading coefficient, free interfacial energy, and dynamic interfacial energy. They reported that when PMMA was the major (matrix) phase, the core–shell morphology together with the discrete PS droplets were observed in the matrix. The core–shell morphology was also observed in the state of having PP as the matrix, but when PS formed the matrix phase, the morphology was separated dispersions. Experimental data were supported by the free interfacial energy model for the first state. Predictions of both spreading coefficient and dynamic interfacial energy models were correct for most samples but none of the models could predict and justify the existence of pure PS droplets in the PMMA matrix. To understand the mechanism of morphology formation, Valera *et al.* investigated the behavior of the system during mixing time. They observed that by feeding PMMA to the binary PP/PS system, PMMA fibrils were formed. Then they were broken into smaller droplets and diffuse toward the PS phase; consequently, the core–shell morphology with a PMMA shell was observed. The core phase may break into smaller droplets.

Reignier *et al.* [47] studied the encapsulated morphologies in blends of PS/HDPE/PMMA, where the matrix was PMMA. It was shown that the calculation of the spreading coefficient and surface free energy predicted enveloping PMMA drops by the phase of PS. However, in the case of highly viscous samples of PS and PMMA, these predictions were not met: the data of microscopy recorded enveloping of the PS drops by the phase of PMMA. Interestingly, the "correct" arrangement of the internal phases was recovered after annealing of the blends. It was suggested that the cause of the observed effect could be the dynamic interfacial tension $\gamma'_{ij}$ in the flow upon mixing, the value of which differ from the static values $\gamma_{ij}$, used in the calculations. Therefore, the model by Guo *et al.* [43]

was modified by replacing a static with a dynamic interfacial tension from the VanOene theory (Eq. (6.9)):

$$
\left( \sum A_i \gamma_{ij} \right)_{\text{PS+PMMA}} = 4\pi R_i^2 \left[ \gamma_{\text{PS/PMMA}} + \frac{R_i}{6}(N_{1.\text{PS}} - N_{1.\text{PE}}) \right]
$$
$$
+ 4\pi R_i^2 \left[ \gamma_{\text{PS/PMMA}} + \frac{R_i}{6}(N_{1.\text{PMMA}} - N_{1.\text{PE}}) \right],
$$
$$
\left( \sum A_i \gamma_{ij} \right)_{\text{PS/PMMA}} = 4\pi R_e^2 \left[ \gamma_{\text{PS/PE}} + \frac{R_e}{6}(N_{1.\text{PS}} - N_{1.\text{PE}}) \right]
$$
$$
+ 4\pi R_i^2 \left[ \gamma_{\text{PMMA/PS}} + \frac{R_i}{6}(N_{1.\text{PMMA}} - N_{1.\text{PS}}) \right], \tag{6.22}
$$
$$
\left( \sum A_i \gamma_{ij} \right)_{\text{PMMA/PS}} = 4\pi R_e^2 \left[ \gamma_{\text{PMMA/PE}} + \frac{R_e}{6}(N_{1.\text{PMMA}} - N_{1.\text{PE}}) \right]
$$
$$
+ 4\pi R_i^2 \left[ \gamma_{\text{PS/PMMA}} + \frac{R_i}{6}(N_{1.\text{PS}} - N_{1.\text{PMMA}}) \right].
$$

This model, which takes into account the contribution of the component elasticity to the value of the interfacial tension, was in good agreement with the observed effect.

Mohammadigoushki *et al.* [48] presented an attempt to predict the morphology development in PA6/PP/PS ternary blends using the dynamic interfacial tension of the blend components evaluated from the Palierne's viscoelastic model in conjunction with the spreading coefficient theory. As was evident by the SEM micrographs, the blend developed a predicted morphology in which composite PP/PS core/shell droplets were dispersed in the PA6 matrix. The further researchers used both the spreading coefficients and the minimal free energy surface model to predict the morphologies of polymer blends [47–55].

A different approach to predict the phase morphology of the ternary polymer blends, offering a spreading coefficient calculation using solubility parameters, was proposed by Koseki *et al.* [56]. The widespread use of the above methods for predicting the morphology of multiphase polymer blends is significantly hampered by the absence of reliable values of the interfacial tensions for many polymer pairs. This is due to the serious challenges associated with the experimental determination of this parameter in polymer systems at high temperatures. This is especially true for elastomeric materials for which experimental values of interfacial tensions are still a rarity. According to the mean-field theory the interfacial tension is proportional to the square root of the interaction Flory–Huggins parameter $\chi_{ij}$:

$$
\gamma_{ij} = \frac{kT}{b^2} \sqrt{\frac{\chi_{ij}}{6}} \sqrt{a^2 + b^2},
$$

where $b$ is the effective length of monomer unit, $k$ is the Boltzmann constant, and $T$ is the absolute temperature. For mixtures of nonpolar or weakly polar molecules, $\chi_{ij}$ is proportional to the difference in the solubility parameters of blend components:

$$\chi_{ij} = \frac{V_r}{RT} \left(\delta_i - \delta_j\right)^2,$$

where $R$ is the gas constant, $V_r$ is the reference molar volume, and $\delta_i$ is the solubility parameter of component $i$. If the assumptions behind the above two equations are valid, one can approximate spreading coefficients $\lambda_{31}$ and $\lambda_{13}$ as follows:

$$\begin{aligned}
\lambda_{31} &\approx |\delta_1 - \delta_2| - |\delta_3 - \delta_1| - |\delta_2 - \delta_3|, \\
\lambda_{13} &\approx |\delta_2 - \delta_3| - |\delta_1 - \delta_2| - |\delta_3 - \delta_1|.
\end{aligned}$$

(6.23)

The adequacy of Eq. (6.23) was verified by analyzing the phase morphologies in 9 ternary mixtures based on acrylonitrile-butadiene rubber (NBR) containing 16, 33, or 46 wt% of AN, chlorinated polyethylene (CPE), containing 29, 35, or 44 wt% Cl and ethylene-propylene rubber (EPR, 45 wt% ethylene). According to the TEM data, morphologies of mixtures of fixed composition NBR/CPE/EPR 10/70/20 corresponded to the predicted structures with only one exception that was attributed to the viscous effects. Thus, in the lack of values of the interfacial tension, this method can provide an alternative way for prediction of morphology since the values of the solubility parameters for polymers are available in many references. In particular, values of the solubility parameter for polymers are available in the handbook by Van Krevelen [57] and can be quite easily measured experimentally.

Most authors agree that the method of prediction of the phase morphology based on the spreading coefficient is quite adequate. However, some researchers have reported the opposite.

Harrats *et al.* [58] studied encapsulation of the PP phase between the PS and PA6 phases (Figure 6.11a and b). They found that this contradicts the prediction of the Harkins equation, whereas the interfacial tension values of $\gamma_{PA6/PS} = 13.72$ mN m$^{-1}$, $\gamma_{PA6/PP} = 13.64$ mN m$^{-1}$, and $\gamma_{PP/PS} = 2.26$ mN m$^{-1}$ demonstrated partial wetting case.

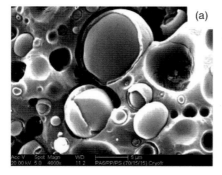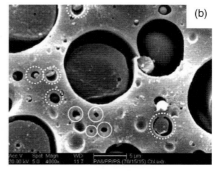

**Figure 6.11** SEM micrograph of PA6/PP/PS 70/15/15 blends: (a) cryofractured and (b) chloroform extracted surfaces [58], with permission © 2005 Elsevier Ltd.

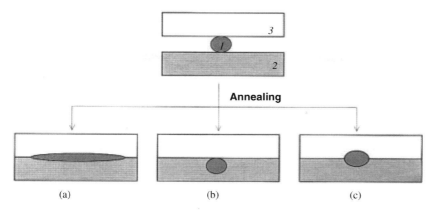

**Figure 6.12** Possible location and shapes of the floating polymer. (a) Thin phase of polymer *1* spread between polymers 2 and 3; (b) droplet of 1 buried in 2; (c) droplet of 1 located between 2 and 3 [59].

They noted that the spreading coefficient theory fails in their case because "the development of the phase morphologies in polymer melt depends also on other key factors such as the viscosity and elasticity of the blend. The use of the interfacial tension alone for the prediction of the type of phase morphologies in immiscible polymer melt can be successful in blend components having closer melt viscosities." This work illustrates one of the few bad prediction of the morphology on the basis of the spreading coefficient. One has to keep in mind that in this case an incorrect prediction was not associated with inadequate model; evidently, the mixing time was not enough for attaining equilibrium blend morphology. Ohishi [59] reported phase morphology in the ternary blends composed of PET, ethylene buten rubber (EBM) and partially neutralized ethylene–methyl methacrylic acid ionomer (EMAA). For prediction of the type of morphology, he used a droplet-sandwich experiment allowing for qualitative analysis of wetting phenomena in ternary polymer systems proposed by Nakamura and Inoue [60]. As Figure 6.12 shows, depending on the balance of the interfacial tensions, there are three possible locations of the droplet 1 placed between two layers of polymers 2 and 3. In a liquid state, droplet 1 (a) can spread between phases 2 and 3, (b) be buried in phase 3 (or in phase 2), and (c) locate between phases 2 and 3 without spreading. One can admit that, by analogy with Figure 6.10, cases (a) and (b) in Figure 6.12 should be referred to complete, and the case (c) to the partial wetting. It was shown [59] that the EMAA droplet sandwiched between the EBM and PET layers widely spread between these phases. When the EBM droplet was placed initially between PET and EMAA, it was encapsulated by the EMAA phase. These findings imply that EMAA will encapsulate EBM in the matrix of PET. This prediction agreed well with the result of the TEM analysis. It is believed that in the absence of data on the interfacial tension, the droplet-sandwich technique may represent a real alternative to the spreading coefficient approach.

In concluding this section, it can be stated that the above analyzed methods of prediction of the phase morphology of multicomponent blends of incompatible polymers are represented by accurate and thermodynamically based models that do not require the introduction of any additional assumptions. However, the accuracy of prediction of the phase structure can be very low due to the low reliability of the values of interfacial tension, which may differ from each other by 30% or more. This is because of the low precision and reproducibility of the existing methods of measuring $\gamma_{ij}$ in melts and the fact that polymers from different manufacturers may contain various impurities and additives, influencing the surface properties of materials. In this regard, the need to support the experimental data and calculations with the microscopic studies is clear [61].

### 6.3.3
### Encapsulated Morphologies: Influence of Different Factors

Ternary systems with the encapsulated inner phases have attracted a great deal of attention in the recent literature. As it was mentioned earlier, fully encapsulated morphologies are formed provided one of the spreading coefficients ($\lambda_{31}$ or $\lambda_{13}$) is positive while $\lambda_{21} < 0$ (Figures 6.9b and c). When all the three coefficients are negative, partial engulfing (Figure 6.9d) takes place. The influence of different factors on the phase morphology of such blends is summarized below.

#### 6.3.3.1 Blend Composition
Varying the composition is one of the main methods to control the phase morphology of polymer blends. Throughout the history of the development of polymer blends one of the principal efforts was directed on the study of dependence of structure and properties of blends on their composition. Numerous studies published since then show [24,28,57,63] that increasing the content of the dispersed phase A in the matrix B is accompanied by a sharp increase in the domain sizes due to increasing the frequency of collisions of drops leading to coalescence. Finally, the fusion of droplets of the dispersed phase occurs so often that they form a continuous network phase B interlaced with the continuous phase A. (Figure 6.13). In this case, one can speak about the formation of the morphology of the two co-continuous phases. These blends have the potential of opening particular application fields where the presence of interconnected structures are a necessary feature (as in separation phenomena, electrical conductivity, tissue engineering scaffolds, and drug delivery devices).

It is believed that polymer blends with the co-continuous morphologies are considered to be essential for practical use because, at this structural geometry, each phase makes a maximal contribution to a property of a blend. Unfortunately in spite of numerous attempts, there is still no satisfactory theory for predicting the percolation threshold in polymer blends, that is, the concentration at which the dispersed phase becomes a continuous phase under given properties of the components and mixing conditions. Till now, researches are guided mainly by a simplest empirical rule, which states that the lower the

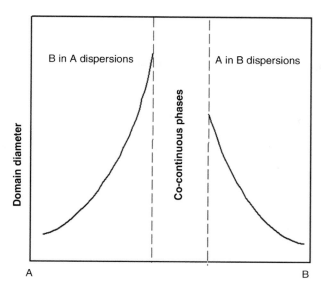

B in A dispersions

A in B dispersions

Co-continuous phases

Domain diameter

A

B

**Figure 6.13** Schematic dependence of domain diameter on composition in blends of polymers A and B.

viscosity and elasticity of the dispersed phase compared with a matrix, the lower the concentration, at which it goes into the continuous phase [28,30].

Many interesting studies on the structure and properties of co-continuous phase structures in binary polymer blends have been published. The authors [62] report very low percolation thresholds for polyethylene in polystyrene binary blends of very high molecular weight. In ternary blends, the main attention is also paid to co-continuous morphologies. Here, the results of research are expected to be even more interesting, since the continuous phases can be developed by two or all three phases. Also, the influence of the component ratios in the composite domains on the phase sizes is of particular interest and importance.

For example, Luzinov *et al.* [64] studied the morphology of the ternary synthetic styrene-butadiene rubber (SBR)/PS/PE blends in relation to the weight ratio of the minor phases SBR and PE, forming the composite dispersed droplets in PS. It was demonstrated that whatever the content of PE, it always formed core structures in the SBR phase, as expected from the spreading coefficients theory. The authors showed that the morphology of the PE/SBR composite domains corresponds to phase morphology of binary blends. When the PE/SBR ratio in this binary blend is lower than the percolation threshold, the PE phase forms multiple subinclusions in the SBR domains (Figure 6.14a). It is interesting that when PE content is increased beyond the percolation threshold, the phase inversion, nevertheless, does not occur because this structural transformation is unfavorable thermodynamically. Instead, a single core morphology forms where large cores of PS are enveloped by the SBR phase (Figure 6.14b). The authors

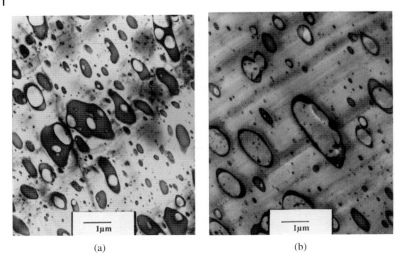

(a)  (b)

**Figure 6.14** Transmission electron micrographs of SBR–2/PS/PE blends of different PE contents in the mixed SBR-2/PE dispersed phase: (a) 15% and (b) 67%. Transition from a multicore (a) to single core (b) composite domains is observed [64], with permission © 1998 Elsevier Science Ltd.

observed also that at high PE contents, small SBR domains could be found in the PE cores (Figure 6.14b). They suggested that this was due to the tendency of PE to encapsulate SBR at PE loadings larger than the theoretical phase inversion point. Also, when the SBR content in the ternary blends was maintained constant at 25 wt% while the PE/PS ratios were in the range from 40/60 to 60/40, the formation of three co-continuous phases was observed [65].

Omonov *et al.* [66] observed similar effects of composition ratios on the development of co-continuous morphology in the PP/polyamide 6 (PA6)/PS blends prepared in a twin-screw extruder. While 15/70/15 compositions were constituted from PP/PS core–shell composite droplets dispersed in the PA6 matrix, the 40/30/30 blends formed three co-continuous phases.

A similar system on the basis of the same polymers but with PP as the matrix phase was studied in the work by Li *et al.* [67] and Wang *et al.* [68]. The blends presented a core/shell morphology with PS as shell and PA6 as core in the PP matrix. The thickness of the PS shell decreased with decreasing the component ratio PS/PA6 or PS content. At the same time, the type of morphology of the ternary blends was unaffected by the core/shell component ratio and depended on the signs of the spreading coefficients only. For the composition PP/PS/PA6 40/30/30, the formation of a triple-continuous morphology was observed [68].

Hemmati *et al.* [45] have reported on the influence of compositional effects on the encapsulated phase morphologies of ternary blends composed of the PP matrix phase and the HDPE/EPDM core/shell minor phases. It was shown that the increasing HDPE content increased both the sizes of the HDPE cores and the HDPE/EPDM composite domains as a whole. Increasing the EPDM content at

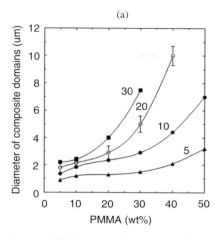
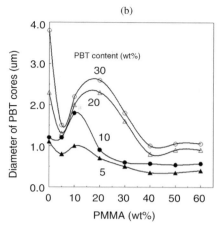

**Figure 6.15** Influence of the content of the PMMA enveloping phase on the average equivalent volumetric diameters of the PBT + PMMA composite domains (a) and the PBT cores only (b) in ternary PMMA/PS/PBT blends. Parameter is the PBT content (wt%) [69], with permission © 2015 Taylor & Francis.

constant HDPE concentration resulted in the increase of the composite domains sizes and reduction of the HDPE cores. The authors concluded that composition affected the domain sizes, while having no noticeable impact on the type of morphology of blends.

Letuchii *et al.* [69,70] systematically studied ternary blend morphologies PMMA/PS/poly(butylene terephthalate) (PBT) in the composition range where concentrations of PBT and PMMA minor phases were limited to those beyond which a complete phase inversion could take place. In order to obtain a full panorama of morphologies developed in this concentration range, the composition of the blends was varied in a systematic manner. Figure 6.15a shows the dependences of the average equivalent volumetric diameters of the composite core/shell PBT/PMMA domains on concentration of the PMMA shell-forming phase. Each curve on these graphs is plotted for the blend with particular fixed concentration of the PBT phase shown by numbers. The composition ratios in ternary blends at any experimental point are easily obtainable in the following way. If, for example, the PBT content is 30% (the upper curve) and the PMMA concentration is 20% (abscissa), then the PS matrix content is 100 − (30 + 20) = 50% and the composition ratio is the PBT/PS/PMMA 30/50/20. As expected, Figure 6.15a illustrates an enlargement of the total sizes of the composite domains PBT + PMMA on the increase of the content of the inner phases.

The next features, which are worth bringing into focus, concern experimental results shown in Figure 6.15b, depicting influence of the content of the PMMA enveloping phase on the diameters of PBT subinclusions. Note that the experimental points on the ordinate of Figure 6.15b belong to average volumetric equivalent diameters of the PBT domains in binary PBT/PS blends. One can observe that subsequent addition of 5 wt% of PMMA to binary 30/70 PBT/PS compositions (upper curve in Figure 6.15b) results in a more than twofold decrease of the PBT subinclusion sizes in the ternary blends. This dispersive effect is probably due to substitution of the PBT/PS interface with the interfacial tension of 3.5 mN m$^{-1}$ for the PMMA/PS interfaces with much lower interfacial tension (0.45 mN m$^{-1}$, at 235 °C). As a consequence, the interfaces become more deformable in flow and favor the formation of smaller domains of the dispersed phase during melt blending. Such behavior is well established theoretically [71,72] and proved experimentally elsewhere [73].

Further increase of the content of PMMA results in increasing PBT domain sizes until the component ratio of binary PBT/PMMA reaches a critical value of percolation threshold under which the multiple PBT subinclusions begin to form [64]. Since in that case an increase of the PMMA shell phase is followed by a corresponding decrease of the PS matrix content, it leads simultaneously to the increase of the dispersed phase concentration and total sizes of the composite domains. As far as, according to thermodynamic demand, PBT must fill in the PMMA "reservoirs" the increase of PMMA may result in corresponding enlargement of the PBT inclusions. As Figure 6.15b shows, when PBT/PMMA ratio becomes lower than the critical value, the multiple PBT subinclusions are formed and their sizes reduce with PMMA content.

Data in Figure 6.15 can be illustrated by microphotograps published by Luzinov *et al.* [64] who studied the morphology of ternary blends PS/SBR–2/PE. The authors used constant content of the major component (PS; 75 wt%) while changing the weight ratio of the two minor constitutive polymers – styrene butadiene rubber, SBR-2 (42.7 wt% bound styrene) and LDPE. The micrographs presented show the development of phase morphology of blends on a gradual increase of the content of the rubber phase. One can observe that if the content of the core-forming phase PE does not exceed 30% of the total content of the disperse phase (PE + SBR), then the composite domains contain more than one core (Figures 6.16a,b). When this content is 50% or higher, the composite droplets consist of single core (Figures 6.16c,d). This phenomenon can be understood easier by imagining that the ternary encapsulated blend is composed from two independent binary blends. In the case under consideration there are PS/SBR and PE/SBR. Since the PS matrix contacts with the SBR shell only and does not "see" the PE phase at all, this binary blend must follow the rules for the ordinary binary mixtures. For example, the larger the content of the composite domains (PE + SBR) the larger is the domain sizes.

The second blend PE/SBR representing the composite droplets in which SBR and PE serve as the matrix and dispersed phase has one unusual feature. Based on requirements of the interfacial forces the phase inversion may not take place here. Since otherwise it is also an ordinary binary mixture, then the smaller the content of the dispersed phase PE with respect to the matrix SBR the smaller is PE-core dimensions. That is what can be observed in Figures 6.15 and 6.16.

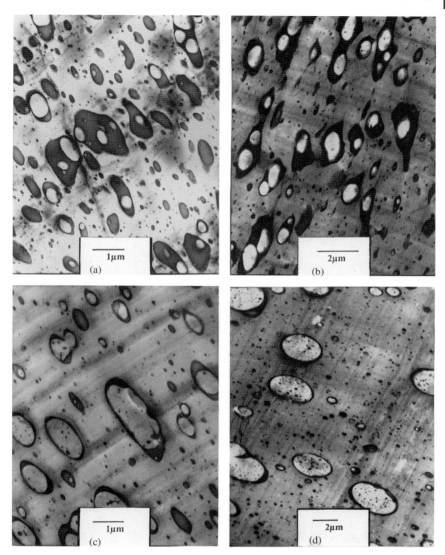

**Figure 6.16** Transmission electron micrographs of PS/SBR-2/PE blends of different PE contents in the mixed SBR–2/PE dispersed phase: (a) 15%; (b) 32%; (c) 50%; (d) 67% [64], with permission © 1998 Elsevier Science Ltd.

### 6.3.3.2  Kinetic Factors

Studies on the encapsulated morphologies in multiphase, mainly ternary, immiscible polymer blends have become very popular in the recent years. Expressing a consolidated opinion about the validity of the spreading coefficient theory and the key role of interfacial forces in development of the encapsulated morphologies, most of the authors also reported on the importance of kinetic factors. Indeed, although the encapsulation itself is a thermodynamically driven phenomenon, the rate of approach to the equilibrium morphology is a time-dependent process controlled by phase viscosity (and elasticity) ratios, the type of flow in a mixer, the intensity and sequence of mixing, and so on. Note that the term "equilibrium morphology" refers here to the dynamic equilibrium which is reached when the phase morphology is no longer changes with a further mixing, at other conditions remaining unchanged.

The effect of the viscosity ratio is still controversial among the authors. MacLean [74] and Minagawa and White [75] reported that low viscosity melts tend to encapsulate high viscosity melts during flow. Nemirovski *et al.* [76] have shown that even though encapsulation is predicted by a positive spreading coefficient it may be hindered by high viscosity of the spreading polymer. They studied the PP/LCP/PS ternary system and claimed that if both thermodynamic conditions, that is, positive spreading coefficients and kinetics factors, that is, phase viscosity ratio ($\eta_{PS}/\eta_{LCP} < 1$) were favorable for the encapsulated behavior, then PS would encapsulate LCP. In cases where these effects were opposing, kinetics could hinder the development of predicted morphology. In agreement with these authors, Koseki *et al.* [56] believe that the viscosity effect might be an important factor in resisting the thermodynamic driving force for encapsulation and thus making it difficult to predict equilibrium morphology from simple thermodynamic considerations. In contrast, Gupta and Srinivasan [77] observed that in PC/hydrogenated styrene-b-butadiene-b-styrene triblock copolymer (SEBS)/PP blends, when SEBS had a higher viscosity than PC, SEBS formed a boundary layer around the PC phase, and otherwise separated disperse morphology was observed. Kim and Kim [78] mentioned that in a ternary polyolefin system, when the two minor phases have equal compositions, the dispersed phase with lower viscosity forms a shell around the other phase.

Le Corroller and Favis [79] have shown that in the PE/PP/PS blends in which the interfacial driving forces for partial wetting were weak, the PE viscosity could have a strong impact on the phase morphology. The viscosity directly affected the quantity and size of PS droplets at the PE/PP interface during annealing. Luzinov *et al.* [64] claimed that the core diameter is determined by the viscosity ratio of core to shell and also the composite droplet size is affected by the viscosity ratio of the shell to matrix. Hemmati *et al.* [44] modified this assumption and claimed that in ternary polymer systems, the average viscosity ratio of minor phases ($\eta_{av}$) to the matrix determines the droplet size, which is easily calculated by the mixture law:

$$\eta_{av} = \eta_1 x_1 - \eta_2 x_2,$$

where $x_1$ and $x_2$ are volume fractions, and $\eta_1$ and $\eta_2$ are viscosities of the dispersed phase.

Hemmati *et al.* [14] working with various ternary blends of 70/15/15 PP/EPDM (or EPR)/HDPE showed that variation of the rubber/HDPE torque (viscosity) ratio from 0.86 to 82 produced no perceptible effect on the type of blend morphology. In agreement with spreading coefficients, the rubber phase invariably encapsulated the HDPE drops. Lusinov *et al.* [65], Reignier *et al.* [47], Letuchii *et al.* [69,70] and Miroshnikov [80] in their studies on different ternary blends composed of the dispersed phases with different core/shell viscosity ratios arrived at the same conclusions.

Considering all of the results reported above, it can be admitted that the effect of viscosity ratio in controlling the morphology of ternary blends compared to interfacial tension is of a smaller importance.

As far as mixing kinetics for the ternary blends with encapsulated morphologies is concerned, again, data from different sources differ in some aspects. Nemirovski *et al.* [76], for a PP/liquid crystalline polymer (LCP)/PS ternary blend with PP as the matrix phase, reported a gradual time-dependant encapsulation process. Complete engulfing of the LCP cores by the PS layers was reached after 15 min of mixing. Note that the viscosity of the PS shell was three times greater than that of the LCP core. Tchomakov *et al.* [55] have shown that the PS/HDPE/PMMA blends processed through a twin-screw extruder retain stable core–shell morphology in spite of large variations in the minor phase viscosities and processing conditions. Letuchii and Miroshnikov [81] working with the PMMA/PS/PBT ternary blends in which PMMA encapsulated PBT cores reported that in spite of the PMMA phase viscosity was eight times higher than PBT, no detectable sign of the enveloping retardation was observed. Almost all the drops of the PBT were engulfed by the PMMA after 2–3 min of mixing. Reignier and Favis [53] for the HDPE/PS/PMMA 70/15/15 ternary blends in which the less viscous PS enveloped around the PMMA domains (PS/PMMA viscosity ratio was 0.24) detected very fast engulfing. The encapsulation was already completed after 2 min of mixing and further processing of this system up to 15 min produced only minor changes of morphology. In fact, the results of these works, which reflect the kinetic factors, probably, do not contradict each other since the enveloping behavior depends not only on the phase viscosities but also on corresponding interfacial tensions as well.

As far as the dependence of the blend morphology on the mixing sequence is concerned there is no common point of view among the authors. Li *et al.* [67], for uncompatibilized ternary blends, have not observed any noticeable influence of the mixing sequence on the blend morphology. Huang *et al.* [82] studied PP/ethylene/1-octene copolymer (rubber)(EOR)/EOR-g-MA and claimed that the order of mixing, namely simultaneous mixing, master batch preparation, or premixed method, does not affect the particle size. They report that the mixing sequence seems to cause a negligible difference in the average size of the particles or their polydispersity for the blends with a unimodal particle size distribution regardless of the matrix polymer used; however, for the blends having a bimodal particle size distribution, the order of mixing seems to affect the dispersed rubber particle size a little more but still not significantly.

On the other hand, Shokoohi *et al.* [83] working with multiphase polymer blends have indicated that the mixing sequence had a strong impact on the final morphology. Letuchii and Miroshnikov [81] observed several interesting phenomena in the study of the kinetics of mixing of the PMMA/PS/PBT and

PMMA/PS/PC 20/60/20 ternary blends in which components were labeled 1, 2, and 3. The numbers refer to the primary dispersed phase, matrix, and secondary dispersed phase, respectively. In agreement with the spreading coefficients theory, phase 1 (PMMA) encapsulated phases 3 (PBT or PC with the development of the composite domains dispersed in matrix 2 (PS). The apparent viscosities of the polymers in blends PMMA/PS/PBT and PMMA/PS/PC were, respectively, (kPas) 1.6/0.4/0.2 and 1.6/0.4/5.4.Three mixing sequences used in this study, denoted MS1, MS2, and MS3, represent the following order of addition of the polymer components upon blending respectively: $(1+2)+3$, $(1+3)+2$, and $(2+3)+1$. In other words, the adopted procedure involved a two-stage process: (i) preblending of the corresponding binary compositions for 3 min and (ii) addition of the corresponding third polymer and further mixing of the ternary composition for up to 40 min.

The kinetic curves of blending for both MS1 (curves 1, 1′) and MS2 (curves 2, 2′) mixing procedures for the PMMA/PS/PBT blends are shown in Figure 6.17. Experimental points on this graph represent volume average equivalent diameters of the composite PBT/PMMA core/shell domains (curves 1, 2), and those for the PBT cores only (dotted curves 1′, 2′) in the ternary blends as determined by TEM.

There are several aspects that attract attention. The first and rather ambiguous one, at first glance, concerns the inverse kinetics of mixing. In some cases, up to a sevenfold increase of average domain sizes with mixing time was observed. Second, MS1 and MS2 result in very different degrees of blend dispersion,

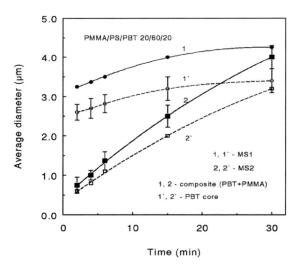

**Figure 6.17** Kinetic curves of mixing for ternary PMMA/PS/PBT 20/60/20 blends prepared at 235 °C according to MS1 (curves 1, 1′) and MS2 (curves 2, 2′) procedures. Solid lines 1 and 2 represent outer diameters of the composite PBT/PMMA core/shell domains while dotted lines 1′ and 2′ refer to those of PBT cores only.

especially during initial and intermediate stages of mixing. Third, MS1 produces minor change of dimensions of both PBT cores and composite domains throughout the blending cycle, while MS2 (curves 2, 2') resulted in a significant enlargement of domain sizes with time. Fourth, the differences between domain sizes obtained with the MS1 and MS2 procedures gradually diminish with blending time. Besides, one can observe that about 20 min of mixing was enough to nearly reach the equilibrium domain sizes with MS1 while with MS2 the equilibrium morphology was not attained even after 30 min of mixing.

SEM microphotographs depicted in Figure 6.18 show that at the very beginning of mixing of a ternary blend, that is, after 1 min (Figure 6.18a), the minor

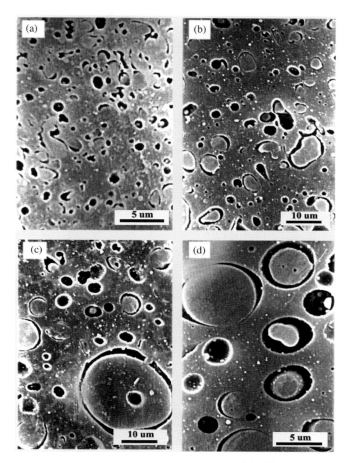

**Figure 6.18** Evolution of phase morphology in PMMA/PS/PBT 20/60/20 blend prepared with MS2 procedure (PBT/PMMA + PS). Heavy dark areas belong to the PMMA shell phase which were leached with cyclohexane. Time was counted from the moment of addition of the PS phase: (a) 1 min; (b) 10 min; (c) 30 min; (d) 30 min (at a higher magnification). Compare domain sizes after 1 (a) and 30 (d) min of blending at similar magnification.

components are dispersed rather finely and predominantly separately: most hollows left after leaching of the PMMA phase have no PBT cores. At the same moment, enveloping also starts and proceeds with mixing time until the PMMA drops "find" their PBT counterparts. A relatively high degree of dispersion of the ternary mixture in the initial stages of mixing can be due to the fact that until the internal phases engulfed each other, the component ratio of PMMA/PS and PBT/PS is 20/60 and, therefore, favors the formation of small drops of PMMA and PBT. On proceeding of phase self-assembly and encapsulation, the composite domains are formed, the volume of which is equal to the sum of volumes of both the dispersed phases. As a result, the dispersed phase content doubles PS/ (PMMA+ PBT) = 60/40 leading to intense coalescence and growth of the particle sizes.

Qualitatively similar results were obtained with the other studied system. Detailed quantitative information on the mixing kinetics for ternary PMMA/PS/ PC 20/60/20 blends prepared with the use of MS1, MS2, and MS3 procedures is given in Figure 6.19a. These data show again that the different blending procedures resulted in different phase dimensions. At the same time, the inverse character of the blending kinetics remained unvaried.

It was important to find out whether the observed phenomena would take place with other component ratios. To this end, the total content of the dispersed phases of the previous system was reduced to 15 wt% and the core/shell PC/PMMA ratio was changed to 1/2. Detailed kinetic curves of mixing for ternary PMMA/PS/PC 10/85/5 blends were obtained separately for the core, shell, and composite domain sizes, using the MS2 procedure. Figure 6.19b shows that the variation of component ratio did not alter the inverse character of the kinetic curves. Quantitative changes compared to Figure 6.19a concerned only the

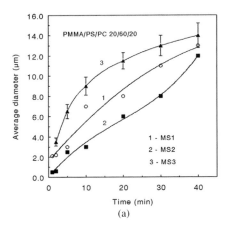
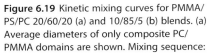
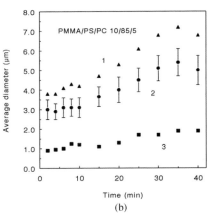

**Figure 6.19** Kinetic mixing curves for PMMA/ PS/PC 20/60/20 (a) and 10/85/5 (b) blends. (a) Average diameters of only composite PC/ PMMA domains are shown. Mixing sequence: 1 – MS1 (PMMA/PS + PC); 2 – MS2 (PMMA/PC + PS) and 3 – MS3 (PC/PS + PMMA). (b) MS2; 1 – composite PC/PMMA domains; 2 – the PMMA shell phase; and 3 – PC cores only.

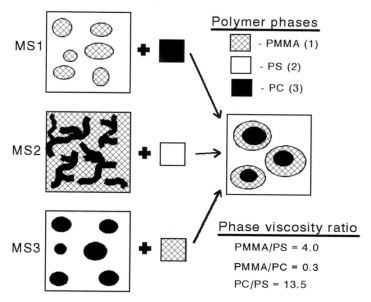

**Figure 6.20** Illustrations showing schematically transition from binary to ternary PMMA/PS/PC morphologies for the three mixing sequences.

smaller domain sizes, apparently due to the lower content of the dispersed phases. SEM images for this blend (not shown) also did not differ much from those shown in Figure 6.18.

The error bars depicted in Figure 6.19b and previous graphs indicate large variations of the domain sizes which are not unusual for polymer blends. Keeping this fact in mind, it was difficult to give a distinct answer concerning the time of development of equilibrium phase morphologies in the studied blends. The only possible conclusion that could be made on this matter was that mixing for 40 min resulted in attaining of nearly equilibrium phase morphologies.

Discussing the influence of the mixing sequences on the domain sizes one has to keep in mind that, according to MS1, MS2, and MS3, different phase morphologies of the starting binary blends composed of different polymers were formed (see sketch in Figure 6.20). On the other hand, the equilibrium phase morphology should not to be dependent on the mixing sequence. Therefore, it is reasonable to suggest that different phase dimensions of blends at intermediate mixing times are due to different pathways from starting to final morphologies.

### 6.3.3.3 Morphological Types
Ravati and Favis [84] and Ravati [85] examined the complete range of morphological types in ternary blends of high-density polyethyleneHDPE,PS, and PMMA prepared by melt mixing over the entire composition variation. HDPE, PS, and PMMA are selected as a model system showing a positive spreading coefficient of PS over PMMA. Thus, in all cases the PS phase separates HDPE

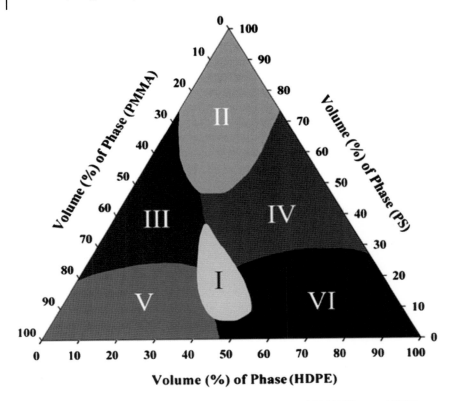

**Volume (%) of Phase (HDPE)**

**Figure 6.21** Triangular concentration diagram showing the various regions of the morphological states for ternary HDPE/PS/PMMA. I – tri-continuous (HDPE/PS/PMMA); II – matrix PS/two dispersed HDPE and PMMA; III – by-continuos PS and PMMA/dispersed HDPE phase; IV – by-continuos PS and HDPE/dispersed PMMA phase; V – matrix PMMA/core (HDPE) – shell(PS) dispersed phase; VI – matrix HDPE/core(PMMA)–shell(PS) dispersed phase.

and PMMA. Four subcategories of morphologies were identified, depending on the composition of phases (Figure 6.21), including: (a) matrix/core–shell dispersed phase (V, VI); (b) tri-continuous (I); (c) bi-continuous/dispersed phase (II, IV); and (d) matrix/two separate dispersed phases (II) morphologies. The phases in these submorphologies are identified and illustrated qualitatively by electron microscopy as well as a technique based on the combination of focused ion beam irradiation and atomic force microscopy.

Letuchii *et al.* [69,70] also studied the effect of composition on the formation of different types of morphology in the PMMA/PS/PBT blends. Figure 6.22 shows photomicrographs of phase structures in ternary PMMA/PS/PBT 40/40/ 20 blend. TEM micrograph of ultrathin section in Figure 6.22a shows signs of a continuous phase of PMMA (white zones), but these data required direct confirmation. Therefore, the complete selective extraction of the PS matrix with cyclohexane was made. As a result, the sample retained its integrity. This means that

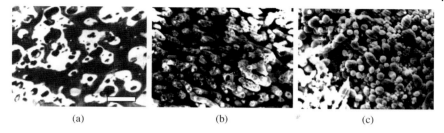

(a)  (b)  (c)

**Figure 6.22** Phase morphologies in the PMMA/PS/PBT 40/40/20 blends: (a) ultrathin section (TEM); (b) the sample after extraction of the PS matrix phase (SEM) and (c) particulate PBT phase fragments (SEM) after selective dissolution of PS and PMMA phases with toluene. Scale bar is 2 μm for all images.

the PMMA and/or PBT continuous phase has been developed (Figure 6.22b). To answer the question which of the polymers formed co-continuous phase, the selective extraction of both PMMA and PS phases with toluene was undertaken. It was found that this procedure led to the loss of integrity of the sample and formation of a suspension of discrete particles of PBT phase shown in Figure 6.22c. Thus, these data confirm the development of the co-continuous PMMA phase at this component ratio.

At a higher component ratio of PBT/PMMA (PMMA/PS/PBT 20/40/40), the co-continuous structures are formed by all three phases (Figure 6.23). The fact that such morphologies are not identified in the ultrathin sections (Figure 6.23a) shows the limited capacity of TEM and the need of independent methods. Using the selective dissolution of the components illustrated in the previous example (Figure 6.22), made it possible to separate the PBT and PMMA phases and unambiguously prove their continuity (Figures 6.23b and c). It is understood that since the PMMA phase envelops PBT, the transition of the latter to the continuous phase automatically results in a co-continuity of the PMMA shell phase.

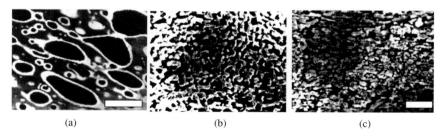

(a)  (b)  (c)

**Figure 6.23** Phase morphologies in the PMMA/PS/PBT 20/40/40 blends: (a) ultrathin section (TEM); (b) co-continuos PMMA + PBT phases after extraction of the PS matrix (SEM) and (c) PBT co-continuos phase after selective extraction of PS and PMMA phases with toluene(SEM). Scale bar is 2 μm (a) and 5 μm (b,c).

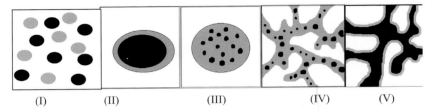

**Figure 6.24** Morphological types in ternary polymer blends with phase encapsulation.

Different structures formed by the core and shell phases placed into a matrix phase define a variety of morphologies in these systems. They have been classified [69] in five types of morphologies and characterized as follows (Figure 6.24): I – separated dispersed phases; II – single core–shell; III– multiple cores–shell; IV – dispersion of the separate domains in one of the bicontinuous phases; V – all three phases are co-continuous. (Morphology of partial encapsulation (partial wetting) is not considered.)

Reignier and Favis [53,54] and Reignier *et al.* [47] have made a systematic study on the morphology evolution in ternary PS/HDPE/PMMA blends in which PMMA droplets enveloped by the PS phase were dispersed in the HDPE matrix. As far as the morphology–composition relationships for the composite domains are concerned the results by Luzinov *et al.* [64] have been confirmed. Experimental data indicated whether upon melt mixing or quiescent annealing the inner phases coalesce separately [53,54]. It was found also that the composite domains merge similar to pure PS droplets provided the shell PS layer is thicker than 0.2 μm. In ternary blends with the HDPE matrix at fixed content of 80 vol.%, the increase of the PS shell phase resulted in significant reduction of the composite domain total sizes. This effect was attributed to the reduction of both the interfacial tension and composite domain viscosity with increasing shell thickness [53].

It was shown [46] that in the presence of a perfectly segregated PMMA core/PS shell in the composite droplets, the thinnest possible molecular-scale shell thickness can be developed. The authors declared that by optimizing the viscous and interfacial effects it is possible to prepare completely segregated core–shell dispersed morphology with a controlled shell thickness ranging from 40 to 300 nm.

### 6.3.3.4 Multiple Percolated Structures

Ravati [85] and Zhang *et al.* [86] reported on a potentially interesting way to achieve low percolation threshold materials by combining a co-continuous structure with a composite droplet type of morphology. They prepared ternary percolated co-continuous systems in HDPE/PS/PMMA blends in which HDPE and PMMA formed two continuous networks, while the PS build up a continuous sheath structure at the HPDE/PMMA interface and studied the effect of composition on the continuity of PMMA/PS/HDPE ternary blends. The continuity of the blends was evaluated by selective solvent extraction of the PMMA and PS phases, respectively, in acetic acid and cyclohexane. Mass loss

(a)

(b)

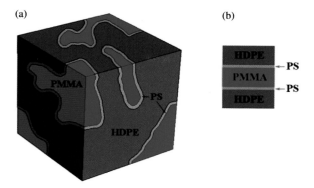

**Figure 6.25** Three dimensional view of the blend structure, showing the multiple percolated structure: thin PS layer at the HDPE/PMMA interface [86], with permission © 2007, American Chemical Society.

measurements were carried out for 1 week and were used to calculate the extent of continuity of the PMMA and PS phases as

$$\% \text{ Continuity} = \frac{W_{\text{initial}} - W_{\text{final}}}{W_{\text{initial}}(i)}. \tag{6.24}$$

Here $W_{\text{initial}}$ is the initial mass of the sample, $W_{\text{final}}$ is the final mass of the sample, and $W_{\text{initial}}(i)$ is the mass of polymer 1 contained in the sample before selective extraction, calculated by considering the blend as homogeneous. This multiple percolated, interfacial tension driven structure (Figure 6.25) resulted in a dramatic decrease in the percolation threshold volume fraction of the encapsulating PS component. As little as 3% polystyrene is shown to generate a continuous PS network (Figure 6.26). The encapsulation effect follows the prediction based on the spreading coefficient theory. According to the authors, this approach could be used as a route to generate novel co-continuous structures and also as a technique to significantly reduce percolation thresholds in multiphase blends.

Using methods of the selective phase extraction and electron microscopy (TEM and SEM) the concentration corresponding transition of the PMMA and PBT former dispersed phases to co-continuous ones (i.e., percolation thresholds) were identified with up to 1–2 wt% error [70]. Figure 6.27 illustrates the threshold lines for the PBT core (curve 1) and PMMA shell (curve 2) phases. The ternary compositions situated above each line correspond to co-continuous phases, those below each line correspond to particulate (discrete) morphologies. The dashed lines in Figure 6.27 show how to evaluate blend compositions at the percolation threshold. The PMMA phase can reach percolation threshold at 20 wt% provided the content of PBT and the PS will be 30 wt% and $100 - (20 + 30) = 50$ wt%, respectively.

Clearly, the percolation threshold for the composite domains may be exceeded in excess of the total concentration of the dispersed phases. This occurs as a

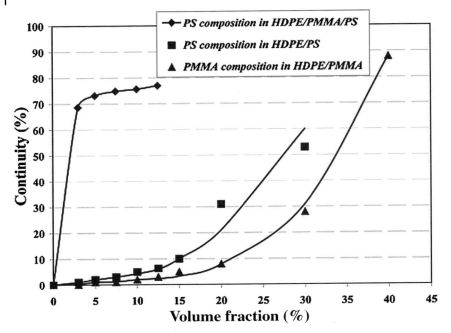

**Figure 6.26** Ultralow percolation threshold in ternary co-continuous polymer blends of PS/HDPE/PMMA [86], with permission © 2007, American Chemical Society.

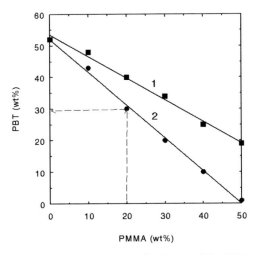

**Figure 6.27** Concentrations of transition of the PBT (curve 1) and PMMA (curve 2) phases in ternary blends PMMA/PS/PBT to co-continuous structures (percolation thresholds). Compositions above and below each lines denote, respectively, co-continuous and discrete phases.

result of increasing the concentration of one or the other of the inner phases, which invariably leads to an increase in the total volume of the composite domains. However, if the content of the core phase is low then the increase of the shell phase will result in the co-continuity of the latter phase only. If, however, the percolation threshold is reached for the core phase, both phases become continuous. This is because, even at very low concentrations of PMMA, this phase, in accordance with the thermodynamic demands, should cover (envelope) the core phase (PBT). In this example, the minimum recorded percolation threshold for the PMMA was 10%. Other lower compositions simply have not been studied. It is believed, however, that careful selection of the ternary blend components, their composition, and viscoelastic and surface properties can, as it was mentioned earlier, achieve much lower percolation threshold values.

It may be concluded that composition have strong impact on the phase morphologies of immiscible ternary blends with encapsulated inner phases. By appropriate choice of the content of constituent polymers, polymer blends with two or three co-continuous phases can be prepared. Increase of concentration of one of or both dispersed phases result in increasing total sizes of the composite domains. Mutual disposition of the core and shell phases in the composite domains is developed in agreement with thermodynamic demand of minimization of surface free energy. Therefore, increasing the concentration of the core phase will never result in phase inversion phenomena. Instead, at high core phase content, the single core composite droplets are formed while at low content – the multiple cores are developed.

### 6.3.3.5 Partial Wetting Morphology

Partial wetting in a ternary polymer blend is the thermodynamic state where all three phases meet at a three-phase line of contact. Pickering emulsions, where solid particles situate at the interface of two other phases, is a classic example of this state. Most of the papers published in recent years on ternary polymer blends have studied morphology formation related to complete wetting, while the partial wetting has attracted much less attention [51,87,88]. Debolt and Robertson [87] studied the morphologies of a ternary blend of Nylon 66 and polystyrene in a polypropylene matrix with and without compatibilization and have found "an association of the polystyrene with the Nylon in the compatibilized blends" (i.e., partial wetting). Horiuchi *et al.* [51] observed partial encapsulation in ternary blends of polyamide-6 (PA6)/PC/PS and PA-6/PC/styrene-(ethylene-butylene)-styrene triblock copolymer (SEBS). They were able to switch to complete encapsulation by using reactive species of PS and SEBS grafted with maleic anhydride (PS-*g*-MA and SEBS-*g*-MA).

Le Corroller and Favis [79] examined the effect of polyethylene viscosity on the partial wetting of ternary PE/PP/PS systems displaying a boundary state of partial to complete engulfing behavior. This work demonstrates that viscous effects can play an important role in the development of a partial wetting morphology at the boundary condition when the driving forces for partial wetting are weak. The high interfacial mobility generated during static annealing in the

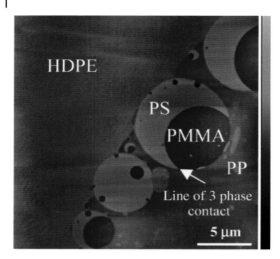

**Figure 6.28** Topographical images of blend PMMA/PP/PS/HDPE displaying partial wetting between the components [42]; with permissions © 2009, American Chemical Society.

low viscosity PE (PE1) system facilitates the disengagement of larger droplets from the interface. However, when the mobility is reduced through the use of a high viscosity polyethylene (PE2) a closely packed PS droplet array at the PE/PP interface is generated and lateral coalescence promotes the PS droplet size growth. A ternary PE1/PP/PC system was also investigated. Such a system displays a strong driving force for partial wetting as suggested by the spreading theory and leads to all PC droplets being located at the PE/PP interface. The influence of the viscous effects noted for PE1/PP/PS is not apparent for the PE1/PP/PC system and the latter displays an almost pure partial wetting behavior.

In addition, a few papers have used the Neumann triangle method to measure interfacial tensions between different polymer pairs [42,89–93]. The blends under consideration all must be of the partial wetting type of morphology, which is a necessary condition for the Neumann triangle method to be applicable (Figure 6.28).

In that method, a droplet of polymer 1 has to be placed between two films of polymer 2 and 3, as shown in Figure 6.12. After melting, the droplet of polymer A reaches equilibrium geometry, and the three phases meet along a common line of contact. For this method to be applicable, the three spreading coefficients must be negative, which corresponds to partial wetting, as illustrated in Figure 6.9. Virgilio *et al.* [41] observed a self-assembling close-packed droplet array at the interface in ternary blends 45/45/10 HDPE/PP/PS featuring partial wetting.

Self-assembly [94] is the fundamental principle that generates structural organization on all scales from molecules to galaxies. It is defined as reversible processes in which pre-existing parts or disordered components of a pre-existing system form structures of patterns. Examples of self-assembling system include

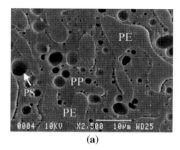 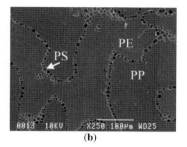

(a)     (b)

**Figure 6.29** SEM micrographs of morphology of HDPE/PP/PS 45/45/10 vol % blends immediately after melt processing with (a) 0% SEB and (b) after 120 min of annealing with 1% SEB modifier. The PS phase has been selectively extracted using cyclohexane [41], with permissions © 2009, American Chemical Society.

weather patterns, solar systems, histogenesis (formation of different tissues from undifferentiated cells), and self-assembled monolayers. The most well-studied subfield of self-assembly is molecular self-assembly (see Wikipedia: http://en.wikipedia.org/wiki/Self-assembly).

The authors [41] have found that the initial mixing of the components in the Brabender mixer (200 °C, 50 rev/min for 8 min) resulted in the formation of the morphology, characterized by a dispersion of droplets PS, located mainly at the HDPE/PP interface with a preference for the PP side (Figure 6.29a). Such an arrangement of the phases is well consistent with the negative values of all three spreading coefficients. In order to further clarify the picture, a quiescent annealing of the blend at 200 °C for 15 to 120 min in the presence of 1% SEB diblock copolymer was undertaken. One can observe that after this procedure PS particles were localized mainly at the HDPE/PP interface (Figure 6.29b).

It is shown that the affinity of the PP droplets to the HDPE/PP interface is predicted by the spreading coefficients theory and can be further controlled by the addition of a diblock copolymer SEB. Even at a high concentration, the PS drops did not coalesce to form a continuous layer on the HDPE/PP interface. Such highly organized Pickering-type structures [94–97] are developed in the case of partial wetting of the HDPE/PP interface by the drops of the PS phase. In the classical Pickering emulsion, the emulsion is stabilized by the adsorption of small colloidal particles at the oil–water interface. It is believed that a significant perfection of the morphology due to the annealing process is associated with a "sweep and grab" effect induced by the coarsening of the HDPE/PP co-continuous morphology. It is believed that this work opens promising prospects for the development of new, complex microstructures in polymer blends.

Ravati and Favis [98] analyzed tunable morphologies in ternary blends comprising biodegradable polymers: poly(butylene succinate) (PBS), poly(lactic acid) (PLA), poly(butylenes adipate-co-terephthalate) (PBAT) and polycaprolactone (PCL). These blends represented a special case when the components had very low interfacial tensions and all the spreading coefficients were close to zero. As a result, replacing one component with another can change the sign of the spreading coefficient and lead to development of a new morphology. Three different morphological types were detected in this study: partial wetting for PBS/PLA/PCL in which PLA droplets self-assembled at the PBS/PCL interface (Figures 6.30a,f). Complete wetting for PBS/PBAT/PLA (Figures 6.30b,h) occurs when the PBAT developed an interlayer between PBS and PLA phases, and an

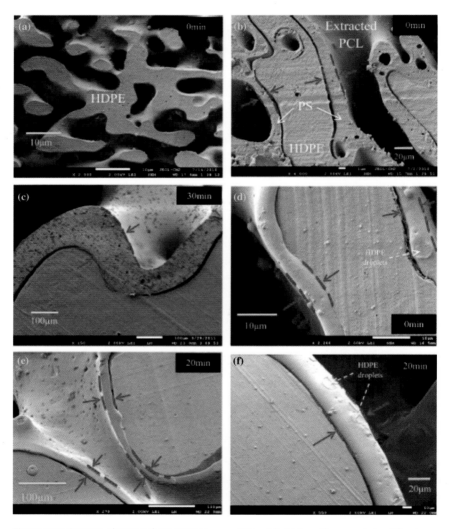

**Figure 6.30** Pictures, showing three wetting morphologies for the ternary blends A/B/C: (a), (f) partial wetting; (b), (h) complete wetting; (c), (g) combined partial and complete wetting; (d), (e) how they are derived from the binary blend form [98], with permission © 2013 Elsevier Ltd.

unusual morphology combining both partial and complete wetting cases for the PBS/PBAT/PCL blend (Figures 6.30c,g).

### 6.3.4
### Ternary Blends with Separated Dispersed Phases

The final morphology of binary systems, provided a sufficiently long time of blending, is developed as a result of multiply repeated events of breakup and coalescence of droplets of the dispersed phase. The situation becomes even more complicated when three immiscible components are blended. Indeed, as outlined above, introduction of the second dispersed phase into a binary system may not only lead to the formation of a definite phase structure, but it is expected that interaction between two systems of dissimilar droplets may also give rise to some new effects as well. As it follows from Figure 6.9a, the formation of the individual separated dispersions of the inner phases 1 and 3 in the matrix polymer 2 takes place provided negative signs of the $\lambda_{31} < 0$ and $\lambda_{13} < 0$ and positive sign of $\lambda_{21} > 0$.

#### 6.3.4.1 Effects of Interaction between Dispersed Phases
Morphological features governed by the interplay of the interfacial tensions and resulting in particular mutual phase disposition seem not to be the only phenomena pertaining to multicomponent polymeric emulsions. Provided that separate dispersions of both the dispersed phases were formed, the addition of the third phase 3 to a binary blend 1–2 was shown to produce a dual effect: substantial growth of the degree of dispersion and formation of co-continuous morphologies [99]. It was shown that these effects had complex origin and strongly depended on blend composition, hydrodynamic interaction between domains of the dispersed phases, and on whether the third phase domains were in liquid or solid state during mixing.

Three ternary systems were studied [69, 99] and shown in Table 6.1: System I was composed of the three liquid phases while the rest two included hard polymeric (PBT) or glass microspheres (GMS).

**Table 6.1** Viscous (at shear rate of $16\,s^{-1}$) and surface properties of blends measured at 190 °C.

| System | Primary dispersed phase1 | Matrix2 | Secondary (or third) dispersed phase3 | Apparent viscosities of phases (kPa s) 1/2/3 | Spreading coefficients (mN m$^{-1}$) | | |
|--------|--------------------------|---------|---------------------------------------|----------------------------------------------|-------------------------------------|--------|--------|
| | | | | | $\lambda_{31}$ | $\lambda_{13}$ | $\lambda_{12}$ |
| I | PMMA | PS | PP | 4.9/1.1/0.4 | −7.6 | −2.6 | +0.4 |
| II | PP | PS | PBT | 0.4/1.1/∞ | −11.2 | −11.0 | +3.8 |
| III | PP | PS | GMS | 0.4/1.1/∞ | <0 | <0 | >0 |

#### 6.3.4.1.1 **System I**

Negative signs of spreading coefficients $\lambda_{31}$ and $\lambda_{13}$ as well as positive $\lambda_{21} = +0.4$ for System I predict (Table 6.1) the formation of separate dispersions by both inner phases in the PS matrix. The SEM photographs of the PMMA phase morphologies in binary 45/55 PMMA/PS and ternary 45/54/1 PMMA/PS/PP blends are shown in Figure 6.31a and b, respectively. The binary blend sample disintegrated in cyclohexane (solvent for PS) leaving discrete PMMA domains (Figure 6.31a). These data provides clear evidence that no dual phase formation at this component ratio takes place. The micrograph in Figure 6.31b shows the effect that was produced by the addition of only 1 wt% of PP to this binary blend. It is seen that this ternary 45/54/1 PMMA/PS/PP blend sample disintegrated in toluene while retained its shape and integrity in cyclohexane indicating development of the PMMA co-continuous phase. Thus, the addition of only 1.0 wt% of the PP phase to a binary 45/55 PMMA/PS blend provoked the transition of PMMA from the discrete to co-continuous phase.

TEM micrographs of ultrathin sections from the same blends are shown in Figure 6.32. Note that PMMA domains are seen as the light areas on these images. One can clearly observe a strong dispersive effect produced by addition of the tiny portion of the PP phase. While Figure 6.32a for the binary blend demonstrates the presence of noticeable portions of large PMMA domains, Figure 6.32b for the ternary blend shows a much finer dispersion of the PMMA phase. Obviously, no indication of encapsulation of the inner phases is detectable on these (and many other) images proving the prediction made with Eq. (6.9). It has to be emphasized that the light zones in Figure 6.32b which could be identified visually as the discrete ones, represent in fact the fragments of the PMMA co-continuous phase. Having no acceptable alternative to quantify such morphologies it was decided to measure these visible portions of the PMMA co-continuous phase on TEM images as discrete domains. Eventually, the sizes of these

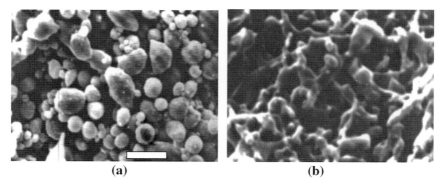

**(a)**          **(b)**

**Figure 6.31** SEM micrographs of PMMA phase morphologies in binary 45/55 PMMA/PS (a) and ternary 45/54/1 PMMA/PS/PP (b) blends. The PS matrix component was leached completely with cyclohexane leaving the discrete PMMA domains in binary (a) or co-continuous PMMA network in ternary (b) compositions. Scale bar is 5 μm.

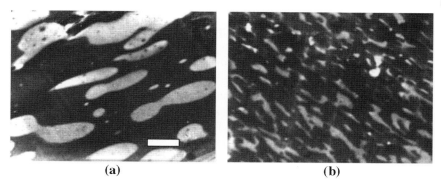

**Figure 6.32** TEM images of 45/55 PMMA/PS binary (a) and 45/54/1 PMMA/PS/PP ternary (b) blends. Scale bar is 1.0 μm.

zones on micrographs can, generally, characterize the degree of dispersion of a system.

It was interesting to further extend these experiments to ascertain how concentrations of each of the dispersed phases affect the morphologies of the ternary blends. The influence of the content of the PP phase on the average PMMA domain sizes in ternary PMMA/PS/PP blends at different concentrations of the PMMA phase (5.0, 20.0, 40.0, and 60.0 wt%) is demonstrated in Figure 6.33. To avoid overlapping, the error bars for the only one of the curves

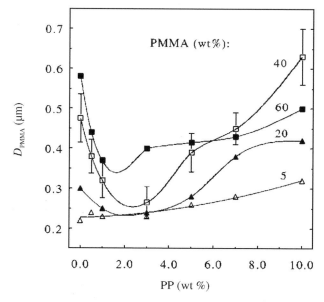

**Figure 6.33** Influence of the PP phase concentration on the average equivalent diameter of the PMMA phase fragments in System I. Parameter is the content of the PMMA primary dispersed phase.

are shown. Note that any variation of the contents of both inner phases was made at the expense of concentration of the PS matrix phase. For instance, the component ratio for ternary blend comprising 40 wt% of PMMA (parameter) and 4.0 wt% of PP (abscissa) is 40/56/4 PMMA/PS/PP.

In Figure 6.33, the experimental points at zero PP concentration represent the PMMA volume average equivalent domain diameters in binary PMMA/PS blends at the indicated PMMA contents. In spite of the complicated character of this graph, a few important phenomena can be visualized. In particular, one can see a trivial effect of coarsening of the PMMA dispersed phase with its content in the blends. As it was mentioned earlier, very small (up to 1–3 wt%) additition of the PP minor dispersed phase to the binary blends produced a pronounced dispersive action.

What is even more intriguing is that this dispersive action of PP gradually diminished with its concentration. One can observe that at 10 wt% of PP, the PMMA fragment sizes in ternary blends grow markedly and become comparable with those in binary compositions. Note also that when the content of the PMMA primary dispersed phase was lower than 5.0 wt%, an addition of the third PP component produced minor, if any, dispersive effect. In contrast, at larger PMMA dosages, a dispersive action of the PP phase was enhanced. It is important to stress again that only small portions of PP were really effective, at least, in the studied systems. Further increase of the PP content diminished its efficacy and led to coarsening PMMA phase in ternary blends.

Then the minimal concentrations of PMMA at which it forms co-continuous morphologies (or percolation threshold, $C^*_{PMMA}$), in presence of different portions of PP were estimated. The technique based on the selective extraction of the PS matrix phase with cyclohexane was used in these experiments since, as it was mentioned earlier, PP did not form co-continuous phases at its loadings below 10 wt%. Experimental data obtained were plotted in Figure 6.34 in the form of a boundary (threshold) curve separating blend compositions with co-continuous and discrete PMMA morphologies. The compositions situated above the curve represent ternary blends with co-continuous PMMA/PS phases, while those below the curve have particulate morphologies.

This graph helps to associate data in Figure 6.33 with particular phase morphologies developed in the blends. For example, Figure 6.34 shows that the PMMA co-continuous phase cannot be formed unless its content in ternary blends is higher than 18–20 wt%. Consequently, the ternary compositions throughout the lower curve in Figure 6.33 (5.0 wt% of PMMA) all have discrete PMMA domains at any PP loadings used. Ternary blends containing 20 wt% of PMMA have mostly similar type of morphology. Concerning blends with 40 wt% of PMMA, the particulate PMMA dispersions were formed provided the PP contents were either below 0.5 wt% or higher 6 wt%. In other cases, as Figure 6.34 demonstrates, the co-continuous phases of PMMA are developed. Finally, all blends containing more than 53 wt% of PMMA, including a binary one, had bicontinuous morphologies.

The foregoing data can be summarized as follows. In binary PMMA/PS blends used in this study about 53 wt% of PMMA was needed to form its own co-

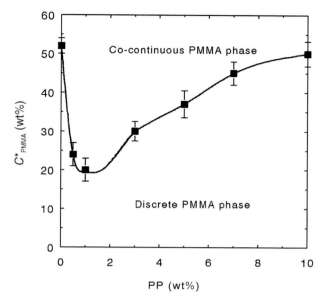

**Figure 6.34** Dependence of minimal concentrations (percolation threshold) C*$_{PMMA}$, at which PMMA develops a co-continuous phase, on the content of the PP third phase in ternary blends PMMA/PS/PP of System I.

continuous phase, while in presence of only 0.5–1.0 wt% of the PP phase, this concentration decreased to about 20 wt%. On further increase of the content of the PP phase, as Figure 6.34 shows, the concentration at which PMMA develops its own co-continuous structures becomes progressively higher. It is important to emphasize that the PMMA phase produces similar effect with respect to the PP dispersed phase. It was observed that in binary PP/PS mixtures, PP forms its own continuous phase at 23 wt%, while in ternary blends this occurs at 13 wt%. These data imply that at appropriate component ratios, the three co-continuous phases can be formed.

Figure 6.32 shows that formation of co-continuous phase is accompanied by the "dispersive" action of the third phase. When concentration of the PMMA primary dispersed phase exceeded 17–20 wt%, an addition of small portions of PP makes the discrete PMMA droplets to coalesce and develop its own co-continuous phase with very high interfacial area. The latter means that the PMMA network bridges become very thin. Further increase of PP concentration did not alter the type of morphology but resulted in its progressive coarsening. In additional experiments a well-known mixing rule [30] has been confirmed. Namely, the lower the apparent viscosity and elasticity of a dispersed phase polymer, the lower is the content at which it forms co-continuous phase in ternary blend, at other conditions being equal.

The next question to be answered is why the efficacy of the third liquid PP phase decreases on increasing its content as Figure 6.34 shows. It is well known [100,101] that the increase of concentration of the dispersed phase unavoidably results in the increase of its particle sizes due to coalescence. Therefore, one can speculate that progressive coarsening of the PP droplets with concentration can be responsible for this effect. To verify this suggestion ternary Systems II and III in which minor dispersed phases were represented by solid, nondeformable and, therefore, noncoalescing particles were studied.

### 6.3.4.2 Ternary Systems with One Solid Phase – Proof of the Mechanism of Phase Interaction

There are not so few publications devoted to morphology and properties of immiscible binary blends containing dispersed solid particles as the third phase. Huitric *et al.* [102] reported on the effect of organically modified layered silicate on the rheological, morphological, and structural properties of immiscible polyethylene/polyamide (PE/PA) blends. It was observed that the dispersed phase sizes decreased with increasing clay content up to 2% and tended to stabilize at higher clay fractions. In blends with PE matrix, clay particles were located predominantly at the PE/PA interface with its thickness increasing with clay content. For PA matrix blends with 2% of clay, the interphase thickness was stabilized at 11 nm while further addition of clay resulted in its dispersion in the PA phase.

Fisher *et al.* [103] studied ternary system composed of PP, ethylene-vinyl alcohol copolymer (EVON) and glass beads (GB). They found that in the PP matrix GBs were encapsulated by EVON with some of the minor EVON component separately dispersed in the matrix. Modification of the interfaces resulted in complete encapsulation of GBs. Wu *et al.* [104] observed a significant change of the phase morphology of PBT/PE blends in the presence of clay. When PE was the matrix (PBT/PE 40/60), the addition of more than 2% of clay changed the morphology to co-continuous. In contrast, the addition of clay in PBT/PE 60/40 with PBT as the matrix, significantly reduced the PE domain sizes. In this case, clay was concentrated in the PBT phase.

In the studies [69,80], solid particles of PBT for ternary PP/PS/PBT System II were prepared *in situ* using binary blends PBT/PS and ordinary mixing procedures at 235 °C. Since the melting point of PBT is 220 °C, both phases were liquid at this temperature. Varying the blend composition and/or shear rate, four PBT/PS master batches with average particle diameters of the PBT phase of 1.6, 3.2, 4.2, and 10.5 μm have been prepared. Figure 6.35 shows that the particles of PBT have near spherical shapes.

These premixtures were then diluted with appropriate amounts of the PS and PP phases to prepare ternary blends of desired compositions. The latter operation was performed at 190 °C when PBT particles obtained in the previous stage were solid and nondeformable. Ternary compositions PP/PS/GMS were prepared by conventional melt mixing. TEM and SEM electron microscopies were used to prove the absence of envelopment around solid particles.

A solvent extraction experiments described above were used to measure concentrations of PP corresponding to the onset and completion of co-continuous

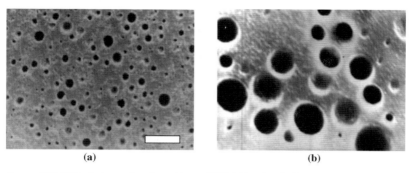

**Figure 6.35** SEM photographs of two binary PBT/PS blends, which were used for preparation of PBT particles. Average PBT domain (dark portions) diameters are 1.6 μm (a) and 10.5 μm (b) Scale bar is 8 μm.

phase formation (i.e., a percolation threshold). Quantitative results obtained with System II and III are summarized in Figure 6.36. Curve 1 corresponds to the onset of the formation of co-continuous PP phase at PBT domain diameter of 1.6 μm while curves 2 and 3 represent the completion of this process in presence of PBT particles of different (1.6 and 10.5 μm) sizes. Three effects are worth to be outlined here. Firstly, the addition of only 1–2 wt% of solid particles of PBT to PP/PS binary blends is enough to reduce PP concentration at which it starts to form co-continuous phase from 20 to 5 wt% (Figure 6.36, curve 1).

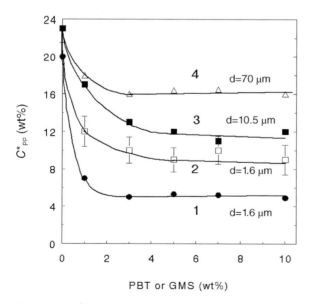

PBT or GMS (wt%)

**Figure 6.36** Influence of the content of PBT (1–3) and GMS (4) solid particles on concentration, $C^*_{PP}$, at which PP starts (1) and completes (2–4) co-continuous phase formation. Average PBT particle diameters are 1.6 μm (1, 2); 10.5 μm (3) and diameter of GMS is 70 μm (4).

Accordingly, solid PBT particles produce similar but even stronger effect compared to liquid droplets in System I (Figure 6.34). Secondly, the curves in Figure 6.36 demonstrate no noticeable reverse effects on increasing PBT or GMS content. Thus, one can attribute such effects in System I to coalescence of the droplets of PP resulting in coarsening of its phase and diminishing its "dispersive" efficiency. Finally, curves 1 to 4 show an obvious tendency of the third PBT or GMS phases toward the increase of the $C^*_{PP}$ on increasing particle sizes.

### 6.3.4.2.1 Mechanism

Phenomenological mechanism based on an idea of free space of matrix (FSM) has been proposed [69,70]. Adopting the simpler case represented by Systems II or III, assume a matrix polymer 2 filled with a large enough portion of liquid primary dispersed phase 1. The latter, nevertheless, still does not form a co-continuous structure and consists of nearly spherical droplets, as sketched in Figure 6.37. On subsequent addition of the solid dispersion of phase 3 (black spheres), the overall concentration of both inner phases becomes higher. Due to negative values of both $\lambda_{31}$ and $\lambda_{13}$, the contacts between domains 1 and 3 are unfavorable and their encapsulation is prohibited. Since there is not enough room in the matrix for arranging spherical particles 3 and 1, the latter are forced to deform during blending and to occupy the vacant volumes between solid spheres 3. Spatial deformation of adjacent liquid drops results in collision and fusion of its tentacle-like ends followed by formation of co-continuous phase, on the one hand, and increase of the specific interface area, that is, the degree of dispersion of this phase, on the other hand.

It is clear that when both inner phases are liquid, they accommodate their shapes simultaneously and stimulate each other to develop their own continuous phases at proper concentrations. An attempt has been made to formalize this mechanism with the use of a simple geometrical approach.

### 6.3.4.2.2 Modeling [69]

Consider an assumed mechanism of interaction between the dispersed phases in ternary PP/PS/PBT blends, as a simple example, composed of the PS matrix and

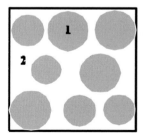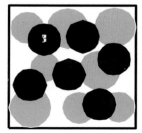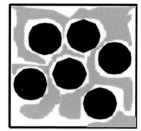

**Figure 6.37** Schema illustrating double action of solid particles of phase 3: transformation of the primary dispersed phase 1 to the co-continuous one and simultaneous increase of the degree of dispersion of a ternary system.

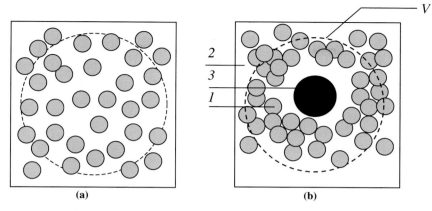

**Figure 6.38** Schema that describes a proposed mechanism of action of solid domains of the PBT phase (3) added to the binary PP/PS emulsion. Dotted circle represents a microvolume $V$ through which collisions and coalescence of the liquid PP domains (2) could take place.

the PP primary dispersed phase both being liquids upon mixing, and solid particles of the PBT secondary dispersed phase. In accordance with the common practice, designate the primary dispersed phase (PP), matrix (PS), and the third phase (PBT) by numbers 1, 2, and 3, respectively, used further as subscripts.

In agreement with the percolation theory [105], a particular threshold volume fraction exists for the domains of different shapes beyond which they begin to contact each other. This volume fraction for the monodisperse spheres packed into the simple cubic lattice is 0.52. The more complex the particle shapes, or rather the larger their radius of gyration, the lower is the percolation threshold.

Since added solid PBT domains occupy some volume of the PS matrix that becomes inaccessible for the PP drops, the partial concentration of the latter per the rest matrix volume can increase beyond the percolation threshold. To accommodate the PP droplets in the reduced free matrix space, they deform in between the PBT domains, collide, coalesce, and develop their own continuous phase.

Let us consider a simplified picture of the discussed phenomenon. Imagine some statistical microvolume of a binary PP/PS melt blend where the equisized PP drops are dispersed through the PS matrix phase, as depicted in Figure 6.38a. An addition to this microvolume of the solid PBT spheres will result in the compaction of the PP drops and their coalescence, especially, in the vicinity of the foreign domains (Figure 6.38b). It is clear that the effects of compaction and coalescence will suffer a gradual decay with the distance from the PBT sphere and, finally, will damp completely outside a microvolume $V$ labeled by a dashed circle in Figure 6.38b.

A final equation enabled the prediction of the percolation threshold $C^{*}_{PP}$ for the liquid dispersed phase (PP) in presence of different amounts of the solid particles:

$$0.52 = \frac{1}{1 - \varphi_3} \left[ \left(1 - e^{-k_2\varphi_3}\right)\varphi_1 k_1 + \varphi_1 e^{-k_2\varphi_3} \right].$$

(6.25)

Here $\varphi_i$ is the volume fraction of $i$-component (i.e., dispersed phases 1 and 3); $k_1$ and $k_2$ are the constants. The value of the constant $k_1$ depends on both the number of primary PP drops fused and the shapes of new domains formed. For the sake of simplicity, coalescence was considered by convention to be the formation of clusters, that is, the drops that just touched each other without merging. Coefficient $k_2 = V/V_3$ denotes the ratio of the microvolume $V$ to the volume of the PBT sphere $V_3$. It was shown by best fitting of experimental data for the ternary PP/PS/PBT blends to Eq. (6.25), that $k_1 = 9$ and $k_2 = 89$. The last number means that the diameter of the microsphere $V$ (Figure 6.38) is 4.5 times larger than that of the PBT domain.

One can realize that the approach used for the development of this model was based on the experimental observations [69] and pure geometrical considerations. The consequence of such formalism is the absence of material variables characterizing viscoelastic and surface properties of polymer phases in the final equation (Eq. (6.25). A detailed description and mathematical background of this approach can be found in [70].

A comparison of experimental and calculated curves, which denote the percolation thresholds for ternary PP/PS/PBT blends, is shown in Figure 6.39. The main conclusion that can be made from this figure is that Eq. (6.25) describes

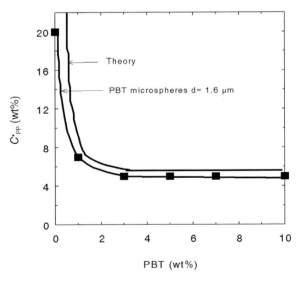

**Figure 6.39** Experimental and theoretical (Eq. 6.16) dependences of the percolation threshold $C^*_{PP}$ for the PP phase in ternary PP/PS/PBT blends on concentration of the PBT solid particles.

the exponential shape of the experimental curves fairly well. Moreover, the model predicts leveling off of the dependence at higher concentrations of the PBT particles. These data provides the evidence in favor of general adequacy of the proposed geometrical model. Almost quantitative correlation between the calculated and experimental curves is not surprising since coefficients $k_1$ and $k_2$ were used actually as fitting parameters.

## 6.4
## Properties of Ternary Polymer Blends

The basis for the estimation of the mechanical properties of the polyblends is the property – composition curve. Usually such a curve is obtained experimentally, which allows one to see how the viscosity of melts, modulus, tensile and impact strength, and other properties of the mixture vary with composition. Property–composition curve gives the opportunity to immediately see if there is a synergy in mutual influence of polymers and how large is the level of synergies. For this reason, numerous attempts were made to theoretically calculate the property–composition curve based on knowing properties of the polymer components and based on some models. The first who proposed a mechanical model of the two-phase polymer blend was M. Takayanagi [106]. Works for calculating two-phase systems modulus, including polymers containing filler are summarized in reviews [107,108].

Model representations in predicting the properties of polymer blends are generally not widely accepted for two reasons. First, during the mixing of polymers there is always a partial miscibility taking place within a few percent or more. This changes the mechanical properties of the starting components and, therefore, quantitative estimation of these changes is practically impossible. Secondly, during the mixing of the polymers in a wide range of compositions, the phase inversion occurs such that instead of a conventional dispersed structure the two coexisting continuous phases are formed. The concentration range in the compositions corresponding to the phase inversion is difficult to predict, since it is defined by the polymers viscosity ratio during mixing and intensity of the intermolecular interaction between the selected pair of polymers. Both of these reasons do not allow one to calculate the property– composition curve of a binary blend on the basis of some model. The calculation of the triangular diagram property–composition of a ternary blend is generally impossible because in addition to the phase inversion, the ternary blend formed area of incompatibility even when mixed polymer pairs are miscible or partially miscible. Often particles of one polymer are encapsulated within the particles of another polymer. Thus, the matrix base polymer becomes the same as polymer of outer layer in a core–shell particle.

For these reasons, the study of the mechanical properties of ternary mixtures are related to their applications or have a goal to find their relation to location on the triangular diagram of areas miscibility or immiscibility. The works are

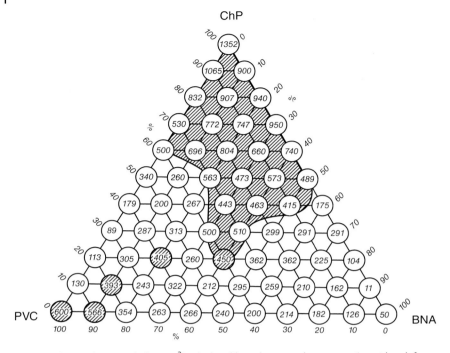

**Figure 6.40** Tensile strength (kg cm$^{-2}$) calculated based on sample cross-section at break for ternary blend of PVC–polychloroprene–poly(butadiene-co-acrylonitrile) 28 mol.%) [109].

rarely found that would establish the dependencies between certain parameters of the phase morphology and modulus, tensile strength, toughness, and other mechanical properties of ternary blends.

Shape of the property–composition curve in the triangular diagram of ternary polymer blends obviously must depend on the intermolecular interaction of the polymers and their miscibility. Figure 6.40 shows the results of determination of the tensile strength of ternary polymer blend: PVC, polychloroprene (CR) and poly(butadiene-co-acrylonitrile) 28 mol. %) (NBR). The last two polymers are elastomers. Blends were prepared by mixing on a mill at 140–150 °C, followed by pressing at 150–160 °C, and cooling in a press. With an interval of 10% of the composition, the tensile strength was determined. Its value was calculated based on cross section of the sample at break. The resulting values reflect simultaneously the tensile strength and elongation at break. This resulted in that the maximum strength for CR-elastomer, because this elastomer has the greatest elongation at break.

Figure 6.40 shows that the NBR, which is well compatible with PVC, has a plasticizing effect and the tensile strength of mixtures PVC–NBR is greatly reduced. The triangular diagram indirectly indicates a possible increase of the miscibility in ternary mixtures and a large region of high strength in the middle of the chart is seen. It should be noted that the determination of miscibility in this work was not carried out [109].

In some cases, the state of miscibility in the ternary blend does not appear, but the particle size is significantly reduced compared with a binary blend. This also leads to improved mechanical properties. So adding into PC the polyviniliden-fluoride (PVDF)/PMMA mixture significantly improves the elongation at break and impact strength. Improving these parameters observed with increasing content of PMMA in admixture with PVDF up to 20–40%. Improvement of the properties of the ternary blend was compared to the corresponding properties of the binary mixture of PC–PVDF. It is in this area of compositions a significant decrease of sizes of dispersed particles occurs in the blend [110].

Growth of intermolecular interactions at the interface in a ternary mixture can lead to a significant improvement in mechanical properties compared with binary mixture. It can be seen from the following example. The ternary blends of polypropylene – styrene-b-ethylene-co-butylene-b-styrene triblock copolymer (SEBS) and high-density polyethylene or polystyrene were investigated. A mixture of PP–SEBS–HDPE exhibits substantially better mechanical properties than PP–SEBS–PS. At the same time, the properties of the ternary mixture are superior compared to binary mixtures of PP–SEBS and PP–HDPE [111].

There is an example of work in which an attempt was made to establish relationship between the phase morphology of a ternary mixture and its properties. It was found that the disperse particles of methacrylated butadiene-styrene copolymer (MSBR) during the mixing process migrate to the interphase boundary of a ternary mixture PS–SAN–MSBR [112]. This copolymer is core–shell latex particles in which methacrylated component forms outer layer. A similar phenomenon was observed in the mixture of high impact polystyrene (HIPS) and acrylonitrile butadiene styrene (ABS) [113]. These works were continued on mixtures of PC–MSBR–brittle polymer (BP). BP was PS, SAN, or PMMA. MSBR and PS were simultaneously added to the PC matrix, or a mixture of BP–MBS was first prepared, and then this mixture was added to PC [114]. The microstructure of mixtures showed localization of core–shell particles of MSBR on the interphase boundary of PC–BP. This resulted in a higher level of toughness and reduced notch sensitivity without any loss of the modulus relative to PC for some of ternary blends. Significant toughening of PC blends was only in blends with PMMA or SAN, but not with PS. Most interesting result was that blends prepared by two-step mixing sequence usually exhibited better properties than those prepared by mixing simultaneously.

One would assume that in the two-step mixing when the MSBR particles are all added to BP, their concentration in the phase of BP is high and more particles are trapped on the interphase surface. During simultaneous mixing, MSBR particles are distributed throughout the volume of the mixture and their concentration in the total volume is less which reduces their location at the interface and does not lead to significant improvement in properties. Thus, the properties of ternary blends are superior to those of binary mixtures, such as PC–BP and PC–MSBR. It is important to note that the best results are achieved when the content of AN in copolymer is 25%, which provides its best interaction with PMMA and, therefore, with PMMA shell in core–shell particles. Work on the addition of compatibilizers in ternary blends showed that their action in these blends is similar to the action in binary blends. It should be noted that the value of the interface surface in the ternary mixtures, in principle, should be more than that in binary mixtures and modification of such a surface is an important contribution to the formation of material properties of the blends.

One can also note that the toughening mechanism in ternary blends is similar to toughening in impact resistant polymers. In the PBT–PC rubber blend, the pair PC–rubber in the process of melt blending forms the core–shell particles with rubber being inside. When determining the impact resistance, cavitation in rubber particles and plastic cracking are usually observed with the participation of both PC shell and PBT matrix [115]. In ternary blends, the presence of miscibility usually has a positive effect on their properties. This regularity is related to the fact that the study of turbidity, as well as the determination of phase or relaxation transitions, reveals two-phase structure only up to a certain particle size of the dispersed phase. It is known that with decreasing particle size to less than 10 nm, substance loses its phase properties and also particles become much smaller than the wavelength of light and do not scatter light. Macromolecules and microvolumes of short-range order are sufficiently large in size and the presence of two-phase structure in polymer–polymer systems may not be detected even at a sufficiently high heterogeneity. Thus, in a mixture 85 : 15 of PVC–butadiene–acrylonitrile copolymer (40%) 10 nm particles are seen on slices, but the study of the properties of the mixture certainly points out to the presence of one-phase structure indicating the complete miscibility [116].

In miscible one-phase blend of PS–poly(phenyleneoxide), microheterogenity exists with size of particles being larger than 3 nm [117].

Thus, an important factor in improving properties of polymer blends is sufficiently small particle size. Miscibility provides the necessary size, but that does not mean the presence of molecular mixing.

It is important that the molecular mixing in the polymer blends, if it happens, influences properties of mixtures. This is especially typical for mixtures with the presence of crystalline or crystallizable polymer. For example, a mixture of natural rubber (NR), which has an ability to crystallize under tension, and SBR can achieve state of almost molecular mixing, if interpolymer of these rubbers is present in the mixture. The transparency of mixture also indicates its one-phase structure. In vulcanizate of mixture of NR/SBR at a ratio of 50/50, NR can not crystallize under tension and tensile strength strongly decreases [118].

In binary mixtures, the efficiency of polymer mixing, that is, gain in the properties, is measured based on the deviation of the experimentally determined value from the value calculated on the basis of the additivity (rule of mixtures). This additive value is calculated relative to the values of individual components of the mixture. There is great difference in the structure of the components: single-phase individual polymers and heterophase polyblend. Therefore, the agreement between the experimental and calculated values is extremely rare.

The properties of the ternary mixture can be also assessed by comparing the performance properties of the individual polymer with an additive index – the average of the performance properties of the three components of the mixture. Also, a comparison of the properties of homogeneous polymers with heterophase mixture can be made. However, it is not a meaningful comparison. More justified is to compare the properties of the ternary blend with the properties of binary blends from which a ternary blend of given composition could be obtained.

Often, the experimental values of the properties of the ternary blends are much closer to those calculated by the additivity properties of the corresponding binary blends [119].

## 6.5
## Conclusions and Future Development

Blending has been used as the most versatile, economic method to produce materials with specific qualities that are able to satisfy complex demands for performance. This approach is very important not only for the development of polymers with new properties, but also for recycling of greener materials. The use of polymer blends and alloys has grown so fast compared with other polymeric materials system mainly because of their low cost and their acceptable performance. Multiphase polymer blends can be considered as a logical development of binary compositions to further improve versatility and extend the range of properties of the materials. It should be noted that attempts to use the multiphase polymer blends began long ago. For example, Alexeenko and Mishustin [108] in 1959 published a systematic study on the structure and properties of ternary mixtures PVC–polychloroprene–poly(butadiene-co-acrylonitrile). In subsequent years, various authors have developed a number of ternary mixtures of elastomers for heavy tires. Thus, in 1970, Speransini and Drost [120] published a paper in which they designed not only ternary but quaternary elastomeric blends for the tire sidewalls. Much more information about multiphase polymer blends during this period was collected in the patent literature. However, in these years, the choice of components for such complex systems was rather intuitive, so the development of such compositions was performed mainly by the method of trial and errors.

For many years a lack of information on the phase morphology and the principles of its formation, which largely determines the final properties of multicomponent polymer blends, was a crucial factor. Finally, in 1988, Hobbs *et al.* [40] published a method for predicting the phase morphology of multiphase mixtures on the basis of calculating the coefficients of spreading. (In fact, this theory in a more complete version was published by Torza and Mason [39] as early as 1970). This work gave a strong boost to research in the field of morphology of multiphase polymer systems, most of which were analyzed in this review. Calculation of three spreading coefficients based on the values of interfacial tension in polymer pairs forming composition enables one to predict accurately three types of phase morphology (Figure 6.9): separate dispersion of the dispersed phases in the matrix, a full, or partial encapsulation of one dispersed phase by another. Prediction can be performed not only for the ternary, but also for multiphase systems (a practical example for the quaternary polymer system is given in [40]).

Multiphase systems possess enormous potential for the development of new advanced materials. For example, in the case of morphology with individual dispersed phases (Figure 6.9a), two or all three phases can be made co-continuous. Furthermore, each of the phases can be "treated" individually: any additives such as fillers, plasticizers, and so forth can be added to any phase selectively. One should note that the addition of dispersed particles as described above may result in a transfer of the dispersed phase to continuous. In the case of blends of elastomers, each of the phases may selectively be vulcanized such that crosslink density may be varied in each phase. The encapsulated polymer blends provide similar, if not wider, capabilities. Here, apart from the listed options, there are ways to form a continuous phase at very low content of additives in the mixture. This can be very useful in the development of different conductive materials.

Discussing the formation of the above three types of phase morphologies and their potentials for creating new materials, one should not forget the well-known powerful methods from the experience of working with binary mixtures which have strong impact on the phase structure and properties of compositions. It can be stated that all the known patterns of influence of component concentrations, their viscoelastic and surface properties on the structure of binary blends (percolation threshold, the degree of dispersion, particle shape, etc.) are generally valid for polyphase systems. For example, the smaller the ratio of the viscosity of the disperse phase to the viscosity of the matrix $\mu = \eta_1/\eta_2$ (and/or ratio of the elasticity of phases, $\psi$), the lower is the content of the dispersed phase at which it transfers to a continuous phase. Values of $\mu$ and $\psi$ close to unity favor melt mixing and formation of the smallest particle sizes of the dispersed phase.

According to the interest of researchers to the multicomponent systems, one can expect rapid development of these blends in the nearest future. Among the reasons arising such interest, the most important is, perhaps, that polymer blends offer a fast and cheap way to obtain new polymeric materials. It is envisioned that the number of blend components will increase. Using the theory of spreading coefficients one can already purposefully create phase morphology of such systems in relation to a particular field of use. Thus, the researcher turns to an engineer designer, who develops multicomponent material with the desired qualities.

Finally, the global challenge of the future is the development of methods for predicting the properties of multicomponent polymer blends. Solving this task will require a large amount of information on the impact of morphology on various properties of such materials. Currently, scientists are just beginning to search for approaches to solving these problems.

# References

1 Utracki, L. (1998) *Commercial Polymer Blends*, Chapman & Hall, London.
2 Kuleznev, V.N. and Surikov, P.V. (2012) Phase equilibrium in ternary polymer blends. *Polym. Sci.*, **A54**, 833–840.
3 Prigogine, I. and Defay, R. (1954) *Chemical Thermodynamics*, Longmans, London.
4 Min, K.E., Chiou, J.S., Barlow, J.W., and Paul, D.R. (1987) A completely miscible ternary blend: Poly(methyl metacrylate)-poly(epichlorohydrin)-poly(ethylene oxide). *Polymer*, **28**, 1721–1728.
5 Brannock, G.R. and Paul, D.R. (1990) Phase behavior of ternary polymer blends composed of three miscible binaries. *Macromolecules*, **23**, 5240–5250.
6 Fowler, M.E., Barlow, J.W., and Paul, D.R. (1987) Effect of copolymer composition on the miscibility of blends of styrene-acrylonitrile copolymers with poly (methyl methacrylate). *Polymer*, **28**, 1177–1184.
7 Voskov, A.L. and Voronin, G.F. (2010) A universal method for calculating isobare-isothermal sections of ternary system phase diagrams. *Russ. J. Phys. Chem.*, **84** (4), 525–533.
8 Hildebrand, J.H. and Scott, R.L. (1950) *Solubility of Nonelectrolytes*, Reinhold, NewYork.
9 Flory, P.J., Eichinger, B.E., and Orwoll, R.A. (1968) Thermodynamics of mixing polymethylene and polyisobutylene. *Macromolecules*, **1**, 287–288.
10 Duffy, D.J., Stidham, H.D., Hsu, S.L., Sasaki, S., Takahara, A., and Kajiyma, T. (2002) Effect of polymer structure on the interaction parameters and morphology development of ternary blends: Model of high performance adhesives and coatings. *J. Mater. Sci.*, **37**, 4801–4809.
11 Ruzette, A.V.G. and Mayes, A.M. (2001) A Simple free energy model for weakly interacting polymer blends. *Macromolecules*, **34**, 1894–1907.
12 Ruzette, A.V.G., Banerjee, P., Mayes, A.M., and Russell, T.P. (2001) A simple model for baroplastic behavior in block copolymer melts. *J. Chem. Phys.*, **114**, 8205–820913.

13 Askadskii, A.A. (2001) *Computational Materials Science of Polymers*, Cambridge International Science Publishing, Cambridge.
14 Gonzales-Leon, J.A. and Mayes, A.M. (2003) Phase behavior prediction of ternary polymer mixtures. *Macromolecules*, **36**, 2508–2515.
15 Sanches, I.C. and Panayiotou, C.C. (1993) *Models for Thermodynamic and Phase Equilibria Calculations* (ed. S.I. Sandler), Marcell Dekher, NewYork.
16 Rodgers, P.A. (1993) Pressure–volume–temperature relationships for polymeric liquids: A review of equations of state and their characteristic parameters for 56 polymers. *J. Appl. Polym. Sci.*, **48**, 1061–1080.
17 Van Krevelen, D.W. (1976) *Properties of Polymers*, Elsevier, Amsterdam.
18 Kim, C.K., Kim, J.J., and Paul, D.R. (1994) Phase stability of ternary blends. *Poly. Eng. Sci.*, **34**, 1788–1798.
19 Hsu, W.P. (2007) A closed-loop behavior of ternary polymer blends composed of three miscible binaries. *Termochim. Acta*, **454**, 50–56.
20 Landry, C.J.T., Yang, H., and Machell, J.S. Miscibility and mechanical properties of a ternary polymer blend: polystyrene-polycarbonate- tetramethyl polycarbonate. *Polymer*, **32**, 44–52.
21 Goh, S.H. and Ni, X. (1999) A completely miscible ternary blend system of poly(3-hydroxybutirate), poly(ethylene oxide) and polyepichlorhydrin. *Polymer*, **40**, 5733–5735.
22 Huges, L.J. and Britt, G.E. (1961) Compatibility studies on polyacrylate and polymethacrylate systems. *J. Appl. Polym. Sci.*, **5**, 337–348.
23 Woo, E.M., Mao, Y.J., and Wu, P.L. (2003) Ternary miscibility in blends of three polymers with balanced binary interactions. *Polym. Eng. Sci.*, **43**, 543–557.
24 Utracki, L.A. (ed.) (1990) *Polymer Alloys and Blends*, Hanser, Munich, Chapter 1.
25 Boudevine, A., Ibos, L., Candau, Y., and Thomas, S. (eds) (2011) *Handbook of Multiphase Polymer Systems*, John Wiley & Sons Ltd, Chichester, p. 193.

26 Utracki, L.A. (ed.) (2002) *Polymer Blends Handbook*, Kluwer, Dorchester, p. 1–96, Chapter 1.

27 Paul, D. R. (1978) Fundamentals and prospects, in *Polymer Blends*, vol. **1** (eds D.R. Paul and S. Newmann), Academic Press, New York, Chapter 1.

28 Favis, B.D. (2000) Factors influencing the morphology of immiscible polymer blends in melt processing, in *Polymer Blends*, vol. **1**, Formulations (eds D.R. Paul and C.B. Bucknall), John Wiley & Sons Inc., New York, Chapter 16.

29 VanOene, H. (1972) Modes of dispersion of viscoelastic fluids in flow. *J. Colloid Interface Sci.*, **40**, 448–467.

30 Paul, DR. and Barlow, J.W. (1980) Polymer blends. *J. Macromol. Sci., Rev. Macromol. Chem.*, **18**, 109–168.

31 Willemse, R.S., Posthuma de Boer, A., van Dam, J., and Goisis, A.D. (1998) Co-continuous morphologies in polymer blends: A new model. *Polymer*, **39**, 5879–5887.

32 Lyngaae-Jørgensen, J. (1993) *Polymer Blends and Alloys* (eds M.J. Folkes and P.S. Hope), Blackie Academic & Professional, London, Chapter 4.

33 de Gennes, P.G. (1985) Wetting: Statics and dynamics. *Rev. Mod. Phys.*, **57** (3), Part 1: 827–863.

34 Dobbs, H. and Bonn, D. (2001) Predicting wetting behavior from initial spreading coefficients. *Langmuir*, **17**, 4674–4676.

35 Erbil, H.Y. (2006) *Surface Chemistry of Solid and Liquid Interfaces*, Blackwell Publ. Ltd, Oxford, p. 194.

36 Geoghegan, M. and Krausch, G. (2003) Wetting at polymer surfaces and interfaces. *Prog. Polym. Sci.*, **28**, 261–302.

37 Bonn, D., Eggers, J., Indekeu, J., Meunier, J., and Rolley, E. (2009) Wetting and spreading. *Rev. Mod. Phys.*, **81**, 739–805.

38 Harkins, W.D. and Feldman, A. (1922) Films. The spreading of liquids and the spreading coefficient. *J. Am. Chem. Soc.*, **44**, 2665–2685.

39 Torza, S. and Mason, S.G. (1970) Three-phase interactions in shear and electrical fields. *J. Colloid Interface Sci.*, **33**, 67–83.

40 Hobbs, S.Y., Dekkers, M.E.J., and Watkins, W.H. (1988) Effect of interfacial

41 Virgilio, N., Marc-Aurel, C., and Favis, B.D. (2009) Novel self-assembling close-packed droplet array at the interface in ternary polymer blends. *Macromolecules*, **42**, 3405–3416.

42 Virgilio, N., Desjardins, P., L'Esperance, G., and Favis, B.D. (2009) In situ measure of interfacial tensions in ternary and quaternary immiscible polymer blends demonstrating partial wetting. *Macromolecules*, **42**, 7518–7529.

43 Guo, H.F., Packirisamy, S., Gvozdic, N.V., and Meier, D.J. (1997) Prediction and manipulation of the phase morphologies of multiphase polymer blends: 1. Ternary systems. *Polymer*, **38**, 785–794.

44 Hemmati, M., Nazokdast, H., and Shariat Panahi, H. (2001) Study on morphology of ternary polymer blends. I. Effect of melt viscosity and interfacial interaction. *J. Appl. Polym. Sci.*, **82**, 1129–1137.

45 Hemmati, M., Nazokdast, H., and Shariat Panahi, H. (2001) Study on morphology of ternary polymer blends. II. Effect of composition. *J. Appl. Polym. Sci.*, **82**, 1138–1146.

46 Valera, T.S., Morita, A.T., and Demarquette, N.R. (2006) Study of morphologies of PMMA/PP/PS ternary blends. *Macromolecules*, **39**, 2663–2675.

47 Reignier, J., Favis, B.D., and Heuzey, M.-C. (2003) Factors influencing behavior in composite droplet-type polymer blends. *Polymer*, **44**, 49–59.

48 Mohammadigoushki, H., Nazockdast, H., and Mostofi, N. (2009) Morphology development and melt linear viscoelastic properties of (PA6/PP/PS) ternary blend system. *J. Elastom. Plast.*, **41**, 339–351.

49 Guo, H.F., Gvozdic, N.V., and Meier, D.J. (1997) Prediction and manipulation of the phase morphologies of multiphase polymer blends: II. Quaternary systems. *Polymer*, **38**, 4915–4923.

50 Hara, M. and Sauer, J.A. (1998) Synergism in mechanical properties of polymer/polymer blends. *J. Macromol. Sci. – Rev. Macromol. Chem.*, **C38**, 327–362.

51 Horiuchi, S., Matchariyakul, N., Yase, K., and Kitano, T. (1997) Morphology

development through an interfacial reaction in ternary immiscible polymer blends. *Macromolecules*, **30**, 3664–3670.

52 Nauman, E.B. and He, D.Q. (2001) Nonlinear diffusion and phase separation. *Chem. Eng. Sci.*, **56**, 1999–2018.

53 Reignier, J. and Favis, B.D. (2000) Control of the subinclusion microstructure in HDPE/PS/PMMA ternary blends. *Macromolecules*, **33**, 6998–7008.

54 Reignier, J. and Favis, B.D. (2003) Core-shell structure and segregation effects in composite droplet polymer blends. *AIChE J.*, **49**, 1014–1023.

55 Tchomakov, K.P., Favis, B.D., Huneault, M.A., Champagne, M.F., and Tofan, F. (2004) Composite droplets with core/shell morphologies prepared from HDPE/PS/PMMA ternary blends by twin-screw extrusion. *Polym. Eng. Sci.*, **44**, 749–759.

56 Koseki, Y., Lee, M.S., and Macosko, C.W. (1999) Encapsulation in ternary elastomer blends. *Rubber Chem. Technol.*, **72**, 109–118.

57 Van Krevelen, D.W. (1997) *Properties of Polymers*, 3rd edn, Elsevier, Amsterdam.

58 Harrats, C., Omonov, T.S., and Groeninckx, G. (2005) Co-continuous and encapsulated three phase morphologies in uncompatibilized and reactively compatibilized polyamide 6/polypropylene/polystyrene ternary blends using two reactive precursors. *Polymer*, **46**, 12322–12336.

59 Ohishi, H. (2004) Phase morphology and compatibilization mechanism in yernary polymer blend systems of polyethylene terephthalate, polyolefin rubber, and ionomer. *J. Appl. Polym. Sci.*, **93**, 1567–1576.

60 Inoue, T. (2003) Morphology of polymer blends, in *Polymer Blends Handbook* (ed. L.A. Utracki), Kluwer Acad. Press, Dordrecht, pp. 547–576. Chapter 8.

61 Gayla, A. and Salaun, S. (2012) Microscopy as a tool to control predicted morphology and/or dispersion of a binary and ternary compounds in polymeric particles and fibre, in *Current Microscopy Contributions to Advances in Science and Technology* (ed. A. Mendez-Vilas), Formatex, Badajoz, Spain.

62 Lyngaae-Jorgensen, J. and Utracki, L.A. (2003) Structuring polymer blends with bicontinuous phase morphology. II. Tailoring blends with ultralow critical volume fraction. *Polymer*, **44**, 1661–1669.

63 Potschke, P. and Paul, D.R. (2003) Formation of co-continuous structures in meltmixed immiscible polymer blends. *J. Macromol. Sci.-Polym. Rev.*, **43**, 87–141.

64 Luzinov, I., Xi, K., Pagnoulle, C., Huynh-Ba, G., and Jerome, R. (1999) Composition effect on the core–shell morphology and mechanical properties of ternary polystyrene/styrene–butadiene rubber/polyethylene blends. *Polymer*, **40**, 2511–2520.

65 Luzinov, I., Pagnoulle, C., and Jerome, R. (2000) Ternary polymer blends with core-shell dispersed phases: effect of the core-forming polymer on phase morphology and mechanical properties. *Polymer*, **41**, 3381–3389.

66 Omonov, T.S., Harrats, C., and Groeninckx, G. (2005) Co-continuous and encapsulated three phase morphologies in uncompatibilized and reactively compatibilized polyamide 6/polypropylene/polystyrene ternary blends using two reactive precursors. *Polymer*, **46**, 12322–12336.

67 Li, Y., Wang, D., Zhang, J.-M., and Xie, X.-M. (2011) Influences of component ratio of minor phases and charge sequence on the morphology and mechanical properties of PP/PS/PA6 ternary blends. *Polym. Bull.*, **66**, 841–852.

68 Wang, D., Li, Y., Xie, X.-M., and Guo, B.-H. (2011) Compatibilization and morphology development of immiscible ternary polymer blends. *Polymer*, **52**, 191–200.

69 Letuchii, M.A. and Miroshnikov, Yu.P. (2015) Interaction of domains in ternary polymer melt blends with separate dispersions of the inner phases. *J Macromol. Sci., Phys.*, **54**, 433–449.

70 Letuchii, M.A., Klepper, L.Ya., and Miroshnikov, Yu.P. (2015) Percolation thresholds in ternary polymer melt blends with separate dispersions of the inner phases: mathematical model. *J. Macromol. Sci., Phys.*, **54**, 393–400.

71 Taylor, G.I. (1934) The formation of emulsions in definable fields of flow. *Proc. R. Soc. (London) A*, **146**, 501–523.

72 Goldsmith, H.L. and Mason, S.G. (1967) The microrheology of dispersions, in *Rheology: Theory and Applications*, vol. **4** (ed. F.R. Eirich), Academic Press, New York, pp. 85–250.

73 Miroshnikov, Yu.P. and Voloshina, Yu.N. (2000) Effect of interphase interactions on the dispersity of binary blends of incompatible polymers. *Polym. Sci., Ser. A*, **42**, 170–176.

74 MacLean, D.L. (1973) A theoretical analysis of bicomponent flow and the problem of interface shape. *Trans. Soc. Rheol.*, **17**, 385–400.

75 Minagawa, N. and White, J.L. (1975) Coextrusion of unfilled and TiO2-filled polyethylene: Influence of viscosity and die cross-section on interface shape. *Polym. Eng. Sci.*, **15**, 825–830.

76 Nemirovski, N., Siegmann, A., and Narkis, M. (1995) Morphology of ternary immiscible polymer blends. *J. Macromolec. Sci., Phys.*, **B34**, 459–475.

77 Gupta, A.K. and Srinivasan, K.R. (1993) Melt rheology and morphology of PP/SEBS/PC ternary blend. *J. Appl. Polym. Sci.*, **47**, 167–184.

78 Kim, B.K. and Kim, M.S. (1993) Viscosity effect in polyolefin ternary blends and composites. *J. Appl. Polym. Sci.*, **48**, 1271–1278.

79 Le Corrolle, P. and Favis, B.D. (2011) Effect of viscosity in ternary polymer blends displaying partial wetting phenomena. *Polymer*, **52**, 3827–3834.

80 Miroshnikov, YuP. (2012) Prediction and design of phase morphology of polymer blends. *Vestnik MITHT*, **6** (5), 53–64 (In Russ).

81 Letuchii, M.A. and Miroshnikov, Yu.P. (2013) Phase self-assembly driven inverse kinetics of mixing in ternary encapsulated polymer blends. *J. Macromol. Sci., Phys.*, **52**, 530–544.

82 Huang, J.J., Keskkula, H., and Paul, D.R. (2006) Elastomer particle morphology in ternary blends of maleated and non-maleated ethylene-based elastomers with polyamides: Role of elastomer phase miscibility. *Polymer*, **47**, 624–638.

83 Shokoohi, S., Arefazar, A., and Naderi, G. (2011) Compatibilized polypropylene/ethylene-propylene-diene-monomer/polyamide6 ternary blends: Effect of twin screw extruder processing parameters. *Mater. Design*, **3**, 1697–1705.

84 Ravati, S. and Favis, B.D. (2010) Morphological states for a ternary polymer blend demonstrating complete wetting. *Polymer*, **50**, 4547–4561.

85 Ravati, S. (2010) Novel conductive polymer blends. PhD Thesis. Ecole Polytechnique de Montreal. Montreal, Canada, 262 p.

86 Zhang, J., Ravati, S., Virgilio, N., and Favis, B.D. (2007) Ultralow percolation threshold in ternary cocontinuos polymer blends. *Macromolecules*, **40**, 8817–8820.

87 Debolt, M.A. and Robertson, R.E. (2006) Morphology of compatibilized ternary blends of polypropylene, nylon 66, and polystyrene. *Polym. Eng. Sci.*, **46**, 385–398.

88 De Freitas, C.A., Valera, T.S., De Souza, A.M.C., and Demarquette, N.R. (2007) Morphology of compatibilized ternary blends. *Macromol. Symp.*, **247**, 260–270.

89 Hyun, D.C., Jeong, U., and Ryu, D.Y. (2007) A simple approach to determine the equilibrium shape and position of sandwiched polymer droplets. *J. Polym. Sci., B: Polym. Phys.*, **45**, 2729–2738.

90 Kim, J.K., Jeong, W.-Y., Son, J.-M., and Jeon, H.K. (2000) Interfacial tension measurement of a reactive polymer blend by the neumann triangle method. *Macromolecules*, **33**, 9161–9165.

91 Zhang, X. and Kim, J.K. (1998) Interfacial tension measurement with the Neumann triangle method. *Macromol. Rapid Commun.*, **19**, 499–504.

92 Virgilio, N., Desjardins, P., L'Esperance, G., and Favis, B.D. (2010) Modified interfacial tensions measured in situ in ternary polymer blends demonstrating partial wetting. *Polymer*, **51**, 1472–1484.

93 Zhang, X. and Kim, J.K. (1998) Interfacial tension measurement with the Newman triangle method. *Macromol. Rapid Commun.*, **19**, 499–504.

94 Uskoković, Y. (2008) Isn't self-assembly a misnomer multi-disciplinary arguments

in favor of co-assembly. *Adv. Colloid Interface Sci.*, **141**, 37–47.

95 Joris, W.O., van Heck, S.J., and Klumperman, B. (2010) Steric stabilization of Pickering emulsions for the efficient synthesis of polymeric microcapsules. *Langmuir*, **21**, 14929–14936.

96 Zoppe, J.O., Venditti, R.A., and Rojas, O.J. (2012) Pickering emulsions stabilized by cellulose nanocrystals grafted with thermo-responsive polymer brushes. *J. Colloid Interface Sci.*, **369**, 202–209.

97 Marku, D., Wahlgren, M., Rayner, M., Sjцц, M., and Timgren, A. (2012) Characterization of starch Pickering emulsions for potential application in topical formulations. *Int. J. Pharm.*, **428**, 1–7.

98 Ravati, S. and Favis, B.D. (2013) Tunable morphologies for ternary blends with Poly(butylene succinate): Partial and complete wetting phenomena. *Polymer*, **54**, 3271–3281.

99 Miroshnikov, Yu.P., Letuchii, M.A., Lemstra, P.J., Govaert-Spoelstra, A.B., and Engelen, Y.M.T. (2000) Morphology of multiphase polymer blends: continuous phase formation in ternary systems. *Polym. Sci., Ser. A*, **42**, 795–805.

100 Janssen, J.M.H. (1993) Dynamics of liquid–liquid mixing, Ph.D. thesis, Eindhoven University of Technology, The Netherlands.

101 Egorov, A.K., Egorova, M.V., and Miroshnikov, Yu.P. (2011) Measuring of coalescence in polymer melt blends flowing through convergent channels. *J. Appl. Polym. Sci.*, **120**, 2724–2733.

102 Huitric, J., Ville, J., Mederic, P., Moan, M., and Aubry, T.J. (2009) Rheological, morphological and structural properties of PE/PA/nanoclay blends: Effect of clay weight fraction. *J. Rheol.*, **53**, 1101.

103 Fisher, I., Siegmann, A., and Narkis, M. (2002) The effect of interface characteristics on the morphology, rheology and thermal behaviour of three-component polymer alloys. *Polym. Composites*, **23**, 34.

104 Wu, D., Zhou, C., and Zhang, M. (2006) Effect of clay on immiscible morphology of poly(butylene terephthalate)/ polyethylene blend nanocomposites. *J. Appl. Polym. Sci.*, **102** (4), 3628.

105 Stauffer, D. and Aharony, A. (1994) *Introduction to Percolation Theory*. Taylor & Francis, Bristol, London.

106 Uemura, S. and Takayanagi, M. (1966) Application of the theory of elasticity and viscosity of two-phase systems to polymer blends. *J. Appl. Polym. Sci.*, **10**, 113–125.

107 Paul, D.R. and Bucknell, C. (eds) (2000) *Polymer Blends*, John Wiley & Sons Inc., NY.

108 Kuleznev, V.N. (1980) *Polym. Blends*, Chemie Publishing, Moscow.

109 Alekseenko, V.I. and Mishustin, I.U. (1959) Studies of compatibility in ternary polymer blends. *Russ. J. Phys. Chem.*, **33** (4), 757–763.

110 Moussaif, N. and Jérôme, R. (1999) Compatibilization of immiscible polymer blends (PC/PVDF) by the addition of a third polymer (PMMA): Analysis of phase morphology and mechanical properties. *Polymer*, **40**, 3919–3932.

111 Gupta, A.K. and Purwar, S.N. (1985) Studies on binary and ternary blends of polypropylene and SEBS, PS and HDPE. *J. Appl. Polym. Sci.*, **30**, 1799–1814.

112 Fowler, M.E., Keskkula, H., and Paul, D.R. (1989) Distribution of MBS emulsion particles in immiscible polystyrene/SAN blends. *J. Appl. Polym. Sci.*, **37**, 225–232.

113 Keskkula, H., Paul, D.R., McCreedy, K.M., and Henton, D.E. (1987) Methyl methacrylate grafted rubbers as impact modifiers for styrenic polymers. *Polymer*, **28**, 2063–2069.

114 Cheng, T.W., Keskkula, H., and Paul, D.R. (1992) Property and morphology relationships for trnary blends of polycarbonate, brittle polymers and impact modifier. *Polymer*, **33**, 1606–1619.

115 Okamoto, M., Shinoda, Y., Kojima, T., and Takashi, I. (1993) Toughening mechanism in a ternary polymer alloy: PBT/PC/rubber system. *Polymer*, **34**, 4868–4873.

116 Matsuo, M., Nozaki, C., and Jyo, Y. (1969) Fine structures and fracture processes in plastic/rubber two -phase polymer systems. I. Observation of fine

structures under the electron
microscope. *Polym. Eng. Sci.*, **9**, 197–205.

**117** Dickinson, L.C., Yang, H., Chu, C.W.,
Stein, R.S., and Chien, J.C.W. (1987)
Limits to compatibility in poly(x-
methylstyrene)/poly(2,6-
dimethylphenylene oxide) blends by
NMR. *Macromolecules*, **20**, 1757–1760.

**118** Kuleznev, V.N. and Dogadkin, B.A.
(1962) On relation between mechanical
properties and composition of polymer

blend. *Colloidnii J. (Colloid J.)*, **24**,
632–633.

**119** Miroshnikov, Yu.P. and Kuleznev, V.N.
(1972) Principle of binary additivity in
estimating properties of heterophase
polymer blends. *Kolloid. Zh.*, **34**,
884–888 (Russ.)

**120** Speransini, A.H. and Drost, S.J. (1970)
EPDM and chlorobutyl blends for tire
sidewalls. *Rubber Chem. Technol.*, **43**,
482–491.

# 7

# Morphology and Structure of Polymer Blends Containing Nanofillers

*Hossein Nazockdast*

*Department of Polymer Engineering, Amirkabir University of Technology, Tehran 15875-4413, Iran*

## 7.1
## Introduction

Polymer blending has been used as one of the successful methods of developing new polymeric materials with improved or new synergy of many specific properties for diverse applications. Although some polymer blends are completely or partially miscible, most polymer blends are immiscible and exhibit multiphase morphology due to their high molecular weight and unfavorable interaction. It is well known that morphology plays a significant role in determining rheological behavior, and physical and mechanical properties of polymer blends. This is the reason why during the last three decades a numerous research work has been devoted to develop strategies for compatibilization of polymer blend constituents in order to control and stabilize the morphology of polymer blend systems. Adding multifunctional block copolymers and reactive polymers in melt blending and employing co-solvent in solution blending are among the most common routes used for compatibilization Nowadays there are many books, and numerous articles and reviews dealing with morphology development, rheology, and the compatibilization of polymer blends [1–6]. Reinforcement of polymers through incorporation of inorganic filler is another successful and well-established route to produce a variety of polymer composites with enhanced thermal and mechanical performance for different applications. Nevertheless, experiences gained from polymer composite technology have demonstrated that for many prescribed properties the enhancing efficiency of adding microscale filler is not necessary determined by filler concentration but is mainly defined by the filler aspect ratio or more specifically by the filler–matrix interfacial area. This concept has opened revolutionary possibilities in the development of new nanoscale-filler-reinforced polymer composites with dramatically enhanced properties at much lower filler loadings. The main reason behind increasingly

*Encyclopedia of Polymer Blends: Volume 3: Structure*, First Edition. Edited by Avraam I. Isayev.
© 2016 Wiley-VCH Verlag GmbH & Co. KGaA. Published 2016 by Wiley-VCH Verlag GmbH & Co. KGaA.

growing interest in polymer nanocomposite is that nanoparticle addition can dramatically change various properties of polymer matrix including electrical and thermal conductivity, dielectric and magnetic permeability, gas barrier properties, and mechanical performances [3,7–14].

More recently, a combination of polymer blending and nanoscale filler reinforcement has received special attention from both academia and industry. This interest stems from the fact that the addition of nanofillers into multiphase polymer blends has proved to be an efficient strategy to develop a new family of polymer nanocomposites with a great tailoring potential for producing products with a combination of prescribed properties. That is the reason why there have been several reviews devoted to the nanoparticles containing polymer blends [15,16].

## 7.2
### Type of Nanofiller Used in Polymer Nanocomposite

Silica nanoparticles (hydrophilic and hydrophobic), layered silicate, surface modified nanosilicates (organoclays), single- and multiwalled carbon nanotubes (CNTs), and graphene are most common nanofiller used in producing polymer nanocomposite and nano-containing polymer blends.

### 7.2.1
### Structure and Characteristic of Layered Silicate

The most common layered silicate used for producing polymer nanocomposites belong to family of 2:1-layered silicate or phyllosilicates. Their crystal lattice consists of a two-dimensional very thin layer of a central octahedral aluminum oxide sheet sandwiched between two tetrahedral coordinated silicon atoms as illustrated in Figure 7.1.

The layer thickness is around 1 mm and the total dimensions may vary from 30 nm to several microns or even larger depending on the source of the layered silicate. The clays prepared by milling typically have lateral platelet dimensions of approximately 0.1–1 mm. Therefore, the aspect ratio of these layers (ratio length/thickness) is very high with values greater than 1000 [17–19]. When in the original pyrophyllite structure the trivalent of cation in the octahedral layer is partially substituted by the divalent Mg ($Mg^{++}$) cation, the structure of montmorillonite clay is formed, which is the best known element of the group of clay minerals called "smecfile clay." In this case, the overall negative charge is balanced by sodium and calcium ion silicates in a regular van der Waals gap between the layer called interlayer or gallery. A particular feature of this type of layered silicate (montmorillonite) is that the layers are bound together by actively weak surface charge known as the cation exchange capacity. This enables water, and polar molecules can readily enter between the unit layers [10,20,21].

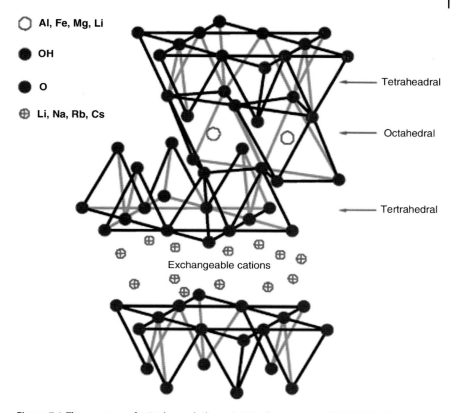

**Figure 7.1** The structure of a 2:1 layered silicate [20] (with permission © 2008 Elsevier).

### 7.2.1.1 Surface Modification of Layered Silicates

It is well recognized that the degree of success in achieving optimum enhancement in nanoparticle-filled composites is strongly determined by the effective aspect ratio of the particles as well as the particle matrix interfacial interaction. On the other hand, layered silicates in their pristine state are only compatible and can, therefore, be readily intercalated leading exfoliated morphology in hydrophilic polymers. Hence, in order to render them with other polymers, it is essential to exchange the alkali counter ions with a cationic–organic surfactant as shown in Figure 7.2. Alkyl ammonium ions are the most commonly used surfactants in layered silicates surface modification.

The organic cations lower the surface energy of the silicate surface and enhance their affinity with the polymer matrix. Moreover, the long organic chains of these surfactants with positively charged ends enable them to be tethered to the surface of the negatively charged silicate layers, resulting in an increase in the gallery spacing [23]. It then becomes possible for the polymer matrix to diffuse between the layers and ultimately separate them into tactoids or platelets, resulting in intercalated or exfoliated morphology. Thus, the surface

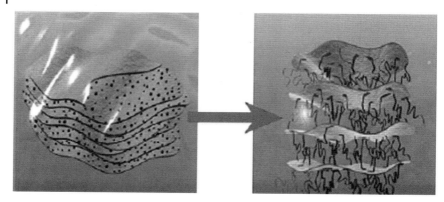

**Figure 7.2** Schematic picture of an ion-exchange reaction. The inorganic, relatively small (sodium) ions are exchanged against more voluminous organic ammonium cations. This ion-exchange reaction has two consequences: firstly, the gap between the single sheets is widen, enabling polymer chains to move in between them, and secondly, the surface properties of each single sheet are changed from being hydrophilic to hydrophobic [20,22] (with permission © 2008 Elsevier).

modification both increase the gallery spacing of layered silicates and serve as a compatibilizer between the hydrophilic clays and hydrophobic polymers [24]. It is worth noting that the capability of layered silicates to exchange ions, which can be quantified by specific properties known as the cation-exchange capacity (CEC), is strongly dependent upon the nature of the soil where the clay was formed. Based on the CEC measurements, the alkyl ammonium content of organo-modified layered silicates (organoclay) is about 35–45 wt%. The interlayer distance is increased by the chain length of the surface modifier and the charge density of the clay [25,26], although the ammonium ion chain arrangement in the organoclay can also play a role [14,25]. It should be noted that the final nanoscopic structures of layered silicate-polymer nanocomposites are defined by thermodynamic affinity and hence compatibility between the type of organically modified nanoclay and the polymer matrix, on the one hand, and preparation methods and related conditions, on the other. Characteristics of various nanoclays are presented in Table 7.1.

**Table 7.1** Chemical structure of commonly used 2:1 phyllosilicates [a] [14,20].

| 2:1 phyllosilicates | General formula |
| --- | --- |
| Montmorillonite | $M_x(Al_{4-x}Mg_x)Si_8O_{20}(OH)_4$ |
| Hectorite | $M_x(Mg_{6-x}Li_x)Si_8O_{20}(OH)_4$ |
| Saponite | $M_xMg_6(Si_{8-x}Al_x)O_{20}(OH)_4$ |

a)  M: monovalenr carion; $x$: degree of isomorphous substitution (between 0.5 and 1.3).

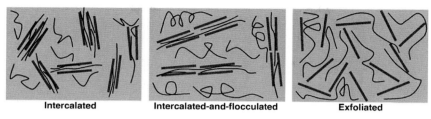

|  |  |  |
|---|---|---|
| **Intercalated** | **Intercalated-and-flocculated** | **Exfoliated** |

**Figure 7.3** Schematic illustration of three different types of thermodynamically achievable polymer/layered silicate nanocomposites [21] (with permission © 2003 Elsevier).

### 7.2.1.2 The Structure of Polymer-Silicate Nanocomposites

Depending on the preparation method and the nature of the components used including polymer matrix and organically modified layered silicate, three types of polymer-layered silicate nanocomposites can be obtained. These three types of nanocomposites are depicted in Figure 7.3. Intercalated nanocomposites are formed when a few extended polymer chains are intercalated between the silicate layers. This results in a well-ordered multilayered structure of alternating polymeric and inorganic layers with a repeated distance (less than 20–30 Å) between them [14,20]. On the other hand, exfoliated or delaminated structures are obtained when the layer clays are well separated and individually dispersed in the polymer matrix [14,20,27]. In this case, the polymers separate the clay platelets by 80–100 Å or more, which is comparable to the radius of gyration of the polymer matrix [27]. The exfoliated nanostructure is of particular interest since it provides the maximum surface area available for the optimum polymer–organoclay interfacial interaction, which, in turn, provides most possible chance to achieve desirable enhancements. Silicate layers are sometimes flocculated due to the hydroxylated edge–edge interaction of the silicate layers.

### 7.2.2
### Structure and Characteristic of Carbon Nanotube

CNTs are the third allotropes of carbon next to diamond and graphite and are constituted by hexagonal networks of carbon atoms, forming cylinders of about 1 nm diameter and 100–10 000 nm length, thus a very high aspect ratio. CNTs are long cylinders of covalently bonded carbon atoms, which look somewhat like a graphene sheet that has been rolled up into seamless tubes [28]. A graphene sheet can be rolled more than one way, producing different types of CNTs. Thus, CNTs can be categorized by their structures: single-walled nanotubes (SWNTs) and multiwalled nanotubes (MWNTs) (see Figure 7.4).

SWNTs have a diameter of ~1 nm, with a tube length that can be many millions of times longer. MWNTs are the nanotubes consisting of up to several tens of graphitic shells with adjacent shell separation of 0.34 nm, diameters of 1 nm, and a high length/diameter ratio [29,30]. The interlayer

**Single-walled CNT**　　　　　　**Multi walled CNT**

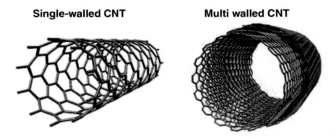

**Figure 7.4** Artistic representation of single and multiwalled carbon nanotube.

distance in MWNT is close to the distance between graphene layers in graphite, ~3.3 A. The way the graphene sheet is wrapped is represented by a pair of indices ($n$, $m$), called the chiral vector. The integers $n$ and $m$ denote the number of unit vectors along two directions in the honeycomb crystal lattice of graphene. If $m = 0$, the nanotubes are called "zigzag," which is named for the pattern of hexagons as we move along the circumference of the tube. If $n = m$, the nanotubes are called "armchair," which describes one of the two confirmers of cyclohexene, a hexagon of carbon atoms. Otherwise, they are called "chiral," in which the $m$ value lies between zigzag and armchair structures. The word chiral means handedness and it indicates that the tubes may twist in either direction (see Figure 7.5).

CNTs are the strongest and stiffest materials yet discovered in terms of tensile strength (150–180 GPa) and elastic modulus (640 GPa–1 TPa). This strength results from the covalent $SP^2$ bonds formed between the individual carbon atoms. Young's modulus value of SWNTs is estimated as high as 1–1.8 TPa depending on

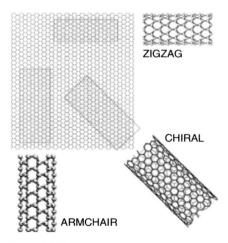

**Figure 7.5** Rolling up a graphene sheet to produce zigzag, armchair, and chiral SWCNTs.

the diameter and chirality. However, in the case of MWNTs, it correlates with the amount of disorder in the sidewalls. For MWNTs, experiments have indicated that only the outer graphitic shell can support stress when the tubes are dispersed in a polymer matrix, and for SWNT bundles (also known as ropes), it has been demonstrated that shearing effects due to the weak inter tube cohesion gives significantly reduced moduli compared to individual nanotube [31].

CNTs have many structures, differing in length, thickness, and in the type of helicity and number of layers. Although they are formed from essentially the same graphite sheet, their electrical characteristics differ depending on these variables, acting either as metals or as semiconductors. Thus, some nanotubes have conductivities higher than that of copper, while others behave more like silicon.

All nanotubes are expected to be very good thermal conductors along the tube, exhibiting a property known as "ballistic conduction," but good insulators laterally to the tube axis. It is predicted that CNTs will be able to transmit up to $6000\,\mathrm{W\,m^{-1}\,K^{-1}}$ at room temperature. The temperature stability of CNTs is estimated to be up to $2800\,°C$ in vacuum and about $750\,°C$ in air. The thermal expansion of CNTs will be largely isotropic, which is different than conventional graphite fibers, which are strongly anisotropic. This may be beneficial for carbon–carbon composites. It is expected that low defect CNTs will have very low coefficients of the thermal expansion.

The chemical reactivity of CNT compared with a graphene sheet is enhanced as a direct result of the curvature of the CNT surface. This curvature causes the mixing of the $\pi$- and $\sigma$-orbital, which leads to hybridization between the orbitals. The degree of hybridization becomes larger as the diameter of SWNT gets smaller. Hence, the reactivity of CNT is directly related to the $\pi$-orbital mismatch caused by an increased curvature. Therefore, a distinction must be made between the sidewall and the end caps of a nanotube. For the same reason, a smaller nanotube diameter results in increased reactivity. Covalent chemical modification of either sidewalls or end caps has shown to be possible. For example, the solubility of CNTs in different solvents can be controlled by this way. However, the covalent attachment of molecular species to fully $SP^2$-bonded carbon atoms on the nanotube sidewalls has proved to be difficult. Therefore, nanotubes can be considered as usually chemically inert.

### 7.2.2.1 Surface Modification of Carbon Nanotube

There has been an immense effort to establish the most suitable conditions for the transfer of either the mechanical load or the electrical charge to individual nanotubes in a polymer nanocomposite. A prerequisite for achieving such a target is the efficient dispersion of individual nanotubes and the establishment of a strong interfacial interaction between CNTs and surrounding polymer matrix. Various chemical modification methods have been successfully utilized in introducing functional moieties to CNTs in order to achieve efficient thermodynamic wetting of nanotubes with polymer matrices, which, in turn, results in better nanotube dispersion [27].

Chemically bonded whole polymer chain is expected to have a greater influence on the nanotube properties and their affinity to polymer matrices as compared to that of low molecular weight functionalities [28]. The modification of CNTs by polymers is divided into two main categories, depending on whether the chemical bonding of polymer chain to the nanotube surface involves either "grafting to" or "grafting from" strategies.

In the "grafting to" approach, polymer chains are grafted to the surface of pristine, oxidized or pre-functionalized CNTs. The main approaches exploited in this functionalization strategy are radical or carbanion additions as well as cyclo-addition reactions to the CNT double bonds [32].

In the "grafting from" approach, the grafting is carried out through polymerization of monomers from surface-derived initiators on either MWCNTs or SWCNTs. These initiators are covalently attached using the various functionalization reactions developed for small molecules, including acid-defect group chemistry and sidewall functionalization of CNTs [33].

The advantage of the "grafting from" approach is that the polymer growth is not limited by steric hindrance, allowing high molecular weight polymers to be efficiently grafted. In addition, nanotube–polymer composites with quite high grafting density can be prepared. However, this method requires strict control of the amounts of initiator and substrate as well as accurately controlled conditions required for the polymerization reaction. To demonstrate this approach, the well-known living polymerization approaches have been employed in recent years because of their tolerance to a wide variety of functional groups in the monomer units.

### 7.2.3
### Structure and Characteristics of Silica Nanoparticles

Silica dioxide ($SiO_2$) under the trademark of Aerosil is highly dispersed amorphous, very pure silicate that is produced by high temperature hydrolysis of silicone tetrachloride in an oxyhydrogen gas flame. It is a white, fluffy powder consisting of spherically shaped primary particles. Aerosil-OX50 has the largest average primary particle size of 40 nm. Aerosol-300 and Aerosol-380 have the same size of 7 nm. The primary particles in the flame interact with developed aggregates which join together reversibly to form agglomerates. Figure 7.6 shows a transmission electron microscopy (TEM) micrograph of an Aerosil-300, in which the primary particles, aggregates, and agglomerates can be observed. Siloxane and silanol groups are situated on the surface of the silica particles. The silanol group is responsible for the hydrophilic behavior of the untreated silica nanoparticles. The surface of silica nanoparticles can be chemically modified by the reaction of the silanol groups with various silanes and silazenes, resulting in hydrophobic silica. For example, Aerosil-R972 and Aerosil-R812 are made hydrophobic with dimethylsilyl or trimethylsilyl groups. For the past decades, fumed silica nanoparticles have been used successfully as thickening agents and thixotrope in liquid systems. Also, silica nanoparticles have been widely used to control the rheology of silicon rubber, plastics, coating, and adhesives.

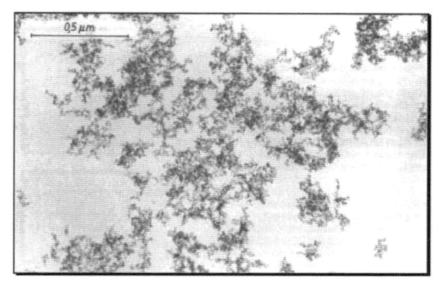

**Figure 7.6** TEM micrograph of Aerosil-300 showing particles in aggregates (courtesy of Degussa).

## 7.3
## Nanostructural Characterization

### 7.3.1
### X-Ray Diffraction

The X-ray diffraction (XRD) is the easiest and most widely used technique to probe the nanostructures of layered silicate filled nanocomposites. This technique can be used to determine the interlayer distance of nanoclays by utilizing Bragg's law expressed as $\sin(\theta) = n\lambda/2d$, where $\lambda$ is the wavelength of the X-ray radiation used, $d$ is the spacing between the diffraction of lattice plans, and $\theta$ is the measured diffraction angle [14,20]. By monitoring the position, shape, and intensity of the basal reflections from the distributed silicate layers, the nanocomposite structures can be identified. In the case of layered silicate–polymer nanocomposites with an intercalated structure, an increase in the interlayer spacing due to polymer intercalation can be determined from a shift of diffraction peak toward lower diffraction angle according to Bragg's law. The greater shift to the lower value of $2\theta$ would be an indication of the higher extent of polymer intercalation and probably better dispersion of layered silicate (see Figure 7.7) [21]. The increase in interlayer spacing induced by polymer intercalation also decreases the periodicity, hence reduces the intensity of the clay peak. In an exfoliated or delaminated structure, because of the disruption of the coherent layer stacking induced the extensive polymer intercalation the basal peaks associated with the clay are no longer visible in the XRD pattern (Figure 7.7). This may be caused by

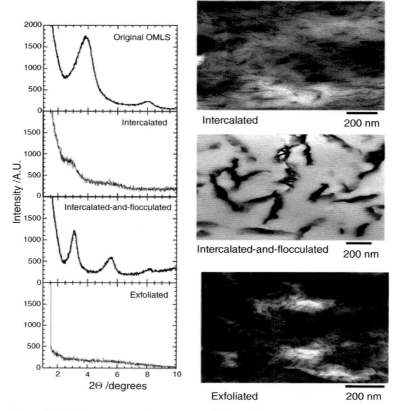

**Figure 7.7** WAXD patterns and TEM images of three different types of nanocomposites [21] (with permission © 2003 Elsevier).

either much too large spacing between the layers (i.e., exceeding 8 nm in ordered exfoliated structure) or complete disordering of the platelets in the nanocomposites [14,20,27,34]. The effect of polymer intercalation on the ordering of organically modified silicate layers can also be determined through variation in the full width at half-maximum (FWHM) and the intensity of the basal reflections. A decrease in the degree of coherent layer space stacking (more disordering) results in peak broadening and loss of its intensity, while an increase in the degree of coherent layer space stacking (more ordered system) results in a relative decrease in the FWHM of the basal reflections upon hybrid formation.

## 7.3.2
### Transition Electron Microscopy (TEM)

Although XRD is a reliable and powerful technique for determining the interlayer spacing of some nanoparticles and the intercalated nanocomposites, it provides little information about the long-range distribution and orientation of the silicate layer in the composites. On the other hand, the TEM enables one to directly obtain a great insight into the qualitative understanding of the inherent nano-scopic structures through observation of the localized area [20,35,36]. Figure 7.7 shows TEM micrographs obtained for intercalated, exfoliated, and flocculated nanocomposites. It is worth mentioning that in addition to these well-defined structures there are other nanocomposites that present both intercalated and exfoliated structures. In this case, the XRD pattern becomes broaden and the applicability of the XRD technique in evaluating the extent of intercalation is lost and, therefore, one must rely on the TEM observation to define the overall structure. Thus, usually XRD and TEM can be utilized as two main complimentary techniques to characterize the layered silicate containing nanocomposites.

### 7.3.3
### Differential Scanning Calorimetry

Differential scanning calorimetry (DSC) can also be employed to gain a further insight into the interaction between the nanoparticles and polymer matrix. This may be achieved by monitoring the nanoclay-induced changes in the thermal transition of amorphous as well as crystalline polymer matrices. Figure 7.8, for example, shows the DSC thermogram of polystyrene/organoclay (PS/organoclay) mixture and an intercalated PS/organoclay nanocomposite. As is clearly seen, while in the physical mixture of PS/organoclay, the characteristic peak

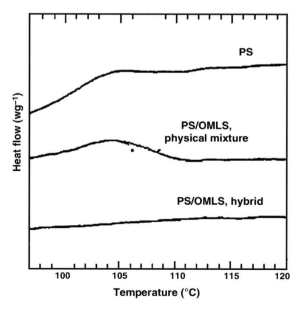

**Figure 7.8** DSC traces of pure PS, a physical mixture of PS/OMLS, and PS-intercalated OMLS [20,39] (with permission © 2000 WILEY-VCH).

corresponding to $T_g$ of PS remained intact, it disappeared or shifted to higher temperature (not covered in this figure) in the intercalated nanocomposite sample. These results reveal that disappearing or shifting the $T_g$ peak of the polymer matrix, which is an indication of retarding the polymer chain motion, can be considered as a measure of the polymer–nanoparticle interfacial interaction. Furthermore, since this effect is highly dependent on the effective surface area of the nanoparticles, DSC can provide additional information about the state of dispersion of nanoparticles. A similar argument is true on the applicability of DSC in determining the polymer–nanoparticle interfacial interaction for crystalline matrices, through evaluating the nanoparticle influence on crystallinity behavior of the nanocomposite. Of course, this is only possible if the effect of the polymer–nanoparticle interaction on the crystallinity is strong enough to be determined by DSC. The addition of organoclays to crystalline polymers may also influence their crystallinity when nanoparticles act as nucleation agents. More importantly, DSC can be employed to study the relationship between flow-induced anisotropic nano-orientation and crystallinity of matrix as one of the challenging issues in the field of polymer nanocomposites [13,37,38].

### 7.3.4
### The Linear Rheological Measurements

The linear and nonlinear viscoelastic behavior of nanocomposites have received considerable attention from both academia and industries [40–43]. This stems from two remarkably important points of view: (1) to take advantage of the sensitivity of rheology to nanostructure to gain the fundamental understanding of the role of nanoparticles in determining microstructure and (2) to establish the microstructure-processing relationship in order to monitor the processing and hence to produce products with optimum properties.

In the linear viscoelastic measurements, a small amplitude oscillatory strain $\gamma(t)$ of the form of $\gamma(t) = \gamma_0 \sin(\omega t)$ and the resulting time-dependent linear shear stress are recorded and interpreted as $\sigma(t) = \gamma_0(G' \sin(\omega t) + G'' \cos(\omega t))$, where $G'$ and $G''$ are storage and loss modulus, respectively. Here $\gamma_0$ is the strain amplitude and its values are kept as low as possible in order to keep the quiescent state microstructure unchanged during the experiment. The other important rheological functions derived include the complex modulus $G^* = (G'^2 + G''^2)^{1/2}$, complex viscosity $\eta^* = \frac{G^*}{\omega}$, and loss tangent, $\tan \delta = \frac{G''}{G'}$.

The linear viscoelastic measurements have shown that, depending on the aspect ratio and interfacial interaction between nanoparticles and polymer matrix there is a concentration of nanoparticles ($\varphi$) above which the nanocomposites exhibit a liquid–solid transition (LST) (called rheological percolation threshold) at a low frequency range [43]. Different criteria have been used to identify the LST. Among these is the Van Gurp plot of the phase angle, $\delta(\omega, \varphi)$, versus the absolute value of the complex modulus ($G^*$). For neat polymer melt and the low content of nanoparticle-filled melts, $\delta(\omega, \varphi)$ tends 90°. As concentration $\varphi$ of nanoparticles increases to a critical value, a drastic decrease in $\delta(\omega, \varphi)$

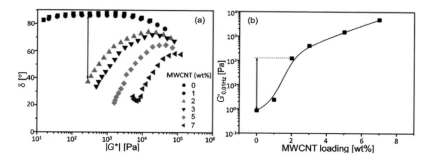

**Figure 7.9** (a) van Gurp–Palmen plot and (b) $G'$ measured at 0.01 Hz as a function of MWCNT loading for PBT nanocomposite melts at 240 °C [44] (with permission © 2007 Wiley Periodicals, Inc.).

at low values of $G'(\omega, \varphi)$ occurs, indicating LST. For example, as shown in Figure 7.9, a rheological percolation threshold or LST occurs at $\varphi = 2\%$ for MWCNT-filled poly(butyl terephathalate) (PBT) [44].

The Cole–Cole plot of the imaginary part of complex viscosity $\eta''(\omega, \varphi)$ versus the real part of complex viscosity $\eta'(\omega, \varphi)$ has also been used as a criterion to determine the LST (see Figure 7.10). This plot shows one arc corresponding to the matrix relaxation at low values of $\varphi$, while it exhibits a pronounced tail at high viscosity side for the nanocomposite when $\varphi$ reached a certain concentration ($\varphi_{\text{percolation}}$) [43,45].

Another criterion is the Han plot $G'(\omega, \varphi)$ versus $G''(\omega, \varphi)$. As can be seen in Figure 7.11 for MWCNT-filled polylactide, the plot of low-filled melt may overlap with that of the matrix, while a further increase in nanoparticle concentration leads to a gradual decrease in the slope in the terminal region [43,46].

The log–log plot of the storage modulus versus frequency ($G'$ versus $\omega$) can also be used to gain some insight about the LST. While nanoparticle-filled melts at low concentrations exhibit characteristic homopolymer-like terminal behavior expressed by the power-law $G' \cong \omega^2$ and $G'' \cong \omega$, the polymer nanocomposite

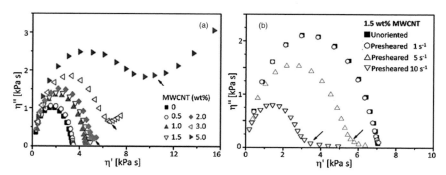

**Figure 7.10** Cole–Cole plot for (a) PCL melts containing different MWCNT loadings and (b) 1.5 wt% MWCNT-filled PCL melts presheared at different strain rates [45] (with permission © 2007 Wiley Periodicals, Inc.).

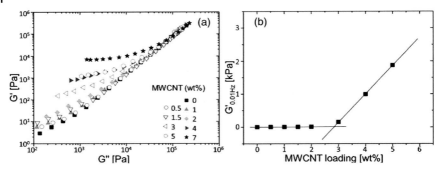

**Figure 7.11** Han plots for carboxylic MWCNT-filled polylactide melts and (b) measured at 0.01 Hz as a function of MWCNT loading [46] (with permission © 2008 Elsevier).

slope of $G'$ versus $\omega$ is much smaller than 2, and $G'$ reached to a plateau at a low frequency range at certain nanoclay concentration (rheological percolation). Recently, the nonlinear viscoelastic measurements have been utilized as a complimentary to linear viscoelastic tests to probe the microstructure evolution in nanocomposites [41,42,47,48]. It was shown that, while the samples containing low concentration of organoclay exhibit a transient behavior similar to that of the matrix, high-concentration filler-loaded samples show a pronounced stress overshoot whose extent increases with increasing the nanofiller content. A similar behavior has been reported for other nanoparticles including CNTs, and graphene-filled nanocomposites. The observed stress overshoot has been interpreted in terms of a solid-like response due to a three-dimensional (3-D) network formed by the particle–particle and/or particle–polymer interaction which is broken up as the convective forces become a dominant parameter.

The results of this experiment together with the results obtained from the flow reversal test can be used to provide valuable information about the nanoscopic structure and structural recovery of nanoparticle-filled polymers. For anisotropic nanoparticle (CNT or nanoclay)-filled polymers, the stress-induced nanoparticle orientation may also have a contribution to the stress overshoot values [49–52].

## 7.4
### Partially Miscible Polymer Blends Containing Nanoparticle

During the past two decades, a combination of polymer blending and nanoparticle incorporation has proved to be an effective strategy with a great tailoring potential to develop a variety of new polymer composites for a wide range of prescribed applications [9,15,24,50,53–63]. It has been found that the mechanism by which nanoparticles influence the phase-separation behavior and separated morphologies in partially miscible blends is significantly deferent compared to that in immiscible blends. Therefore, it would be more reasonable to discuss them in two separated categories.

7.4.1

**The Effect of Nanoparticles on Phase Separation of Partially Miscible Polymer Blends**

Recently, there has been a considerable interest in studying the influence of nanoparticles on the phase-separation behavior of partially immiscible polymer blends. From earlier studies, it was recognized that the addition of nanoparticles to the partially miscible polymer blends can strongly influence the behavior and kinetics of phase separation and resulted morphologies. This subject has been extensively studied by several research groups utilizing both theoretical and experimental approaches [15,64]. Recently, a comprehensive book chapter has been devoted to theoretical and modeling aspects of the phase-separation behavior of polymer blends containing nanoparticles by Ginzburg [65]. Therefore, these subjects have not been covered in this chapter. It has been recognized that the determination of the phase-separation diagram or phase-separation temperature is a prerequisite for studying the effect of nanoparticles on the thermally induced phase separation of nanoparticle-filled polymer blends.

Small angle light scattering (SALS) and optical microscopy are the most commonly used methods to study the phase-separation behavior of polymer blends. However, the validity of results obtained by these two techniques is strongly influenced by the transparency and thickness of the samples. Therefore, due to multiple scattering and high turbidity, these techniques are not applicable for nanoparticle-filled polymer blends. On the other hand, viscoelastic measurements have been successfully used as an alternative method in probing the phase-separation behavior of nanoparticle-filled polymer blends. That is because the linear viscoelastic response of polymer blends has proved to be very sensitive to microstructural changes taking place during the phase separation regardless of transparency and refractive indices of the blend components [55,66–68]. Time-temperature superposition (TTS) principle [63,67], Cole–Cole plot [67], plots of ($\eta''(\omega, \varphi)$ versus $\eta'(\omega, \varphi)$), and Han plot ($G'(\omega, \varphi)$ versus $G''(\omega, \varphi)$) are among most frequently used protocols to study the phase-separation temperature. By utilizing TTS, it is possible to distinguish whether the blend is in a homogeneous or heterogeneous region. The Cole–Cole plot of a blend shows a single arc when it is in a homogeneous region, whereas a tail or one additional arc with the original arc is observed when the blend is in the phase-separated state. The Cole–Cole plot has been found to be more sensitive and provide more insight compared to TTS for studying the phase-separation behavior. It should be noted that since the TTS method utilizes the viscoelastic results through the frequency sweep experiments performed at various temperatures, the accuracy of determining the phase separation is strongly affected by the temperature interval used in those experiments. That is, the smaller the temperature interval used for linear viscoelastic measurements, the more accurate results for the phase separation will be obtained. More recently, the isochronous dynamic temperature ramp experiment ($G'$ and $G''$ versus temperature) has been used as a more appropriate method to determine the critical phase-separation temperature [63,69]. In this experiment, usually storage modulus $G'$ is measured as a function of temperature at a constant low frequency at which the linear viscoelastic behavior is prevailed. In the case of lower critical solution temperature (LCST) blends, the storage modulus ($G'$) decreases as the temperature increases in the homogenous region due to promoted chain mobility. But when the temperature further increases and reaches the phase separation, the concentration fluctuation and domain-interface-induced stress will have extra contribution to the storage modulus. This will, in turn, compensate decreasing of the storage modulus ($G'$) caused by the promoted chain mobility. It has been shown that in strongly dynamically asymmetric blends such as PS/poly(vinyl methyl ether) (PEMV) blends the latter effect dominates and consequently a pronounced upturn in $G'$ appears as the temperature increases [54]. In these blends, the temperature at the lowest value of $G'$ in the curve of $G'$ versus temperature was denoted as the binodal temperature (BT). On the other hand, in the weakly dynamic asymmetric blends like poly (methyl methacrylate)/PS (PMMA/PS) blends, due to low contribution of the domain-interface-induced stress, the former effect dominates and, therefore, only small change in the decreasing rate of $G'$ with an increase in temperature can be detected.

The BT also called the rheological phase-separation temperature can be quantitatively measured through the onset and the inflection point of the upturn in $G'$ respectively [60]. On the other hand, the spinodal decomposition temperature (ST) can be determined by the theoretical treatment introduced by Fredricsson and Larson's based mean-field theory which was modified for the partially miscible polymer blends by Ajji and Choplin [70].

Recently, a few research studies have taken advantage of strong correlation between morphology and viscoelastic responses of polymer blends to study the morphological changes taking place during the phase separation [69,71,72]. Accordingly, the time evolution of dynamic modulus ($G'$) may show an increase or decrease depending on the changes on morphology during the phase separation. A comparison between these results with those predicted from the viscoelastic models can provide more insight into understanding the phase-separation behavior.

Over the years, the phase separation induced by the variation of temperature, pressure, and/or composition has been interpreted by nucleation and growth (NG) or spinodal decomposition (SD) mechanisms. However, recently Tanaka found a new mechanism of phase separation that could hardly be explained by these two classical mechanisms (NG and SD) [69,72]. According to Tanaka in this novel type of phase separation, in addition to initial diffusive and final hydrodynamic regimes, which are known for classical phase separation, there exists an intermediate viscoelastic region where elastic-force balances instead of interfacial tension-induced domain morphology. This type of phase separation is called viscoelastic phase separation (VPS). The main feature of VPS in the polymer blends is the induction of a percolated network of a more viscoelastic phase even if it is the minor phase. It should be noted that strong dynamic asymmetry between blend components is the key to VPS.

7.4.2
**The Effect of Silica Nanoparticles on Phase Separation of PMMA/Polyvinyl Acetate**

The effect of the addition of fumed silica on phase separation of PMMA/ poly-vinyl acetate (PVAc) with the same absorption tendency at the surface of the fumed silica was studied by Lipatov and Nestrove [73] and VanderBeek *et al.* [74]. They observed filler-induced change in the cloud point curve with either a decrease or increase in the phase separation with filler content at a component ratio of 30/70 or 20/80 respectively. These observations were explained in terms of a selective adsorption of one of the blend components at the filler interface. The selective adsorption of one component at the filler interface changes the ratio of this component both in the border layer and in the matrix bulk. In their latest study on the same system, Nestrov and coworkers confirmed the compatibilizing effect of the nanofillers by establishing $(\lambda_{AB})_{\text{filled}} < (\lambda_{AB})_{\text{unfilled}}$ through measuring the $T_g$ values and calculating the component blend ratio of the PVAc-enriched phase and the PMMA-enriched phase, $(\lambda_{AB})_{\text{filled}} < (\lambda_{AB})_{\text{unfilled}}$ [75]. In order to provide more insights into understanding the real mechanism, Lipotov investigated the effect of fumed silica on the phase-separation behavior of the chlorinated polyethylene/ethylene vinyleacetate (CPE/EVA) copolymer as an LCST system in which the blend components have different affinities to the solid surface of the fumed silica [76]. They demonstrated that the addition of fumed silica (at 5% or 10%) led to a dramatic change in the phase-separation temperature and the slope of the cloud point curve (see Figure 7.12). More specifically, they found that the phase-separation temperature

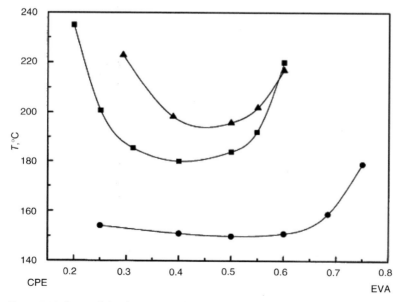

**Figure 7.12** Curve of cloud point for unfilled (■) and filled with 10% (D) and 5% (●) aerosol [76] (with permission © 2001 Published by Elsevier).

of the filled blends may either increase or decrease depending on the filler concentration. According to the authors, for 10% of the fumed silica, the thickness of the layer between two particles is low enough that both components transit into the state of the boarder layer. In this case, the variation of the shape and position of the bimodal curve of the phase diagram, induced by the nanofiller, is mostly controlled by the variation of the free energy of the turnery system and nanoparticles act as a compatibilizer. This obviously leads to an increase in the phase-separation temperature. At lower concentration of nanofiller, the possible effect is the redistribution of the blend components according to the molecular masses between the filler surface and in the bulk that may diminish the phase-separation temperature.

### 7.4.3
### Effect of Nanosilica on Phase-separation Behavior of PMMA/SAN Blends

The effect of the addition of nanoparticle on the phase-separation behavior of PMMA/styrene-co-acrylonitrile (PMMA/SAN) blends as an LCST system have been studied by many researchers [77–78]. Huang *et al.* examined the effect of the addition of 5% of untreated nanosilica on the phase-separation behavior of PMMA/SAN 70/30 blends by utilizing the viscoelastic measurements [62]. The filled and unfilled blends of various compositions were prepared by solution casting. They showed that, while below 190 °C, the TTS principle in the form of $G'$ versus $a_T\omega$ and HAN plots ($G'$ versus $G''$) is applicable for the PMMA/SAN filled with 5% of the unthreatened nanosilica, the TTS of the same sample failed when the temperature was increased above 195 °C. Also, as can be observed from Figure 7.13, below 190 °C, the sample shows only one single arc, whereas above 195 °C, the Cole–Cole plots show an extra drift tail, which is an indication of the phase-separation-induced elastic response.

In order to determine the phase-separation temperature more accurately, Hung constructed the $G' - T$ curve by picking up values of $G'$ at the same frequency (0.1 rad s$^{-1}$) and at various temperatures. As can be seen in the $G' - T$ curve, shown in Figure 7.14, at low temperatures where the blend is in the homogenous region, the $G'$ decreases linearly with increasing the temperature. This is due to the promoted chain mobility associated with an increased distance from the $T_g$ of the blend. With a further increase in temperature to the vicinity of phase separation (193 °C), the slope of the $G' - T$ curve decreases as the result of the enhanced elasticity induced by thermodynamic domination over chain mobility. The temperature corresponding to the change in the slope of the $G' - T$ curve (193 °C) was considered as the rheological phase-separation temperature (see Figure 7.15.)

Figure 7.15 compares the phase diagram of unfilled PMMA/SAN and nanosilica-filled PMMA/SAN/5% blends obtained through the linear viscoelastic measurements. As can be seen, the critical temperature and critical composition are estimated to be 184 °C and 56% respectively. Figure 7.15 also shows that the addition of 5% nanosilica shifts the phase diagram vertically with respect to the

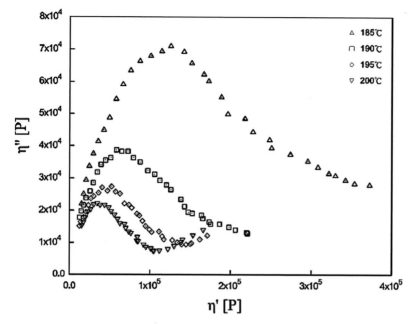

**Figure 7.13** Cole–Cole plot for PMMA/SAN(70/30)/SiO$_{2-5}$ mixture showing the occurrence of a tail beyond 195 °C [77] (with permission © 2005 Acta Materialia Inc.(Elsevier)).

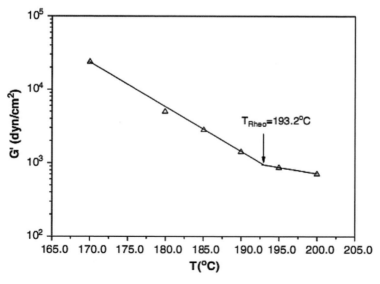

**Figure 7.14** Temperature dependence of *G'* in the mixture of PMMA/SAN (70/30)/SiO$_{2-5}$ on temperature. The data are picked at 0.1 rad s$^{-1}$. $T_{Rheo}$ signifies the phase temperature determined from rheological methods [77] (with permission © 2005 Acta Materialia Inc.(Elsevier)).

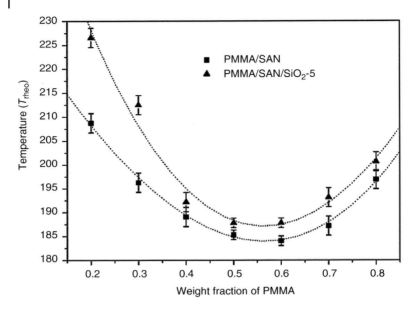

**Figure 7.15** Phase diagrams of PMMA/SAN/SiO$_{2-5}$ and PMMA/SAN mixtures as determined by shear rheology: $T_{Rheo}$. The dotted lines are obtained by the least-squares fitting for a polynomial. (Equation [77], with permission © 2005 Acta Materialia Inc. (Elsevier)).

unfilled blend, and the amount of this shift highly depends on the blend composition. More specifically, the nanoparticle-induced shift is much larger for the blends in which PMMA is the minor component ($\phi_{PMMA} < 5\%$) compared to the blends in which PMMA is the major component ($\phi_{PMMA} > 56\%$). On the basis of the above-described results and calculated Flory–Hugins interaction parameters, with taking into account the composition change caused by the addition of the nanofiller, Hung proposed a mechanism for explaining the selective adsorption of the PMMA chain on the silica surface. Accordingly, the adsorption of longer chain fraction of PMMA on the surface of the nanoparticles would lead to a decrease in the average molecular weight of PMMA, which is in favor of PMMA/SAN miscibility.

The same research group used a combination of dynamic rheology and optical microscopy methods to provide more insight into the effect of hydrophilic nanosilica on the phase-separation behavior with focus on determining the binodal and the ST in PMMA/SAN blend [63]. It was shown that the TTS principle works quite well for the neat components (PMMA and SAN) and nanoparticle-filled components which confirmed that neither the blend component nor the filled components contribute to the elasticity of the phase-separated blends. The TTS master curves of PMMA/SAN 60/40 and PMMA/SAN/SiO$_2$ (60/40/3) showed that, while for the PMMA/SAN blend TTS is valid at low temperature (below 170 °C), the TTS breaks down at temperatures above 170 °C. However, for PMMA/SAN/SiO$_2$ 60/40/3, there was no obvious failure of the TTS except for the

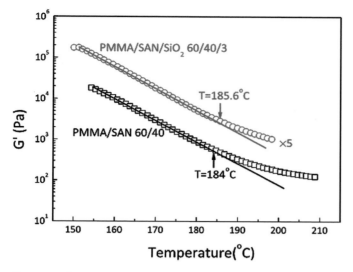

**Figure 7.16** Dynamic temperature ramp for blends PMMA/SAN 60/40 and PMMA/SAN/ SiO$_2$ (60/40/3). The straight lines denote the apparent linear fit with the storage modulus at low temperature. $G'$ of the particle-filled blend are shifted to avoid overlapping [63] (with permission © 2012 Elsevier).

deviation of low frequencies. The observed failure of TTS could be attributed to the appearance of concentration fluctuation in PMMA/SAN weakly dynamic asymmetric blends, where the failure temperature is usually lower than the BT. These results suggested that the TTS principle not only can be used to estimate the phase-separation temperature of PMMA/SAN blend, but it is also sensitive enough to detect the possible effect of the incorporation of nanoparticles.

The isochronous dynamic temperature $(G' - T)$ is among the other adopted rheological methods used to determine the BT in unfilled and nanoparticle-filled partially miscible blends. In this method, deviation from the temperature dependence of $G'$ in the homogenous regime is considered as the BT. Figure 7.16, for example, shows the temperature ramp results reported for the unfilled PMMA/ SAN 60/40 blend and PMMA/SAN/SiO$_2$ (60/40/3 wt%) blend nanocmposite.

In this method, the deviation point from the apparent line at low frequencies is usually defined as the phase-separation temperature. It follows that this is about 184 °C and 185 °C for the unfilled and filled PMMA/SAN/SiO$_2$ blends respectively. This temperature is much higher than the temperature (160–170 °C) obtained through the deviation from the terminal behavior suggesting that at 180 °C, the blend has already changed from homogenous to heterogeneous.

It is well known that partially miscible polymer blends undergo two different thermodynamic-induced phase-separation mechanisms depending on blend composition. The blend with near-critical composition experiences spinodal decomposition (SD), resulting in co-continuous morphology in the initial stage, while that with off-critical compositions undergo the nucleation and growth (NG) mechanism, forming droplet-matrix morphology. Obviously, these two phase-separated morphologies are expected to exhibit different viscoelastic behavior. The phase-separated-induced droplet matrix is expected to exhibit a linear viscoelastic-type behavior similar to that predicted by the Palierne model introduced for the polymer blend with droplet-matrix morphology [79]. Accordingly, in the frequency sweep experiment ($G' - \omega$), a shoulder in $G'$ due to droplet-deformation-induced elasticity is observed at low frequency followed with a liquid-like terminal behavior. On the other hand, co-continuous morphology shows a power-law dependence of $G'$ on frequency at the low frequency range, where the liquid-like terminal behavior could hardly be observed. Thus, by comparing the results obtained for the unfilled and the nanosilica-filled blends, Gao *et al.* reached the conclusion that the addition of 3% nanosilica did not make an appreciable change in low-frequency response of the PMMA/SAN blends [63]. Therefore, they could justify the results, as shown in Figure 7.16, for the phase-separation temperature results obtained from the temperature ramp method for the PMMA/SAN blends.

An attempt has also been made by Gao and coworker to obtain the BTs of near-critical compositions by employing the gel-like criterion, which is a characteristic behavior of polymer blends with co-continuous morphology, during spinodal decomposition [63]. The gel-like behavior refers to a phenomenon that the dependence of $G'$ and $G''$ on frequency follows the same power-law relation in the low frequency range. In this case, the loss angle tangent ($\tan \delta$) would be a constant and independent of frequency at sufficiently low frequencies. From the results obtained from the above-described method for the near-critical composition of the unfilled and filled PMMA/SAN blends, they found very close BT for the unfilled (165.8 °C) and filled (167.7 °C) blends. This suggested that the addition of nanosilica at a concentration of 0.03 does not have an appreciable effect on BT of PMMA/SAN blends. It should be noted that due to possible errors in low-frequency rheological measurements, $\tan \delta$ might be converted to a single point. That is why Gao *et al.* used the arithmetic of three points of intersection as the phase-separation temperature where the standard deviation can be regarded as the error in the determination of the bimodal temperature.

In the same work, they utilized the Cole–Cole plots to determine the phase-separation temperature for off-critical compositions PMMA/SAN 50/50, 40/60, and corresponding nanosilica-filled blends, which exhibited a shoulder in the $G' - \omega$ plot. At 160 °C, the Cole–Cole plot appeared as a self-circular arc, indicating a homogenous state in the blend. When the temperature was increased to 170 °C, a slight deviation at high $\eta'$ corresponding to low frequencies with no change in the rest of the Cole–Cole plot was observed. This behavior was related to the elastic contribution of concentration fluctuation [63]. At higher temperature, the Cole–Cole plot showed an additional small half-circle or drift tail which became larger with a further increase in temperature. By considering these results, it was suggested that the concentration-fluctuation-induced elasticity

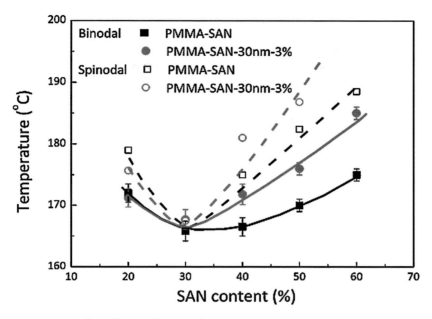

**Figure 7.17** The binodal phase diagram of PMMA/SAN (solid circles) and PMMA/SAN/SiO$_2$ (solid squares) by combining the gel-like method and Cole–Cole plot. The spinodal points of PMMA/SAN (hollow circles) and PMMA/SAN/SiO$_2$ (hollow squares) are also shown. The lines are adopted from [63] (with permission © 2012 Elsevier).

can be separated from that caused by phase-separated domains. Accordingly, the BT for the off-critical PMMA/SAN 40/60 blends and that for the corresponding nanosilica-filled blend were found to be 175 °C ± 1 °C and 185 °C ± 1 °C respectively. Figure 7.17 shows the binodal phase diagram of PMMA/SAN and PMMA/SAN/SiO$_2$obtained by using a combination of the gel-like criterion and the Cole–Cole plot method. By comparing these results, it was concluded that the addition of nanosilica significantly shifts the phase diagram of PMMA/SAN blends, especially for blends in which PMMA was a minor phase. This was attributed to the selective location of nanosilica in PMMA which was confirmed by scanning electron microscopy (SEM) micrographs. They also introduced a new method by integrating Ajji–Choplin's modified mean filled theory and the Doi–Edwards model which could successfully be used to quantitively determined ST. This new method could also be extended to nanosilica-filled PMMA/SAN blends. It should be noted that in the new method the fluctuation temperature elastic contribution to the storage modulus $G'$ was used instead of the whole dynamic modulus.

In a recent study, Huang and coworkers investigated the effect of the selective location of nanosilica on the phase-separation behavior of PMMA/SAN blends by using a combination of the rheological method and optical microscopy [62]. The selective location of nanosilica was controlled by varying the chain length of

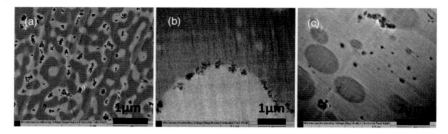

**Figure 7.18** TEM images of (a) PMMA/SAN/ SiO$_2$–OH 70/30/3 annealed at 172.5 °C for 2 h, (b) PMMA/SAN/SiO$_2$–PSm 50/50/3 annealed at 200 °C for 12 h, and (c) PMMA/SAN/SiO$_2$–PS1 80/ 20/3 annealed at 200 °C for 12 h. PMMA-rich and SAN-rich domains are observed as the white and dark regions, respectively [62] (with permission © 2012 American Chemical Society).

PS grafted on silica nanoparticles through surface-initiated atom transfer radical polymerization (ATRP). As is shown in Figure 7.18, the bare SiO$_2$−OH particles were selectively located in the PMMA-rich phase, while the particles having medium chain length grafted PS, SiO$_2$−PS$_m$ (as well as short chains grafted not shown here) were selectively located at the interface. However, the selective localization in the PMMA-rich phase occurred again for the particles with longer PS chain length SiO$_2$ − PS$_l$. These results were explained in terms of thermodynamic parameters, especially the interfacial tension values. Based on the power-law analysis of the moduli of components and the shifted Cole–Cole plot, it was shown that the selective location of silica nanoparticles had strong influence on the rheological transient temperature (in BTs) at near-critical as well as off-critical composition for both neat and nanoparticle-filled blends. On the contrary, little effect on optically determined BTs could be detected through changing the selective location of silica nanoparticle. This discrepancy was justified through the TEM observation of blends under different phase-separation conditions (Figure 7.19). According to the authors, nanoparticles-related coarsening on morphology during phase separation and most nanoparticle-induced slowing down were found in off-critical compositions with nanosilica particles preferentially located in the interface.

On the other hand, selectively locating nanoparticles in the minor phase could act as a nucleating site but decreased the total number of nuclei. The difference in the rheological transition temperature was attributed to the effect of nanosilica particles or the viscoelasticity of the components and the morphology during the phase separation. As is shown in Figure 7.20, the rheological transition temperature is much higher than that of the optically determined cloud temperature; the authors attributed this difference mainly to the spatial resolution of the two methods and not necessarily to a difference in the time scale of the two methods.

The influence of the selective location of silica nanoparticles on phase-separation behavior and the kinetics of phase separation of PMMA/SAN blends were also studied via a high-throughput (HTP) as a new approach which combines a composition and temperature gradient. The selective location of nanosilica

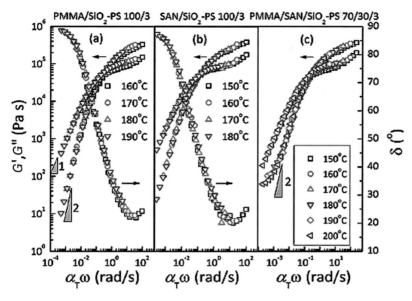

**Figure 7.19** Time–temperature superposition of polymer/nanoparticle composites (a) PMMA/SiO$_2$–PS 100/3, (b) SAN/SiO$_2$–PS 100/3, and (c) PMMA/SAN/SiO$_2$–PS 70/30/3. The reference temperature ($T$ ref) is 160 °C for all the nanocomposites [62] (with permission © 2012 American Chemical Society).

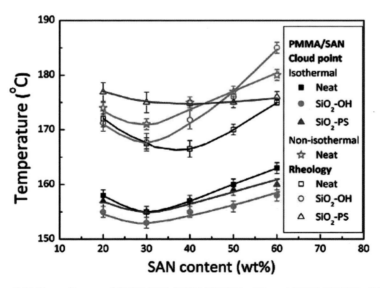

**Figure 7.20** Phase diagram of PMMA/SAN, PMMA/SAN/SiO$_2$–OH, and PMMA/SAN/SiO$_2$–PS blends obtained via rheology and optical microscopy. The content of nanoparticles is 3 wt% in the latter two blends [62] (with permission © 2012 American Chemical Society).

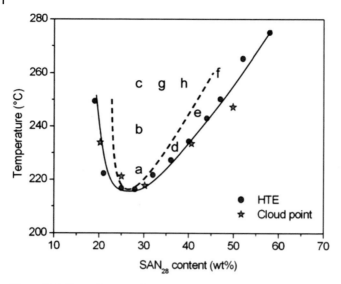

**Figure 7.21** Phase diagram of PMMA/SAN obtained from $\varphi$–$T$ gradient films as prepared with the HTP setup. The cloud points are reproduced [78] (with permission © 2013 WILEY-VCH).

particles was controlled by choosing two different types of treated surface nanosilica: hydrophilic and hydrophobic [78]. The evolution of the phase separation was studied by optical microscopy, small-angle neutron scattering (SANS) and TEM. Figure 7.21 presents the phase diagram obtained by the HTP experimentation and cloud point. The optical microscope pictures taken from different regions during the phase transition (as marked in phase diagram Figure 7.21) are also shown in Figure 7.22.

They demonstrated that the HTP experimentation can be used as a suitable tool to study the phase behavior of the blend. The SALS results obtained for the spinodal decomposition showed a slow-down effect on coarsening for both hydrophobic and hydrophilic silica particles. For the hydrophobic nanosilica, which is selectively located in the PMMA-rich phase, the induced slow-down of coarsening was attributed to the reduction of mobility of PMMA chains, whereas the reduced coarsening observed for the hydrophilic nanosilica was related to minimizing the coalescence effect caused by the preferential localization of the silica nanoparticles at the PMMA/SAN interface.

The effect of organically modified layered silicates (Closite 30B) on the phase-separation behavior of PMMA/SAN blends was studied by Shumsky [80], through the viscoelastic and light-scattering method at below, near, and above the phase-separation temperature. From the light-scattering results, it was shown that for unfilled PMMA/SAN blends of all the compositions, the phase-separation rates are the same and the phase-separation process are distinctive only by the time of onset of this process. It was demonstrated that linear

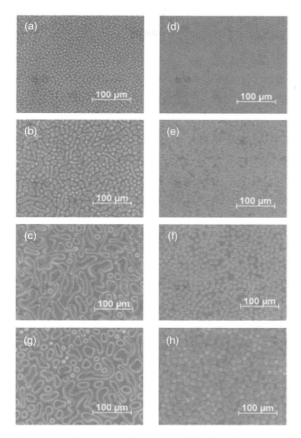

**Figure 7.22** OM images of PMMA/SAN blends in the region of spinodal decomposition (a–c and g–h) and binodal decomposition (d–f) [78] (with permission © 2013 WILEY-VCH).

viscoelastic measurements are sensitive enough to detect the effect of the addition of nanoclays on changing of the interaction between the components and the interphase that promoted the change of the free mixing energy. As shown in Figure 7.23, the addition of the silicate layer results in an increase about 7 °C in the phase-separation temperature. At the isothermal condition, the phase separation begins earlier and proceeds whit a higher rate as compared with the unfilled blends. It was also found that the phase-separation temperature determined by the rheological method was about 10 °C lower than the cloud-point temperature [80].

The effect of the addition of thermally reduced graphene (TRG) on the phase-separation behavior of poly[($\alpha$-methyl styren)-co(acrylonitrile)]/ PMMA (P$\alpha$MSAN/PMMA) blends as an LEST system has been studied through melt rheology and electrical conductivity spectroscopy by Shumsky *et al.* [80].

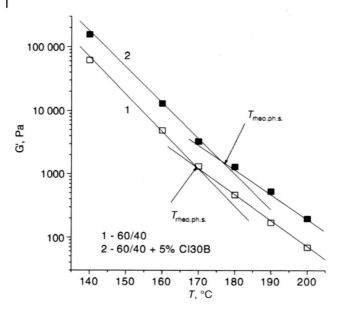

**Figure 7.23** Determination of the phase-separation temperature by the rheological method [80] (with permission © 2010, Springer-Verlag).

### 7.4.4
### Effect of Addition of Nanoparticles on Phase-separation Behavior of PS/PVME

The blends of PS/poly(vinyl methyl ether) (PS/PVME) with exhibiting LEST behavior and strong dynamic asymmetry have been used as an ideal and attractive system for studying the thermodynamics of the phase diagram, phase-separation kinetics, and resulting morphologies at the different conditions. In addition to the concentration-fluctuation-induced phase separation, under the quiescent condition, the phase separation under shear flow has also been investigated [68,81–85]. Moreover, one of the main and distinct features of PS/PVME is that this system undergoes a phase separation through the viscoelastic mechanism rather than the normal concentration fluctuation mechanism. In this context, an increasing activity has been directed toward the effect of nanoparticles on phase-separation behavior and resulting morphologies of PS/PVME blend systems. Yurekli and coworker, for example, investigated the effect of the disk diameter of anisotropic organically modified layered silicates on the phase-separated morphology of near-critical PS/PVME blends through the temperature gradient combination method, and by employing atomic force microscopy (AFM) [84,85]. Films with comparable amounts of thermodynamically equivalent silicates, but varying only in lateral disk diameter, were examined through the temperature gradient. It was shown that small disk diameters (30 nm and 0.5 nm) had a stunning effect on the kinetics of the spinodal decomposition and significantly change the separated morphologies, resulting in smaller and more circular domains. On the other hand, the addition of the larger diameter disk silicate layers did not have noticeable influence on the separated morphologies and only accelerated the phase-separation kinetics. The slowing down of the kinetics of phase separation by layered silicates was explained by a pinning mechanism similar to that proposed by Balazs and coworkers' computer simulation method [64]. The observed difference in the influence of layered silicate disk diameter on the phase-separated morphologies of the blends was attributed to more exfoliated and better dispersion for silicates with smaller lateral dimension. The extent of pinning was found to become more pronounced as the layered silicate diameter decreases and concentration goes up. The authors argued that these effects must also depend on the relative wetting of the surfaces by the polymers and for the case examined that the silicates are probably only marginally more compatible with the PVME than PS, and these effects would be significantly enhanced for cases where there is a large preferential attraction of one of the polymer components to the silicate surface.

In their later research, Krishnamorti's group examined the effect of the addition of highly anisotropic organically modified layered silicate (montmorillonite) on the phase-separation behavior of deuterated polystyrene (DPS)/PVME blends by utilizing a combination of SANS and a recently developed two-dimensional combinatorial method based on the light-scattering method. It was found that the incorporation of the layered silicates up to a volume fraction of 0.04 leaves the phase diagram unchanged. In other words, while at temperatures far below the LEST the Flory–Huggins parameter $\chi$ values were dependent on the concentration of added layered silicate, at temperature close to the LEST, where concentration fluctuations are large, the thermodynamics are not significantly affected by the presence of the layered silicate. These results indicate that in spite of a significant difference in the polarity of the blend components, there is not a strong preferentiality for either of the components by silicate surfaces. This was evidenced by the XRD measurements that showed only the intercalation of the layered silicate by both components.

During the past decade, PS/PVME blends due to strong dynamic asymmetry induced by the large difference in glass transition temperature ($T_g$) of the components (more than 110 °C) have been considered as a attractive LCST system for studying the newly recognized VPS mechanism [81,82]. Tanaka for the first time proposed that in PS/PVME blends under the deep quench condition, due to a large difference in viscoelastic properties of the coexisting components, the VPS mechanism will control the phase separation, although few researchers did not observe VPS even for much deeper quenches and could only detect the normal phase separation (NPS) [69]. The main feature of VPS is the induction of a percolating network formed between more viscoelastic phase even if volumetrically is a minor phase, which is different from the NPS in which the minor phase is separated in the form of isolated droplets. It is worth mentioning that two kinds of distinct morphologies have been observed in VPS. One is characterized by the formation and subsequent breakup of the percolating network-like structure and the one characterized by the presence of moving droplets. The former type occurs in the polymer blends, whereas the latter type is usually observed for polymer solutions. Since Tanaka's findings, several research studies have been devoted to the VPS of polymer blends by employing different techniques, especially rheology [69,81,82].

Recently, there has been a considerable interest in studying the influence of addition of nanoparticle on the phase-separation behavior of PS/PVME blend [68,83–85]. Charachorlou and Goharpey investigated the effect of hydrophilic nanosilica on the phase-separation behavior of PS/PVME blends by using rheology, DSC, and optical microscopy [68]. Through isochronous dynamic temperature ramp measurements $(G' - T)$, they showed that at low temperatures (homogenous region) the addition of silica nanoparticles did not change the trend of decreasing $G'$ with temperature. However, the addition of 4 wt% of silica nanoparticles shifted the $G'$ to higher values (about 10 degrees) when the temperature reached the transition temperature (rheological phase-separation temperature). This was attributed to the reduction of the concentration-fluctuation-induced elasticity resulted from enhanced miscibility caused by the absorption of PVME on the nanoparticle surface. They supported their suggestion by the slowing effect of nanoparticles on the kinetics of phase separation via the results of $G'$ versus time and optical microscopy observation [68].

Xia and coworkers examined the effect of hydrophilic silica nanoparticles on the morphology of PS/PVME (10/90) in an unstable region, using the optical microscopy and dynamic rheology techniques [81]. For unfilled PS/PVME (10/90) blend annealed at 120 °C, they observed the transition network-like structure formation and subsequent phase inversion. This was considered as an indication of the typical VPS mechanism (see Figure 7.24).

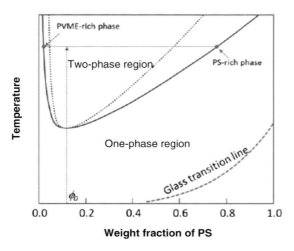

**Figure 7.24** Schematic phase diagram of dynamically asymmetric PS/PVME blends [81] (with permission © 2013, American Chemical Society).

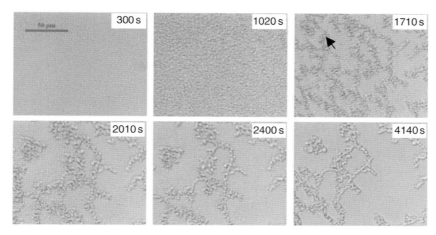

**Figure 7.25** Morphology evolution of PS/PVME (10/90) blend after temperature jump to 120 °C. The phase indicated by the arrow is the PS-rich phase, while the other one is the PVME-rich phase. The scale bar denotes 50 μm [81] (with permission © 2013, American Chemical Society).

However, as can be observed from Figure 7.25, for the nanoparticle-filled blend, a droplet-like structure without the formation of a network-like structure was formed. The significant nanoparticle-induced decrease of dynamic asymmetry, characterized by rheological relaxation time, was proposed to be the primary origin of the disappearing of VPS for the nanoparticle-filled blends. Also, they suggested that, due to the preferential affinity of silica nanoparticles, the PVME phase that is a dynamically fast phase becomes the slower phase, which leads to dynamic inversion compared with those unfilled blends [81].

In their very recent work, Xia and coworkers have studied the morphology change and related molecular mechanism of the PS/PVME blend induced by hydrophilic and hydrophobic silicate nanoparticles by using the optical microscopy and dynamic rheology techniques. In their study, they used the PS/PVME (10/90) blend varying in hydrophilic (A200) or hydrophobic (R972) concentration (1, 2, 4, 8 wt%). The abbreviations of the blends used are given in Table 7.2.

The morphology evaluation of pure PS/PVME (10/90) and that of blend with different hydrophilic nanosilica loading at 120 °C are shown in Figure 7.25 and Figure 7.26, respectively.

**Table 7.2** Abbreviations for hydrophilic and hydrophobic SiO$_2$-filled PS/PVME blends [82].

| Blends | Abbrev. | Blends | Abbrev. |
| --- | --- | --- | --- |
| PS/PVME/A200 (10/90/1) | A1 | PS/PVME/R972 (10/90/1) | R1 |
| PS/PVME/A200 (10/90/2) | A2 | PS/PVME/R972 (10/90/2) | R2 |
| PS/PVME/A200 (10/90/4) | A4 | PS/PVME/R972 (10/90/4) | R4 |
| PS/PVME/A200 (10/90/8) | A5 | PS/PVME/R972 (10/90/8) | R8 |

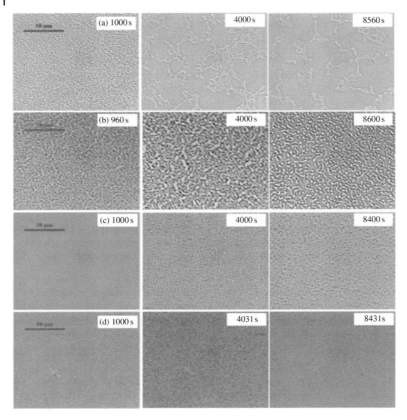

**Figure 7.26** Morphology evolution of blends with different nanoparticle loadings during phase separation at 120 °C: (a) A1, (b) A2, (c) A4, and (d) A8. The scale bar denotes 50 μm [81] (with permission © 2013, American Chemical Society).

As can be observed from Figure 7.25, the neat blends form a sustainable transient network structure displaying the VPS mechanism. From Figure 7.25, it can also be noticed that the blends A1 and A2 form a transition network structure exhibiting similar VPS to pure blends. The more refined network structure and prolonged VPS characteristic observed for the A1 and A2 blends compared to the pure blends were explained in terms of the retarded kinetics of phase separation caused by the preferential affinity of hydrophilic silica for the PVME phase. However, A4 and a4 blends form only the droplet structure after being annealed for 8400 s. From the above-described morphology evaluations that were supported by dynamic rheological measurements, it was concluded that there is a threshold of nanoparticle loading for the network–droplet morphological transition in hydrophilic silica nanoparticle-filled near-critical PS/PVME (10/90) blends. The authors proposed that the origin of the morphological transition can be ascribed to a significantly restrained dynamic of PVME chains due to their preferential interaction with the hydrophilic surface of nanoparticles.

In order to obtain more pieces of evidence for better understanding the role of the nanoparticle–polymer interaction in the morphology formation, the authors investigated the phase separation of PS/PVME (10/90) filled with hydrophobic silica nanoparticles. As can be observed from Figure 7.26, in contrast to the hydrophilic silica nanoparticle-filled blends, the blends with hydrophobic silica nanoparticle still form the network-like structure even at high loading of 8 wt%. Moreover, the PS-rich phase network generated from VPS becomes smaller and can sustain for much longer time under higher hydrophobic silica loading. The enhanced viscoelastic network induced by hydrophobic silica nanoparticle was related to the selective dispersion of nanoparticles in the PS-rich phase, which increased the dynamic asymmetry between the PS and PVME phases. From the above morphology evaluation and rheological measurements, the authors proposed that the network of hydrophilic silica nanoparticles formed in the PVME-rich phase would significantly reduce the dynamic asymmetry of the two phases. This, in turn, leaves the compositional asymmetry to be the dominant factor in controlling the morphology development during phase separation. However, the presence of the hydrophobic silica nanoparticles would lead to an increase in the dynamic asymmetry and reduce the compositional asymmetry because of their selective dispersion in the PS-rich phase, balking the continuous network structure.

## 7.5
## Immiscible Polymer Blends Containing Nanoparticle

### 7.5.1
### Introduction

A century ago, Ramsden and Pickering showed that a small portion of colloidal particles added into low-viscosity emulsions can act as an emulsifying agent in stabilizing the emulsions through selective localization in the interface [86,87]. Recently, it has been recognized that the idea of the solid-particle-induced stabilization of low-viscosity emulsion can also be adopted to stabilize the morphology of immiscible polymer blends consisting of components exhibiting different wetting characteristics [88,89]. On the other hand, it has been well established that the incorporation of nanofillers in the polymer can be utilized as a new method of producing polymer composites with significantly enhanced properties. This can be achieved with much lower filler loadings compared to conventional micro-sized filler composites. In more recent years, many research groups have taken advantages of both compatibilizing and reinforcement functionalities of nanofillers for producing nanofilled polymer blends as a new generation of polymer composites.

### 7.5.2
### Parameters Determining Localization of Nanoparticles

It has been demonstrated that the morphology of a nanoscopic structure and the final properties of nanofilled polymer blends are significantly controlled by the localization of nanofillers in one or two phases or at the interface. It should be noted that the localization of nanoparticles is usually established during the mixing stage, even though the resulting localization may undergo some changes during the melt processing. It is generally accepted that in low-viscosity emulsions, nanoparticle localization is predominantly controlled by thermodynamic preferential attraction between nanoparticles and blend constituents. In high-viscosity immiscible blends, however, due to a highly reduced Brownian diffusion coefficient and hence minimized thermodynamic induced localization contribution, the kinetic parameters such as the blend component viscosity differences and sequence of feeding can also play competitive roles. As a consequence, in the majority of nanofilled polymer blends, the nanoparticles are distributed unevenly between the blend components. Very recently, some increasingly research activities have been directed toward studying the role of these two competitive groups of parameters. However, due to difficulties concerning the quantitative evaluation of the interfacial tension between the components, in particular, polymer–nanoparticle interfacial tension and mixing parameters, the exact role of these parameters in determining the nanoparticle localization has not been fully understood.

### 7.5.2.1 Thermodynamic Parameters (Wetting Parameters)

Thermodynamic preferences of nanoparticles for selective localization in polymer blends can be predicted by estimating the interaction between polymer pairs as well as nanoparticles with each blend component. The balance of interfacial energies can be used to calculate the wettability parameter $W_{AB}$ according to Young's equation. The wettability $W_{AB}$ determines the ability of the particle to be wetted by liquid A and B and directly related to the contact angle:

$$-1 < W_{AB} < 1 = \cos\theta = \frac{\gamma_{\text{Particle B}} - \gamma_{\text{Particle A}}}{\gamma_{AB}},$$

where $\theta$ represents the contact angle, $\gamma_{\text{Particle B}}$ and $\gamma_{\text{Particle A}}$ are the interfacial tensions between the particle and polymer B and that between the particle and polymer A respectively, and $\gamma_{AB}$ is the interfacial tension between the two polymers A and B. If $W_{AB} > 1$, the nanoparticles are localized only within polymer A, if $W_{AB} < -1$, the nanoparticles are selectively distributed within polymer B, and if $-1 < W_{AB} < 1$, the particles are concentrated at the interface. The third case is more likely to occur in polymer blends, exhibiting a high degree of incompatibility when the differences in the particle–polymer interaction are small [16].

The interfacial tension between the blend components can be determined by the harmonic mean equation introduced by Wu, which is typically used for materials with low interfacial tension energies [89–91]:

$$\gamma_{AB} = \gamma_A + \gamma_B - 4\left(\frac{\gamma_A^d \gamma_B^d}{\gamma_A^d + \gamma_B^d} + \frac{\gamma_A^p \gamma_B^p}{\gamma_A^p + \gamma_B^p}\right),$$

where $\gamma_A$ and $\gamma_B$ are the surface tensions of polymers A and B, and superscripts $d$ and $p$ denote the dispersive and polar components of the surface energy respectively. The surface tension of each polymer component can be determined by different techniques including the breaking thread, imbedded fiber retraction, the retraction of the deformed drop, the pendant drop, and rheological or linear viscoelastic measurements [92]. Xing *et al.* examined all these methods to determine the interfacial tension for PS/PA blends [93]. Although the values obtained from these different techniques were not consistent, the variations were found to be in an acceptable range (20%). However, it has always been difficult to experimentally determine the surface tension of nanoparticles and interfacial tension between polymer and nanoparticles. Nevertheless, many attempts have been made by several research groups [93] to predict the nanoparticle localization in polymer blends, by using Owens–Wendt [94] or Girifalco–Good [95,96], and Wu equations [97]:

$$\gamma_{AB} = \gamma_A + \gamma_B - 2\left(\sqrt{\gamma_A^d \gamma_B^d} + \sqrt{\gamma_A^p \gamma_B^p}\right)$$

$$\gamma_{AB} = \gamma_A + \gamma_B - 2\left(\sqrt{\gamma_A \gamma_B}\right)$$

$$\gamma_{AB} = \gamma_A + \gamma_B - 4\left(\frac{\gamma_A \gamma_B}{\gamma_A + \gamma_B}\right)$$

The Sumita group, for example, performed such a prediction by utilizing the Owens–Wendt equation for PB/PMMA, PP/PMMMA, and PB/PP blends, all filled with carbon black (CB) [98,99]. They checked their predictions with experimental results obtained by electrical conductivity measurements and electron microscopy. It was shown that, as long as the viscosities of the two polymers are comparable, the interfacial energy is the main parameter determining the uneven distribution of the CB filler in polymer blend matrices. Zaikin *et al.* also focused their study on the prediction of the redistribution conditions for CB particles from bulk to the interface, in blends of a number of different polymers [98,99]. In their new approach, they evaluated the affinity of the filler with polymers, in terms of the enthalpy of the wetting of the fillers with the selected low molecular weight liquid analogs as an indication of polymer–filler interaction energy. They also estimated the bounding strength between the filler and the polymer by measuring the increment of the peeling force [100,101]. These authors demonstrated that the filler redistribution and preferential localization at the interface are strongly dependent upon the order of the component mixing. In other words, the redistribution and localization of CB particles can only proceed if the filler is first mixed with a polymer component that has lower capability of wetting the filler surface. This dependence was suggested to be associated with the nonequilibrium location of CB particles at the interface between the polymeric phases and is due to a high energy of disruption of macromolecules from the solid surface.

An attempt was also made by Thareja and Velankar to quantitatively evaluate the contribution of the polymer–filler interaction in determining the particle distribution in polymer blends [102]. For this purpose, they conducted experiments with several particle types incorporated into two pairs of model polymers:

polyisoprene/polyisobutylene (PI/PIB) and polyisoprene/polydimethyl siloxane (PI/PDMS), which had more chemical difference than PI/PIB. They also used the Girifalco–Good theory as a mean to predict the interfacial activity of the particles. The solid surface energy required by this theory was assumed to be equal to the critical surface tension of the particles needed for interfacial activity. According to the Girifalco–Good prediction, only the PTFE particle was expected to locate at the interface while experimentally all types of fillers used were observed at the PI droplet surface. This suggests that the partial wettability of particles by both phases is responsible for their interfacial adsorption. To explain the disagreement between theory and experiment, the authors proposed that the Girifalco–Good theory is not capable of predicting the interfacial activity of particles and a more elaborate theoretical approach is necessary to capture the relevant surface energies.

Elias and coworkers studied the capability of the Owens–Wendt equation in predicting the polymer–particle interfacial tension and nanoparticle localization in PP/PS (70/30) blends [103]. They used two types of pyrogenic nanosilica: hydrophilic nanosilica with a specific area of $200 \, \text{m}^2 \text{g}^{-1}$ and hydrophobic one having a specific area of $150 \, \text{m}^2 \text{g}^{-1}$.

The authors compared their predictions with morphology observation on electron microscopy (TEM and SEM). SEM and TEM image analyses proved that the hydrophilic silica tends to locate inside the PS phase, while the hydrophobic one was located at the PP/PS interface and in the PP phase. Interestingly, they demonstrated that a quantitative analysis of the linear viscoelastic measurements based on the Palierne model, extended to filled immiscible blends, can be utilized as an improved method for determining the effective interfacial tension.

Attempts were also made by Fudazi and Nazockdast to predict the thermodynamic equilibrium localization of the two types of nanosilica (hydrophilic and hydrophobic) in blends of polypropylene (PP) and liquid crystal polymer (LCP), PP/TLCP (80/20) using the Owens–Wendt theory [104,105]. The predicted results showed that the hydrophobic silica was expected to preferentially be located in the LCP phase, while the hydrophilic one was concentrated at the surface. The TEM micrographs showed that the hydrophobic silica particles were mostly concentrated at the interface, which was in agreement with the predicted results. However, in the hydrophilic silica-filled blends, a fraction of nanoparticles were also distributed in the PP matrix. Such a discrepancy could be explained in terms of the concentration effect, the competitive role of kinetic parameters, and/or temperature effect on the surface tension determination of the components.

Thus, for many polymer blends, in particular those polymer pairs that have a close affinity with nanofillers, the preferred nanoparticle localization predicted through the selected theories, based on the interfacial energies, could hardly be supported experimentally. This has been mainly related to the difficulties concerned with the quantitative determination of the surface tension of the different components, especially at evaluated temperatures and relatively strong interfacial interaction created by the adsorption of polymer chains on the nanoparticle surface. On the other hand, it has been shown that in high-viscosity polymer blends, in contrast to low-viscosity emulsions, the Brownian diffusion time required for nanoparticles to attain their thermodynamic preferred localization is significantly longer compared to the mixing time which is usually around few minutes. This allows the kinetic parameters to play dominating roles in controlling the nanoparticle localization.

### 7.5.2.2  Kinetic Parameters (Dynamic Processes)

It is now well recognized that the final nanofiller localization in high-viscosity polymer blends may be controlled through selecting an appropriate order of the feeding of different components (polymers and nanofillers). It has also been found that the effectiveness of this approach is highly dependent on other parameters such as the viscosity of the blend components, mixing time, and the type and concentration of the nanofiller. On the other hand, the nanoparticle distribution has proved to have a significant effect on determining morphology and, hence, the final properties of these ternary systems. This has motivated increasing activities toward producing polymer blend containing nanoparticles.

### 7.5.2.3  Effect of Feeding Sequence

The addition of nanofillers into polymer blends to produce nanofilled polymer blends may be carried out through three different feeding routes. In the first feeding route, also known as direct feeding, all the components (polymers and nanofillers) are fed into mixer simultaneously. In this mixing route, due to the interaction of different parameters, it would be very difficult to decouple the effect of individual parameters on determining the nanofiller localization. Since if the polymer pair have significantly different melting temperature, the nanofiller will be first localized in the lower melting polymer phase, even if the nanofiller has greater affinity toward the second polymer. In the later stage of mixing, the nanofillers may tend to transfer to the more thermodynamically preferred molten polymer phase. Thus, by knowing this and the fact that the morphology evolution is taking place all along the mixing process, one can realize that it is very difficult to distinguish the real role of individual parameter in determining the final nanofiller localization in this mixing route. In the second route of feeding, the polymer components are first melt blended for an appropriate time period and nanofillers are added afterward while the blending is still running. In the third route, the nanofiller is first mixed with one polymer and the resulting premixed compound is mixed with the second polymer.

Thus, depending on the feeding sequence, the filler may need to follow a different path or transfer from one phase to other in order to reach their thermodynamic equilibrium or preferred localization. By knowing this and the fact that the melt mixing period is usually well below the Brownian diffusion time, one will be able to control the final nanoparticle localization and morphology in the final ternary system product. Gubbels and coworkers conducted a series of research studies, focused on preparing CB-filled electric conductive composites based on PE/PS blends through controlling CB particle localization and with

(a) 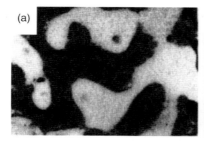 (b)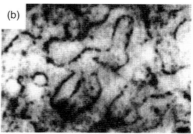

**Figure 7.27** Optical micrographs of a PE/PS 45/55 blend filled with 1 wt% of carbon black. (a) Carbon black localized in the PE phase and (b) carbon black localized at the interface by taking advantage of the transfer of the filler from the PS phase toward the PE phase [107] (with permission © 1994, American Chemical Society).

considering the concept of the double-percolation threshold. The so-called percolation threshold refers to the lowest concentration of the conducting filler, at which continuous conducting chains or networks are formed. They compared four situations where CB was dispersed in a (1) PS, (2) PE, (3) PS/PE (45/55) blend in which the CB was confined inside the PE phase, and (4) in the same PS/PE blend but turning the CB particles to be localized in the interface [106–108]. According to the authors, the CB-filled PE showed a greater electric conductivity than the filled PS, because in the case of the filled PE, the CB particles were forced to segregate in the amorphous phase, resulting in a lower percolation threshold. The third sample (PS/PE 45/55) displayed much higher electric conductivity because a double percolation could be achieved through turning the appropriate blend ratio (PS/PE 45/55) for a co-continuous morphology and percolation of the CB inside the PE phase (see Figure 7.27).

The electrical percolation threshold found for this sample was about 3 wt% of CB, which was well below the value (5 wt%) that was reported for homogeneously dispersed CB in a polymer matrix. Interestingly, the authors obtained much lower percolation threshold (0.4 wt%) for the PS/PE (45/55) blend (fourth sample) in which the CB particles were allowed to migrate from less interacting phase (PS) toward more favorite interacting phase (PE). Accordingly, this can be achieved by the kinetic control of the thermodynamically driven transfer of the fillers from the PS phase (the less preformed one) to the PE phase, and stop the blending process at an adequate time at which most of the fillers have concentrated at the interface. This is evidenced by the results shown in Figure 7.28, which clearly shows that the resistivity passes through a minimum when CB particles have concentrated at the interface and progressively increases as the particles cross the interface and accumulate in the PE phase.

Similar results have been reported for elastomer blends filled with CB [88,109,110]. This phenomenon has also been studied by Zaikin *et al.*, who varied the sequence of feeding of their various polymer pairs with CB. They observed much enhanced electrical conductivity when the CB particles had crossed the interface [100,101]. Elias *et al.* examined this phenomenon on the

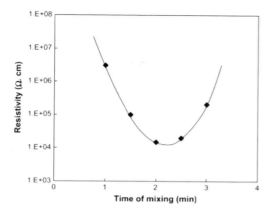

**Figure 7.28** Electrical resistivity versus mixing time of PE/PS co-continuous (45/55) blend filled with 1 wt% of carbon black. From left to right CB is localized in the PS phase, then at the interface (minimum of the resistivity), and finally in the PE phase [107] (with permission © 1994, American Chemical Society).

PP/PS/silica turnery system [103]. In their study, they used a sequence of mixing in which the silica was first mixed with PP and the resulting premixed composite was then with PS in the later mixing stage. With the help of the SEM and TEM images, they showed that almost all the hydrophilic silica particles are transferred from the PP which is a less favorite phase than the PS which is the preferred phase (see Figure 7.29) [103].

In another research, Elias *et al.* studied the effect of mixing sequence on the filler localization in the PP/EVA/silica 80/20/3 wt% ternary system [111]. In this study,

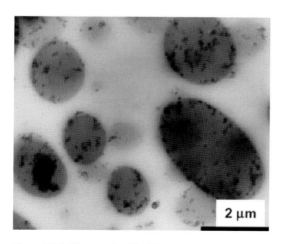

**Figure 7.29** Micrograph of PP/PS 70/30 wt% containing 3 wt% hydrophilic silica. The nanocomposite was obtained by premixing the silica with PP before adding the PS. The silica particles have totally migrated from the PP matrix to the PS phase which has better affinity [103] (with permission © 2007 Elsevier).

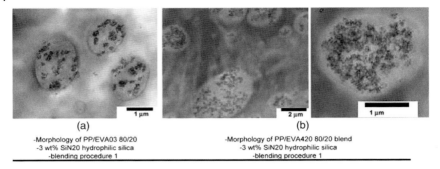

(a)

-Morphology of PP/EVA03 80/20
-3 wt% SiN20 hydrophilic silica
-blending procedure 1

(b)

-Morphology of PP/EVA420 80/20 blend
-3 wt% SiN20 hydrophilic silica
-blending procedure 1

**Figure 7.30** Morphology of PP/EVA80/20 blend prepared with blending procedure 1 (the three components added simultaneously in the extruder). (a) PP/EVA03 with 3 wt% Si N20 hydrophilic silica. (b) PP/EVA420 with 3 wt % Si N20 hydrophilic silica [111] (with permission © 2008 Elsevier).

they used two types of silica (hydrophilic and hydrophobic) and two EVA grades with different $M_w$ (EVA03 $M_w = 53,500$ and EVA420 $M_w = 12\,000$). Their TEM and SEM observations proved that, when all the three components were added to the extruder simultaneously, the hydrophilic silica was preferentially located in the EVA droplets (see Figure 7.30). However, with the same feeding sequence, the hydrophobic silica was concentrated at the interface, although in contrary to the PP/PS/silica system, a distinct layer of the silica particles at the interface could not be observed. On the other hand, when the hydrophobic or the hydrophilic silica particles were first compounded with PP and resulting premixed compound was mixed with EVA, the silica particles were transferred from PP, accumulated in the EVA, and were close to the interface (see Figure 7.31).

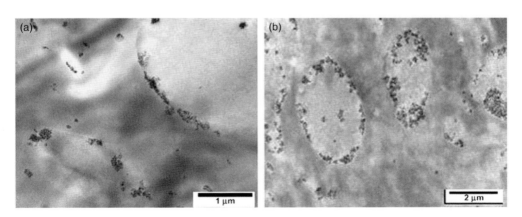

**Figure 7.31** Morphology of PP/EVA80/20 blends prepared with blending procedure 2 (the silica was pre-compounded with PP at 200 °C for 5 min. Then the obtained material was blended with the EVA 03 during a second extrusion step). (a) PP/EVA420 80/20 blend filled with 3 wt% Si H20 hydrophobic silica. (b) PP/EVA420 80/20 blend filled with 3 wt% Si N20 hydrophilic silica [111] (with permission © 2008 Elsevier).

#### 7.5.2.4 **Effect of Viscosity**

The effect of the viscosity difference of the blend polymer pairs on the nanofiller localization in immiscible blends has been studied by a few researchers [108,112,113]. Persson *et al.*, for example, studied this effect on PE/PIB blends filled with the aluminum borate whisker using a different feeding sequence in which the polymer pairs had different viscosities but a comparable interaction with the aluminum whisker [113]. Their observation evidenced that the whisker accumulates in the higher viscosity polymer phase. They purposed that the blend organizes itself to minimize its dissipative energy during mixing which could be attributed to the thermodynamic force that tends to minimize the dissipative energy of the blend during mixing. They supported their hypothesis through estimating the viscosities calculated based on appropriate models with the assumption that the filler can be localized in one polymer or the other. In their similar study on the PA/SAN/whisker system, they observed that in contrary to the PE/PIB/whisker, the filler particles were concentrated in the PA phase, although it had lower viscosity compared to SAN. By comparing these findings, they reached a conclusion that the effectiveness of the viscosity ratio is weak and can dominate only when the filler has comparable affinity toward the polymers. By contrast, when one of the polymers has much greater thermodynamic interaction with the filler (PA in the PA/SAN blend), the thermodynamic effect will dominate the viscosity ratio effect.

In another interesting work, Feng *et al.* focused on the effect of the viscosity ratio on the CB localization in the PP/PMMA/CB system with using the same PP grade but three types of PMMA with different molecular weight [114]. The samples were prepared by the simultaneous feeding procedure. The authors, with the help of TEM observation, proved that when the viscosities of PP and PMMA were comparable, the CB particles were located in the PMMA droplets (see Figure 7.32). This observation was found to be in agreement with Sumita's prediction based on the wetting parameter.

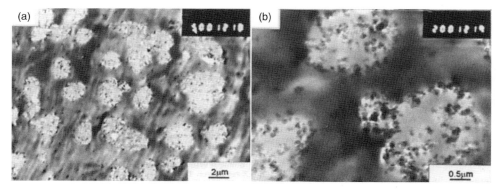

**Figure 7.32** TEM micrographs of the PP/PMMA blend. (a) magnification, 5000X. and (b) magnification 20 000× [114] (with permission © 2003 Society of Plastics Engineers).

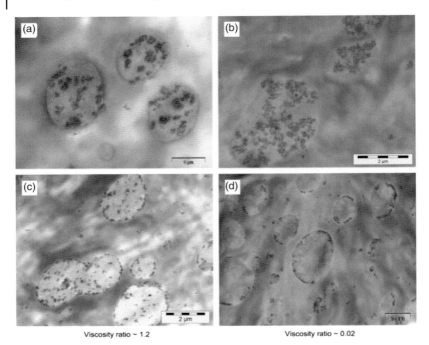

**Figure 7.33** TEM images of PP/EVA/silica blends. The polymer components and silica particles were added simultaneously in the extruder (procedure 1). (a) PP/EVA03/ hydrophilic silica, (b) PP/EVA420/hydrophilic silica, (c) PP/EVA03/hydrophobic silica, and (d) PP/EVA420/hydrophobic silica [112] (with permission © 2008 Wiley Periodicals, Inc.).

As the viscosity of PMMA increases, the CB was found to be located in the interface. On further increase in PMMA viscosity, the CB particles were located in the PP matrix. In a recent work, Elias and coworkers studied this effect in the PE/EVA/silica system [112]. They used one PP grade with two EVA grades of varying viscosities denoted as EVA03 ($M_w = 53,500$) with viscosity close to PP and EVA420 ($M_w = 12,000$) with viscosity lower than PP. For preparing the samples, the three compounds were fed to extruder simultaneously. The TEM result showed that the hydrophilic silica particles were located in the EVA droplets, regardless of the EVA viscosity. On the other hand, the hydrophilic silica particles accumulated at the interface only in the case of low-viscosity EVA (see Figure 7.33).

As the authors stated, in the simultaneous feeding method, the EVA is melted before PP and, therefore, the hydrophilic filler is expected to remain in the preferred EVA phase, all along the mixing. However, the hydrophobic silica particles can migrate to their preferred PP phase if the thermodynamic interaction dominates the viscosity effect, and this can only occur in the case of low-viscosity EVA. The viscosity effect was also studied by Feng *et al.* [114] and Zhou *et al.* [115] for CB particles. They showed that under normal processing conditions, CB particles tend to be preferentially incorporated into the lower viscosity polymer. They also suggested that the interfacial energy of particle–polymer can only dominate the CB localization when the viscosity ratios of both polymers tend to 1.

From the above results, one can realize the fact that the viscosity ratio of the two polymers can play an influential role in determining the nanoparticle localization, especially in systems where the thermodynamic wetting parameter is not strong enough to play a predominating part. However, due to the complexity of the mixing process, it is often very difficult to clearly distinguish the viscosity effect from the effects of the thermodynamic infraction and other kinetic parameters. Another problem is that, in many cases, the TEM or SEM examinations performed on the solid sample do not necessary represent the real picture of the equilibrium nanoparticle distribution reached during mixing and before solidification.

Moreover, the mechanism by which nanoparticles are transferred from one phase to the other, passes the interface, or more generally move inside the blend in the molten state and during the mixing process is rarely discussed in the literature.

The above-mentioned situations become even more complicated for the case of blends containing nanofillers such as layered silicates, graphene, and CNTs. Because in these blend nanocomposites, the dynamics (kinetics of transport) and transfer of nanoparticles from one phase to the other phase highly depend on the aspect ratio of the filler and hence the extent of intercalation (for nanoclay and graphene) or the state of dispersion (for CNT). Moreover, for some selected polymer blends, neither of the two polymer phases is capable of intercalating organoclay or graphene. There are also some cases where due to comparable interfacial tension between the organoclay and the two polymer phases, it is difficult to localize the organoclays in a describable phase. One way of overcoming these problems is to add another polymeric compatibilizer with the ability of localizing the organoclays in one phase where it can also promote the melt intercalation process.

Zhu *et al.*, for example, showed that the location of nanoclay can be switched from one phase to another in PP/PS blends by tailoring the polarity of PP with the addition of PP-G-MA, or by sulfonate groups to PS [117]. According to the authors, this promotes the migration of clay particles from less polar to high polarity blend components of PP/PS blends. Tiwari *et al.* also demonstrated that the addition of PP-G-MA facilitated the transfer of nanoclay from the PS to PP phase, and such a change in the nanoclay localization was found to be more efficient when PS was dispersed in the PP matrix [118]. This improved the microfibrill formation of the PA-dispersed phase during the melt-spinning process. The concept of double percolation and, hence, the exclusively localization of nanofillers in one of the two phases in immiscible blends has received tremendous attention over the last two decades. Among the nanofillers, due to their unique atomic structure and properties such as the high aspect ratio, high strength, and high thermal and electrical conductivity, CNTs have received special attention. Potschke *et al.* studied the localization and double percolation for

MWCNT-filled PC/PE blends in which MWCNT was first melt mixed with PC and then blended with PE, in order to obtain a co-continuous morphology [119,120]. They showed that MWCNT remained in the PC phase. The authors also investigated the blends of PC containing MWCNT with PP containing nanoclay and found that while MWCNTs remained in the PC phase, the nanoclays migrated from the PP phase to the interphase [121]. A similar study was carried out by Wu *et al.* on PET containing MWCNT blended with PVDF, PP, PA6:6, and HDPE [122,123]. In the case of the PET/PVDF and PET/PA blends, MWCNTs stayed in the PET phase, whereas in the case of the PET/PP and PET/HDPE blends, MWCNTs mainly accumulated at the interface. Li *et al.* in their study on MWCNT-filled PA/PVDF blends showed that MWCNTs were exclusively in the PA phase [124]. It was shown by Meincke *et al.* that in polyamide/acrylonitrile butadiene styrene (PA/ABS) blends, MWCNTs remained in the PA phase and created a triple continuous structure [125]. In all these studies, the selective localization of MWCNTs could be explained based on the concept of the wetting parameter, and MWCNTs concentrated in the more polar phase. Recently, Potschke *et al.* showed that in both PE/PA and PE/PC blends, MWCNTs migrated from the PE phase to the PA or PC phase and a lower electrical percolation threshold could be achieved with the help of the direct feeding method [126]. In their more recent work on MWCNT-filled PE/ABS blends, they found that even when MWCNTs were first melt mixed with ABS, MWCNTs were exclusively localized in the PC phase, in spite of the minor difference in the wetting behavior between the two phases. Goldel *et al.* from their study on the localization of MWCNTs in the co-continuous PC/SAN blend also showed that MWCNTs were exclusively in the PC phase and this was independent of the feeding sequence [127]. The complete transfer of CNTs from the SAN to PC phase with minor difference in the wetting behavior of the two polymer phases could be explained in terms of the high aspect ratio of CNTs which facilitated their relatively fast transfer from the SAN phase to the more preferable PC phase.

It has been suggested that the percolation threshold with the lowest CNT loading can be achieved if CNTs are localized at the interface of polymer blends. However, this ideal localization of CNTs can hardly be obtained since the pristine CNT rarely has equal affinity with different polymer components. Furthermore, as mentioned above (see also Section 7.5.4.1), the localization of nanoparticles with a high aspect ratio such as CNTs at the interface is unstable and can easily penetrate into the preferred phase. Nevertheless, attempts have been made by some researchers to localize the CNTs at the interface by using either an appropriate polymeric compatibilizer or functionalized CNTs. Bose *et al.* used styrene-maleic anhydride copolymer (SMA) to localize MWCNTs at the interface in PA6/ABS blends [128,129]. This interfacial localization was attributed to the chemical coupling between CNTs and SMA compatibilizer already located at the interface.

Wu *et al.* used functionalized MWCNTs in order to perform the interfacial localization of MWCNTs in the poly(ε-caprolactone)/poly(L-lactic acid) (PCL/

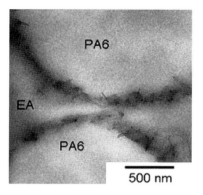

**Figure 7.34** TEM micrographs of a blend of 10 wt% PA6/90 wt% EA+ 2 wt% MWNTs [131] (with permission © 2010 Elsevier).

PLA) blend [130]. They suggested that, a good affinity of the carboxylic groups on the MWCNT surface with both the PCL and PLA phases was responsible for the observed interfacial localization. In more recent years, Baudouin *et al.* examined the possibility of the localization of unfunctionlized MWCNTs at the interface in immiscible polymer blends of PA6 or PA12 and ethylene-acrylate (EA) copolymers [131]. The effect of the mixing sequence was also investigated. They found that when MWCNTs were first mixed in PA6, most of them migrated to the interface, even though aggregates of some of CNT reminded in the PA phase. For the PA12/EA blend, which was prepared using the same feeding sequence, the most of the MWCNTs were remained in the PA12 phase.

However, when MWCNTs were first mixed with the EA copolymer or feeding all compounds simultaneously, a large part of the MWCNTs migrated to the interface whatever PA type used (see Figure 7.34).

When PA12 was used, some parts of the MWCNTs penetrated inside the PA droplets and some of the MWCNTs remained in the EA phase.

By a combination of thermogravimetry analysis (TGA) and separation techniques, they demonstrated that the first polymer that contact with MWCNTs during melt mixing was adsorbed irreversibly by noncovalent adsorption. According to the authors, the resulting modification of interfacial thermodynamics could explain the observed nanoparticle confinement. In a similar study on ternary blends of copolyamide 6/12 and copolymer of ethylene methylacrylate (EMA) containing unfunctionlized MWCNT, Baudouin *et al.* showed that when MWCNTs were premixed with the copolyamide, they remained inside this phase after blending with EMA. This behavior was not in agreement with the thermodynamic prediction. When MWCNTs were first mixed with EMA or for simultaneous blending, they were selectively localized at the interface which was in agreement with the prediction based on wetting parameters [131]. They concluded that the localization of the unfunctionlized MWCNTs is governed by either the thermodynamics or on the long-term partial adsorption (or wrapping) of polymer on MWCNTs depending on the sequence of mixing.

### 7.5.3
### Rheology of Immiscible Polymer Blends Containing Nanoparticle

As was discussed in Section 7.3.4, rheology, in particular the melt linear visco-elastic measurements, can be successfully utilized for the nanostructure characterization of polymer nanocomposites. This arises from the great sensitivity of viscoelastic properties of nanocomposites to the state of dispersion and distribution of nanoparticles. It has been well recognized that depending on the aspect ratio and interfacial interaction between nanoparticles and polymer matrix, there is a concentration of nanoparticles ($\varphi$) above which the nanoparticles exhibit a liquid–solid-like transition (LST) in a low frequency range. Such a solid-like behavior has been considered as an indication of a 3-D percolated network formed due to the particle–particle and/or particle–matrix interactions. The LST behavior of polymer nanocomposites can be characterized by different criteria including the plot of the storage modulus ($G'$) versus frequency, the Van Gurp plot of phase angles versus absolute complex modulus ($G^*$), the Cole–Cole plot of imaginary ($\eta''$) versus real ($\eta'$) part of the complex viscosity and the Han plot of storage modulus ($G'$) versus loss modulus ($G''$) [43].

On the other hand, it is now well established that the melt rheology can be used as a powerful tool to study the morphology development of polymer blends. The strong correlation between the melt rheology and morphology and the fact that the morphologies of polymer blends essentially determine their physical and mechanical properties have motivated a large number of publications including a comprehensive book chapter by Utracki on the rheology of polymer blends [1]. The Palierne model has extensively been used to describe the linear viscoelastic behavior of immiscible polymer blends with matrix-disperse morphology [132–138]. This model expresses the complex modulus of an immiscible blend as

$$G^*(\omega) = G_m^*(\omega)\frac{1 + 3\phi H(\omega)}{1 - 2H(\omega)}$$

$$H(\omega) = \frac{4\left(\dfrac{\alpha}{R}\right)\left[2G_m^*(\omega) + 5G_d^*(\omega)\right] + \left[G_d^*(\omega) - G_m^*(\omega)\right]\left[16G_m^*(\omega) + 19G_d^*(\omega)\right]}{40\left(\dfrac{\alpha}{R}\right)\left[G_m^*(\omega) + G_d^*(\omega)\right] + \left[2G_d^*(\omega) + G_m^*(\omega)\right]\left[16G_m^*(\omega) + 19G_d^*(\omega)\right]}$$

where $G_m^*$ is the complex modulus of the matrix, $G_d^*$ is the complex modulus of the dispersed phase (droplet), $\alpha$ is the interfacial tension, $\phi$ is the volume fraction of the dispersed phase, and $R$ is the volume average radius of the droplets. It is worth pointing out that in order to use the Palierne model, which is originally derived for emulsion, for immiscible polymer blends, it is assumed that the interfacial tension does not depend on local shear deformation nor does it depend on the variation of the interfacial area. In other words, the interfacial properties are only characterized by scalar interfacial tension. The Palierne model can predict a pronounced increase in the melt elasticity at a low frequency range and a long relaxation time, both of which are related to the deformability of dispersed droplets and their recovery and, therefore, to the interfacial tension. Also, this model has proved to be highly effective in describing the interphase formed in immiscible blends. Moreover, it has been shown that by fitting the melt linear viscoelastic data to the Palierne model, it is possible to distinguish the matrix-disperse morphology from the co-continuous one. From the above, one may realize that the Palierne model can provide a great insight into understanding the governing mechanisms and the role of the parameters affecting the morphology evolution, including droplet breakup and coalescence processes in immiscible polymer blends.

More recently, several researchers have taken the above-mentioned advantages of the melt rheology to study the state of dispersion, nanoparticle localization, and its effect on the morphology development and stabilization of nanofilled polymer blends. These subjects are discussed in the following sections.

### 7.5.3.1 Polymer Blends Containing Nanosilica

Vermant *et al.* studied the linear viscoelastic behavior of the silica-filled PMDS/PIB system [139]. They showed that in the filled blends, in contrast to the unfilled blend sample, the elastic response of PIB droplets to the pre-shearing (detected by the shoulder in $G'$ versus frequency) was reduced upon nanosilica loading and the shear rate dependence was completely suppressed for the blend filled with 1% silica. Such morphology stability or compatibilization effect was mostly attributed to the suppression of the coalescence of PIB droplets. This was confirmed by the change in the interfacial properties obtained by fitting the experimental data to the Palierne model. It is worth mentioning that the Palierne model can successfully be used to determine the volume average dispersed particle radius $R$ from the $G'$ versus frequency plot, since the interfacial tension $\gamma_{12}$ values are readily available from an independent measurement. However, due to the difficulty in measuring the interfacial tension for the filled blends, Vermant *et al.* used the ratio $R/\gamma_{12}$ in the Palierne model.

The ratio $R/\gamma_{12}$ was calculated from the following relation:

$$\tau = \frac{R_{\eta_a}}{4\gamma_{12}} \times \frac{(19p + 16)(2p + 3 - 2\varphi(p - 1))}{10(p + 1) - 2\varphi(5p + 2)},$$

where $\eta_a$ is the viscosity of the matrix phase, and $\tau$ is the relaxation time determined by the crossover frequency between $G'$ and $G''$, $p$ is the viscosity ratio, and $\varphi$ is the volume fraction of the dispersed phase.

The linear viscoelastic behavior of similar PDMS and PIB and the same grade of fumed silica were studied by Thareja and Velankar [140]. The silica-filled blend samples exhibited a pronounced viscosity upturn or a gel-like behavior at low frequencies, a behavior which is significantly deviated from the Palierne model. Their direct visualization showed distinct clusters of selected PIB particles (Figure 7.35) which was suggested to be induced by the nanoparticle bridging mechanism.

Accordingly, these clusters were responsible for the gel-like behavior. Moreover, the viscoelastic behavior of the filled blends was found to be sensitive to the blending procedure (pre-shearing). Obviously, there was an apparent discrepancy between these results and those reported by Vermant *et al.* for which the authors

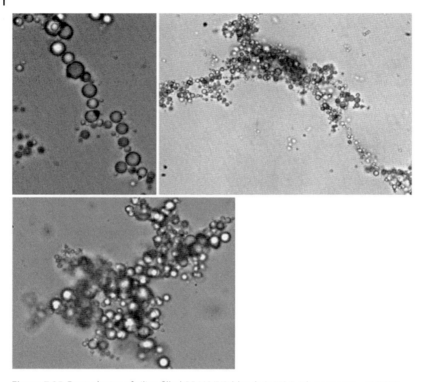

**Figure 7.35** Drop cluster of silica-filled PDMS/PIB blends [140] (with permission © 2006, Springer-Verlag).

did not provide a reason. Vermant *et al.*, however, in their later study on the same ternary system also found a gel-like behavior for the filled blend samples after being subjected to a long resting period [139]. Nevertheless, the stresses involved are small and could not be detected in blends of technical relevant (high temperature) polymers. Finally, the authors came to the conclusion that both coalescence and droplet breakup processes are affected by the presence of silica nanoparticles.

Fudazi and Nazockdast used the linear and nonlinear melt rheology in order to compare the compatibilizing role of the nanosilica particles with that of the elastic compatibilizers such as styrene-butadiene-styrene (SBS) and styrene-ethylene-butylene-styrene (SEBS) [104,105]. The most linear viscoelastic results as expressed by the $G'$ versus frequency plot and Cole–Cole plot showed that while hydrophobic silica-filled samples showed a behavior similar to that of the unfilled blend samples, the hydrophilic silica-filled one exhibited a pronounced elastic response in low frequencies. Direct observations on TEM showed that the hydrophobic silica particles were mostly located at the interface and a fraction of particles in the form of small aggregates were distributed in the PP matrix, whereas the hydrophilic silica particles were distributed all over the sample with greater concentration inside the nematic LCP-dispersed phase. The large aggregates of the selected hydrophilic particles were suggested to be responsible for the high elastic response of the filled blends. By comparing these results, the authors concluded that the hydrophobic nanosilica could play a compatibilizing role in the PP/LCP/silica system, although not as much effective as the elastic compatibilizer used in their study.

An attempt was made by Elias *et al.* to quantitatively analyze the melt rheological behavior of silica-filled PP/PS blends on the framework of the Parlierne model, extended to filled immiscible blends, and calculating the $R/\gamma_{12}$ ratio [111]. However, as the authors pointed out, due to the localization of the silica particles in two phases and its effect on the viscosity ratio, limited cases could be investigated. For a sample in which the hydrophilic silica particles were assumed to be totally localized in the PS phase, the interfacial tension was calculated to be reduced from 2.15 to 0.085 mN m$^{-1}$. They also found that the hypothesis that the hydrophilic silica is preferentially localized in the PS droplets and that the hydrophobic silica dispersed in the PP matrix was much closer to the actual situation. From these findings and the TEM observation, the authors came to the conclusion that the stabilization mechanism of the PP/PS blend by hydrophilic silica is the reduction in the interfacial tension, whereas the hydrophobic silica as a rigid layer preventing the coalescence of PS droplets. In their later research, Elias *et al.* attempted to use a similar approach for determining more accurately the effective interfacial tension based on the melt rheological measurements and the Palierne model for silica-filled PP/EVA blends [111]. However, in some cases, the droplet relaxation time was found to be too similar to the one of the matrices. The two peaks corresponding to each relaxation process were much convoluted and, therefore, the determination of the relaxation times became inaccurate (for example, see Figure 7.36).

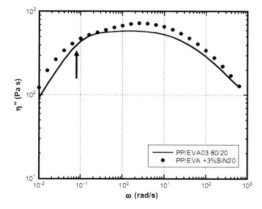

**Figure 7.36** Frequency dependence of the imaginary part of complex viscosity for PP/EVA03 blend filled with 3 wt% silica particles. The silica particles have been simultaneously dispersed in PP/EVA03 blend during mixing. (C) Blend with hydrophilic silica (N20) and full line: experimental virgin blend. The arrow indicates the relaxation time of the EVA droplets [111] (with permission © 2008 Elsevier).

In order to overcome this difficulty, the author introduced an improved route by which the ratio of $R/\gamma_{12}$ could be determined by isolating the dispersed phase relaxation contribution from the complex modulus, using a modified Palierne model. Consequently, the Palierne model can be divided into two contributions:

$$G^*_{\text{Palierne}} = G^*_{\text{Compositon}} + G^*_{\text{Droplet}},$$

where $G^*_{\text{Compositon}}$ is the complex shear modulus of the blend without any interfacial effects (Kerner model), $G^*_{\text{Droplet}}$ captures the droplet-induced interfacial effect and more specifically the extra elasticity brought by the droplet deformability. The storage part of the complex shear modulus ($G^*$) for a blend of two viscoelastic fluids can be presented as follows:

$$G'_{\text{Palierne}}(\gamma_{12}, G^*_m, G^*_d) = G'_{\text{Palierne}}(0, G^*_m, G^*_d) + G'_{\text{Palierne}}(\gamma_{12}, \eta_m, \eta_d)$$
$$= G'_{\text{Kerner}} + G'_{\text{Palierne}}(\gamma_{12}, \eta_m, \eta_d),$$

where $G^*_m$ and $G^*_d$ are the complex shear moduli of the matrix and the dispersed phase, respectively, $\eta_m$ and $\eta_d$ are the Newtonian viscosity of the matrix and that of the dispersed phase, respectively. From these equations, the contribution of the dispersed phase relaxation that only depends on the interfacial tension ($\gamma_{12}$) could be determined by subtracting the composition effects from the experimental data of $G^*$. From their findings, they concluded that even if silica mainly plays a thermodynamic role in the stabilization mechanism by decreasing the effective interfacial tension, the influence of particular rheological conditions and the structure of the phases must also be taken into account for a complete understanding of the final morphology of the system. It is worth mentioning that what is discussed above regarding the case of the Palierne model is only applicable for those silica-filled blends which are below their percolation and do not exhibit a nonterminal storage modulus of the low frequency range.

### 7.5.3.2 Polymer Blends Containing Nanoclay

Due to interesting physical and mechanical properties and continuously growing industrial applications of polymer blends containing nanoclay, the melt rheology of these materials has received special attention from the scientific and industrial communities. The melt rheology and, more specifically, the melt linear viscoelastic measurements can provide a great insight into the degree of dispersion and exfoliation of nanoclay in blend components as well as the localization of exfoliated (platelets) or intercalated (tactoids) in the filled blends. Moreover, the melt rheology has proven to be capable of determining the blend composition needed for reaching co-continuous morphology and double percolation structure in nanoclay-filled polymer blends.

Koshkava *et al.* used the melt rheology to study the extent of melt intercalation and nanoclay localization in nanoclay-filled PA/PE blends with composition as tabulated in Table 7.3 [141].

They used two different organically modified nanoclays (organoclays): Closite 30B and Closite 15A. It was shown that the blends containing 1 wt% more polar

**Table 7.3** Composition of prepared samples [141].

| Sample | PA6 (wt%) | LDPE (wt%) | Qrganocaly (phr) |
|---|---|---|---|
| Ln2 | 0 | 100 | 2 |
| Pn2.3 | 100 | 0 | 2.3 |
| P85L15 | 85 | 15 | 0 |
| P85L15n0.5 | 85 | 15 | 0.5 |
| P85(Ln5)[a] | 85 | 15 | 0.5 |
| P85L15nl | 85 | 15 | 1 |
| P85Ll5n2 | 85 | 15 | 2 |
| P85(Ln11)[a] | 85 | 15 | 2 |
| P85L15n4 | 85 | 15 | 4 |
| P70L30 | 70 | 30 | 0 |
| P70L30nl.6 | 70 | 30 | 1.6 |

a) Prepared in different order feeding, that is, first preparation of nanocomposite-based LDPE then mixed it with PA6 as a dispersed phase.

organoclay (Closite 30B) exhibited a pronounced nonterminal storage modulus at a low frequency range, whose plateau magnitude was increased with increasing the organoclay loadings (see Figure 7.37).

According to the authors, such a pseudosolid-like behavior implies that the exfoliated (platelets) or the highly intercalated (tactoids) organoclays are exclusively localized in the form of a percolated network in the PA matrix. This was evidenced by SEM micrographs which showed much smaller PE-dispersed particle size in the filled blends compared to that in the unfilled blends. This was suggested to be mainly due to the suppression of the coalescence of droplets by the organoclays dispersed in the PA matrix. However, Khatua *et al.* who

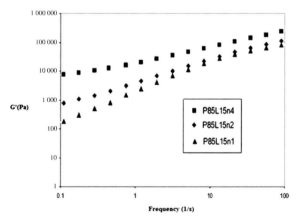

**Figure 7.37** Storage modulus versus frequency at 240 °C for P85Ll5n4, P85Ll5n2, and P85Ll5n1 [141] (with permission © 2011 Wiley Periodicals, Inc).

observed a similar result from the organoclay-filled PA/ethylene propylene rubber (PA/EPR) blend suggested that the presence of the exfoliated organoclay at the interface could play the same role [142]. Interestingly, the magnitude of the storage modulus ($G'$) at the plateau region for the blends was found to be higher compared to that of the PA matrix filled with equal amount organoclay content, assuming that in the filled blends the platelets are exclusively located in the PA phase. It should be noted that the authors calculated the PE droplet and interface elastic contribution and found it negligible compared to the measured $G'$ of the filled blends at the plateau region. Thus, they concluded that the presence of the PE droplets in the filled blends can rearrange the platelets to form a new and stronger percolated network structure. Consequently, this was mainly attributed to the suppression of the coalescence of the PE droplets by the platelets distributed in the matrix. Khatua *et al.* who observed similar results for the organoclay-filled PA/EPR blends believed that the presence of organoclay at the interface could also lead to the reduction of EPR-dispersed size. It is worth mentioning that the presence of the organoclay inside the dispersed droplets is not expected to cause an appreciable change in the storage modulus of filled blends.

Chow *et al.* carried out a similar melt rheological experiment on PA/EPR blends containing organoclay (Nanometer 1.30 TC) [143]. For these ternary systems, they also found a strong nonterminal storage modulus ($G'$), whose increasing trend with increasing organoclay was measured in terms of reducing the slope of $G'$ versus frequency range (see Figure 7.38). In a similar manner to the polymer nanocomposites, they attributed these results to a pseudosolid-like behavior, due to the nano-reinforcing or the percolating effect of exfoliated clay in the PA matrix. The authors suggested that the higher value and the smaller slope of $G'$ at the low frequency range obtained for the filled blend compared to those of the filled PA could be attributed to the stronger interaction between the

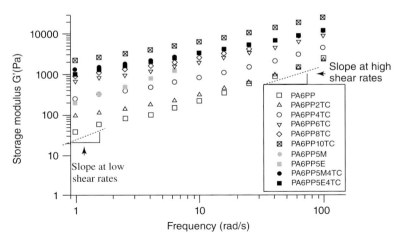

**Figure 7.38** Variation of the storage modulus as a function of frequency for the PA6/PP nanocomposites [144] (with permission © 2005 WILEY-VCH Verlag).

droplets and their tendency to form a 3-D superstructure. Some of the filled blends, in which PP/G-MA or EPR/G-MA were added as a compatibilizer, exhibited stronger pseudosolid-like behavior compared with the uncompatibilized samples. This was related to the improved degree of dispersion of organoclay.

There have been many cases where neither of the blend compounds have enough capability for intercalating organoclay. There are also some cases where the two polymer pairs have comparable thermodynamic tendency toward organoclay and, therefore, it will be difficult to reach a thermodynamic preference for the selective localization in one of the two phases. In such cases, the melt rheology can be used as a quick and yet reliable technique for selecting the appropriate compatibilizer to either assist the melt intercalation of the organoclay or to select the appropriate compatibilizer. As will be seen in the later sections, there are many other important reasons for using the compatibilizer in nanoparticle-filled polymer blends. Tiwari *et al.*, for example, studied the effect of polypropylene-glycidil methacrylate (PP-GMA) as a compatibilizer on the extent of dispersion and distribution of organoclay and their influences on morphology and mechanical properties in PP/EOR (ethylene-co-octan copolymer 70/30) blends [118]. By performing the melt viscoelastic experiment, they showed that the addition of PP-GMA, due to its greater compatibility with PP, could not only facilitate the preferred localization of organoclay in the PP matrix but also significantly improved the extent of organoclay dispersion (see Figure 7.39).

The plateau region observed in $G'$ versus frequency (Figure 7.39) and stress relaxation versus time (not shown) which are indications of a percolated network formed in the PP matrix are clear evidence of the above-mentioned compatibilizer actions.

Bigdeli *et al.* used the melt linear viscoelastic measurements to study the organoclay (Closite 30B) distribution in PP/PBT blends [145]. The filled blend samples were prepared first by melt mixing the organoclay with PBT and then the

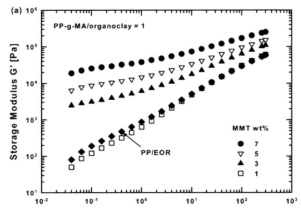

**Figure 7.39** Storage modulus of PP/PP-GMA/MMT/EOR nanocomposites determined at different MMT contents and at a fixed PP-GMA/organoclay ratio of 1.0 [118] (with permission © 2012 Wiley Periodicals, Inc.).

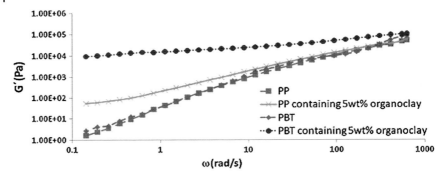

**Figure 7.40** Melt viscoelastic results of PP and PBT containing 5 wt% organoclay [145] (with permission © 2012, Springer Science + Business Media Dordrecht).

resulting premixed compound was compounded with PP in the later stage of mixing. From the results shown in Figure 7.40, one may notice that the extent of nanoclay dispersion in PBT is much greater than that in PP, which was considered to be an indication of a higher thermodynamic tendency of the organoclay toward the PBT phase with respect to the PP phase.

From the viscoelastic results, it was suggested that, while at the low concentration of organoclay most platelets remain inside the more favorable PBT droplet, or at the interface, a fraction of organoclays could transfer to the PP matrix at a higher organoclay content. This effect was found to be intensified when PP-GMA was used as a compatibilizer (see Figure 7.41). The authors supported the rheological arguments by more pieces of evidence obtained from the morphological study using SEM and TEM.

A similar experiment was conducted by Abdolrasouli *et al.* on organoclay-filled PLA/PE blends using the PEG-MA compatibilizer [146]. These authors also found compatibilizer assistance in transforming the organoclay, but in this

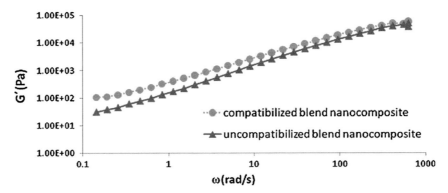

**Figure 7.41** Comparison of melt viscoelastic results of PP/PBT compatibilized and uncompatibilized blend nanocomposites containing 3 wt% organoclay [145] (with permission © 2012, Springer Science + Business Media Dordrecht).

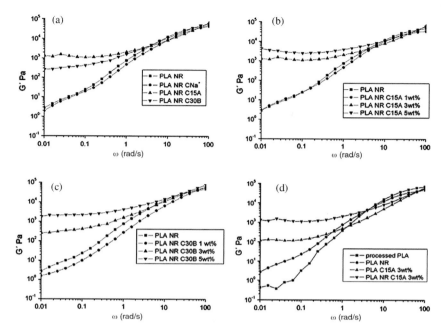

**Figure 7.42** Influence of (a) nanoclay type, (b) concentration of C15A, (c) C30B, and (d) composition of the blends over rheological properties of the composite [147] (with permission © 2012 Elsevier).

case from a more preferred PLA matrix to PE droplets, leading to lowering the elasticity due to the contribution of the percolated network, in the PLA matrix. This was checked by SEM, which showed a larger PE particle size in compatibilized filled blend samples compared to the uncompatibilized one. Bilinis *et al.* employed the melt rheology in order to elevate the effect of the nanoclay interaction in determining the extent of nanoclay dispersion in PLA/NR blends [147]. For this purpose, they used the unmodified clay (CNa$^+$) and two different organically modified nanoclays denoted as Closite 15A and Closite 30B. They found a typical liquid-like behavior at low frequencies for both PLA- and NR-containing unmodified clay. This was an indication of a poor dispersion and the lack of interaction of the clay with any of the polymers. However, a drastic change in the viscoelastic response (a solid-like behavior) was observed by the addition of the organoclays C15A and C30B for both PLA and NR (see Figure 7.42). This suggested that neither of the two organoclays can selectively localize in one of the two phases (polymers) but are preferentially located at the interface.

Hutric *et al.* investigated the linear viscoelastic and steady shear properties of PE/PA 80/20 and 20/80 blends containing different amounts of the same organoclay grade (Closite 30B) [148]. They found that the blends in which PE was the matrix irrespective of organoclay weight fraction exhibited a behavior similar to that of PE/PA 80/20-unfilled blend at low frequencies (low shear rates). For the blends with the PA matrix containing lower than 2 wt% organoclay, the behavior and interpretation of the ternary blends were similar to those observed for the PE matrix blends, whereas the blends with the weight fraction of organoclay higher than 4 wt%, the linear viscoelastic and steady shear properties showed a solid-like behavior similar to that of sufficiently filled nanocomposites. Interestingly, the authors found good agreement between the rheological measurements and morphological analysis.

Hong *et al.* used the linear viscoelastic measurements to study the effect of the presents of nanoclay on the interfacial tension of nanoclay-filled PET/PE 10/90 blends in which the platelets were preferentially located at the interface [149]. From the comparison of the measured storage modulus ($G'$) versus frequency and uniaxial extensional force of the blends measured as a function of the Hencky strain ($\epsilon$) for several compositions, they verified the organoclay-induced reduction of the interfacial tension. This was found to be in agreement with the interfacial tension reduction which was predicted from the extensional force measured based on the Levitt *et al.* method [150]. From these results, they showed that at organoclay loading of 1 wt% the platelets are selectively located at the interface and act as a compatibilizer. Kontopoulou *et al.* investigated the melt rheological properties of EPR-G-MA/PP 70/30 and 30/70 blends both containing equal amounts of organoclay ($NR_4T_{MM}$) [151]. With the help of TEM, they found that irrespective of the ratio of the blend components, the exfoliated organoclays (platelets) were exclusively localized in the EPR-G-MA phase. By fitting the experimentally measured storage modulus $G'$ versus the frequency of the filled EPR-G-MA/PP (70/30) blend to the Palierne model, they found a fitted value of $\sigma = 1$ which was very close to that of $\sigma = 1.1$ for the unfilled counterpart (see Figure 7.43). From this finding, the authors concluded that the presence of organoclays did not affect the interfacial properties of this blend sample.

Recently, an attempt was made by the Ville research group to investigate and characterize the interphase in PE/PA blends with nodule morphology (PE or PA) filled with the same organoclay grade (Closite 30B), using the rheological and

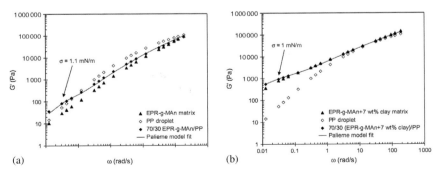

**Figure 7.43** (a) Palierne model fit for EPR-g-MA/PP 70/30 blend; (b) Palierne model fit for model blend, which comprises an EPR-g-MA matrix containing 7 wt% NR [151] (with permission © 2007 Elsevier).

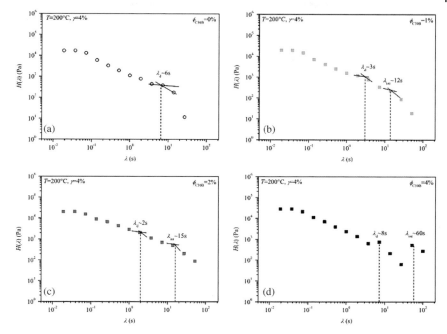

**Figure 7.44** Relaxation spectra for PE matrix blends at a clay weight fraction of: (a) 0 wt%, (b) 1 wt%, (c) 2 wt%, and (d) 4 wt% [152] (with permission © 2012 Elsevier).

structural experimental techniques [152]. In order to investigate more thoroughly the viscoelastic contribution of the interphase, relaxation spectra, $\lambda H(\lambda)$ were computed from $G'$ versus $G''$ plots (reported in their previous work) and using the program described by Honerkemp and Weese [153]. The results of $H(\lambda)$ versus time obtained for the unfilled and filled blends are shown in Figure 7.44.

The relaxation of the unfilled blend showed only one peak at $\lambda_d = 6$ s, which was attributed to the form relaxation of the PA nodule, whereas the filled blends showed an additional peak ($\lambda_{int}$) at longer time which was suggested to be due to the interphase contribution. The form relaxation $\lambda_d$ was decreased with increasing the organoclay content up to 2 wt% ($\lambda_d = 2$ s), while it raised again and reached to 8 s which was higher than that obtained for the unfilled blends. According to the authors, the decrease in the value of $\lambda_d$ at low organoclay concentration was due to the decrease in the PA nodule relaxation droplet size caused by the presence of the organoclay, while increasing in the value of $\lambda_d$ was attributed to the accumulation of organoclay inside the PA nodules and, consequently, their greater resistance to deformation. More interestingly, the relaxation time ($\lambda_{int}$) corresponding to the interphase was about 12 s at 1 wt% organoclay and 15 s at 2 wt% and reached to 60 s when the organoclay content was increased to 4 wt%. The increase in the interphase relaxation time was related to the increase in the average interphase thickness. However, as the authors pointed out, using only average interphase thickness to characterize the interphase is certainly insufficient to a great extent because of the presence of numerous interphase defects which were evidenced by TEM micrographs. Finally, in contrast to PE matrix blends, the relaxation spectra of PA matrix blends above 2 wt% organoclay did not exhibit any well-marked peak which was observed for PA/organoclay nanocomposites at 5 wt% organoclay. The percolated network structure formed in the PA matrix was the reason why the relaxation time spectra of PA matrix blends at high organoclay content did not show any relaxation time peak. In other words, the percolation network of clay particles and, hence, its remarkably high elastic response in the PA matrix masks the viscoelastic contribution of both PE nodules and interphase.

In their later work, Aubry's research group used a similar rheological approach combined with the TEM and SEM study, in order to compare the interphase characteristics of PE/PA blends which were compatibilized with organoclay (Closite 30B) and PE-g-MA (Orevac) [154]. The critical strain ($\gamma_c$), which defines the extent of the linear viscoelastic regime, is plotted as a function of organoclay or Orevac concentration, as shown in Figure 7.45. While the value of $\gamma_c$ of the blend compatibilized with Orevac is independent of the compatibilizer content, it drastically decreased for the organoclay-filled blends. These results indicated very significant differences between the two types of the compatibilized blends; this was attributed to major differences in the interphase nature ($\phi_{\text{orevac}} = \phi_{\text{clay}}$). The results shown in Figure 7.45 also imply a transition fraction at $\phi = 1\,\text{wt}\%$ which is suggested to being due to a change in morphology for the organo-filled ternary blends, whereas the morphology of the blends compatibilized by Orevac remained unchanged over the whole range of the Orevac concentration considered (up to 6 wt%).

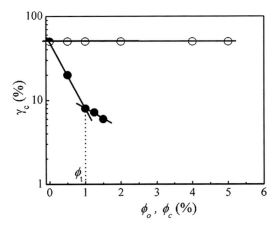

**Figure 7.45** Critical strain $\gamma_c$ as a function of Orevac or clay volume fraction (°) blends compatibilized with Orevac and (•) PE/PA blends filled with clay [154] (with permission from AIP Publishing LLC).

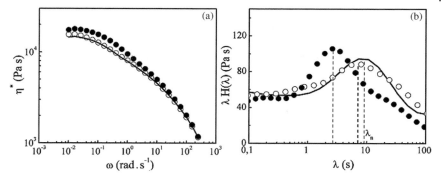

**Figure 7.46** (a) Complex viscosity as a function of frequency, (b) weighted relaxation spectra for (—) PE/PA, (°) PE/PA/O 0.5%, and (•) PE/PA/C 0.5% [154] (with permission from AIP Publishing LLC).

The effects of addition of 0.5 wt% and 1 wt% of organoclay or Orevac on $\eta^*$ versus frequency and $[\lambda H(\lambda)]$ of the blends are shown in Figures 7.46 and 7.47. The results of the similar experiments obtained for the blends containing 1 wt% Orevac or organoclay are also shown in Figures 7.46 and 7.47 for comparison. It should be noted that in plots of $[\lambda H(\lambda)]$ versus $\lambda$, presented in these figures, the peak of the PE matrix has been eliminated from the results. The authors suggested that there are some analogy between the Orevac-compatibilized blends and the organoclay-filled blends at low organoclay content (up to transition fraction $\phi_t = 1$ wt%). In other words, in these organoclay-filled blends with low organoclay content, the exfoliated organoclays (platelets) are accumulated at the interface and act similar to the organic compatibilizer.

### 7.5.3.3 Polymer Blends Containing Carbon Nanotubes
The rheology of CNT-filled polymer blends has received special attention mostly due to its capability of distinguishing the feature of co-continuous morphology and the so-called double-percolation threshold in the field of electro-conductive

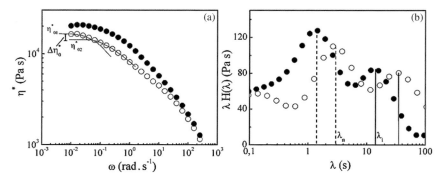

**Figure 7.47** (a) Complex viscosity as a function of frequency. (b) Weighted relaxation spectra for (°) PE/PA/O 1% and (•) PE/PA/C 1% [154] (with permission from AIP Publishing LLC).

nanocomposites. The concept of the rheological percolation threshold was first introduced by Potschke for CNT-filled polycarbonate (PC) [155]. They observed a remarkable viscosity up-turn and nonterminal or solid-like behavior at low frequencies for CNT-filled PC having greater than 2 wt% CNT. This was ascribed to a rheological percolation threshold, due to a high aspect ratio of CNT, which also correlated well with the electrical percolation threshold obtained for the same PC/CNT nanocomposites. Later, a more effective method called double-percolation threshold was used in which a two-phase polymer blend was used as a matrix. In this type of polymer blend nanocomposite, either one of the two polymer phase is continuous and the conducting filler is exclusively localized (in the form of a percolated network) in the continuous phase or two polymers are co-continuous and the fillers are preferentially localized in one phase or at the interface.

The application of the melt linear viscoelastic measurements for studying the nanoscopic structure, in particular the particulate network structure (LST) of polymer nanocomposites, was described in Section 7.5.3. Furthermore, the melt rheology of the polymer blends containing CNT has many common features with the silica or organoclay-filled polymer blends which were discussed in detail in previous sections. Therefore, here we focus on some specific rheological aspects of CNT-filled polymer blends.

Over past few decades, attempts have been made by several researchers to study the correlation between the percolation threshold obtained from the rheological behavior and the electrical properties of immiscible polymer blends containing CNT [106,156–158]. Zondere *et al.*, for example, investigated the effect of different mixing sequences (simultaneous PA12 masterbatch and PE masterbatch) on the rheological behavior of PA/PE 75/25 and 25/75 blends containing different amounts of CNTs [159]. It was shown that, in contrary to the poor dispersing ability of PE, PA had enough capability in dispersing the CNT due to its polar nature, which resulted in the rheological and electrical percolation threshold around 1.4 wt% CNT. The blend in which the PA was a matrix exhibited a solid-like behavior at a low frequency range, which was related to the well-dispersed CNT in the PA matrix and the formation of the percolated network structure.

Among various mixing sequences, the PA/PE 75/25 blend nanocomposite prepared from PE masterbatch showed slightly higher elastic response at a low frequency region compared to the blend nanocomposite prepared from simultaneous mixing and PA masterbatch. By means of SEM images, and knowing that the viscosity of PE was higher than that of PA, this was attributed to the formation of the surface and volume percolation network where the CNT-coated PE domains were interconnected through the interphase. In this way, the CNT trapped in the interphase could act as centers or a junction to connect the few CNTs that were dispersed through the PA matrix. The localization of the CNTs at the interface intensified the solid-like behavior. This was explained in terms of increased elasticity due to the retardation of the shape relaxation of PE droplets and/or clustering of the finer PE domains.

An attempt was made by Tao *et al.* to quantitatively determine the fraction of CNTs trapped at the interface and PA droplets in the PA/EA copolymer EA(10/90) immiscible blends by fitting the melt linear viscoelastic data to the Palierne model and calculating the interfacial tension [157]. The CNT-filled PA/EA 10/90 blends were prepared by first mixing 2 wt% CNT with EA and the premixed compound was then blended with 10 wt% PA in the later mixing stage. They found that of the total 1.8 wt% CNT content in the blend, 0.7 wt% CNT remained inside the EA matrix, while 1.1 wt% CNT migrated to the interface. It should be noted that the CNTs trapped inside the EA droplets could be neglected from both mass balance and rheological standpoints. From their analysis, they found that the interfacial energy is affected by the interfacially trapped CNTs. They could also estimate the interfacial area coverage by the CNTs. The calculated value was high (about 40%) which was found to be consistent with the morphology of the blend. Moreover, they demonstrated that the analysis of relaxation spectra $[\lambda H(\lambda)]$ can provide some insights into the molecular processes occurring as a consequence of the blending and especially the effect of the CNT localization on the interface. From these findings, they concluded that the surface of the droplets is only partially covered so that the uncovered surface relaxes faster than the covered surface.

Taheri *et al.* studied the linear viscoelastic behavior of the PC/ABS blends by varying the blend ratio and the CNT content [160]. They found that the addition of the CNT had a greater effect on increasing the storage modulus ($G'$) at low frequencies for ABS than that of PC. This was attributed to the higher affinity of the CNT to ABS compared to PC. They also found more enhancing effect of the CNT on the low frequency storage modulus ($G'$) for PC/ABS 70/30 blend nanocomposite in comparison with the PC/ABS 30/70 blend. These results were consistent with TEM images which showed that while in the former nanocomposites CNTs were localized at the interface and ABS droplets in the later one, most of the CNTs remained inside the ABS matrix. A similar research work was carried out by Abbasi *et al.* on CNT-filled PA/PP 60/70 and 50/50 blends with co-continuous morphology [161]. The blend nanocomposites were prepared by first mixing the CNT with PA, and PP was added to the mixer in the later mixing stage. As evidenced by the linear viscoelastic results, CNTs, due to the high interfacial interaction with the polar PA, were well dispersed and, hence, resulted in a strong percolated network in PA. In contrast, CNTs were poorly dispersed and induced a weak interconnectivity in PP. The CNT-filled blends exhibited similar rheological behavior to that of the PA nanocomposite, indicating that most CNTs are localized in the PA phase. These results were verified by TEM images (see Figure 7.48).

Recently, an attempt was made by Rostami *et al.* to use the melt rheology in order to determine the CNT localization and its role on the morphology development of PMMA/PS/PP-ternary blends which originally had preferred core-shell-type morphology. From the rheological results, they found that in contrast to the unfilled blends, whose core-shell morphology was fully controlled by the thermodynamic parameters, in the case of CNT-filled blends, the kinetic parameters could also play an effective role [162].

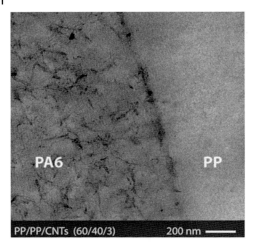

**Figure 7.48** TEM images of PA6/PP/CNTs (60/40/3) nanocomposites at different magnifications [161] (with permission © 2014 Wiley Periodicals, Inc.).

### 7.5.4
### The Role of Nanoparticles on the Morphology Evolution of Nanofilled Polymer Blends

#### 7.5.4.1 Nanoparticle Migration (Dynamic and Transfer of Nanoparticles)

As discussed in the previous sections, there have been a considerable number of reports dealing with experimental evidences for nanoparticle localization in immiscible polymer blends. However, the mechanisms by which the nanoparticles are transferred from one phase to the other, pass the interface, or more generally move inside the blend at the molten state have been discussed very rarely. Fenouillot *et al.* were the first to identify and discuss at least qualitatively these mechanisms [16]. They postulated that during mixing, convection shear forces are high enough to shift the adsorption/disruption equilibrium by extracting the particle from the interface. According to the authors, the transfer of nanoparticles between the two phases of co-continuous immiscible blends is provided firstly by the contact between nanoparticles and blends interface, and secondly by transfer from one phase to the other through passing the interface. The transport of the nanoparticles toward the interface can be provided by a combination of the particle diffusion (Brownian induced motion) and the shear-induced collisions of the particles with the blend interface (blend domains). By simplifying calculations performed on a molten polymer model (PP) and spherical particles, they showed that the order of magnitude of Brownian motion time (7500 s) was very large and consequently not comparable with mixing time (a few minutes). In other words, the migration of nanoparticles can hardly occur under static conditions in molten polymers. In the case of blends with matrix/droplet morphology, these mechanisms can only be applicable when nanoparticles are initially located and dispersed in the matrix phase. While nanoparticles are selectively embedded in the droplets, the transfer of nanoparticles must be governed by other mechanisms which have not been identified. It is worth mentioning that the coalescence of polymer droplets could also play a role in the transfer of nanoparticles from one phase to the other. However, due to difficulties concerning with decoupling the effects of the parameters such as blend composition, droplet deformability, and the type and extent of the convective (shear) forces during melt mixing, it is difficult to declare whether this latter mechanism is significant or not. Furthermore, the complete transfer of nanoparticles from one phase to the other occurs when nanoparticles can cross the interface. This implies that a polymer chain adsorbed on the particle surface must desorb progressively to be regulated by the other polymer chains (the one with better affinity). As the authors acknowledged such a competitive adsorption of polymer melt blends on a solid particle, which can play an important role in the transfer of particles, has been rarely discussed in the literature.

An attempt was made by Goldel *et al.* to study the kinetics of the transfer of CNT from pre-compounded SAN to more favorable an initially unfilled PC phase during melt mixing in a continuous twin-screw extruder [163]. As the authors pointed out due to the complex mechanisms and kinetics involved with the nanoparticle transfer during the melt mixing, they had to use several simplifications for explaining the governing mechanism. They demonstrated that similar to discontinuous microcompounder, complete transfer of nanoparticles could occur in a twin-screw extruder but within much shorter residence time (60–90 s) compared to 5 min of mixing time in the discontinues mixer. They also found that the transfer rate of particles is strongly coupled with the development of the interface area between the blend phases. Thus, the premixing of CNT in one of the two polymer phases could be a determining parameter for the final dispersion of CNTs in the blend. From their findings, Goldel *et al.* proposed similar mechanisms for the transport of nanoparticles to reach and contact with the blend interface [163]. They also studied the kinetics of the subsequent transfer through the blend interface into the favorable (better wetting) phase. The probability that this mechanism can transfer CNT within the available time scale could be expressed and correlated with the main influencing factors like the blend composition and deformation rate of the mixing process. However, quantitative evaluation was not possible since the time scales reached for the formation of the wetting angle and the subsequent transfer of individual CNTs were still unknown. More recently, an attempt was made by Goldel *et al.* to study the effect of the nanoparticle aspect ratio on the transfer dynamic and the stability of nanoparticles at the blend interface [164]. These authors on the basis of the pure theoretical consideration of Krasovitski *et al.* found that the transfer of the high aspect ratio fillers is much faster and more efficient than that of low aspect ratio particles [165].

The high aspect ratio fillers were transferred much faster and more efficiently as compared to low aspect ratio nanoparticles. In other words, the probability of nanoparticle localization in both phases or at the interface after mixing decreased with increasing aspect ratio. This was attributed to the change of blend interface curvature during the nanoparticle transfer process. They addressed this difference as the "slim-fast mechanism" (see Figure 7.49).

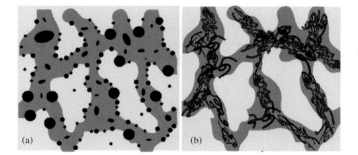

**Figure 7.49** Typical localization scenarios for low (a) and high (b) aspect ratio fillers that were predispersed within a poor wetting blend phase (yellow) and subsequently melt blended for the same mixing times with a more favorable polymer (blue). (a) Spheres and ellipsoidal particles are distributed in between the blend interface and the preferred (blue) blend phase according to their interfacial stability. (b) Slim CNTs are highly selectively localized in the preferred blend phase (blue) after melt blending [164] (with permission © 2011, American Chemical Society).

This was found to be a reasonable explanation for their previous experimental findings which showed fast and complete transfer of CNT from the less favorable SAN phase to the more favorable wetting PC phase. Further verification was made by comparing PC/SAN blends filled with either CNT or CB which showed that after 5 min of melt mixing, CNTs were mostly transferred and localized in the PC phase, while a fraction of CB was trapped at the interface.

Zhang *et al.* examined the transfer behavior of three different nanoparticles, organoclay (Closite 6A), tungsten disulfide nanotube, and tungsten disulfide multilayer onion-shaped (W2S) particle in PS/PMMA (50/50) blends. They showed that Closite 6A was more efficient in reducing PMMA domain size compared to other two types of nanoparticles. They explained this in terms of a higher aspect ratio of organoclay platelets and, hence, their greater ability to saturate the interface between matrix and droplets [166].

The above-discussed findings on the role of the geometry and size of nanoparticles on their displacement kinetics and their transfer from one phase to the other clearly indicates that the extent of the intercalation of layered silicates and graphene, and the state of dispersion of CNT obtained during melt mixing can play important roles in determining the nanoparticle localization. This, once again, emphasizes the significant role of kinetic parameters such as melt-mixing history and the viscosity of the blend polymer components in determining nanoclay compatibilization and polymer morphology of blends containing organoclay, grapheme, and CNT-type nanoparticles.

### 7.5.4.2 The Compatibilizing Effect of Nanoparticles
It is well known that compatibilization is the key strategy to manipulate the interfacial properties of most immiscible polymer blends in order to refine their morphology for optimum physical and mechanical properties. Compatibilization is usually carried out by using a third compound called a

compatibilizer which is often a macromolecule such as block/graft or star copolymer. The presence of a compatibilizer in an immiscible blend plays two major roles: (1) it lowers the interfacial tension, thus promoting the breakup of droplet during blending (emulsification role) and (2) it minimizes the subsequent coalescence of the droplets and, hence, stabilizes the blend morphology (coalescence suppression role) during blending as well as the latter processing steps. Moreover, and equally important, the addition of a compatibilizer to immiscible blends can enhance the interfacial adhesion and, hence, improve the mechanical properties, especially the ultimate strength of finished products. This type of compatibilizing method has been successfully used to produce a variety of industrial polymer blends with a wide range of prescribed properties.

On the other hand, more recently a new compatibilization strategy based on the addition of nanoparticles has been explored to control the morphology of immiscible blends. Indeed, it has been found that added particles such as CB, silica, clay and $TiO_2$ nanoparticles into polymer blends exhibiting different wetting characteristics can play a role similar to block copolymers. However, while block copolymers are absorbed at the interface due to their dual chemical nature, the tendency of nanoparticles to accumulate at the interface to form a highly stable monolayer depends on the wettability, size, and initial location of nanoparticles as well as nanoparticles' interaction.

The compatibilizing effect of nanoparticles in polymer blends has been studied by many investigators. For example, Gubbels *et al.* for the first time observed the compatibilizing effect of nanoparticles in immiscible polymer blends. They evaluated the coalescence suppression in a co-continuous CB-filled PE/PS blends by performing static annealing (200 °C for a few minutes) [106,107]. The authors found that the CB particles were localized at the interface in order to minimize the interfacial energy of the binary blend. It was also shown that the compatibilizing effect of the CB on the coalescence suppression and coarsening of both phases were dependent upon the CB content. The moderate compatibilizing effect was obtained with only 1 wt% added CB, whereas for a 5 wt% CB-filled blend sample, the average size of both PS and PE phases remained unchanged even after 24 h at 200 °C.

In another work, Vermant *et al.* investigated the effect of the addition of the hydrophobic nanosized silica particles on the flow-induced morphology of the PDMS/PIB blend by using a well-defined flow protocol followed by linear viscoelastic measurements in conjunction with the Palierne model [59]. The effect of nanoparticle concentration on the droplet coalescence and the shear sensitivity of the flow-induced morphology were evaluated in terms of the ratio between the volume average droplets radius over the interfacial tension versus time (see Figure 7.50).

They found that the addition 1 wt% of nanoparticle completely slowed down or suppressed the coalescence of the PIB droplets. This was evidenced by TEM observation which showed a distinct layer of silica nanoparticles formed at the interface (see Figure 7.51).

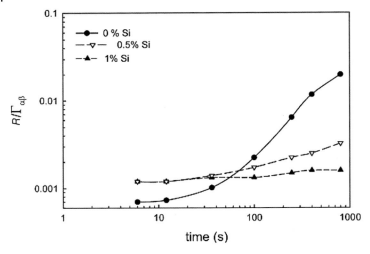

**Figure 7.50** Evolution as a function of time for a step-down experiment depicted of $R/\Gamma$ for a 70/30 PDMS/PIB blend containing different concentrations of silica ($T = 20\,^{\circ}$C) [59] (with permission © 2004, Springer-Verlag).

In more recent years, Vandebril *et al.* used the same approach as used by Vermant *et al.* to study the role of two geometrically different (spherical and spindle-like) nanoparticles but with the same tailored surface chemistry on the suppression coalescence of PIB droplets in the PDMS/PIB blends [58]. In this study, it was demonstrated that the coalescence of the PIB droplets was slowed down or even completely suppressed when the nanoparticles were located at the interface. It was also found that an isotropic nanoparticle tends to stabilize the blend morphology more efficiently than the spherical counterparts. The authors, based on their findings obtained from a combination of microscopy and

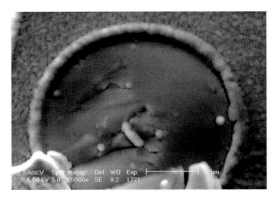

**Figure 7.51** Cryo SEM freeze fracture images of a 30 wt% PIB in PDMS blend containing 1 wt% of silica particles: detail showing the particles at the interfaces [59] (with permission © 2004, Springer-Verlag).

interfacial rheology, concluded that the nanoparticles stabilize the morphology mainly by changing the surface rheological properties, whereas the polymeric compatibilizer can also strongly influence the interfacial tension.

Tong *et al.* studied the effect of surface nature (hydrophilic and hydrophobic) and the concentration of nanosilica on the morphology of PDMS/PIB 90/10 blends by means of the optical-shear technique [167]. It was shown that the coalescence behavior of PIB droplets under the low shear rate was determined not only by the surface nature but also by the nanoparticle concentration. The hydrophobic nanosilica suppressed the droplet coalescence and stabilized the blends by forming a nanoparticle layer at the interface. On the contrary, the hydrophilic nanoparticles prohibited the coalescence through the bridging effect of particle aggregates and the formation of droplet clusters at low nanoparticle concentrations. However, the coalescence of PIB droplets was found to be promoted more when hydrophilic nanosilica was added. The authors believed that this phenomenon could be attributed to the easy migration of hydrophilic particle aggregates toward the PDMS matrix at higher added nanoparticles which led to the coalescence of PIB droplets.

In another work, Li *et al.* investigated the compatibilizing effect of alcohol-treated $TiO_2$ (T) on PP/PET blend morphology both in the absence and presence (up to 6 wt%) of PP-gr-MA(MA) [168]. The PP was mixed with $TiO_2$, and the resulting nanocomposites were melt blended with PET in the absence or presence of 0.3 and 6 wt% PP-G-MA. The TEM micrographs showed that in both PP/PET filled with 2 wt% $TiO_2$ (PP/PET/2T) and PP/PET filled with 4 wt% $TiO_2$ (PP/PET/4T) nanocomposites, the $TiO_2$ nanoparticles were mostly migrated to the interface and to less extent transferred to the PET phase (see Figure 7.52).

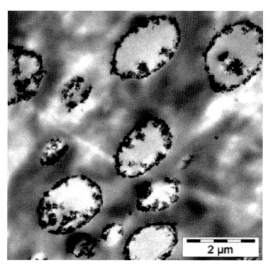

**Figure 7.52** TEM image of PET/PP/2T nanocomposite [168] (with permission © 2009 Wiley Periodicals, Inc.).

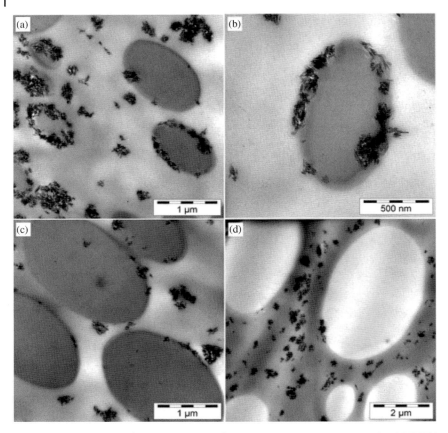

**Figure 7.53** TEM images of PET/PP/3MA/2T and PET/PP/6MA/2T nanocomposites: (a–c) PET/PP/ 3MA/2T, (d) PET/PP6MA/2T [168] (with permission © 2009 Wiley Periodicals, Inc.).

The addition of PP-G-MA to the blend nanocomposites was found to greatly reduced the migration rate of the $TiO_2$ nanoparticles in such a manner that in the PP/PET/6MA/2T containing 6% PP-G-MA, almost all $TiO_2$ particles preferentially remained in the PP matrix (see Figure 7.53). This was suggested to be due to enough maleic anhydride functional groups in the PP matrix to react with the hydroxyl group of $TiO_2$. Figure 7.54 schematically explains the two links occurring in the PP matrix (between MA and $TiO_2$) and the PET-dispersed phase (between PET and $TiO_2$).

As can be clearly noticed from Table 7.4, the most efficient compatibilizing effect expressed in terms of the reduction in PET-dispersed phases can be achieved for blend nanocomposites, in which the $TiO_2$ nanoparticles were exclusively located at the interface.

The compatibilizing role of organoclay on the morphology of immiscible blends has been studied by many research groups. Khatua *et al.*, for example,

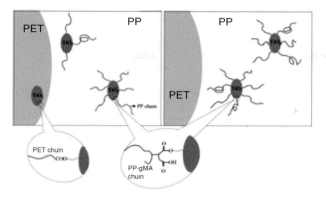

**Figure 7.54** Schematic of (a) PET/PP/3MA/2T and (b) PET/PP/6MA/2T nanocomposites [168] (with permission © 2009 Wiley Periodicals, Inc.).

investigated the effect of organoclay (Closite 20A) on the morphology of three nanocomposite blends of PA6/EPR 80/20, 20/80, and 50/50. From the SEM and TEM micrographs, it was found that for the PA6/EPR 80/20 blend, the EPR average volume diameter was reduced drastically with a small amount (0.5 wt%) of added organoclay [142]. Also, the EPR domain size in this blend did not change upon annealing at the elevated temperature, which was an indication of morphology stability. On the contrary, the presence of organoclay in the PA6/EPR 20/80 blend did not reduce the dispersed PA phase appreciably. On the other hand, WAXD results revealed that while PA almost exfoliated the organoclays, the EPR could only intercalate them unevenly. Furthermore, the extent of the reduction of the EPR domains of the organoclay-filled blends was found to be higher compared to the blend in which an *in situ* formed EPR-GMA was used as a compatibilizer. From these results, the authors concluded that the added organoclay can play a compatibilizing role if the platelets are finely dispersed in the PA matrix.

**Table 7.4** Number and volume average diameter of PET droplets in PET/PP and PET/PP/TiO$_2$ [168].

| Materials | PET droplets | |
| --- | --- | --- |
| | Number average diameter (μm) | Volume average diameter (μm) |
| PET/PP | 5.4 | 7.2 |
| PET/PP/MA | 3.4 | 5.1 |
| PET/PP/2T | 1.4 | 3.6 |
| PET/PP/4T | 1.1 | 3.0 |
| PET/PP/3MA/2T | 2.0 | 4.3 |
| PET/PP/6MA/2T | 3.8 | 5.5 |

A similar study was carried out on the organoclay-filled PA/PE 80/20 blend by Koshkava *et al.* by using a combination of XRD, TEM, SEM, and rheological measurements [141]. They also suggested that the hindrance effects of the platelets on minimizing the coalescence of EPR droplets in the PA matrix had the main contribution in the compatibilizing role of nanoclay. Gelfer *et al.* proposed that the decrease in the PS domain size in the PS/PMMA blend upon adding organoclay was due to the compatibilizing function of the excessive surfactant present in the organoclays, as well as the increased viscosity [169].

An attempt was also made by Wang *et al.* to determine how the organoclay compatibilizes the PS/PP blend [170]. The PS domain was found to be gradually reduced by the addition of the organoclay. They attributed this to the coexistence of the two immiscible polymer chains between the intercalated clay platelets. Thus, the two chains play a role similar to block copolymer acting as a compatibilizer. The compatibilizing role of organoclay in PCL/PEO blends was studied by Fang *et al.* using a combination of XRD, TEM, and melt viscoelastic measurements [171]. They demonstrated that when organoclay-filled blends were prepared through a two-step mixing, in which the layered silicates were first premixed with PEO or with PCL, the platelets were migrated from the PEO phase or the PCL phase to the interface. They compared the compatibilizing role of the platelets with a diblock copolymer as schematically represented in Figure 7.55.

For a diblock copolymer, block A and block B are compatible with phase A and phase B, respectively, and their connection link is found at the interface. In the case of the organically modified clay, both A and B polymers can have a strong interaction with the surface of the platelets that are localized at the interface. In this way, the organoclay plays the role of the coupling agent.

From the above, it is clear that the compatibilizing effect of the added organoclay highly depends on the particle's initial location or the phase where intercalation occurs because this determines the extent of melt intercalation and, hence, the effectiveness of the aspect ratio of the nanoparticles on the kinetics of

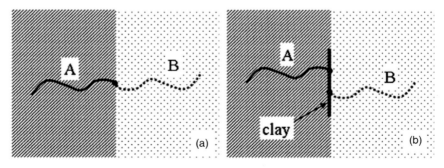

**Figure 7.55** Comparison of the compatibilization mechanisms of (a) a diblock copolymer and (b) a clay platelet [171] (with permission © 2007 Wiley Periodicals, Inc.).

transport and resulting localization. This will, in turn, determine the main compatibilizing mechanism of the platelets in determining the final morphology of the blend nanocomposites.

Si *et al.* compared the compatibilizing effect of organoclay in immiscible blends with different intercalation behavior [172]. They chose two model blend systems: PS/PMMA in which the organoclay was expected to exfoliate in only one phase (PMMA) and PC/SAN24 where organoclay was known to exfoliate in both phases. They found an increase in compatibility and the consequent localization of the platelets at the interface in both blend systems. This compatibilization was attributed to the large surface area of the exfoliated organoclay platelets which enabled the formation of *in situ* graft on the organoclay surfaces during the melt mixing in both blend systems. The effect of a combination of organoclay (Closite 20A) and some other polymeric-type compatibilizers differing in their polar functionalities on the morphology of PP/SEBS blends was studied by Martin *et al.* [173]. They demonstrated that the organoclay (Closite 20A) could be selectively located in one or other polymer phases, depending on the ability of the compatibilizers to react with the polymer components. The compatibilization effect of PP-G-MA was found to be more efficient compared to a $N_2$-plasma surface-treated PP (PP*). With the addition of 5 wt% PP-G-MA, the organoclay (Closite 20A) exhibited a tendency to locate at the PP–SEBS interface. The increase in PP-G-EPDM up to 15 wt% resulted in fine exfoliated platelets located at the interface and in the PP matrix, which led to a dramatic decrease in the interfacial tension and, hence, in the SEBS domain size, whereas when PP* was used as a compatibilizer, there was little improvement even though PP* intercalated the nanoclay. According to the authors, because of the affinity of the PP* to both organoclay and SEBS droplets, an enclosing effect on the organoclay and SEBS droplets hindered the suitable dispersion of organoclay and its contribution to the coalescence of SEBS droplets.

As described above, the dispersed particle size reduction in immiscible polymer blends containing nanoparticles is governed by complex processes occurring during melt compounding, where the droplet's brakeup mechanisms are opposed to coalescence. The parameters involved with these processes and their role in determining the morphology were summarized and described by Fenouillot *et al.* [16] as (1) a reduction of the interfacial energy, (2) the inhibition of coalescence by forming a solid barrier around the droplets, (3) the viscosity ratio changing due to the presence of nanofillers in one of the two phases, (4) the formation of a physical network of nanoparticles (when percolation is reached) that inhibits the droplets or the matrix motion, and (5) the adsorption interaction of macromolecules at the surface of the solid particles.

As mentioned previously, Hong *et al.* showed that in nanoclay-filled PBT/PE blends, the size reduction is obtained when the nanoclays are selectively located in one of the two phases [149]. At low nanoclay content, the added nanoparticles (platelets) are concentrated at the interface and suppress the coalescence, whereas at high content, the viscosity change would be responsible for reducing the droplet size. In some organoclay-filled polymer blends, the added organoclays (platelets) may be localized at the interface as well as in one of the two phases. In such situations, it is difficult to identify the origin of the nanoparticle-induced compatibilization (the reduction in droplet size and morphology stability). An attempt was made by Huitric *et al.* to provide some more insight into this problem by calculating the ratio between platelets located in the matrix and at the interface in PE/PA6 blend systems [148]. They showed that when PE is the matrix, all the platelets were accumulated at the interface and, hence, the size reduction could be only originated from coalescence suppression through the solid-like barrier effect of the platelets. In the PE/PA 20/80, however, the platelets were localized in both the interface and the PA matrix.

Several authors utilized other techniques rather than direct observation by TEM or SEM to identify the origin of droplet-size reduction. Sinha *et al.* for example, estimated the contribution of the viscosity ratio on the droplet-size reduction in PC/PMMA 30/70 blend nanocomposites containing Closite 20A by using the Taylor equation [174]:

$$\mathrm{Re} = \frac{0.5\gamma}{\eta_m \dot{\gamma}} \times \frac{16P + 16}{19P + 16},$$

where $\eta_m$ is the viscosity of the matrix, $P$ is the viscosity ratio, $\dot{\gamma}$ is the shear rate, and $\gamma$ is the interfacial tension. They suggested that the reduction in the particle size, due to the added nanoclay, was not only originated by change in the viscosity ratio but other mechanism may also play roles. More importantly, as discussed previously (see Section 7.5.4.1), several research groups have utilized the linear viscoelastic measurements in conjunction with the Palierne model to elucidate the emulsification role of added nanoparticles in reducing the dispersed size in nano-filled blends.

Among nanofillers, CNTs have received tremendous attention since they combine a high aspect ratio with exceptional electrical, mechanical, and thermal properties. Recently, special attention has been paid to study the compatibilizing effect of CNTs and nanographene on the morphology of immiscible polymer blends. The compatibilizing effect of the functionalized MWCNTs on morphology stabilization in PA/PE blends was studied by Baudouin *et al.* using the TEM micrograph together with iTEM software (Olympus soft imaging system) to evaluate the change in the droplet diameter [175]. The addition of MWCNTs was found to enhance both phase dispersion and stability of the dispersed phase along mixing time for very low filler content (0.5 wt% MWCNT). This was proved to be due to the interfacial localization of the MWCNTs. The main stabilization mechanism proposed by the authors was the formation of the deformable barrier network at the interface, which added as a mechanical barrier against the coalescence of the droplet.

Yong *et al.* studied the compatibilization effect of graphene oxide (GO) on the morphology development of PA/PVDF (90/10) blends. The size of the PVDF-dispersed particle was drastically reduced and became more uniform, revealing a well-defined compatibilization effect of the amphiphilic GO. They found a marked reduction in the PVDF-dispersed particle, which was an indication of a well-defined compatibilization effect of the GO.

On the basis of Fourier transform and Raman spectra finding, they proposed a mechanism for compatibilization. The edge polar groups of GO can form hydrogen bands with the PA chain which can interact with electron-withdrawing fluorine on PVDF chains, leading to the so-called charge transfer C–F bonding. In this case, GOs could exhibit favorable interactions with both PA and PVDF phases, thus stabilizing the interface during GO migration from the PVDF to PA phase.

The localization of graphene and GO in immiscible blends was investigated by few researchers. As mentioned before, similar to nanoclay, the extent of intercalation and, hence, the phase where the graphene is situated (come into contact) can play an important role in determining their location. In other words, in addition to thermodynamic parameters (wetting parameters), the sequence of mixing and viscosity of the polymer can also have effective contributions in determining the graphene nanoplatelet localization [86,176–178]. Liebscher *et al.*, for example, showed that the PC/SAN/graphene 59/40/1 blend prepared by simultaneous melt mixing procedure exhibited very poor intercalation or dispersion of graphenes, and the fillers could hardly be found localized selectively in one phase [179]. However, a much better dispersion of graphene platelets was found for the blends in which the nanographene was first melt mixed with either PC or SAN and then blended with the polymer matrix for longer time and at higher shear rates. When nanographene was premixed with PC, they remained in the PC phase after blending with SAN. However, in the blend where graphene was first melt mixed with SAN, the nanographene was located partially in the SAN phase or at the interface, and some smaller sized particles were transferred to the PC phase.

Cao *et al.* also showed that the addition of 0.5 wt% of graphene oxide sheets (GOS) to the PA/polyphenylene oxide (PPO) 90/10 blend resulted in a dramatic decrease in the PPO-dispersed size. This was attributed to the strong interaction of GOS with both PA and PPO phases [180]. In another research, Cao *et al.* found a similar compatibilization effect for PP-grafted-GOS in the same blend system. This was related to the fact that PP-g-GOS could adsorb PPO on their basal planes while exhibiting intermolecular interactions with PP through their grafted PP chains [177].

## References

1 Isayev, A.I. and Palsule, S. (2011) *Encyclopedia of Polymer Blends*, vol. **2**, Processing, vol. **2**, John Wiley & Sons, Inc., Hoboken, NJ.

2 Utracki, L.A. and Wilkie, C.A. (2014) *Polymer Blends Handbook*, Springer, New York.

3 Utracki, L.A. (2004) *Clay-Containing Polymeric Nanocomposites*, vol. **1**, iSmithers Rapra Publishing, London.

4 Paul, D.R. (2012) *Polymer Blends*, vol. **1**, Elsevier, New York, NY.

5 Folkes, M.J. and Hope, P.S. (1993) *Polymer Blends and Alloys*, Springer, New York.

6 Koning, C., Van Duin, M., Pagnoulle, C., and Jerome, R. (1998) Strategies for compatibilization of polymer blends. *Prog. Polym. Sci.*, **23** (4), 707–757.

7 Koo, J.H. (2006) *Polymer Nanocomposites*, McGraw-Hill Professional Pub, New York, NY.

8 Mai, Y.-W. and Yu, Z.-Z. (2006) *Polymer Nanocomposites*, Woodhead Publishing, Sawston, Cambridge.

9 Gao, F. (2012) *Advances in Polymer Nanocomposites: Types and Applications*, Elsevier, New York, NY.

10 Ray, S.S. (2013) *Clay-Containing Polymer Nanocomposites: From Fundamentals to Real Applications*, Elsevier, New York, NY.

11 Winey, K.I. and Vaia, R.A. (2007) Polymer nanocomposites. *MRS Bull.*, **32** (04), 314–322.

12 Moniruzzaman, M. and Winey, K.I. (2006) Polymer nanocomposites containing carbon nanotubes. *Macromolecules*, **39** (16), 5194–5205.

13 Kim, H., Abdala, A.A., and Macosko, C.W. (2010) Graphene/polymer nanocomposites. *Macromolecules*, **43** (16), 6515–6530.

14 Alexandre, M. and Dubois, P. (2000) Polymer-layered silicate nanocomposites: preparation, properties and uses of a new class of materials. *Mater. Sci. Eng. R: Reports*, **28** (1), 1–63.

15 Taguet, A., Cassagnau, P., and Lopez-Cuesta, J.-M. (2014) Structuration, selective dispersion and compatibilizing effect of (nano) fillers in polymer blends. *Prog. Polym. Sci.*, **39** (8), 1526–1563.

16 Fenouillot, F., Cassagnau, P., and Majeste, J.-C. (2009) Uneven distribution of nanoparticles in immiscible fluids: morphology development in polymer blends. *Polymer (Guildf)*, **50** (6), 1333–1350.

17 Kojima, Y., Usuki, A., Kawasumi, M., Okada, A., Kurauchi, T., and Kamigaito, O. (1993) Synthesis of nylon 6–clay hybrid by montmorillonite intercalated with ε-caprolactam. *J. Polym. Sci. A: Polym. Chem.*, **31** (4), 983–986.

18 Anastasiadis, S.H., Karatasos, K., Vlachos, G., Manias, E., and Giannelis, E.P. (2000) Nanoscopic-confinement effects on local dynamics. *Phys. Rev. Lett.*, **84** (5), 915.

19 Hackett, E., Manias, E., and Giannelis, E.P. (2000) Computer simulation studies of PEO/layer silicate nanocomposites. *Chem. Mater.*, **12** (8), 2161–2167.

20 Pavlidou, S. and Papaspyrides, C.D. (2008) A review on polymer–layered silicate nanocomposites. *Prog. Polym. Sci.*, **33** (12), 1119–1198.

21 Ray, S.S. and Okamoto, M. (2003) Polymer/layered silicate nanocomposites: a review from preparation to processing. *Prog. Polym. Sci.*, **28** (11), 1539–1641.

22 Fischer, H. (2003) Polymer nanocomposites: from fundamental research to specific applications. *Mater. Sci. Eng. C*, **23** (6), 763–772.

23 Kim, G.-M., Lee, D.-H., Hoffmann, B., Kressler, J., and Stöppelmann, G. (2001) Influence of nanofillers on the deformation process in layered silicate/polyamide-12 nanocomposites. *Polymer (Guildf)*, **42** (3), 1095–1100.

24 Zerda, A.S. and Lesser, A.J. (2001) Intercalated clay nanocomposites: morphology, mechanics, and fracture behavior. *J. Polym. Sci. B: Polym. Phys.*, **39** (11), 1137–1146.

25 LeBaron, P.C., Wang, Z., and Pinnavaia, T.J. (1999) Polymer-layered silicate nanocomposites: an overview. *Appl. Clay Sci.*, **15** (1), 11–29.

26 Wang, K.H., Choi, M.H., Koo, C.M., Choi, Y.S., and Chung, I.J. (2001) Synthesis and characterization of maleated polyethylene/clay nanocomposites. *Polymer (Guildf)*, **42** (24), 9819–9826.

27 Giannelis, E.P. (1996) Polymer layered silicate nanocomposites. *Adv. Mater.*, **8** (1), 29–35.

28 Meyyappan, M. and Srivastava, D. (2000) Carbon nanotubes. *IEEE Potentials*, **19** (3), 16–18.

29 Ganesh, E.N. (2013) Single walled and multi-walled carbon nanotube structure, synthesis and applications. *Int. J. Innov. Technol. Explor. Eng.*, **2** (4), 2278–3075.

30 Saito, R., Dresselhaus, G., and Dresselhaus, M.S. (1998) *Physical Properties of Carbon Nanotubes*, vol. **4**, Imperial College Press, London, England.

31 McClory, C., Chin, S.J., and McNally, T. (2009) Polymer/carbon nanotube composites. *Aust. J. Chem.*, **62** (8), 762–785.

32 Spitalsky, Z., Tasis, D., Papagelis, K., and Galiotis, C. (2010) Carbon nanotube–polymer composites: chemistry, processing, mechanical and electrical properties. *Prog. Polym. Sci.*, **35** (3), 357–401.

33 Qin, S., Qin, D., Ford, W.T., Resasco, D.E., and Herrera, J.E. (2004) Functionalization of single-walled carbon nanotubes with polystyrene via grafting to and grafting from methods. *Macromolecules*, **37** (3), 752–757.

34 Beyer, G. (2002) Nanocomposites: a new class of flame retardants for polymers. *Plast. Addit. Compound.*, **4** (10), 22–28.

35 Ma, J., Xu, J., Ren, J.-H., Yu, Z.-Z., and Mai, Y.-W. (2003) A new approach to polymer/montmorillonite nanocomposites. *Polymer (Guildf)*, **44** (16), 4619–4624.

36 Morgan, A.B. and Gilman, J.W. (2003) Characterization of polymer-layered silicate (clay) nanocomposites by transmission electron microscopy and X-ray diffraction: a comparative study. *J. Appl. Polym. Sci.*, **87** (8), 1329–1338.

37 Nazari, B., Nazockdast, H., and Katbab, A.A. (2014) The role of flow-induced microstructure in rheological behavior and nonisothermal crystallization kinetics of polyethylene/organoclay nanocomposites. *Polym. Eng. Sci.*, **54** (8), 1839–1847.

38 Wang, H. and Qiu, Z. (2012) Crystallization kinetics and morphology of biodegradable poly(l-lactic acid)/graphene oxide nanocomposites: influences of graphene oxide loading and crystallization temperature. *Thermochim. Acta*, **527** (2012), 40–46.

39 Zanetti, M., Lomakin, S., and Camino, G. (2000) Polymer layered silicate nanocomposites. *Macromol. Mater. Eng.*, **279** (1), 1–9.

40 Abbasi, S., Carreau, P.J., Derdouri, A., and Moan, M. (2009) Rheological properties and percolation in suspensions of multiwalled carbon nanotubes in polycarbonate. *Rheol. Acta*, **48** (9), 943–959.

41 Khalkhal, F., Carreau, P.J., and Ausias, G. (2011) Effect of flow history on linear viscoelastic properties and the evolution

of the structure of multiwalled carbon nanotube suspensions in an epoxy. *J. Rheol.*, **55** (1), 153–175.

42 Solomon, M.J., Almusallam, A.S., Seefeldt, K.F., Somwangthanaroj, A., and Varadan, P. (2001) Rheology of polypropylene/clay hybrid materials. *Macromolecules*, **34** (6), 1864–1872.

43 Song, Y. and Zheng, Q. (2015) Linear rheology of nanofilled polymers. *J. Rheol.*, **59** (1), 155–191.

44 Wu, D., Wu, L., and Zhang, M. (2007) Rheology of multi-walled carbon nanotube/poly (butylene terephthalate) composites. *J. Polym. Sci. B: Polym. Phys.*, **45** (16), 2239–2251.

45 Wu, D., Wu, L., Sun, Y., and Zhang, M. (2007) Rheological properties and crystallization behavior of multi-walled carbon nanotube/poly (ε-caprolactone) composites. *J. Polym. Sci. B: Polym. Phys.*, **45** (23), 3137–3147.

46 Wu, D., Wu, L., Zhang, M., and Zhao, Y. (2008) Viscoelasticity and thermal stability of polylactide composites with various functionalized carbon nanotubes. *Polym. Degrad. Stab.*, **93** (8), 1577–1584.

47 Nazockdast, E. and Nazockdast, H. (2011) Rheological Modeling of Polymer/layered silicate nanocomposites. *Appl. Rheol.*, **21** (2), 25434.

48 Nazockdast, E., Nazockdast, H., and Goharpey, F. (2008) Linear and nonlinear melt-state viscoelastic properties of polypropylene/organoclay nanocomposites. *Polym. Eng. Sci.*, **48** (7), 1240.

49 Abbasi, S., Carreau, P.J., and Derdouri, A. (2010) Flow induced orientation of multiwalled carbon nanotubes in polycarbonate nanocomposites: rheology, conductivity and mechanical properties. *Polymer (Guildf)*, **51** (4), 922–935.

50 Somani, R.H., Hsiao, B.S., Nogales, A., Fruitwala, H., Srinivas, S., and Tsou, A.H. (2001) Structure development during shear flow induced crystallization of i-PP: in situ wide-angle X-ray diffraction study. *Macromolecules*, **34** (17), 5902–5909.

51 Vermant, J. and Solomon, M.J. (2005) Flow-induced structure in colloidal suspensions. *J. Phys. Condens. Matter*, **17** (4), R187.

52 Ranjbar, B. and Nazockdast, H. (2015) Shear flow-induced orientation and structural recovery of multiwalled carbon nanotube in poly(ethylene oxide) matrix. *J. Appl. Polym. Sci.*, **132** (15), 1–9.

53 Ajji, A. and Choplin, L. (1991) Rheology and dynamics near phase separation in a polymer blend: model and scaling analysis. *Macromolecules*, **24** (18), 5221–5223.

54 El-Mabrouk, K., Belaiche, M., and Bousmina, M. (2007) Phase separation in PS/PVME thin and thick films. *J. Colloid Interface Sci.*, **306** (2), 354–367.

55 Hashimoto, T., Itakura, M., and Shimidzu, N. (1986) Late stage spinodal decomposition of a binary polymer mixture: II. Scaling analyses on Qm ($\tau$) and Im ($\tau$). *J. Chem. Phys.*, **85** (11), 6773–6786.

56 Jinnai, H., Hasegawa, H., Hashimoto, T., and Han, C.C. (1991) Time-resolved small-angle neutron scattering in intermediate-and late-stage spinodal decomposition of perdeuterated polybutadiene-protonated polyisoprene blends. *Macromolecules*, **24** (1), 282–289.

57 Özdilek, C., Bose, S., Leys, J., Seo, J.W., Wübbenhorst, M., and Moldenaers, P. (2011) Thermally induced phase separation in PαMSAN/PMMA blends in presence of functionalized multiwall carbon nanotubes: Rheology, morphology and electrical conductivity. *Polymer (Guildf)*, **52** (20), 4480–4489.

58 Vandebril, S., Vermant, J., and Moldenaers, P. (2010) Efficiently suppressing coalescence in polymer blends using nanoparticles: role of interfacial rheology. *Soft Matter*, **6** (14), 3353–3362.

59 Vermant, J., Cioccolo, G., Nair, K.G., and Moldenaers, P. (2004) Coalescence suppression in model immiscible polymer blends by nano-sized colloidal particles. *Rheol. Acta*, **43** (5), 529–538.

60 Vleminckx, G., Bose, S., Leys, J., Vermant, J., Wubbenhorst, M., Abdala, A.A., Macosko, C., and Moldenaers, P. (2011) Effect of thermally reduced graphene sheets on the phase behavior, morphology, and electrical conductivity in poly [(α-methyl styrene)-co-

(acrylonitrile)/poly (methyl-methacrylate) blends. *ACS Appl. Mater. Interfaces*, **3** (8), 3172–3180.

61 Fan, Z. and Advani, S.G. (2005) Characterization of orientation state of carbon nanotubes in shear flow. *Polymer (Guildf)*, **46** (14), 5232–5240.

62 Huang, C., Gao, J., Yu, W., and Zhou, C. (2012) Phase separation of poly (methyl methacrylate)/poly (styrene-co-acrylonitrile) blends with controlled distribution of silica nanoparticles. *Macromolecules*, **45** (20), 8420–8429.

63 Gao, J., Huang, C., Wang, N., Yu, W., and Zhou, C. (2012) Phase separation of poly (methyl methacrylate)/poly (styrene-co-acrylonitrile) blends in the presence of silica nanoparticles. *Polymer (Guildf)*, **53** (8), 1772–1782.

64 Balazs, A.C., Emrick, T., and Russell, T.P. (2006) Nanoparticle polymer composites: where two small worlds meet. *Science*, **314** (5802), 1107–1110.

65 Ginzburg, V.V. (2005) Influence of nanoparticles on miscibility of polymer blends. A simple theory. *Macromolecules*, **38** (6), 2362–2367.

66 Niu, Y., Yang, L., Shimizu, K., Pathak, J.A., Wang, H., and Wang, Z. (2009) Investigation on phase separation kinetics of polyolefin blends through combination of viscoelasticity and morphology. *J. Phys. Chem. B*, **113** (26), 8820–8827.

67 Niu, Y.-H. and Wang, Z.-G. (2006) Rheologically determined phase diagram and dynamically investigated phase separation kinetics of polyolefin blends. *Macromolecules*, **39** (12), 4175–4183.

68 Gharachorlou, A. and Goharpey, F. (2008) Rheologically determined phase behavior of LCST blends in the presence of spherical nanoparticles. *Macromolecules*, **41** (9), 3276–3283.

69 Khademzadeh Yeganeh, J., Goharpey, F., and Foudazi, R. (2010) Rheology and morphology of dynamically asymmetric LCST blends: polystyrene/poly (vinyl methyl ether). *Macromolecules*, **43** (20), 8670–8685.

70 Fredrickson, G.H. and Larson, R.G. (1987) Viscoelasticity of homogeneous polymer melts near a critical point. *J. Chem. Phys.*, **86** (3), 1553–1560.

71 Zhang, R., Dong, X., Wang, X., Cheng, H., and Han, C.C. (2009) Nucleation/ growth in the metastable and unstable phase separation regions under oscillatory shear flow for an off-critical polymer blend. *Macromolecules*, **42** (7), 2873–2876.

72 Tanaka, H. (2009) Formation of network and cellular structures by viscoelastic phase separation. *Adv. Mater.*, **21** (18), 1872–1880.

73 Lipatov, Y.S. and Nesterov, A.E. (1992) Effect of filler concentration on the phase separation in poly (vinyl acetate)-poly (methyl methacrylate) mixtures. *Polym. Eng. Sci.*, **32** (17), 1261–1263.

74 VanderBeek, G.P., Stuart, M.A.C., Fleer, G.J., and Hofman, J.E. (1991) Segmental adsorption energies for polymers on silica and alumina. *Macromolecules*, **24** (25), 6600–6611.

75 Nesterov, A.E., Lipatov, Y.S., and Ignatova, T.D. (2001) Effect of an interface with solid on the component distribution in separated phases of binary polymer mixtures. *Eur. Polym. J.*, **37** (2), 281–285.

76 Lipatov, Y.S., Nesterov, A.E., Ignatova, T.D., and Nesterov, D.A. (2002) Effect of polymer–filler surface interactions on the phase separation in polymer blends. *Polymer*, **43** (2002), 875–880.

77 Huang, Y., Jiang, S., Li, G., and Chen, D. (2005) Effect of fillers on the phase stability of binary polymer blends: a dynamic shear rheology study. *Acta Mater.*, **53** (19), 5117–5124.

78 Li, W., Abee, R.M.A., and Goossens, J.G.P. (2013) The control of silica nanoparticles on the phase separation of poly (methyl methacrylate)/poly (styrene-co-acrylonitrile) blends. *Macromol. Chem. Phys.*, **214** (23), 2705–2715.

79 Palierne, J.F. (1990) Linear rheology of viscoelastic emulsions with interfacial tension. *Rheol. Acta*, **29** (3), 204–214.

80 Shumsky, V.F., Getmanchuk, I., Ignatova, T., Maslak, Y., Cassagnau, P., Boiteux, G., and Melis, F. (2010) Effect of nanofillers on the phase separation and rheological properties of poly(methyl methacrylate)/ poly(styrene-co-acrylonitrile) blends. *Rheol. Acta*, **49** (8), 827–836.

81 Xia, T., Huang, Y., Jiang, X., Lv, Y., Yang, Q., and Li, G. (2013) The molecular mechanism of the morphology change in PS/PVME/silica blends based on rheology. *Macromolecules*, **46** (20), 8323–8333.

82 Xia, T., Huang, Y., Peng, X., and Li, G. (2010) Morphological transition induced by nanoparticles in dynamically asymmetric PS/PVME blends. *Macromol. Chem. Phys.*, **211** (20), 2240–2247.

83 Xavier, P. and Bose, S. (2013) Multiwalled-carbon-nanotube-induced miscibility in near-critical PS/PVME blends: assessment through concentration fluctuations and segmental relaxation. *J. Phys. Chem. B*, **117** (28), 8633–8646.

84 Yurekli, K., Karim, A., Amis, E.J., and Krishnamoorti, R. (2004) Phase behavior of PS-PVME nanocomposites. *Macromolecules*, **37** (2), 507–515.

85 Yurekli, K., Karim, A., Amis, E.J., and Krishnamoorti, R. (2003) Influence of layered silicates on the phase-separated morphology of PS-PVME blends. *Macromolecules*, **36** (19), 7256–7267.

86 Pickering, S.U. (1907) CXCVI. —emulsions. *J. Chem. Soc. Trans.*, **91**, 2001–2021.

87 Ramsden, W. (1903) Separation of solids in the surface-layers of solutions and 'suspensions' (observations on surface-membranes, bubbles, emulsions, and mechanical coagulation). Preliminary account. *Proc. R. Soc. London*, **72**, 156–164.

88 Hess, W.M., Scott, C.E., and Callan, J.E. (1967) Carbon black distribution in elastomer blends. *Rubber Chem. Technol.*, **40** (2), 371–384.

89 Walters, M.H. and Keyte, D.N. (1965) Heterogeneous structure in blends of rubber polymers. *Rubber Chem. Technol.*, **38** (1), 62–75.

90 Wu, S. (1982) *Polymer Interface and Adhesion*, M. Dekker, New York, NY.

91 Ross, S. and Morrison, E.D. (1988) Colloidal systems and interfaces.

92 Kamal, M.R., Calderon, J.U., and Lennox, B.R. (2009) Surface energy of modified nanoclays and its effect on polymer/clay

nanocomposites. *J. Adhes. Sci. Technol.*, **23** (5), 663–688.

93 Xing, P., Bousmina, M., Rodrigue, D., and Kamal, M.R. (2000) Critical experimental comparison between five techniques for the determination of interfacial tension in polymer blends: model system of polystyrene/polyamide-6. *Macromolecules*, **33** (21), 8020–8034.

94 Owens, D.K. and Wendt, R.C. (1969) Estimation of the surface free energy of polymers. *J. Appl. Polym. Sci.*, **13** (8), 1741–1747.

95 Good, R.J. and Girifalco, L.A. (1960) A theory for estimation of surface and interfacial energies. III. Estimation of surface energies of solids from contact angle data. *J. Phys. Chem.*, **64** (5), 561–565.

96 Girifalco, L.A. and Good, R.J. (1957) A theory for the estimation of surface and interfacial energies. I. Derivation and application to interfacial tension. *J. Phys. Chem.*, **61** (7), 904–909.

97 Wu, S. (1974) Interfacial and surface tensions of polymers. *J. Macromol. Sci. Polym. Rev.*, **10** (1), 1–73.

98 Asai, S., Sakata, K., Sumita, M., and Miyasaka, K. (1992) Effect of interfacial free energy on the heterogeneous distribution of oxidized carbon black in polymer blends. *Polym. J.*, **24** (5), 415–420.

99 Sumita, M., Sakata, K., Asai, S., Miyasaka, K., and Nakagawa, H. (1991) Dispersion of fillers and the electrical conductivity of polymer blends filled with carbon black. *Polym. Bull.*, **25** (2), 265–271.

100 Zaikin, A.E., Karimov, R.R., and Arkhireev, V.P. (2001) A study of the redistribution conditions of carbon black particles from the bulk to the interface in heterogeneous polymer blends. *Colloid J.*, **63** (1), 53–59.

101 Zaikin, A.E., Zharinova, E.A., and Bikmullin, R.S. (2007) Specifics of localization of carbon black at the interface between polymeric phases. *Polym. Sci. Ser. A*, **49** (3), 328–336.

102 Thareja, P. and Velankar, S.S. (2008) Interfacial activity of particles at PI/PDMS and PI/PIB interfaces: analysis based on Girifalco–Good theory. *Colloid Polym. Sci.*, **286** (11), 1257–1264.

103 Elias, L., Fenouillot, F., Majesté, J.-C., and Cassagnau, P. (2007) Morphology and rheology of immiscible polymer blends filled with silica nanoparticles. *Polymer (Guildf)*, **48** (20), 6029–6040.

104 Foudazi, R. and Nazockdast, H. (2013) Rheology and morphology of nanosilica-containing polypropylene and polypropylene/liquid crystalline polymer blend. *J. Appl. Polym. Sci.*, **128** (6), 3501–3511.

105 Foudazi, R. and Nazockdast, H. (2010) Rheology of polypropylene/liquid crystalline polymer blends: effect of compatibilizer and silica. *Appl. Rheol.*, **20** (1), 12218.

106 Gubbels, F., Blacher, S., Vanlathem, E., Jérôme, R., Deltour, R., Brouers, F., and Teyssie, P. (1995) Design of electrical composites: determining the role of the morphology on the electrical properties of carbon black filled polymer blends. *Macromolecules*, **28** (5), 1559–1566.

107 Gubbels, F., Jérôme, R., Teyssie, P., Vanlathem, E., Deltour, R., Calderone, A., Parente, V., and Brédas, J.-L. (1994) Selective localization of carbon black in immiscible polymer blends: a useful tool to design electrical conductive composites. *Macromolecules*, **27** (7), 1972–1974.

108 Gubbels, F., Jérôme, R., Vanlathem, E., Deltour, R., Blacher, S., and Brouers, F. (1998) Kinetic and thermodynamic control of the selective localization of carbon black at the interface of immiscible polymer blends. *Chem. Mater.*, **10** (5), 1227–1235.

109 Callan, J.E., Hess, W.M., and Scott, C.E. (1971) Elastomer blends. Compatibility and relative response to fillers. *Rubber Chem. Technol.*, **44** (3), 814–837.

110 Sircar, A.K. and Lamond, T.G. (1973) Carbon black transfer in blends of cis-poly (butadiene) with other elastomers. *Rubber Chem. Technol.*, **46** (1), 178–191.

111 Elias, L., Fenouillot, F., Majesté, J.-C., Alcouffe, P., and Cassagnau, P. (2008) Immiscible polymer blends stabilized with nano-silica particles: rheology and effective interfacial tension. *Polymer (Guildf)*, **49** (20), 4378–4385.

112 Elias, L., Fenouillot, F., Majesté, J., Martin, G., and Cassagnau, P. (2008) Migration of nanosilica particles in polymer blends. *J. Polym. Sci. Part B Polym. Phys.*, **46** (18), 1976–1983.

113 Persson, A.L. and Bertilsson, H. (1998) Viscosity difference as distributing factor in selective absorption of aluminium borate whiskers in immiscible polymer blends. *Polymer (Guildf)*, **39** (23), 5633–5642.

114 Feng, J., Chan, C., and Li, J. (2003) A method to control the dispersion of carbon black in an immiscible polymer blend. *Polym. Eng. Sci.*, **43** (5), 1058–1063.

115 Zhou, P., Yu, W., Zhou, C., Liu, F., Hou, L., and Wang, J. (2007) Morphology and electrical properties of carbon black filled LLDPE/EMA composites. *J. Appl. Polym. Sci.*, **103** (1), 487–492.

116 Clarke, J., Clarke, B., Freakley, P.K., and Sutherland, I. (2001) Compatibilising effect of carbon black on morphology of NR–NBR blends. *Plast. Rubber Compos.*, **30** (1), 39–44.

117 Zhu, Y., Xu, Y., Tong, L., Xu, Z., and Fang, Z. (2008) Influence of polarity on the preferential intercalation behavior of clay in immiscible polypropylene/ polystyrene blend. *J. Appl. Polym. Sci.*, **110** (5), 3130–3139.

118 Tiwari, R.R., Hunter, D.L., and Paul, D.R. (2012) Extruder-made TPO nanocomposites: I. Effect of maleated polypropylene and organoclay ratio on the morphology and mechanical properties. *J. Polym. Sci. B: Polym. Phys.*, **50** (22), 1577–1588.

119 Pötschke, P., Bhattacharyya, A.R., and Janke, A. (2003) Morphology and electrical resistivity of melt mixed blends of polyethylene and carbon nanotube filled polycarbonate. *Polymer (Guildf)*, **44** (26), 8061–8069.

120 Pötschke, P., Bhattacharyya, A.R., and Janke, A. (2004) Carbon nanotube-filled polycarbonate composites produced by melt mixing and their use in blends with polyethylene. *Carbon*, **42** (5), 965–969.

121 Pötschke, P., Kretzschmar, B., and Janke, A. (2007) Use of carbon nanotube filled polycarbonate in blends with montmorillonite filled polypropylene. *Compos. Sci. Technol.*, **67** (5), 855–860.

122 Wu, M. and Shaw, L. (2006) Electrical and mechanical behaviors of carbon nanotube-filled polymer blends. *J. Appl. Polym. Sci.*, **99** (2), 477–488.

123 Wu, M. and Shaw, L.L. (2005) A novel concept of carbon-filled polymer blends for applications in PEM fuel cell bipolar plates. *Int. J. Hydrogen Energy*, **30** (4), 373–380.

124 Li, Y. and Shimizu, H. (2008) Conductive PVDF/PA6/CNTs nanocomposites fabricated by dual formation of cocontinuous and nanodispersion structures. *Macromolecules*, **41** (14), 5339–5344.

125 Meincke, O., Kaempfer, D., Weickmann, H., Friedrich, C., Vathauer, M., and Warth, H. (2004) Mechanical properties and electrical conductivity of carbon-nanotube filled polyamide-6 and its blends with acrylonitrile/butadiene/ styrene. *Polymer (Guildf)*, **45** (3), 739–748.

126 Pötschke, P., Pegel, S., Claes, M., and Bonduel, D. (2008) A novel strategy to incorporate carbon nanotubes into thermoplastic matrices. *Macromol. Rapid Commun.*, **29** (3), 244–251.

127 Göldel, A., Kasaliwal, G., and Pötschke, P. (2009) Selective localization and migration of multiwalled carbon nanotubes in blends of polycarbonate and poly(styrene-acrylonitrile). *Macromol. Rapid Commun.*, **30** (6), 423–429.

128 Bose, S., Bhattacharyya, A.R., Kodgire, P.V., Misra, A., and Pötschke, P. (2007) Rheology, morphology, and crystallization behavior of melt-mixed blends of polyamide6 and acrylonitrile-butadiene-styrene: influence of reactive compatibilizer premixed with multiwall carbon nanotubes. *J. Appl. Polym. Sci.*, **106** (5), 3394–3408.

129 Bose, S., Bhattacharyya, A.R., Kodgire, P.V., and Misra, A. (2007) Fractionated crystallization in PA6/ABS blends: influence of a reactive compatibilizer and multiwall carbon nanotubes. *Polymer (Guildf)*, **48** (1), 356–362.

130 Wu, D., Zhang, Y., Zhang, M., and Yu, W. (2009) Selective localization of

multiwalled carbon nanotubes in poly (ε-caprolactone)/polylactide blend. *Biomacromolecules*, **10** (2), 417–424.

**131** Baudouin, A.C., Devaux, J., and Bailly, C. (2010) Localization of carbon nanotubes at the interface in blends of polyamide and ethylene-acrylate copolymer. *Polymer (Guildf)*, **51** (6), 1341–1354.

**132** Shi, D., Hu, G.-H., Ke, Z., Li, R.K.Y., and Yin, J. (2006) Relaxation behavior of polymer blends with complex morphologies: palierne emulsion model for uncompatibilized and compatibilized PP/PA6 blends. *Polymer (Guildf)*, **47** (13), 4659–4666.

**133** Yang, H., Zhang, H., Moldenaers, P., and Mewis, J. (1998) Rheo-optical investigation of immiscible polymer blends. *Polymer (Guildf)*, **39** (23), 5731–5737.

**134** Vinckier, I., Moldenaers, P., and Mewis, J. (1996) Relationship between rheology and morphology of model blends in steady shear flow. *J. Rheol.*, **40** (4), 613–631.

**135** Vinckier, I., Mewis, J., and Moldenaers, P. (1997) Stress relaxation as a microstructural probe for immiscible polymer blends. *Rheol. Acta*, **36** (5), 513–523.

**136** Graebling, D., Muller, R., and Palierne, J.F. (1993) Linear viscoelastic behavior of some incompatible polymer blends in the melt. Interpretation of data with a model of emulsion of viscoelastic liquids. *Macromolecules*, **26** (2), 320–329.

**137** Bousmina, M., Ait-Kadi, A., and Faisant, J.B. (1999) Determination of shear rate and viscosity from batch mixer data. *J. Rheol.*, **43** (2), 415–433.

**138** Bousmina, M. and Muller, R. (1993) Linear viscoelasticity in the melt of impact PMMA. Influence of concentration and aggregation of dispersed rubber particles. *J. Rheol.*, **37** (4), 663–679.

**139** Vermant, J., Vandebril, S., Dewitte, C., and Moldenaers, P. (2008) Particle-stabilized polymer blends. *Rheol. Acta*, **47** (7), 835–839.

**140** Thareja, P. and Velankar, S. (2007) Particle-induced bridging in immiscible polymer blends. *Rheol. Acta*, **46** (3), 405–412.

**141** Khoshkava, V., Dini, M., and Nazockdast, H. (2012) Study on morphology and microstructure development of PA6/LDPE/organoclay nanocomposites. *J. Appl. Polym. Sci.*, **125** (S1), E197–E203.

**142** Khatua, B.B., Lee, D.J., Kim, H.Y., and Kim, J.K. (2004) Effect of organoclay platelets on morphology of nylon-6 and poly (ethylene-r an-propylene) rubber blends. *Macromolecules*, **37** (7), 2454–2459.

**143** Chow, W.S., Bakar, A.A., Ishak, Z.A.M., Karger-Kocsis, J., and Ishiaku, U.S. (2005) Effect of maleic anhydride-grafted ethylene–propylene rubber on the mechanical, rheological and morphological properties of organoclay reinforced polyamide 6/polypropylene nanocomposites. *Eur. Polym. J.*, **41** (4), 687–696.

**144** Chow, W.S., Ishak, Z.A.M., and Karger-Kocsis, J. (2005) Morphological and rheological properties of polyamide 6/ poly(propylene)/organoclay nanocomposites. *Macromol. Mater. Eng.* **290** (2), 122–127.

**145** Bigdeli, A., Nazockdast, H., Rashidi, A., and Yazdanshenas, M.E. (2012) Role of nanoclay in determining microfibrillar morphology development in PP/PBT blend nanocomposite fibers. *J. Polym. Res.*, **19** (11), 1–10.

**146** Abdolrasouli, M.H., Nazockdast, H., Sadeghi, G.M.M., and Kaschta, J. (2015) Morphology development, melt linear viscoelastic properties and crystallinity of polylactide/polyethylene/organoclay blend nanocomposites. *J. Appl. Polym. Sci.*, **132** (3), 1–11.

**147** Bitinis, N., Verdejo, R., Maya, E.M., Espuche, E., Cassagnau, P., and Lopez-Manchado, M.a. (2012) Physicochemical properties of organoclay filled polylactic acid/natural rubber blend bionanocomposites. *Compos. Sci. Technol.*, **72** (2), 305–313.

**148** Huitric, J., Ville, J., Médéric, P., Moan, M., and Aubry, T. (2009) Rheological, morphological and structural properties of PE/PA/nanoclay ternary blends: effect of clay weight fraction. *J. Rheol.*, **53** (5), 1101–1119.

149 Hong, J.S., Kim, Y.K., Ahn, K.H., Lee, S.J., and Kim, C. (2007) Interfacial tension reduction in PBT/PE/clay nanocomposite. *Rheol. Acta*, **46** (4), 469–478.

150 Levitt, L., Macosko, C.W., and Pearson, S.D. (1996) Influence of normal stress difference on polymer drop deformation. *Polym. Eng. Sci.*, **36** (12), 1647–1655.

151 Kontopoulou, M., Liu, Y., Austin, J.R., and Parent, J.S. (2007) The dynamics of montmorillonite clay dispersion and morphology development in immiscible ethylene–propylene rubber/polypropylene blends. *Polymer (Guildf)*, **48** (15), 4520–4528.

152 Ville, J., Médéric, P., Huitric, J., and Aubry, T. (2012) Structural and rheological investigation of interphase in polyethylene/polyamide/nanoclay ternary blends. *Polymer (Guildf)*, **53** (8), 1733–1740.

153 Honerkamp, J. and Weese, J. (1993) A nonlinear regularization method for the calculation of relaxation spectra. *Rheol. Acta*, **32** (1), 65–73.

154 Labaume, I., Médéric, P., Huitric, J., and Aubry, T. (2013) Comparative study of interphase viscoelastic properties in polyethylene/polyamide blends compatibilized with clay nanoparticles or with a graft copolymer. *J. Rheol.*, **57** (2), 377–392.

155 Pötschke, P., Fornes, T.D., and Paul, D.R. (2002) Rheological behavior of multiwalled carbon nanotube/polycarbonate composites. *Polymer (Guildf)*, **43** (11), 3247–3255.

156 Wode, F., Tzounis, L., Kirsten, M., Constantinou, M., Georgopanos, P., Rangou, S., Zafeiropoulos, N.E., Avgeropoulos, A., and Stamm, M. (2012) Selective localization of multi-wall carbon nanotubes in homopolymer blends and a diblock copolymer. Rheological orientation studies of the final nanocomposites. *Polymer (Guildf)*, **53** (20), 4438–4447.

157 Tao, F., Auhl, D., Baudouin, A., Stadler, F.J., and Bailly, C. (2013) Influence of multiwall carbon nanotubes trapped at the interface of an immiscible polymer blend on interfacial tension. *Macromol. Chem. Phys.*, **214** (3), 350–360.

158 Bose, S., Bhattacharyya, A.R., Kulkarni, A.R., and Pötschke, P. (2009) Electrical, rheological and morphological studies in co-continuous blends of polyamide 6 and acrylonitrile–butadiene–styrene with multiwall carbon nanotubes prepared by melt blending. *Compos. Sci. Technol.*, **69** (3), 365–372.

159 Zonder, L., Ophir, A., Kenig, S., and McCarthy, S. (2011) The effect of carbon nanotubes on the rheology and electrical resistivity of polyamide 12/high density polyethylene blends. *Polymer (Guildf)*, **52** (22), 5085–5091.

160 Taheri, S., Nakhlband, E., and Nazockdast, H. (2013) Microstructure and multiwall carbon nanotube partitioning in polycarbonate/acrylonitrile-butadiene-styrene/multiwall carbon nanotube nanocomposites. *Polym. Plast. Technol. Eng.*, **52** (3), 300–309.

161 Abbasi Moud, A., Javadi, A., Nazockdast, H., Fathi, A., and Altstaedt, V. (2015) Effect of dispersion and selective localization of carbon nanotubes on rheology and electrical conductivity of polyamide 6 (PA6), polypropylene (PP), and PA6/PP nanocomposites. *J. Polym. Sci. B: Polym. Phys.*, **53** (5), 368–378.

162 Rostami, A., Nazockdast, H., Karimi, M., and Javidi, Z. (2015) Role of multiwalled carbon nanotubes localization on morphology development of PMMA/PS/PP ternary blends. *Adv. Polym. Technol.* **35** (1), 80–89.

163 Göldel, A., Kasaliwal, G.R., Pötschke, P., and Heinrich, G. (2012) The kinetics of CNT transfer between immiscible blend phases during melt mixing. *Polymer (Guildf)*, **53** (2), 411–421.

164 Göldel, A., Marmur, A., Kasaliwal, G.R., Pötschke, P., and Heinrich, G. (2011) Shape-dependent localization of carbon nanotubes and carbon black in an immiscible polymer blend during melt mixing. *Macromolecules*, **44** (15), 6094–6102.

165 Krasovitski, B. and Marmur, A. (2005) Particle adhesion to drops. *J. Adhes.*, **81** (7–8), 869–880.

166 Zhang, W., Lin, M., Winesett, A., Dhez, O., Kilcoyne, A., Ade, H., Rubinstein, M., Shafi, K., Ulman, A., and Gersappe, D.

(2011) The use of functionalized nanoparticles as non-specific compatibilizers for polymer blends. *Polym. Adv. Technol.*, **22** (1), 65–71.

167 Tong, W., Huang, Y., Liu, C., Chen, X., Yang, Q., and Li, G. (2010) The morphology of immiscible PDMS/PIB blends filled with silica nanoparticles under shear flow. *Colloid Polym. Sci.*, **288** (7), 753–760.

168 Li, W., Karger-Kocsis, J., and Thomann, R. (2009) Compatibilization effect of TiO$_2$ nanoparticles on the phase structure of PET/PP/TiO2 nanocomposites. *J. Polym. Sci. B: Polym. Phys.*, **47** (16), 1616–1624.

169 Gelfer, M.Y., Song, H.H., Liu, L., Hsiao, B.S., Chu, B., Rafailovich, M., Si, M., and Zaitsev, V. (2003) Effects of organoclays on morphology and thermal and rheological properties of polystyrene and poly (methyl methacrylate) blends. *J. Polym. Sci. B: Polym. Phys.*, **41** (1), 44–54.

170 Wang, Y., Zhang, Q., and Fu, Q. (2003) Compatibilization of immiscible poly (propylene)/polystyrene blends using clay. *Macromol. Rapid Commun.*, **24** (3), 231–235.

171 Fang, Z., Harrats, C., Moussaif, N., and Groeninckx, G. (2007) Location of a nanoclay at the interface in an immiscible poly (ε-caprolactone)/poly (ethylene oxide) blend and its effect on the compatibility of the components. *J. Appl. Polym. Sci.*, **106** (5), 3125–3135.

172 Si, M., Araki, T., Ade, H., Kilcoyne, A.L.D., Fisher, R., Sokolov, J.C., and Rafailovich, M.H. (2006) Compatibilizing bulk polymer blends by using organoclays. *Macromolecules*, **39** (14), 4793–4801.

173 Martín, Z., Jiménez, I., Gómez-Fatou, M. a., West, M., and Hitchcock, A.P. (2011) Interfacial interactions in polypropylene–organoclay–elastomer nanocomposites: influence of polar modifications on the location of the clay. *Macromolecules*, **44** (7), 2179–2189.

174 Ray, S.S., Bousmina, M., and Maazouz, A. (2006) Morphology and properties of organoclay modified polycarbonate/poly (methyl methacrylate) blend. *Polym. Eng. Sci.*, **46** (8), 1121.

175 Baudouin, A.-C., Auhl, D., Tao, F., Devaux, J., and Bailly, C. (2011) Polymer blend emulsion stabilization using carbon nanotubes interfacial confinement. *Polymer (Guildf)*, **52** (1), 149–156.

176 Yang, J., Feng, C., Dai, J., Zhang, N., Huang, T., and Wang, Y. (2013) Compatibilization of immiscible nylon 6/poly (vinylidene fluoride) blends using graphene oxides. *Polym. Int.*, **62** (7), 1085–1093.

177 Cao, Y., Feng, J., and Wu, P. (2012) Polypropylene-grafted graphene oxide sheets as multifunctional compatibilizers for polyolefin-based polymer blends. *J. Mater. Chem.*, **22** (30), 14997–15005.

178 Ye, S., Cao, Y., Feng, J., and Wu, P. (2013) Temperature-dependent compatibilizing effect of graphene oxide as a compatibilizer for immiscible polymer blends. *RSC Adv.*, **3** (21), 7987–7995.

179 Liebscher, M., Tzounis, L., Pötschke, P., and Heinrich, G. (2013) Influence of the viscosity ratio in PC/SAN blends filled with MWCNTs on the morphological, electrical, and melt rheological properties. *Polymer (Guildf)*, **54** (25), 6801–6808.

180 Cao, Y., Zhang, J., Feng, J., and Wu, P. (2011) Compatibilization of immiscible polymer blends using graphene oxide sheets. *ACS Nano*, **5** (7), 5920–5927.

# Index

*Encyclopedia of Polymer Blends: Volume 3: Structure*, First Edition. Edited by Avraam I. Isayev.
© 2016 Wiley-VCH Verlag GmbH & Co. KGaA. Published 2016 by Wiley-VCH Verlag GmbH & Co. KGaA.